Lecture Notes in Computer Science 7001

Commenced Publication in 1973
Founding and Former Series Editors:
Gerhard Goos, Juris Hartmanis, and Jan van Leeuwen

Xuejia Lai   Jianying Zhou   Hui Li (Eds.)

Information Security

14th International Conference, ISC 2011
Xi'an, China, October 26-29, 2011
Proceedings

 Springer

Volume Editors

Xuejia Lai
Shanghai Jiao Tong University
Department of Computer Science and Engineering
800 Dongchuan Road, Min Hang, Shanghai 200240, China
E-mail: lai-xj@cs.sjtu.edu.cn

Jianying Zhou
Institute for Infocomm Research
1 Fusionopolis Way, #21-01 Connexis, South Tower, Singapore 138632
E-mail: jyzhou@i2r.a-star.edu.sg

Hui Li
Xidian University
Key Laboratory of Computer Networks and Information Security
2 South Taibai Road, Xi'an, Shaanxi 710071, China
E-mail: lihui@mail.xidian.edu.cn

ISSN 0302-9743   e-ISSN 1611-3349
ISBN 978-3-642-24860-3   ISBN 978-3-642-24861-0 (eBook)
DOI 10.1007/978-3-642-24861-0
Springer Heidelberg Dordrecht London New York

Library of Congress Control Number: 2011938871

CR Subject Classification (1998): E.3, E.4, D.4.6, K.6.5, C.2, J.1, H.4

LNCS Sublibrary: SL 4 – Security and Cryptology

Typesetting: Camera-ready by author, data conversion by Scientific Publishing Services, Chennai, India

Printed on acid-free paper

Springer is part of Springer Science+Business Media (www.springer.com)

Preface

The 14th Information Security Conference (ISC 2011) was held on October 26–29, 2011 in Xi'an, China. The conference was sponsored by the China Computer Federation (CCF) and co-organized by Xidian University and Shanghai Jiao Tong University. The conference also received partial financial support from the National 111 Program (B08038).

This year the conference received 95 submissions. They were evaluated on the basis of their significance, novelty, technical quality, and practical impact. Each paper was reviewed by three Program Committee members and the reviewing process was "double-blind". After careful reviews and intensive discussions, 25 papers were selected for presentation at the conference, and constitute this Springer volume of proceedings, available at the conference. Beside the regular program, ISC 2011 has two invited speakers, Feng Bao and Dieter Gollmann. The Program Committee chose the paper "Replacement Attacks on Behavior Based Software Birthmark" authored by Zhi Xin, Huiyu Chen, Xinche Wang, Peng Liu, Sencun Zhu, Bing Mao, and Xie Li as the best paper, and the paper "SudoWeb: Minimizing Information Disclosure to Third Parties in Single Sign-On Platforms" authored by Georgios Kontaxis (student), Michalis Polychronakis, and Evangelos Markatos as the best student paper. A few selected papers have been recommended for publication in the ISI-ranked International Journal of Information Security (IJIS).

There is a long list of people who volunteered their time and energy to put together the conference and who deserve special thanks. Thanks to all the 58 members of the Program Committee and the 88 external reviewers for all the hard work they put into evaluating the papers. We are also very grateful to all the people whose work ensured a smooth organization process: the ISC Steering Committee, and Masahiro Mambo in particular, for their advice; Publicity Co-chairs Sara Foresti and Xiaofeng Chen; and Qingqi Pei, Yuanyuan Zuo, Xiaoyan Zhu, and Yun Shi of the Organizing Committee.

Last but certainly not least, our thanks go to all the authors who submitted papers and all the attendees.

August 2011

Xuejia Lai
Jianying Zhou
Hui Li

ISC 2011

14th International Conference on Information Security
Xi'an, China
October 26–29, 2011

General Chair

Pizheng Li   CCF TCCOMM, China
Hui Li   Xidian University, China

Program Chairs

Xuejia Lai   Shanghai Jiao Tong University, China
Jianying Zhou   Institute for Infocomm Research, Singapore

Program Committee

Michel Abdalla	ENS and CNRS, France
Joonsang Baek	KUSTAR, UAE
Feng Bao	Institute for Infocomm Research, Singapore
Alex Biryukov	University of Luxembourg, Luxembourg
Mike Burmester	Florida State University, USA
Levente Buttyan	BME, Hungary
Ee-Chien Chang	Naional University of Singapore, Singapore
Liqun Chen	HP Laboratories Bristol, UK
Xiaofeng Chen	Xidian University, China
Chen-Mou Cheng	National Taiwan University, Taiwan
Sherman Chow	University of Waterloo, Canada
Vanesa Daza	Universitat Pompeu Fabra, Spain
Roberto Di Pietro	Università di Roma Tre, Italy
Claudia Diaz	Katholieke Universiteit Leuven, Belgium
Xuhua Ding	Singapore Management University, Singapore
Josep Domingo-Ferrer	Universitat Rovira i Virgili, Spain
Eduardo Fernandez	Florida Atlantic University, USA
Josep Ferrer	Universitat de les Illes Balears, Spain
Sara Foresti	Università degli Studi di Milano, Italy
Guang Gong	University of Waterloo, Canada
Stefanos Gritzalis	University of the Aegean, Greece
Dawu Gu	Shanghai Jiao Tong University, China
Guofei Gu	Texas A&M University, USA

Steering Committee

Ed Dawson   Queensland University of Technology,
                                 Australia
Hui Li   Xidian University, China
Javier Lopez   University of Malaga, Spain
Spyros Magliveras   Florida Atlantic University, USA
Masahiro Mambo (Chair)   Kanazawa University, Japan
Eiji Okamoto   University of Tsukuba, Japan
Susan Wetzel   Stevens Institute of Technology, USA
Yuliang Zheng   University of North Carolina at Charlotte,
                                 USA

External Reviewers

Man Ho Au   Yuto Kawahara   Adnan Saleem
Josep Balasch   Po-Chun Kuo   Sumanta Sarkar
Filipe Beato   Junzuo Lai   Alessandra Scafuro
Shaoying Cai   Fagen Li   Peter Schwabe
Pengsu Cheng   Jin Li   Francesc Sebe
Kai-Yuen Cheong   Yan Li   Yingbo Song
Cheng-Kang Chu   Jingqiang Lin   Jordi Soria-Comas
Eleni Darra   Shou-De Lin   Georgios Spathoulas
Ning Ding   Wen-Ming Liu   Klara Stokes
Mickael Emirkanian   Zhen Liu   German Saez
Imran Erguler   Giovanni Livraga   Xiao Tan
Xinxin Fan   Flavio Lombardi   William Tankou
Chengfang Fang   Weiliang Luo   Donghai Tian
Oriol Farras   David Mandell-Freeman   Mehdi Tibouchi
Carmen Fernandez-Gago   Jian Mao   Hongbing Wang
Hua Guo   Seiichi Matsuda   Xi Xiong
Sara Hajian   Qixiang Mei   Jia Xu
Jinguang Han   Taher-Adel Mohamed   Zhaoyan Xu
Feng Hao   Pablo Najera   Zhi Xu
Takuya Hayashi   Gregory Neven   Qiang Yan
Kelly Heffner-Wilkerson Takashi Nishide   Bo-Yin Yang
Julio Hernandez-Castro   Kazumasa Omote   Guomin Yang
Yoshiaki Hori   Soyoung Park   Zhenxin Zhan
Honggang Hu   Andreas Pashalidis   Fangfang Zhang
Yupu Hu   Gerardo Pelosi   Jialong Zhang
Xinyi Huang   Gabor Pek   Mingwu Zhang
Yun-Ju Huang   Bo Qin   Jianjie Zhao
Sebastiaan Indesteege   Evangelos Rekleitis   Qingji Zheng
Georgios Kambourakis   Ruben Rios
Ioanna Kantzavelou   Panagiotis Rizomiliotis

Table of Contents

Attacks

Replacement Attacks on Behavior Based Software Birthmark . . . .　1
    Zhi Xin, Huiyu Chen, Xinche Wang, Peng Liu, Sencun Zhu,
    Bing Mao, and Li Xie

Attacking Traitor Tracing Schemes Using History Recording and
Abrupt Decoders. . . . .　17
    Aggelos Kiayias and Serdar Pehlivanoglu

How to Find Short RC4 Colliding Key Pairs. . . . .　32
    Jiageng Chen and Atsuko Miyaji

Protocols

A Formal Approach to Distance-Bounding RFID Protocols . . . .　47
    Ulrich Dürholz, Marc Fischlin, Michael Kasper, and Cristina Onete

MASHA – Low Cost Authentication with a New Stream Cipher . . . .　63
    Shinsaku Kiyomoto, Matt Henricksen, Wun-She Yap,
    Yuto Nakano, and Kazuhide Fukushima

Toward Pairing-Free Certificateless Authenticated Key Exchanges . . . . . .　79
    Hu Xiong, Qianhong Wu, and Zhong Chen

Public-Key Cryptosystems

Security Analysis of an RSA Key Generation Algorithm with a Large
Private Key . . . .　95
    Fanyu Kong, Jia Yu, and Lei Wu

Adaptive Secure-Channel Free Public-Key Encryption with Keyword
Search Implies Timed Release Encryption . . . .　102
    Keita Emura, Atsuko Miyaji, and Kazumasa Omote

The n-Diffie-Hellman Problem and Its Applications . . . .　119
    Liqun Chen and Yu Chen

Network Security

RatBot Anti-enumeration Peer-to-Peer Botnets . . . .　135
    Guanhua Yan, Songqing Chen, and Stephan Eidenbenz

Detecting Near-Duplicate SPITs in Voice Mailboxes Using Hashes   152
  Ge Zhang and Simone Fischer-Hübner

Software Security

Multi-stage Binary Code Obfuscation Using Improved Virtual
Machine   168
  Hui Fang, Yongdong Wu, Shuhong Wang, and Yin Huang

Detection and Analysis of Cryptographic Data Inside Software   182
  Ruoxu Zhao, Dawu Gu, Juanru Li, and Ran Yu

System Security

SudoWeb: Minimizing Information Disclosure to Third Parties in Single
Sign-On Platforms   197
  *Georgios Kontaxis, Michalis Polychronakis, and
  Evangelos P. Markatos*

Hello rootKitty: A Lightweight Invariance-Enforcing Framework   213
  *Francesco Gadaleta, Nick Nikiforakis, Yves Younan, and
  Wouter Joosen*

Opacity Analysis in Trust Management Systems   229
  Moritz Y. Becker and Masoud Koleini

Database Security

On the Inference-Proofness of Database Fragmentation Satisfying
Confidentiality Constraints   246
  Joachim Biskup, Marcel Preuß, and Lena Wiese

Round-Efficient Oblivious Database Manipulation   262
  Sven Laur, Jan Willemson, and Bingsheng Zhang

A Privacy-Preserving Join on Outsourced Database   278
  Sha Ma, Bo Yang, Kangshun Li, and Feng Xia

Privacy

APPA: Aggregate Privacy-Preserving Authentication in Vehicular
Ad Hoc Networks   293
  Lei Zhang, Qianhong Wu, Bo Qin, and Josep Domingo-Ferrer

Assessing Location Privacy in Mobile Communication Networks   309
  *Klaus Rechert, Konrad Meier, Benjamin Greschbach,
  Dennis Wehrle, and Dirk von Suchodoletz*

How Much Is Enough? Choosing ϵ for Differential Privacy............. 325
Jaewoo Lee and Chris Clifton

Digital Signatures

Non-interactive CDH-Based Multisignature Scheme in the Plain Public
Key Model With Tighter Security................................. 341
Yuan Zhou, Haifeng Qian, and Xiangxue Li

An Efficient Construction of Time-Selective Convertible Undeniable
Signatures 355
Qiong Huang, Duncan S. Wong, Willy Susilo, and Bo Yang

Efficient Fail-Stop Signatures from the Factoring Assumption.......... 372
Atefeh Mashatan and Khaled Ouafi

Author Index... 387

Replacement Attacks on Behavior Based Software Birthmark

Zhi Xin[1], Huiyu Chen[1], Xinche Wang[1], Peng Liu[2],
Sencun Zhu[2], Bing Mao[1], and Li Xie[1]

[1] State Key Laboratory for Novel Software Technology,
Department of Computer Science and Technology,
Nanjing University, Nanjing 210093, China
{zxin,maobing}@nju.edu.cn,
{mylobe.chen,xinchewang}@gmail.com
[2] The Pennsylvania State University,
University Park, PA 16802, USA
sxz16@psu.edu, pliu@ist.psu.edu

Abstract. Software birthmarks utilize certain specific program characteristics to validate the origin of software, so it can be applied to detect
software piracy. One state-of-the-art technology on software birthmark
adopts dynamic system call dependence graphs as the unique signature
of a program, which cannot be cluttered by existing obfuscation techniques and is also immune to the no-ops system call insertion attack.
In this paper, we analyze its weaknesses and construct replacement attacks with the help of semantics-equivalent system calls to unlock the
high frequent dependency between the system calls in an original system
call dependence graph. Our results show that the proposed replacement
attacks can destroy the original birthmark successfully.

Keywords: software birthmark, replacement attack.

1 Introduction

Software piracy has been deemed as one of the most serious threats to intellectual
property. A band of companies with high research and development input are
suffering from software counterfeiting. Business software alliance publishes a
study report about illegal copying and unauthorized resale of applications every
year, indicating 51.4 billion of huge loss in 2009 [25]. Microsoft just accused La
Familia, Mexico Drug Cartel, for suspicious piracy of Office 2007 in Mexico, and
this unauthorized business earned $2.2 million dollars every day [24]. To protect
their intellectual property, software vendors have adopted several categories of
approaches, including software watermark [26], tamper-proofing [28], obfuscation
[7], and software birthmark [3,4,5,6,8,15]. Among them, software watermark and
birthmark are both designed against program theft.

There are already a band of software watermarking methods [2,26] proposed in
the past, which embed "copyright notice" inside the original source code to identify the ownership of the software. Also, software birthmarks [3,8] identify the

X. Lai, J. Zhou, and H. Li (Eds.): ISC 2011, LNCS 7001, pp. 1–16, 2011.
© Springer-Verlag Berlin Heidelberg 2011

authorship by unique program characteristics in the event of suspected theft. However, these watermarks and birthmarks are all easy to be radically tampered by semantics-preserving obfuscation [16,32] or compiler's optimization. So researchers present dynamic API based birthmark [4,6,31] and behavior based software birthmark [15], which are considered to be resistant to the existing obfuscation methods. Dynamic API based birthmarks like SCSSB (System call Short Sequence Birthmark) [31] identify a program with the partial trace of its system calls. Furthermore, behavior based software birthmark not only considers the sequence of system calls but also utilizes the dependence relationships between system calls to solidify the uniqueness of birthmark formed as *SCDG* (system call dependence graph). In the graph, each vertex indicates a system call, and each edge means dependence relationship between two system calls. Also, the insight that SCDG captures distinctive program characteristics is accepted by other security tasks, such as behavior based malware detection [9,11]. Compared to the plentiful research works that have been done in birthmark-sabotage obfuscation methods, the methods to evade dynamic API based birthmarks like SCSSB and SCDG seem absent at the present time. Previously, a kind of evasion attack called *mimicry attack* [19], has been designed to defeat a host-based intrusion detection system (IDS). It can evade the IDS's detection model by inserting *no-ops* system calls. Although no-ops system call insertion in spirit could also be used to deceive SCSSB in a similar way, it is incapable of fooling the the SCDG approach, because the inserted system calls would not have any dependency relationship with the existing ones. That is, no-ops system call insertion cannot directly hurt the effectiveness of the SCDG approach.

   In this paper, we focus on designing evasion attacks on the dynamic API based SCDG birthmarks. In particular, we propose a novel system call replacement technique which tangles the existing dependence relationships without altering the original program semantics and then "fool" the SCDG comparison algorithm. Since the birthmark is a graph, the evasion approach must modify the graph significantly and transform the dependence relationships to distinct formats. Following this thought, there are four candidate techniques to achieve it: (1) remove original vertex and edge; (2) insert new vertex and edge; (3) replace original vertex; (4) replace original edge. Based on this observation, we analyze the feasibility with each of these four thoughts. The first one is rarely adopted because only the meaningless system calls can be deleted without alerting the original function of program. Indeed, compiler may leach such useless system calls already. The second method looks working since it certainly modifies the original graph to another one. However, in practice it does not work either because all the insertions only increase the size of the whole graph in the periphery but not the subgraph that indicates the uniqueness of a program. We illustrate this problem in more detail in Section 3. The third one is feasible by replacing a system call with a semantic equivalent one, such as replacing *lseek* with *llseek*. However, even if we can summarize all the mappings following this strategy, a new type of birthmark can be constructed easily to defeat this attack by treating the semantic-equivalent ones the same [15].

Finally, we adopt the fourth thought by replacing an original edge with a new vertex and two new edges. This replacement approach affects the nature of the graph. However, the embedded system calls must establish new dependence relationship with two original system calls and also not generate any side effect to original program semantics. We solve this challenging problem with several elaborate system calls and parameters to make it "no-op" even it has dependence relationship with the existing ones. And we also ensure the new-born dependence relationships are common so that the replacement patterns cannot be recognized and filtered facilely unless very complicated semantic analysis is involved in the birthmark comparison algorithm, which is hard. Our replacement code can be loaded as a specific dynamic library before standard dynamic library like *libc.so* to perform the replacement in runtime. As a summary, the contributions of our work can be enumerated as follows:

1. We designed and implemented a novel system call replacement technique to modify the dependence relationships between system calls as a dynamic library without changing any existing source code.
2. We summarized the rules to select the replacement system calls that have dependence relationships with original system calls with the "no-op" semantic and are also hard to be recognized.
3. To the best of our knowledge, our evasion research is the first one to deceive the dynamic behavior based birthmark, which is the state-of-the-art technology on software birthmark.

Although, this work is launched from a software copyright robber's perspective, we never intend to strengthen software piracy, but rather to explore the limitations of the current software birthmark technology and also encourage the development of better software birthmark techniques.

The remainder of this paper is structured as follows. First, Section 2 presents our system call replacement framework and the rules to select replacement attacks. Then Section 3 explains the effectiveness of replacement attacks. Section 4 details the implementation and Section 5 evaluates our empirical experiments. Section 6 discusses the limitations of our work. In addition, we show the difference between our work and related works in Section 7. Finally, in Section 8, we conclude the whole work.

2   Our Replacement Attack Framework

First, we will present the overview of our attack framework. Then we will illustrate the rationale of our various replacement attacks. Finally we will summarize the rules on selecting proper system calls used for replacement.

2.1   Framework Overview

Our purpose is to deceive the isomorphism algorithms used by SCDG verification. First, we define the system call dependence relationship, which includes two system calls, dependence variable and dependence style.

Definition 1 (SCDR: System Call Dependence Relationship). *The system call dependence relationship between two system calls can be represented as a 4-tuple SCDR = (R,M,S,D) where*

- *R means Relier, whose output can be delivered to the posterior one.*
- *D means Dependent, whose input comes from the previous one.*
- $\alpha : D \to R$ *means Dependent depends on Relier.*
- *M means Medium, which represents the variable passed from relier to dependent.*
- *S means Style, which describes the output method from a relier, including RV (return value) and PA (parameter). In another word, in RV situation, the Medium is a return value from a relier, and relatively in PA situation, the Medium exists as a parameter of the relier, which is filled during system calling.*

As explained before, *we launch replacement attacks by replacing an existing edge with a new vertex and two new edges*, which transforms the current graph into a new graph. An example has been presented in Fig 1. The dependence relationship *read→open* has been replaced into *read→lseek→open*. The system call *lseek* is declared as "off_t output lseek (int *filedes*, off_t *offset*, int *whence*)" in the Unix environment [14], which is used to set the current file offset. In this example, because the third parameter is SEEK_SET, which means it does addressing from the every beginning of a file, the return value *output* should be equal to the second parameter of *lseek* plus zero. In other words, *lseek* merely assigns *fd1* to *output*. As such, the replacement still retains equivalent semantics.

After presenting a replacement attack example, we describe the set of dependence relationships that our replacement technique may attack. As we know, the total number of permutations of all the system calls is enormous, and also

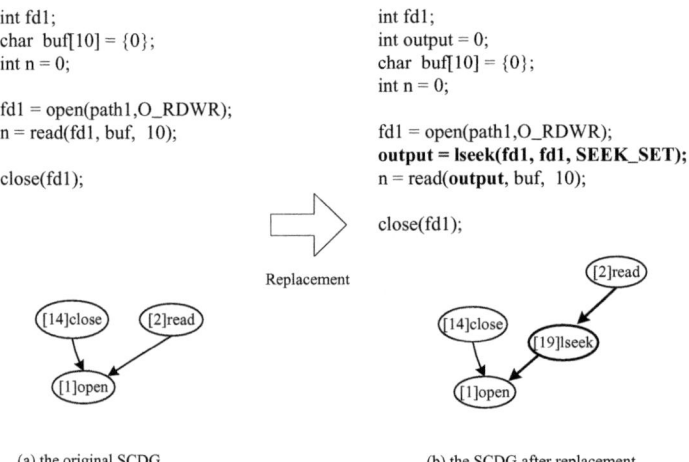

(a) the original SCDG   (b) the SCDG after replacement

Fig. 1. An example of replacement attacks in open→read relationship

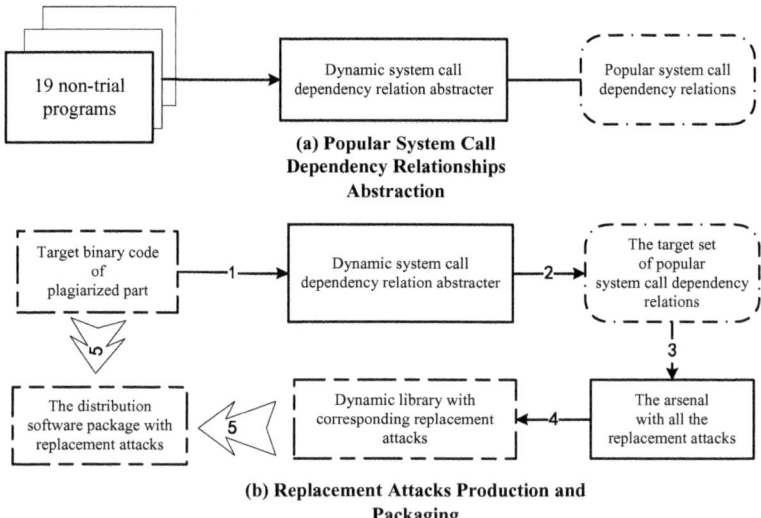

Fig. 2. System overview

every pair of system calls occurs with a distinguishing frequency for each program. We record them from 19 diverse non-trivial programs by abstracting and counting their system call dependence relationships. The whole procedure can be presented in Fig.2 (a), and we get all the system call dependence relationships sorted by their occurrence frequencies. Then we choose the popular ones by setting a threshold *alpha*, which is *1000* in this paper. Finally, we get 11 kinds of popular system call dependence relationships presented in Table 1 as *popular system call dependence relationships* with their names and system call numbers. We believe that as long as our replacement attacks can disrupt these prevalent relationships, they can deceive the SCDG verification algorithm effectively and efficiently.

Also in Fig.2(b), we explain the whole procedure how the replacement attacks deceive the SCDG birthmark. At the first place, we obtain the native binary code of a program whose source code is plagiarized. Then, we feed it to our dynamic system call dependence relationship abstracter, which distills dependence relationships that belong to *popular system call dependence relationships*. Against this target set, our tool assembles the specific dynamic library with replacement attacks. Then we modify the *.dynamic* section of the target binary code file to ensure that the specific dynamic library is loaded before the standard dynamic library is. The details are discussed in Section 4. Finally, this specific dynamic library is packaged into the software package for distribution.

2.2 Various Replacement Attacks

In this section, we exhibit our replacement attack arsenal and also describe the skills that we used to construct the replacement attack. We summarize all our

Table 1. The list of popular system call dependence relationships frequency

Programs Name	stat64 →read 195→3	open →read 5→3	read →open 3→5	close →open 6→5	access →read 33→3	llseek →open 140→5	munmap →mmap 91→192	llseek →write 140→4	llseek →mmap 140→192	mmap →open 192→5	open →write 5→4
					Popular System Call Dependence Relationships						
Flock 2.0.3	2	0	0	1	0	0	0	0	0	0	0
Epiphany 2.22.2	6823	3013	1802	1950	111	13	83	1	6	164	8
Amaya 10	4	0	0	1	0	0	0	0	0	0	0
Opera 9.52	7	0	0	1	0	0	0	0	0	0	0
Galeon 2.0.7	1261	510	582	527	127	15	82	0	12	153	6
AbiWord 2.4.6	62	75	293	299	63	3	122	1	1	213	92
KWord 1.6.3	62	161	566	363	63	149	249	10	82	132	23
LyX 1.5.3	5927	2592	1852	2323	171	130	622	6	90	602	850
Kile 2.0.0	691	279	2478	1339	10840	2110	1218	7	2079	93	104
Gedit 2.22.3	1261	2679	1841	1739	86	42	65	6	9	142	11
Bluefish 1.0.7	5332	2629	1602	1615	95	22	84	4	20	100	100
GNU Emacs 22.2.1	847	250	986	441	119	1914	224	3230	65	78	8
Vim 7.1.138	134	23	112	75	0	20	28	0	18	27	20
Pidgin 2.5.2	5244	2537	2200	1944	100	213	360	3	209	0	8
Kopete 0.12.7	0	0	19	27	0	0	733	11	1	21	2
Kmess 1.5	566	146	423	246	1071	32	154	10	30	88	20
GnoCHM 0.9.9	105	44	69	45	0	0	23	0	0	2	0
Evince 2.22.2	12680	5710	2609	2926	95	11	37	2	6	88	27
Evolution 2.22.3	6457	327	3601	2289	194	344	554	22	272	355	19
SUM	51663	21333	21035	18151	13135	5751	3916	3301	2900	2240	1292

replacement attacks for 11 popular system call dependence relationships to three categories based on the M (medium) type:

1. Handle the "file descriptors" medium. As we declared before, M (medium) is the dependence variable, such as *fd1* and *off_set* in Fig 1. For this category, there are four kinds of popular relationships as presented in Fig 3. We explain the mechanism based on one of them, *read→open*. The original code in the left top of Fig 3 shows this dependence relationship. We construct three types of replacements for this category, which are also effective to all of them:

   (a) "lseek" attack. This attack has been illustrated in Fig 1. *lseek* is used to set the file offset. When we insert *lseek* in the middle position, it breaks the original connection between *open* and *read*. Also, *lseek→open*, *read→lseek* and *read→lseek→open* are all common dependence relationships in programs. So this type of replacement is hard to detect.

   (b) "dup and dup2" attack. "dup" and "dup2" system calls duplicate the file descriptor and also construct corresponding internal data structure for file operations. However, our replacement just transforms dependence from original one into a duplicated one, which also sabotages the graph.

   (c) "fcntl" attack. "fcntl" can change the properties of a file that is already open. The parameter "F_DUPFD" means duplicating an existing descriptor. So it works just like "dup". The return value *fd1* is assigned to *fd2* equally.

2. Handle the "file path buffer" medium. Related to this category, the M (medium) always occurs as a buffer which stores the string of some file path. The S (style) is mostly PA (parameter). Variable *path_buffer* in the right of Fig 3 reveals this situation, which is a parameter of the *read* system call and is filled with data read from the file. It is passed into the system call

stat, which reads the file state in the path of *path_buff*. We offer two kinds of attacks here:

(a) "link" attack. System call "link" creates an extra directory entry point to the target file's i-node, the kernel data structure for a file. Later, the *stat* system call gets the same file state information from another temporary path "_xz_path", which is a temporary file path. Again, this type of new dependence situation is common, so it is hard to detect such replacements.

(b) "rename" attack. System call "rename" changes a file or a directory to a new name, so it also modifies the dependence variable M (medium) from *path_buffer* to _xz_path. It works just like the "link" attack.

3. Handle the "memory address" medium. There are two kinds of popular dependence relationships, which are connected with memory address M (medium), including munmap→mmap2 and llseek→mmap2. System call "mmap2" is declared as "void* mmap (void *addr, size_t *len*, int *prot*, int *flag*, int *filedes*, off_t *off*);". The programmers use this system call to map a given file to a region in memory. Its return value represents the starting address of the mapped area. For this category, we have not found any safe replacement.

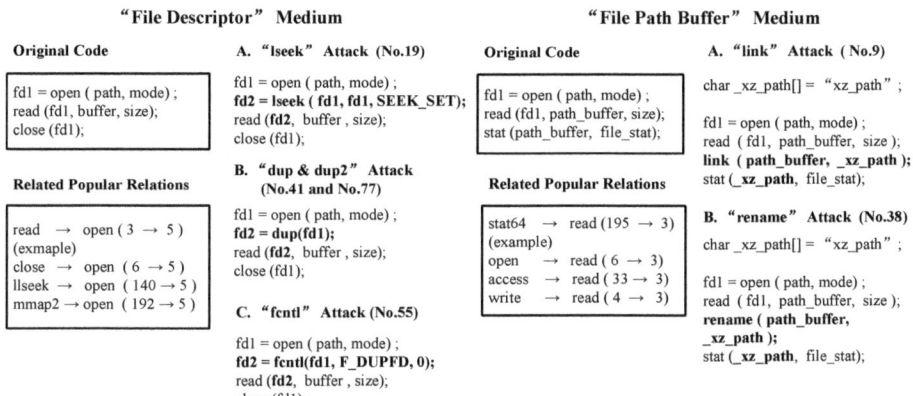

Fig. 3. The replacement attacks related to "File Descriptor" and "File Path Buffer" Medium

2.3 The Rules of Replacement Attacks Construction

From the previous analysis, we can distill skills on how to find these equivalent replacements. First, the system calls with the same data type of parameters and return values are good candidates, just like "dup" with *integer* in both parameters and "link" with two *char** parameters. For this rule, we have to emphasize two things: the "return value" here means the output in semantic, which could be either a real return-value or a parameter; the "same type" could be two types with the same size but not exactly the same language type, just as the return-value of "lseek" with *off_t* and one of its parameters with *integer* which can be considered as the same type. Second, the context after replacement should be as common as possible, which means if there is a novel approach to match this replacement, it cannot easily decide whether this is a replacement or just common code. We evaluate whether some replacement is common by searching it in real code like 19 non-trivial programs. And all of our replacement scenarios can be easily found. Although our skills are presented in the Unix environment, they also can be used for other platforms, like Windows.

3   Effectiveness of Replacement Attacks

In this section, we first compare various graph-subgraph isomorphism algorithms that can be used in SCDG verification and then choose one of them as the representative to discuss the effectiveness of our replacement method by comparing with new vertex and edge insertion attacks.

3.1   Graph-Subgraph Isomorphism Algorithms

Basically, there are three kinds of algorithms for (sub)graph isomorphism detection, including Nauty algorithm [10], Ullman algorithm [17] and VF algorithm [30]. First, McKay's Nauty algorithm reduces matching complexity by transforming the graphs to a canonical form for quick graph isomorphism judgement. However, it asks exponential time for matching. Furthermore, it cannot be used to solve the graph-subgraph isomorphism problem, which makes it unavailable for our experiment [12]. Another two alternative approaches are based on searching and branch pruning of unprofitable paths. One is a backtracking algorithm proposed by Ullmann [17], and the other is referred to as the VF algorithm [30] based on a depth-first search strategy.

The key idea of the Ullmann algorithm is the employment of a procedure called refinement procedure based on the depth-first brute-force search [17]. During the procedure, the algorithm continually checks whether any vertex in G_1 has no corresponding vertex in G_2. If no one is found, the algorithm terminates at its SUCCEED exit. Otherwise, the algorithm jumps to its FAIL exit. The refinement procedure is triggered after entering each node in the tree search, which results in a reduction of the number of successor nodes that must be searched. Relatively, VF has the similar principles to Ullmann, which uses a set of feasibility rules to reduce computational cost of the matching process. The feasibility rules can prune the search space significantly and speed up the matching process. Although both VF and Ullmann algorithms can specify a subgraph isomorphism, according to the experimental results [20], VF algorithm achieves a better performance in matching time, and the Ullmann algorithm is not always able to find a solution. So in our evaluation procedure we choose from the VF family the algorithm VF2, a higher version of the VF algorithm, which is also used by previous SCDG birthmark [15].

3.2   Effectiveness Analysis against VF Algorithm

Then we discuss the effectiveness by comparing the effect of two potential methods against the SCDG birthmark, including new vertex and edge insertion and
our edge replacement. In our case, each node has two attributes, the system call
number and a unique index number. Control dependence relationships between
system calls are denoted as directed edges. We present an example of system call
dependence graph in Fig 4: Fig 4-(a) shows the SCDG birthmark of the plaintiff
program; Fig 4-(b) presents the SCDG of the suspect program. It is obvious that
Fig 4-(a) is a subgraph of Fig 4-(b), and they are graph-subgraph isomorphism.
Then in Fig 5-(b), we insert a node b_8 with the attribute of *close* from the periphery and also replace an edge between *open* and *read* with a node c_8 with the
attribute of *lseek* in the middle and also two other new edges. The VF algorithm
describes the whole procedure with State Space Representation (SSR), which
allows one to simultaneously make syntactic and semantic comparison of the
pairs of nodes to be matched. First, it checks whether the two nodes have the
same system call number in semantics feasibility rules. Then, it checks whether
the node pair meets the syntax feasibility rules defined by the VF algorithm [12].
The internal procedure of the VF algorithm contains three steps. First, for each
intermediate state s (initial state s_0 contains no component), the algorithm computes the candidate set $P(s)$, which is composed by vertexes connected to the
mapped ones in $M(s)$. Second, for each node pair p in $P(s)$, the feasibility rules
are evaluated, which are used to judge whether to accept the new node. Then if
they succeed, state s' would be obtained by adding p. The same operations will
be executed in a recursive way with the input of s'. Otherwise, another node
pair in $P(s)$ will be considered. Until $M(s)$ that records the recognized mapping
nodes pair covers all the nodes of G_1, the algorithm specifies an isomorphism.
Otherwise, the algorithm exits with failure.

The insertion method presented in Fig.5-(b) and our replacement method
affect the isomorphism algorithm in different ways. With new vertex and edge
insertion in Fig 5-(b), after the algorithm accepts the first pair of matched nodes,
it gets the matched node pair set $M(s)=\{ (a_1, b_1) ,(a_2, b_2)\}$. After several similar

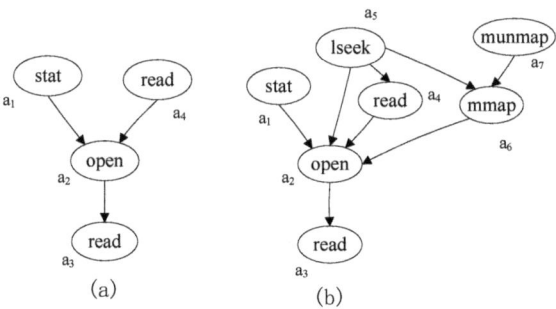

(a)   (b)

Fig. 4. (a) the SCDG birthmark of the plaintiff program;(b) the SCDG of the suspect
program

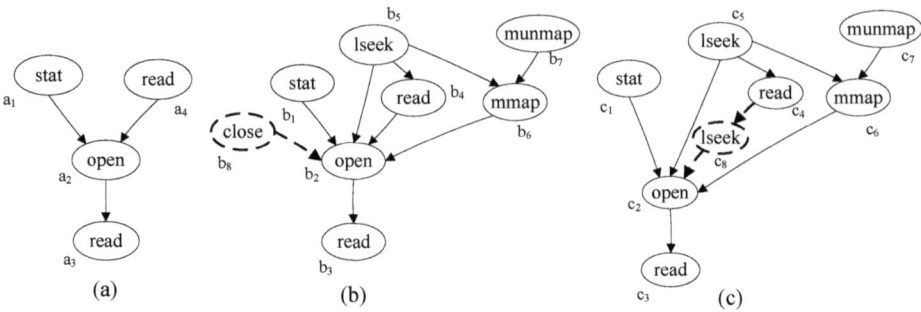

Fig. 5. (a) the SCDG of the plaintiff program;(b) the SCDG of the suspect program with the insertion of b_8;(c)the SCDG of the suspect program with the insertion of c_8

matching steps, it gets $M(s) = \{(a_1, b_1), (a_2, b_2), (a_3, b_3), (a_4, b_4)\}$, which covers all the nodes of Fig 5-(a). Then, the algorithm reports a graph-subgraph isomorphism between Fig 5-(a) and Fig 5-(b). However, with our edge replacement in Fig 5-(c), after the algorithm checks the pairs of (a_1, c_1), (a_2, c_2) and (a_3, c_3), it gets the $P(s) = \{(a_4, c_5), (a_4, c_6), (a_4, c_8)\}$. In other words, a_4 (*read*) has three matching candidates but none of them has the same system call number because both c_5 and c_8 represent *lseek* and c_6 contains *mmap*. For the plaintiff program, there is no node left. So the algorithm exits with a failure. The replacement attack is successful.

To summarize, insertion of new vertices and edges only enlarges the whole graph but does not modify the characteristic-carrying subgraph, so it will not evade the software birthmark evaluation. Relatively, our intermediate replacements break the original structure of the mapping component of the graph, which interferences the search of the VF algorithm and evades the verification.

4   Implementation

Our implementation includes two parts: one is dynamic system call dependence relationship abstracter, which is implemented based on the SCDG construction tool called Hawk [15], and the other is a tool called *SCReplacer* which generates a dynamic library materializing the corresponding replacement attacks.

The Hawk tool is used to capture the system call traces and dependence through machine code instrumentation at runtime. We extend Hawk to record the the inter-system-call dependencies and obtain the statistic results to identify the popular ones among all the system calls. Also, all the statistic results have been presented in Table 1.

The other part of work is mainly about how to construct the dynamic library that materializes the replacement attacks. We build our tool based on the ELF hacking tool ERESI [18], which is a multi-architecture binary analysis framework using a specific domain specific language for reverse engineering and program manipulation. We implement *SCReplacer* with the specific scripts to modify the ELF binary files. First, we review the dynamic linking procedure on ELF files. The *.dynamic* (Dynamic Symbol Table) section (in each ELF file) stores all the symbol information for dynamic linking, which consists of an array of entries. The entries are defined as Elf32_Dyn with two fields: one is called d_tag, which is the type of entry; the other is called d_un, which presents the offset in .dynstr section of the related string name. Inside these symbols, the dynamic linking libraries are all marked as DT_NEEDED in d_tag field. When the runtime linker starts the dynamic linking procedure, it traverses all of these library entries and links them one by one. Basically, there are three steps involved in the replacement attack as shown in Fig 6. First, we add our replacement code as a dynamic linking library and mount it in the *.dynamic* section, listing it as a DT_NEEDED entry by replacing a useless exiting one (e.g., a DT_DEBUG entry) stealthily, just like the entry 12 with *d_tag* as DT_DEBUG in Fig 6. Second, our approach reuses the library path which has existed in the *.dynstr* (Dynamic String Table) section that stores all the strings of dynamic symbols. When we change entry 12 (see Fig 6) from a DEBUG entry to a DT_NEEDED entry, we modify the d_val attribute of the entry into a partial path of an existing one (i.e., reusing an existing library path), so that our modification will be less noticeable, just like *rt.so.1* to *librt.so.1*. We can directly observe that *rt.so.1* can be a substring of *librt.so.1* from the fourth char. Third, we force our dynamic library to be loaded before the standard dynamic runtime library by moving our library closer to the beginning of the startup (library-loading) queue, such as switching the order between entry 00 and entry 12 in Fig 6.

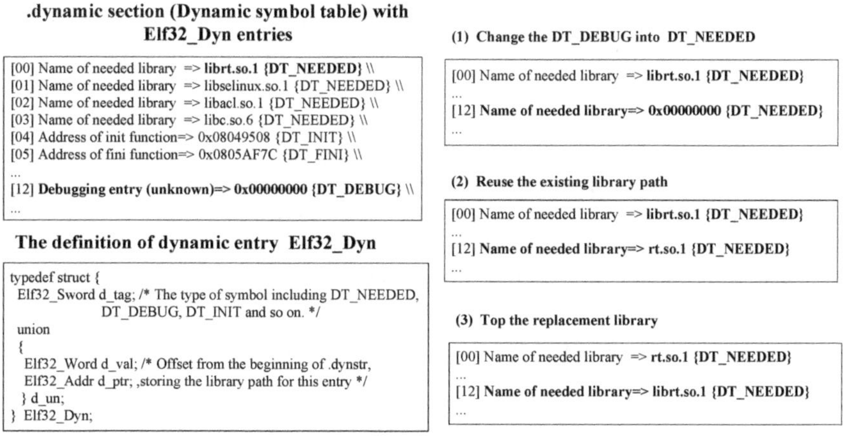

.dynamic section (Dynamic symbol table) with Elf32_Dyn entries

```
[00] Name of needed library  => librt.so.1 {DT_NEEDED} \\
[01] Name of needed library  => libselinux.so.1 {DT_NEEDED} \\
[02] Name of needed library  => libacl.so.1 {DT_NEEDED} \\
[03] Name of needed library  => libc.so.6 {DT_NEEDED} \\
[04] Address of init function=> 0x08049508 {DT_INIT} \\
[05] Address of fini function=> 0x0805AF7C {DT_FINI} \\
...
[12] Debugging entry (unknown)=> 0x00000000 {DT_DEBUG} \\
...
```

The definition of dynamic entry Elf32_Dyn

```
typedef struct {
  Elf32_Sword d_tag; /* The type of symbol including DT_NEEDED,
              DT_DEBUG, DT_INIT and so on. */
  union
  {
    Elf32_Word d_val; /* Offset from the beginning of .dynstr,
    Elf32_Addr d_ptr; ,storing the library path for this entry */
  } d_un;
} Elf32_Dyn;
```

(1) Change the DT_DEBUG into DT_NEEDED

```
[00] Name of needed library  => librt.so.1 {DT_NEEDED}
...
[12] Name of needed library=> 0x00000000 {DT_NEEDED}
...
```

(2) Reuse the existing library path

```
[00] Name of needed library  => librt.so.1 {DT_NEEDED}
...
[12] Name of needed library=> rt.so.1 {DT_NEEDED}
...
```

(3) Top the replacement library

```
[00] Name of needed library  => rt.so.1 {DT_NEEDED}
...
[12] Name of needed library=> librt.so.1 {DT_NEEDED}
...
```

Fig. 6. .dynamic section symbols

5 Evaluation

In this section, we present our experiments in a group of programs with effectiveness and performance overhead. To the best of our knowledge, there is only one SCDG birthmark analysis and production tool previously presented as a research work [15]. We evaluate our replacement attacks against it. We rebuild the VF2 subgraph-graph isomorphism algorithm by Networkx [23], which is the same evaluation algorithm used in SCDG comparison [15] but is much more convenient to check and display the intermediate results. We test our experiments with hardware configuration as Intel Core Duo CPU 2.53GZ and 512M memory and also with software configuration as Ubuntu 9.04 and Linux kernel 2.6.28.

First, we test the effectiveness. The birthmark verification group in previous work [15] include: GNU Aspell spell checker 0.60.5, which has been contained by three other programs, including Bluefish 1.0.7, Kword 1.6.3, and Lyx 1.5.3. First, we run the plaintiff program Aspell to check some spelling problems in a test file and extract the birthmark from all the subtrees. Next, we assemble the corresponding replacement attacks in specific dynamic libraries and run suspicious programs linked with such libraries. Then, all the subtrees of suspicious programs are extracted by using their spelling checking functions with the similar test files. Finally, we verify the birthmark with suspicious program subtrees in VF isomorphism algorithm. Also, three basic conditions about birthmark accelerate the verification procedure: (1) The minimal size of subtrees must be greater than 15; (2) A suspicious subgraph should be at least 0.9 the size of the birthmark; (3) Every node has an attribute of system call number so that it can prune the matching procedure earlier. We call the subtrees which satisfy these conditions as "candidate subtrees". As a result, we filter through the total 83 subtrees of Aspell to get the birthmark with 117 vertices and 123 edges , which is also located in all the suspicious programs. However, after replacements, we can observe the obvious differences in the system call distribution between Aspell birthmark subtree and the relative subtrees in suspicious programs as presented in Fig 7. The relative subtrees are generated by replacement attacks including "lseek", "link" and "dup". All the verification procedures fail indeed after replacement. That is, because of our replacement attacks, software like Bluefish and Lyx can stealthily use Aspell without being charged.

Second, we test the performance overhead of our replacement attacks. We run all the suspicious programs with similar test files. Then every program uses its

Table 2. The effectiveness of replacement attacks

Plaintiff program	Graph profile						Verification interruption
	before replacement			after replacement			
	candidate subtrees	vertexes	edges	candidate subtrees	vertexes	edges	
Bluefish	10	20599	13665	10	26154	13708	✓
Kword	15	41352	3528	15	43376	3645	✓
Lyx	22	47244	29268	22	54192	29315	✓

Table 3. The overhead of replacement attacks

Plaintiff program	Extra overhead							
	before replacement				after replacement			
	$C_1(s)$	$C_2(s)$	$C_3(s)$	AVG_c	$M_1(s)$	$M_2(s)$	$M_3(s)$	AVG_m
Bluefish	7.032	7.809	7.089	7.310	7.180	7.539	7.736	7.485
Kword	4.353	4.404	3.986	4.248	5.041	4.222	3.895	4.386
Lyx	4.322	4.056	4.458	4.289	4.804	4.877	4.662	4.768

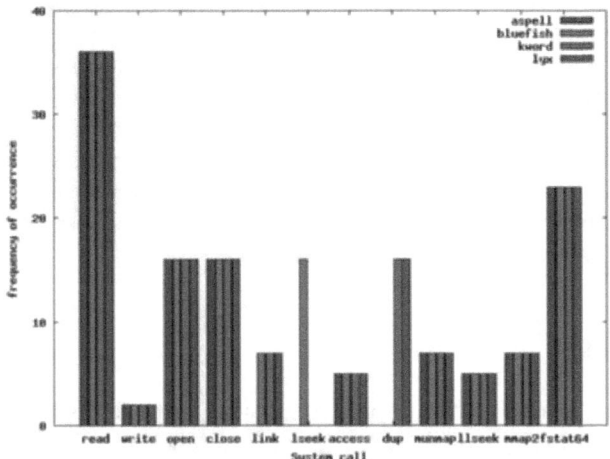

Fig. 7. The system call distribution of birthmark subtree

own spelling check function to capture these bugs and replace them with the right ones. We record the total time of our whole procedure, marked as $C_n(s)$. Also, we repeat the whole process after replacement and collect the total time (marked as $M_n(s)$). We can observe that the average overhead is very low: 2.39% for Bluefish, 0.68% for Kword and 11.18% for Lyx in Table 3. These overhead basically all come from the embedded system calls.

6   Limitations and Future Work

We are aware of several defects of our work. The first thing is that we have not gotten proper replacement attacks for the *memory address* M (medium), such as munmap→mmap2 and llseek→mmap2. But from the current experiment result, it may not be a big problem. The attacks that we already select deceive the SCDG birthmark verification well. Next, our replacement attacks exist as a dynamic library packaged with its original code. Even it can be masqueraded to be a common dynamic library that is required by lots of other commercial off-the-shelf software products, there is still some clue to perceive the library interruption. We believe that the compiler-based system call insertion technique should be a better choice, which transforms the system call replacement into the native code transparently. We take it as a future work.

7   Related Work

Software watermark techniques can be inadvertently sabotaged by semantic-equal compiler optimizations like GNU GCC or be bypassed by a determined attacker [13], such as additive attacks. Also, there are a band of obfuscation methods [16,32] which can damage various birthmarks [3,5,8]. These obfuscation tools mainly clutter the grammar structures like data structure, program control flow and so on, which are all subject to the range of obfuscating transformations [7]. But all of these tools cannot prejudice the birthmark based on dynamic API sequence [21] and also the SCDG [15]. However, there are already dynamic API related evading techniques existing in intrusion detection system field called *mimicry attack* [19,29], which deceives an IDS by no-ops system call insertion. But for SCDG birthmark, to the best of our knowledge, our work is the first attempt toward evading it.

8   Conclusion

As the safeguard against software piracy, software birthmarks have been broadly to evaluate the copyright of software programs. The latest software birthmark technique constructs signature from dynamic system call dependence graph, which cannot be evaded or sabotaged by current obfuscation methods. In this paper, we capture its defect in evaluation procedure and construct the semantic-safe system call replacements to deceive the birthmark successfully with low performance overhead.

Acknowledgments. This work was supported in part by grants from the Chinese National Natural Science Foundation (61073027, 90818022, and 61021062), and the Chinese 973 Major State Basic Program(2009CB320705). And Peng Liu was partially supported by AFOSR FA9550-07-1-0527 (MURI), ARO W911NF-09-1-0525 (MURI), and NSF CNS-0905131.

References

1. Collberg, C., Thomborson, C.: Software watermarking: models and dynamic embeddings. In: POPL 1999: Proceedings of the 26th ACM SIGPLAN-SIGACT Symposium on Principles of Programming Languages. ACM, New York (1999)
2. Collberg, C., Carter, E., Debray, S., Huntwork, A., Kececioglu, J., Linn, C., Stepp, M.: Dynamic path-based software watermarking. SIGPLAN Not. (2004)
3. Myles, G., Collberg, C.S.: Detecting Software Theft via Whole Program Path Birthmarks. In: Zhang, K., Zheng, Y. (eds.) ISC 2004. LNCS, vol. 3225, pp. 404–415. Springer, Heidelberg (2004)

4. Schuler, D., Dallmeier, V., Lindig, C.: A dynamic birthmark for java. In: ASE 2007: Proceedings of the Twenty-Second IEEE/ACM International Conference on Automated Software Engineering. ACM, New York (2007)
5. Tamada, H., Nakamura, M., Monden, A.: Design and evaluation of birthmarks for detecting theft of Java programs,
http://citeseerx.ist.psu.edu/viewdoc/summary?doi=10.1.1.98.7502;
http://se.naist.jp/jbirth/papers/tamada04iasted.pdf
6. Tamada, H., Okamoto, K., Nakamura, M., Monden, A., Matsumoto, K.-i.: Dynamic software birthmarks to detect the theft of Windows applications. In: Proc. Int. Symp. on Future Software Technology 2004 (2004)
7. Collberg, C., Thomborson, C.: A taxonomy of obfuscating transformations. Technical report 148, The University of Auckland (1999)
8. Males, G., Collberg, C.: K-gram based software birthmarks. In: SAC 2005: Proceedings of the 2005 ACM Symposium on Applied Computing. ACM, New York (2005)
9. Christodorescu, M., Jha, S., Kruegel, C.: Mining specifications of malicious behavior. In: ESEC-FSE 2007: Proceedings of the the the 6th Joint Meeting of the European Software Engineering Conference and the ACM SIGSOFT Symposium on The Foundations of Software Engineering. ACM, New York (2007)
10. Garey, M.R.: Practical Graph Isomorphism. Congressus Numerantium, Canberra (1981)
11. Bayer, U., Comparetti, P.M., Hlauschek, C., Kruegel, C., Kirda, E.: Scalable, Behavior-Based Malware Clustering. In: Proceedings of the 16th Annual Network and Distributed System Security Symposium, NDSS 2009 (2009)
12. Cordella, L.P., Foggia, P., Sansone, C., Vento, M.: A (Sub)Graph Isomorphism Algorithm for Matching Large Graphs. IEEE Transactions on Pattern Analysis and Machine Intelligence 26(10) (October 2004)
13. Collberg, C., Thomborson, C.: On the Limits of Software Watermarking,
http://www.cs.arizona.edu/ collberg/Research/Publications/
CollbergThomborson98e/index.html
14. Richard Stevens, W.: Advanced Programming in the Unix Environment. Addison Wesley Longman Inc., Amsterdam (1992) ISBN: 0-201-56317-7
15. Wang, X., Jhi, Y.-C., Zhu, S., Liu, P.: Behavior based software theft detection. In: Proceedings of the 16th ACM Conference on Computer and Communications Security. ACM, New York (2009)
16. Zelix Pty Ltd: The Zelix KlassMaster Java obfuscator,
http://www.zelix.com/klassmaster/
17. Ullmann, J.R.: An Algorithm for Subgraph Isomorphism. Journal of the Association for Computing Machinery (1976)
18. ERESI team, the ERESI Reverse Engineering Software Interface (2011),
http://www.eresi-project.org/
19. Wagner, D., Soto, P.: Mimicry attacks on host-based intrusion detection systems. In: Proceedings of the 9th ACM Conference on Computer and Communications Security. ACM, New York (2002)
20. Foggia, P., Sansone, C., Vento, M.: A Performance Comparison of Five Algorithms for Graph Isomorphism. Journal of the Association for Computing Machinery (1999)
21. Wang, X., Jhi, Y.-C., Zhu, S., Liu, P.: Detecting Software Theft via System Call Based Birthmarks. In: Annual Computer Security Applications Conference, ACSAC 2009, December 7-11, pp. 149–158 (2009)

22. Zhang, X., Tallam, S., Gupta, R.: Dynamic slicing long running programs through execution fast forwarding. In: Processing of 14th ACM SIGSOFT Symposium on Foundations of Software Engineering (2006)
23. Networkx, the Python package for the creation, manipulation, and the study of complex networks (2011), `http://networkx.lanl.gov/`
24. Parrack, D.: Microsoft accuses Mexican drug cartel La Familia of selling bootleg Office software,
    `http://vista.blorge.com/2011/02/05/microsoft-accuses-mexican-drug-cartel-la-familia-of-selling-bootleg-office-software/`
25. International Planning and Research Corporation: Seventh annual BSA and IDC global software piracy study,
    `http://portal.bsa.org/globalpiracy2009/studies/09_Piracy_Study_Report_A4_final_111010.pdf`
26. Zhu, W., Thomborson, C., Wang, F.-Y.: A Survey of Software Watermarking. In: Kantor, P., Muresan, G., Roberts, F., Zeng, D.D., Wang, F.-Y., Chen, H., Merkle, R.C. (eds.) ISI 2005. LNCS, vol. 3495, pp. 454–458. Springer, Heidelberg (2005)
27. Collberg, C.S., Thomborson, C.: Watermarking, Tamper-Proofing, and Obfuscation - Tools for Software Protection. IEEE Transactions on Software Engineering, 735–746 (2002)
28. Aucsmith, D.: Tamper Resistant Software: An Implementation. In: Anderson, R. (ed.) IH 1996. LNCS, vol. 1174, pp. 317–333. Springer, Heidelberg (1996)
29. Forrest, S., Hofmeyr, S., Somayaji, A.: The Evolution of System-Call Monitoring. In: Proceedings of the 2008 Annual Computer Security Applications Conference (ACSAC 2008), pp. 418–430. IEEE Computer Society, Washington, DC, USA (2008)
30. Cordella, L.P., Foggia, P., Sansone, C., Vento, M.: Evaluating Performance of the VF Graph Matching Algorithm. Journal of the Association for Computing Machinery (1999)
31. Wang, X., Jhi, Y.-C., Zhu, S., Liu, P.: Detecting Software Theft via System Call Based Birthmarks. In: Proc. of the 25th Annual Computer Security Applications Conference, ACSAC (December 2009)
32. Collberg, C., Myles, G., Huntwork, A.: SandMark - A Tool for Software Protection Research. IEEE Security and Privacy 1(4) (2003)
33. Garey, M.R., Johnson, D.S.: Computers and Intractability: A Guide to the Theory of NP-Completeness. Freeman & co., New York (1979)

Attacking Traitor Tracing Schemes Using History Recording and Abrupt Decoders

Aggelos Kiayias[1,*] and Serdar Pehlivanoglu[2,**]

[1] Computer Science and Engineering,
University of Connecticut
Storrs, CT, USA
aggelos@cse.uconn.edu

[2] Computer Science and Engineering
Zirve University, Turkey
spehlivan38@gmail.com

Abstract. In ACM-DRM 2001, Kiayias and Yung [19] introduced a classification of pirate decoders in the context of traitor tracing that put forth traceability against history recording and abrupt pirate decoders. History recording pirate decoders are able to maintain state during the traitor tracing process while abrupt decoders can terminate the tracing operation at will based on the value of a "React" predicate. Beyond this original work, subsequently a number of other works tackled the problem of designing traitor tracing schemes against such decoders but with very limited success.

In this work, we present a new attack that can be mounted by abrupt and resettable decoders. Our attack defeats the tracing algorithm that was presented in [19] (which would continue to hold only for deterministic pirate decoders). Thus we show that contrary to what is currently believed there do not exist any known tracing procedures against abrupt decoders for general plaintext distributions. We also describe an attack that can be mounted by history recording (and available) decoders.

Keywords: Traitor Tracing. Abrupt Decoders.

1 Introduction

Traitor tracing is a piracy detection mechanism for a setting where content is illegally redistributed by licensed receivers to unlicensed ones, i.e., entities who were not intended originally to receive the content. The redistribution is achieved through the issuing of a malicious decoder that circumvents the access control system used by the content distribution system. More specifically, a traitor tracing scheme is a type of encryption for the multiuser setting where an authority is

---

* Research partly supported by NSF Awards 0447808, 0831304, 0831306.
** Research conducted in part while at Nanyang Technological University; supported by The Singapore National Research Foundation under Research Grant NRF-CRP2-2007-03

X. Lai, J. Zhou, and H. Li (Eds.): ISC 2011, LNCS 7001, pp. 17–31, 2011.

capable of performing an analysis to any working malicious decoder and recover at least one of the keys that was used in its construction.

Following standard terminology the decoder created by an adversary is called a pirate decoder, the user keys available to the adversary are called traitor keys (and the users that divulge their keys to the adversary are called traitors); the analysis process of the authority is called tracing. Traitor tracing emerged first in the work of Chor, Fiat and Naor [6] as a solution to threats against broadcast encryption shortly after the first non-trivial broadcast encryption scheme was presented by Fiat and Naor [12].

Of particular interest to the present paper is the work of Kiayias and Yung [19] that introduced a classification of stronger adversarial models for the construction of pirate decoders. Since the introduction of traitor tracing schemes it was assumed that the pirate decoder is "resettable" (i.e., it does not maintain state along the tracing process) and available (i.e., the pirate decoder remains available for as long the tracing process wishes to experiment with it). These assumptions were relaxed in [19] where history-recording (i.e., decoders that do maintain state throughout) and abrupt pirate decoders (i.e., decoders that may terminate tracing if they detect it) were considered instead. In the terminology of that paper, history-recording decoders were also called type-1 decoders, abrupt decoders were also called type-2 decoders, while decoders that combined both functionalities were called type-3 decoders, see Figure 1. In contrast, type-0 decoders were the decoders that were assumed in previous works such as that of [6]. Traitor tracing schemes were presented for type-2 and type-3 decoders that, in the latter case, utilized the ability to watermark (see e.g., [7]) the underlying plaintext space. For the case of type-3 decoders a general transformation was outlined (and applied to the scheme of [6]) that showed how type-0 schemes can be "lifted" to the type-3 setting if watermarking is available. Other works that followed up this model include [16,20,21].

Regarding the type-0 decoders, the positive results utilized a technique called "hybrid colorings" that was implicit in previous works and is ubiquitous in almost all traitor tracing schemes. This - almost universal - traitor tracing technique can be summarized in the following fashion: the tracing center prepares a new

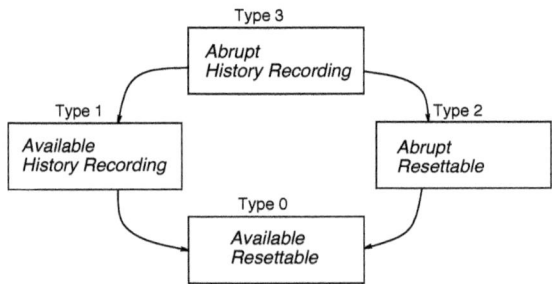

Fig. 1. The types of pirate decoders introduced in [19]

transmission instruction of the multiuser encryption scheme that enables some decoders to decrypt incorrectly, i.e. some receivers may fail to decrypt the transmission, or simply decrypt in a different fashion compared to other receivers. The coloring of the user population corresponds to the number of different ways that the receivers decrypt the special ciphertext that induces the coloring. The tracing center will interact with the pirate decoder and query the transmission that induce such colorings which eventually will enable the recovery of at least one traitor key embedded in the pirate decoder. Such technique is found useful for tracing against type-0 (resettable and available) pirate decoders. The schemes of the area employing this technique include[2,4,5,6,8,11,19,21,22,25].

Tracing abrupt type-2 decoders, in the absence of watermarking, proved to be not an easy task. In the original modeling of the classes of advanced pirate decoders in [19] there was a simplifying assumption that was made : the abrupt decoders utilized a "React" predicate to determine when to terminate tracing that was modeled as a deterministic predicate. Based on such assumption [19] proved that hybrid colorings are useful in tracing such abrupt decoders, at least in a setting where some ambiguity in tracing is permitted[1]. This positive result was then extended to the first proposal for a public-key traitor tracing scheme with sublinear ciphertext length and unambiguous tracing by Matsushita and Imai [21][2].

Our Contribution. While the assumption of a deterministic React predicate might be reasonable in the description of a plausibility result it can be easily argued to be unrealistic. Pirate decoders may behave in arbitrary fashion and not utilize an explicit React predicate that is deterministic and known to the tracer. In the present paper we address this issue. Specifically :

We consider the general case of abrupt decoders when the React predicate is unknown to the tracer and incorporated as part of the program of the pirate decoder. We then show that such abrupt decoders are *untraceable* for the hybrid colorings traitor tracing technique. It is worth noting that in [19] it was stated that the results would apply to probabilistic React predicates; while this holds true it still assumes that the tracer possesses the knowledge of the probabilistic behavior of React and this can be taken into account by the tracing procedure. If this capability is missed, as it may very well be in practice, our new negative result applies. For completeness we also present in full detail the attack that was alluded to in [19] against tracing for available history-recording decoders. Our result for abrupt decoders can in fact be seen as a probabilistic version of the attack with history-recording decoders and from the adversarial point of view can be seen as a tradeoff between memory and error-probability.

Related Work. After the introduction of traitor tracing, a number of subsequent constructions were presented [1,2,4,5,8,10,13,15,18,24,26,27,28] improving either the security modeling or the efficiency parameters of these schemes.

---

[1] In fact it was also claimed that it is possible to extend the results to the probabilistic setting but this turns out to be false as we will see herein.

[2] However, this scheme was broken by [20] and by [16] due to some other issues, i.e. the scheme is not even traceable against type-0 decoders.

Two other tracing strategies based on a variant of hybrid colorings were put forth in [23] called 'binary search tracing' and 'noisy binary search tracing'. The work by [17] gives an improvement on the round complexity of the tracing technique that is the number of rounds of interaction that are required between the tracing authority (or simply the tracer) and a rogue device in order for the tracer to establish the desired identities. A new tracing strategy with such an improvement was presented in [17] which relies on an application of fingerprinting codes superimposed on the tracing process. We note that all these three tracing strategies are succesful against only pirate decoders of type-0.

2   Traitor Tracing: Definitions

A traitor tracing scheme is based on an underlying encryption mechanism called a multiuser encryption scheme (ME) that is a triple (**KeyDist, Transmit, Receive**) of algorithms. The parameter of the scheme is the number of users n, the number of receivers and is associated with three sets $\mathsf{K}, \mathsf{M}, \mathsf{C}$ corresponding to the sets of keys, plaintexts and ciphertexts respectively. We next describe the I/O of these procedures below:

-   **KeyDist.** It is a probabilistic algorithm that on input 1^n, it produces $(tk, ek, sk_1, \ldots, sk_n)$. The decryption key sk_i is to be assigned to the i-th user while ek is the encryption key. The tracing key tk is some auxiliary information to be used for tracing that may be empty.
-   **Transmit.** It is a probabilistic algorithm that given a message $m \in \mathsf{M}$, it prepares an element $c \in \mathsf{C}$. We will write the following to denote the distribution of the output ciphertext for a given message: $c \leftarrow$ **Transmit**(ek, m).
-   **Receive.** It is a deterministic algorithm that on input c sampled from **Transmit**(ek, m) and a user-key sk_i for some $i \in [n]$ where $(tk, ek, sk_1, \ldots, sk_n) \leftarrow$ **KeyDist**(1^n), it either outputs m or fails. Note that **Receive** can also be generalized to be a probabilistic algorithm but we will not take advantage of this here.

The above determine the syntax of the algorithms that define a multiuser encryption scheme ME. We expect from such a scheme to satisfy correctness in the usual sense. In particular we require that: for any $n \in \mathbb{N}$, for any message $m \in \mathsf{M}$ and for any $u \in [n]$, it holds that

$$\mathbf{Prob}[\mathbf{Receive}(\mathbf{Transmit}(ek, M), sk_u) \in m] = 1$$

where $(tk, ek, sk_1, \ldots, sk_n)$ is distributed according to **KeyDist**(1^n). Note that the above may also generalized to hold with overwhelming probability.

A q-ary generalization of the above scheme takes input as a vector of q messages and is called the q-ary multiuser encryption scheme. In this setting the correctness is generalized to return any of the q messages that are given as input to the encryption algorithm (any of them would be considered a valid outcome for the decryption operation).

Security. The security requirement from a multiuser encryption scheme is standard: the scheme is semantically secure to an outsider who does not possess any of the private keys. This is a quite standard notion for which we skip the details of the corresponding semantic security game that is considered between the challenger and the adversary. We say that the scheme is ε-semantically secure if the adversary winning the semantic security game has advantage at most ε.

In an alternative setting where the hybrid encryption approach is employed, the content transmission operates at two levels: first, a one-time content key k is selected and encrypted with the multiuser encryption scheme. Second, the actual message will be encrypted with the key k and will be transmitted alongside the encrypted key. It follows that a minimum requirement would be that the scheme ME should be sufficiently secure to carry a cryptographic key k. As an encryption mechanism this is known in the context of public key cryptography as a "Key Encapsulation Mechanism" [9]. The security model for such formalization approach will focus on the type of security that needs to be satisfied by a multiuser encryption scheme in order to be used as a key encapsulation mechanism. Such a formalization can be found in [17].

Black-Box Traitor Tracing. A black box traitor tracing scheme involves a fourth algorithm **Trace**^{PD} which is an algorithm accessing a pirate decoder PD which can be considered as a probabilistic circuit that takes as input a ciphertext c and outputs some message m or fails. The tracing algorithm **Trace**^{PD} is given as input the tracing key tk, and then queries the pirate decoder PD as an oracle. It finally outputs a set S which is a subset of $\{1, \ldots, n\}$.

In this setting, the tracer has merely black-box access to the pirate decoder. Black-box traitor tracing may in some cases allow tracing to be performed remotely without the physical availability of the pirate decoder.

The major challenge in the black-box traitor setting is to extract information regarding the original keys utilised in the construction of the pirate decoder. The tracer will communicate with the pirate decoder using a set of specially crafted queries. These queries will not be necessarily normal transmissions as the tracing center is allowed to communicate with the decoder in an arbitrary way. The response of the decoder may be equal to the decrypted plaintext, or be simply of binary form, essentially "yes", in case of returning the content in the cleartext form, or "no", in case of responding arbitrarily or jamming.

In our exposition, we will use the threshold σ to impose the adversarial constraint related to the success probability of the pirate decoder in decrypting regular transmissions. This is of particular importance, since tracing would be impossible against a pirate decoder that is not required to operate correctly at least some of the time. More specifically, we say a pirate decoder is σ-pirate decoder if for a randomly chosen m in the finite message space, we have that $\mathbf{Prob}[PD(\mathbf{Transmit}(ek, m) = m] \geq \sigma$. We next give a definition for black box traitor tracing:

Definition 1. *A multiuser encryption scheme* ME $=$ (**KeyDist, Transmit, Receive**) *is a* black box traitor tracing scheme for t-coalitions with success probability $1 - \alpha$ against σ-pirates *if there exists a tracing algorithm* **Trace** *such that any polynomial time adversary $\mathcal{A}$ wins the following game with probability at most α.*

1. *The adversary $\mathcal{A}$ outputs a set $\mathsf{T} = \{u_1, \ldots, u_j\} \subseteq [n]$ of at most t colluding receivers.*
2. *The challenger runs $\mathbf{KeyDist}(1^n)$ and provides $sk_{u_1}, \ldots, sk_{u_j}$ to $\mathcal{A}$. It keeps tk to itself.*
3. *The adversary $\mathcal{A}$ outputs a pirate decoder PD that is a σ-pirate.*
4. *The challenger now runs $\mathbf{Trace}^{PD}(tk)$ to obtain a set $\mathsf{S} \subseteq [n]$. Note that $\mathbf{Trace}$ is only given black-box oracle access to PD.*

We say that the adversary $\mathcal{A}$ wins the above game if the set S is either an empty subset or it is not a subset of T.

One may also consider a more general view of the black box tracing model, that is related to the case that the pirate decoder is a tamper resistant box, such as a music player and the response of the decoder is not the exact decryption of the transmission but rather the actual rendering of the cleartext transmission on a display device. In such case, the tracer can still extract useful information by observing whether the given ciphertext results in music being played or not. It is possible to address such issues in the above definition by adapting the notion of σ-pirate accordingly; i.e. the decoder succeeds if the response of the decoder is considered useful in its application domain as opposed to be plaintext m.

A pirate decoder is said to be resettable if the tracer has the capability to reset the pirate decoder to its initial state and the decoder is available for a new query. This gives the tracer the advantage of asking queries that will be handled independently during the tracing process, i.e., effectively preventing the decoder from using previous querying information submitted by the tracer in order to decide its present action. In contrast, a *history recording* pirate decoder "remembers" the previous queries made by the tracer and because the tracing procedure is public, the history recording capability can be used by the decoder to evade tracing.

A decoder is said to be available if it lacks a self-defensive mechanism, i.e, even if it realizes some abnormality in the content-transmission it is incapable of halting the tracer process. On the other occasion the decoder is called abrupt. Abrupt pirate decoders are those devices that may take some counter-actions against the tracing process and force it to stop. In any case (software aggressive-counter actions or hardware shutting-down mechanisms) we will assume that the tracer wants to avoid the occurrence of any such reaction and if such reaction is triggered it is immediately detectable by the tracer. On the other hand, the pirate decoder does not want such a mechanism to be triggered during normal operation. Since it is not possible to force the pirate decoder not to use such reaction mechanisms if they are available, what is needed to be shown is that there are systems where the usage of such mechanisms is detrimental to the pirate decoder itself (i.e. the triggering of the mechanism leaks some information about the traitor keys or it significantly interferes with the decoder's data reception capabilities).

It is possible to take into account all the above in definition 1 by (i) allowing the oracle PD to maintain a state across invocations or not (this would correspond to history-recording vs. resettable) (ii) allowing the oracle PD to output a special symbol $\perp$ after which *every* query to PD will be answered necessarily by $\perp$ (even in case of a resettable oracle PD). We also stress that our definition of σ-pirate requires that the decoder succeeds in decrypting any proper transmission with at least σ probability regardless of the other capabilities of the decoder. This captures the fact that the decoder has to be useful at least above a threshold σ.

2.1   Hybrid Colorings

The formalism for coloring discussed in [19], is a useful tool in describing how a tracer can interact with the pirate decoder in order to obtain some information about the keys it possesses. A coloring of the user population is a partition $\{U_i\}_{i \in \mathbb{N}}$ of $U = \{1, \ldots, n\}$. The ciphertext space of a multiuser encryption scheme is extended to allow the "coloring" of a user population. More specifically we consider a new transmission instruction of the multiuser encryption scheme that enables some decoders to decrypt incorrectly, i.e. some receivers may fail to decrypt the transmission, or simply decrypt in a different fashion compared to other receivers. The coloring of the user population corresponds to the number of different ways that the receivers decrypt the special ciphertext that induces the coloring.

We denote the new transmission by $\mathbf{Transmit}^{ext}(ek, m)$ and the set of new ciphertext space by C^{ext}. Given some $s \in C^{ext}$, we define the following relation of the set of users $U : a \equiv b$ if and only if $\mathbf{Receive}(s, sk_a) = \mathbf{Receive}(s, sk_b)$ (if the $\mathbf{Receive}$ algorithm is probabilistic, the definition can be amended accordingly while maintaining the properties of the equivalence relation — we will only consider the case of deterministic decryption here). The equivalence classes of $\equiv$ for some $s \in C^{ext}$ define a coloring over the user population. Observe that a valid ciphertext defines only one equivalence class, i.e., all users are colored by the same color.

We say that a multiuser encryption scheme can induce a family of colorings $\{\{U_i^{(j)}\}_i\}_j$ if it is possible to extend the normal transmission such that given 1^j we can produce $s \in C^{ext}$ that induces the coloring $\{U_i^{(j)}\}_i$ over the user population.

Hybrid colorings is a special collection of $n + 1$ partitions of the set of users: $\{\{U_1^{(n,r)}, U_2^{(n,r)}\}\}_{r=0,\ldots,n}$ with,

$$U_1^{(n,r)} = \{1, \ldots, r\} \qquad U_2^{(n,r)} = \{r+1, \ldots, n\}$$

where we set $U_1^{(n,0)} = U_2^{(n,n)} = \emptyset$ by definition. Hybrid colorings can be produced in a trivial traitor tracing scheme that is linear in transmission length and parameterized by an encryption scheme $(\mathbf{E}, \mathbf{D})$. In such a scheme, a message is transmitted as $\langle \mathsf{E}_{k_1}(m), \ldots, \mathsf{E}_{k_n}(m)\rangle$ where k_i is given to the i-th receiver as a secret key. This transmission will be denoted by $\mathbf{Transmit}_L(ek, m)$ with L signifying that transmission length is a linear function in the number of users. We, next, decribe the special tracing transmissions in such scheme that extend the normal transmission into the one that induces hybrid coloring. The tracing queries consist of the special transmission $\mathbf{Transmit}_L^s(ek, m)$ for $s = 0, 1, \ldots, n$ by substituting the first s ciphertexts with a random string.

$$\mathbf{Transmit}_L^s(ek, m) =$$
$$\langle \mathsf{E}_{k_1}(R), \mathsf{E}_{k_2}(R), \ldots \mathsf{E}_{k_s}(R), \mathsf{E}_{k_{s+1}}(m), \ldots \mathsf{E}_{k_n}(m)\rangle$$

where R is a random string of the same length as the message m. Finally, we say a ciphertext $c \in C^{ext}$ if there exists some $s \in \{0, \ldots, n\}$ and $m \in M$ such that $c = \mathbf{Transmit}_L^s(ek, m)$ holds. It is now easy to observe that such extension makes it possible for the linear length scheme to induce the hybrid colorings: a coloring of type $(\mathsf{U}_1^{(n,s)}, \mathsf{U}_2^{(n,s)})$ is induced from the tracing ciphertexts $\mathbf{Transmit}_L^s(ek, m)$. In this case, two colors exist: the users in set $\mathsf{U}_1^{(n,s)}$ corresponds to the ones who fail to decrypt while $\mathsf{U}_2^{(n,s)}$ consists of the users who are capable of decrypting the transmission.

It is shown in [19] that a black-box tracing algorithm exists for a multiuser encryption scheme that can induce hybrid coloring. Such multiuser encryption scheme is proven to be black-box traitor tracing for resettable decoders. A further analysis from [17] shows that a number of $O(\frac{n^3 \ln 1/\varepsilon}{\sigma^2})$ tracing transmissions are needed to query a σ-pirate decoder where ε is the winning probability of the tracer, i.e. the accusation probability of an innocent receiver. We note that hybrid colorings (or sometimes called the linear tracing strategy) is a technique that is implicit in almost all previous work and is ubiquitous in all traitor tracing schemes (cf. a non-exhaustive list of [2,4,5,6,8,10,11,14,16,18,19,21,22,25,28].)

Tracing via hybrid colorings. We recall that the tracing strategy proceeds as follows: for $s = 0, \ldots, n$ transmit $\lambda = O(\frac{n^2 \ln 1/\varepsilon}{\sigma^2})$ ciphertexts of the form $\mathbf{Transmit}_L^s(ek, m)$ and record the way the pirate decoder responds to such special ciphertexts. In the case of available and resettable pirate decoder, there are only two possible cases: (i) the decoder decrypts correctly or (ii) the decoder chooses (or is forced to) not to decrypt correctly.

In the coloring view of the tracing procedure the following important observation can be made : a pirate decoder corrupting t traitors keys will be able to identify only $t + 1$ colorings among all $n + 1$ possible colorings. More specifically, unless the user $s' + 1$ is a traitor, then the two colorings $(\mathsf{U}_1^{(n,s')}, \mathsf{U}_2^{(n,s')})$ and $(\mathsf{U}_1^{(n,s'+1)}, \mathsf{U}_2^{(n,s'+1)})$ are indistinguishable in the views of the traitors, and it also holds that the corresponding tracing ciphertexts are also indistinguishable (assuming the security of the underlying scheme). The response of the pirate decoder cannot be very different between these two cases.

Now to justify that tracing works the following argument is made : the ciphertexts of the form $\mathbf{Transmit}_L^0(ek, m)$ induce a coloring of $(\mathsf{U}_1^{(n,0)}, \mathsf{U}_2^{(n,0)})$ for which the decoder decrypts with at least σ probability: all traitors have the same color of decrypting. On the other hand, the ciphertexts of the form $\mathbf{Transmit}_L^n(ek, m)$ induces a coloring of $(\mathsf{U}_1^{(n,n)}, \mathsf{U}_2^{(n,n)})$ for which the decoder fails to decrypt: all traitors have the same color of not-decrypting. Hence, there will be a point s' for which the decoder switches from decrypting to not-decrypting that can be observed with probability at least $\frac{\sigma}{n}$ due to the triangular inequality. The observed difference implies that the user with index $s'+1$ is a traitor with probability $1-\varepsilon$ where ε is the approximation error that should be allowed since the tracer will statistically recover the switching point by issuing λ ciphertexts (note that ε can be made arbitrarily small). This tracing algorithm that utilizes hybrid colorings will be denoted by $\mathbf{Trace}_{HC}$.

Hybrid Colorings Against History-recording Decoders. It is also stated in [19] (but without presenting an analysis) that a history recording pirate decoder with at least two traitor keys can defeat the hybrid color tracing strategy as follows: upon detecting tracing, the decoder might continue to expose the first plaintext (thus yielding the first color) as if it was given the regular transmission for a random number of trials and then after the counter expires start returning the plaintext that corresponds to the second color (note that both colors are available to the decoder). Such behavior is possible due to the capability of history-recording which means that a state is maintained across tracing tests.

Let us see how this strategy applies to the context of a linear length multiuser encryption scheme: the detection of the tracing procedure is possible when the traitor set is partitioned into two by a transmission of the form $\mathbf{Transmit}_L^s(ek, m)$, $0 < s < n$, i.e., the case that there exists a traitor in both of the intervals $[1, s]$ and $[s+1, n]$. After the detection of tracing which will happen when transmission of type s is passed in the hybrid coloring schema, a history-recording decoder will continue decrypting for some number of ciphertexts and then it will stop decrypting thus effectively switching to the other color. If it happens that the two traitor keys are at locations i, j with $|i - j| > 1$ the pirate decoder can force the tracer to accuse an innocent user with great probability. In the next section, we will give a rigorous description for this attack.

Hybrid Colorings against Abrupt Decoders. Abrupt pirate decoders use a polynomial time predicate $\mathbf{React}$ with domain the set of all partitions of the traitor keys/indices. Upon inputting a ciphertext $c \in \mathsf{C}$ to the pirate decoder, the decoder calculates the partition of the traitor set based on the projection of the coloring induced by c over the traitor keys. If the predicate $\mathbf{React}$ of the decoder returns true, then the decoder activates its self-defensive mechanism. In [19] only deterministic predicates are considered and it is claimed that hybrid colorings imply black box traitor tracing against abrupt decoders. The authors further claim that such result can be easily extended to the general probabilistic case. However, in the next section we will show that this claim is not correct by presenting a probabilistic polynomial-time predicate which defeats the tracing strategy induced by hybrid colorings.

It follows that all traitor tracing schemes (including [4,5,11,21]) that employ tracing via hybrid colorings, (i.e., performing a walk with special transmissions to randomize the ciphertexts one by one) would be susceptible to our attack.

This means that such schemes can be useful for only available and resettable decoders[3].

3   Attack against Hybrid Coloring Based Tracing

The hybrid coloring tracing strategy is used to prove black-box traceability for resettable and available σ-pirate decoders, for some $\sigma > 0$, by querying the decoder with the special tracing ciphertexts as many as some number λ times; this approximates the success probability of decrypting these tracing ciphertexts. The number λ is a function of σ, the number of receivers n and the failure probability ε that is desired in tracing; λ is not necessarily kept secret but we can assume it is not known to the decoder, otherwise the analysis below can be further simplified. We refer to the analysis of [17] and take $\lambda = \frac{75n^2 \cdot \ln(8/\varepsilon)}{\sigma^2}$. However the attacks would work for other values of λ in a straightforward manner. So far in the categorization of the response of the decoder, we argued for a simple interpretation of the response behavior of the pirate: given any ciphertext we said a traitor or a pirate decoder will either decrypt or not. In the algorithm description below we follow the more general interpretation that says that given any ciphertext a pirate decoder will return a plaintext which will be considered as its color (this is without loss of generality).

We next describe the attack on the hybrid coloring tracing that can be mounted by any history recording decoder which does not even need to know the value of λ in advance. Note that it is assumed that the value of λ is stable across tracing (as it is the case for all previous positive results).

Description of the attack for history-recording pirates. The attack utilizes the history-recording capability of the decoder to extract the value λ by observing and maintaining a suitable counter. We assume a traitor coalition of size at least 3 with traitor indices t_1, t_2, t_3. The default plaintext that the decoder responds with is set to be the plaintext of the third traitor. The decoder can count the number of special tracing ciphertexts that randomize only the decryption available to traitor with index t_1 as follows: the counter is initialized to 0. When the traitor t_1 starts returning a different plaintext compared to traitors t_2, t_3 the pirate decoder starts counting and increments the counter with every such ciphertext. When the traitor t_2 starts to decrypt differently than t_3 the value of the counter would be equal to $\lambda \times (t_2 - t_1)$ and hence the value of λ can be computed by the decoder. Now a new counter will be initialized and after another λ transmissions for which t_3 decrypts differently compared to t_1, t_2 the decoder will start responding with the plaintext that the traitor t_1 outputs.

---

[3] We note that from these works only [19] and [21] claim traceability beyond type-0 decoders. Note that [21] was already broken in due to other issues in [16]; Still here we demonstrate that the problem is much deeper as we show that essentially any attempt to trace abrupt decoders through hybrid coloring is ultimately flawed and the only known scheme [19] believed to be secure against abrupt decoders is ultimately also defeated.

$Input :$ Given a ciphertext $c \in \mathsf{C}^{ext}$
$Parameters :$ Traitor identities $\mathsf{T} = \{t_1, \ldots, t_a\}$ with $a > 2$
$Variables :$ A triple of counters α, β and γ all set 0 initially
           A color d, set to empty initially.
Compute the coloring induced by c and let d_i be the color of traitor t_i.
Set d to be d_3
If $d_1 \notin \{d_2, d_3\}$ then increment α by 1.
If $d_1 = d_2$ and $d_3 \neq d_1$ then do the following:

- If $\gamma = 0$ then set $\gamma = \frac{\alpha}{t_2 - t_1}$ and $\beta = 1$.
- Else if $\beta \leq \gamma$ then increment β by 1.
- Otherwise if $\beta > \gamma$ then set d to be the color of d_1.

Respond with the plaintext corresponding to color d.

Fig. 2. The high-level operation of the history-recording pirate decoder that breaks the hybrid coloring strategy

Through out the above decoder-tracer interaction, there is a single switch from the way the decoder responds: from the plaintext/color of the third traitor to the plaintext/color of the first traitor. On the tracer side this happens, provided that λ is correctly computed, when the type of tracing ciphertext/coloring changes from $(\mathsf{U}_1^{(n,t_2)}, \mathsf{U}_2^{(n,t_2)})$ to $(\mathsf{U}_1^{(n,t_2+1)}, \mathsf{U}_2^{(n,t_2+1)})$. Hence the tracer will conclude that $t_2 + 1$ is the traitor index.

Following this logic, Figure 2 describes how a pirate decoder capable history-recording can defeat the hybrid coloring strategy even without knowing the value λ. The decoder, basically, calculates λ and will take the action so that the tracer will accuse the user with index $t_2 + 1$ always.

Theorem 1. *(The attack for history-recording pirates) Consider a black box traitor tracing algorithm* **Trace**$_{HC}$ *based on hybrid coloring technique that has success probability* $1 - \varepsilon$ *against resettable σ-pirates for w coalitions. Let the history-recording decoder $\mathcal{D}$ described in figure 2 corrupts the receivers in set* $\mathsf{T} = \{t_1, \ldots, t_a\}$ *with $w \geq a > 2$ and $t_3 > t_2 + 1$. If the tracing algorithm* **Trace**$_{HC}$ *is applied to the decoder $\mathcal{D}$ then it will accuse the innocent user $t_2 + 1$.*

Proof of Theorem 1:   The correctness of the statement of the theorem relies on the accuracy of the computation of the λ value, that is the number of queries requested by the tracing algorithm **Trace**$_{HC}$ for each type of coloring/tracing ciphertext. As seen in the figure 2, the λ value is computed as $\frac{\alpha}{t_2 - t_1}$. The α counter is incremented whenever the tracing query induces one of the following colorings $\{(\mathsf{U}_1^{(n,t_1)}, \mathsf{U}_2^{(n,t_1)}), \ldots, (\mathsf{U}_1^{(n,t_2-1)}, \mathsf{U}_2^{(n,t_2-1)})\}$. Indeed, all these colorings imply that d_1, the color of the first traitor, is different than the rest of the colors. For the next λ queries, as counted by the counter β, a coloring of type $(\mathsf{U}_1^{(n,t_2)}, \mathsf{U}_2^{(n,t_2)})$ will be induced. Finally, starting from queries that induce a coloring of $(\mathsf{U}_1^{(n,t_2+1)}, \mathsf{U}_2^{(n,t_2+1)})$, the decoder will responds differently thus resulting in the termination of tracing and an accusation to be made.

Recall the rationale used in proving the correctness of the the tracing algorithm: unless $t_2 + 1$ is a traitor, the decoder is not able to distinguish a coloring of type $(\mathsf{U}_1^{(n,t_2)}, \mathsf{U}_2^{(n,t_2)})$ from a coloring of type $(\mathsf{U}_1^{(n,t_2+1)}, \mathsf{U}_2^{(n,t_2+1)})$. According to this logic the tracer will conclude that $t_2 + 1$ is a traitor. Hence, the decoder of figure 2, by using its capability of history-recording, is capable of making the tracer to accuse an innocent receiver.   ∎

We note that the restriction in the above theorem that $t_3 > t_2 + 1$ is superficial and we can easily modify the pirate decoder of figure 2 to obtain a similar theorem where other users would be accused. We note that the value λ in many cases will be public (or can be guessed). In this case the above attack can be simplified to require only two traitors.

Remark. One may consider more sophisticated tracers that try to foil the above attacker by making the calculation of λ harder or by hiding the order of the traitors: specifically, a tracer, (1) may induce the hybrid coloring over a randomly permuted set of indices of users or (2) may not let the traitors know their indices or even (3) query a random value λ ciphertexts for different user locations. These are indeed possible tracing counter-measures against a history-recording decoder of Figure 2. However, our contribution in this section (and of the paper in general as the discussion here applies to abrupt decoders as well) is to show that the standard tracer in the literature that employs the hybrid coloring tracing technique and was believed to be successful in a number of scenarios is foiled because of our attacks. The design of better tracing algorithms as shown by our results has to use more intricate arguments than those believed to be sufficient used so far and such is left to future work. We consider our work an important step forward in understanding traitor tracing of these devices and any potential upcoming improved tracing algorithms would have to circumvent the class of attacks we introduce here.

Description of the attack for abrupt pirates. We now consider an abrupt but resettable decoder. In Figure 3 we describe a pirate decoder that is equipped with a self-defensive mechanism taking advantage of the way the hybrid colorings project themselves on the traitor keys. The decoder will respond to normal transmissions (from its perspective) with $\sigma = 1$ probability while in the case of tracing ciphertexts (that induce a different coloring to the users) it will respond with a valid plaintext with probability $1 - \frac{1}{n^3}$. This will make the decoder stop responding for tracing ciphertexts that induces a coloring of $\{\mathsf{U}_1^{n,s}, \mathsf{U}_2^{n,s}\}$ for some value s. The receiver with index s will then be accused of piracy.

For the sake of ease of presentation we assume a traitor coalition of size at least 2 with $t_2 - t_1 - 1 = \alpha$ for a suitably large α (we make this explicit below). The correctness of the theorem, again similar to the case for history-recording decoder, relies on the tracing algorithm **Trace**$_{HC}$ that queries $\lambda = \frac{75n^2 \cdot \ln(8/\varepsilon)}{\sigma}$ transmissions for each type of coloring. We recall that the hybrid coloring tracing algorithm was believed to be successful by [19,21] and the latter work proposed a

---

Input : Given a ciphertext $c \in C^{ext}$
Parameters : Traitor identities $\mathsf{T} = \{t_1, \ldots, t_a\}$ with $a > 1$
Compute the coloring induced by c and let d_i be the color of traitor t_i.
if **React**$(\{d_1, \ldots, d_a\})$ returns true then then deploy the self-defensive
action with probability $\frac{1}{n^3}$
In any other case respond with the color of d_a.

The predicate **React**: returns true if d_1 has a different color from the
rest of the traitor colors.

Fig. 3. The high-level operation of the abrupt but resettable pirate decoder that breaks
the hybrid coloring

scheme with hybrid colorings as the first sublinear tracing scheme against abrupt
decoders. Our attack presented herein thus fails all known tracing algorithms
against abrupt pirate decoders. We also remark that that repairing the tracer
needs non-trivial improvements to tracing that circumvent our attack; this is an
important open question.

We now present more formally the claim that the hybrid-coloring based standard tracing algorithm fails to be successful against abrupt decoders.

Theorem 2. *(The attack for abrupt pirates) Consider the black box traitor tracing algorithm* **Trace**$_{HC}$ *based on the hybrid coloring technique that has success
probability $1 - \varepsilon$ against resettable σ-pirates for w coalitions. Let the abrupt decoder $\mathcal{D}$ described in figure 3 corrupt the receivers in set $\mathsf{T} = \{t_1, \ldots, t_a\}$ with
$a > 1$ and $t_2 - t_1 - 1 = \alpha > \frac{n}{75}$. If the tracing algorithm* **Trace**$_{HC}$ *is applied
to the decoder $\mathcal{D}$ then it will accuse an innocent user with probability at least
$1 - \frac{\varepsilon}{8} - \frac{75\ln(8/\varepsilon)}{n}$.*

Proof of Theorem 2:   The pirate decoder we designed decrypts any valid transmission which essentially implies that $\sigma = 1$. During the tracing process the
decoder will be given $(\alpha + 1)75n^2 \cdot \ln(8/\varepsilon)$ number of tracing ciphertexts that
induce a coloring of $\{U_1^{n,s}, U_2^{n,s}\}$ for some $s = t_1, \ldots t_2 - 1$. For each such transmission the predicate **React** returns true, hence the decoder will deploy the self
defensive mechanism with probability $\sigma_0 = \frac{1}{n^3}$. If that happens then a receiver
with an index from $\{t_1, \ldots t_2 - 1\}$ would be accused of piracy.

To have the decoder defeat the tracer it should be that the decoder deploys
the self-defensive mechanism for an index from $\{t_1 + 1, \ldots, t_2 - 1\}$ but not t_1.
The decoder succeeds in evading from tracing algorithm with probability $(1 -
1/n^3)^{75n^2 \cdot \ln(8/\varepsilon)}$ as the decoder does not deploy the self-defensive mechanism
with probability $1 - 1/n^3$ for each transmission that induces $\{U_1^{n,t_1}, U_2^{n,t_1}\}$. That
probability is at least $1 - \frac{1}{n^3} \cdot 75n^2 \cdot \ln(8/\varepsilon) = 1 - \frac{75\ln(8/\varepsilon)}{n}$ as it holds that
$(1 + x)^a \geq 1 + ax$ for any real number $x > -1$ and natural number a.

On the other hand, an innocent receiver will be accused with probability
$1 - (1 - 1/n^3)^{75\alpha n^2 \cdot \ln(8/\varepsilon)}$ as the probability of not deploying the self-defensive
mechanism for any of the transmissions that induce coloring $\{U_1^{n,s}, U_2^{n,s}\}$, for

some $s = t_1 + 1, \ldots, t_2 - 1$, is at most $(1 - 1/n^3)^{75\alpha n^2 \cdot \ln(8/\varepsilon)}$. Given that $\alpha > n/75$ the latter probability has an upper bound of $e^{-\ln 8/\varepsilon} = \varepsilon/8$. Hence, the decoder succeeds in accusing an innocent receiver with probability at least $1 - \varepsilon/8$.

Overall, a pirate decoder described in Figure 3 will accuse an innocent receiver with probability at least $1 - \frac{\varepsilon}{8} - \frac{75\ln(8/\varepsilon)}{n}$. ∎

4   Conclusion

In this paper, we propose a new attack against hybrid-coloring based tracing for abrupt decoders. We also present in full detail the attack that was alluded to in [19] against tracing for history-recording decoders. Our new attack exploits the fact that the abrupt decoder may behave arbitrarily and not necessarily follow a deterministic "React" predicate, whenever given a tracing ciphertext. This is a realistic scenario, and given the fact that (in the absence of watermarking) there was no other tracing technique against abrupt decoders other than the hybrid-coloring one, the design of a traitor tracing scheme against history recording and abrupt decoders is left as an open problem.

References

1. Boneh, D., Franklin, M.: An Efficient Public-Key Traitor Tracing Scheme. In: Wiener, M. (ed.) CRYPTO 1999. LNCS, vol. 1666, pp. 338–353. Springer, Heidelberg (1999)
2. Boneh, D., Naor, M.: Traitor tracing with constant size ciphertext. In: ACM Conference on Computer and Communications Security, pp. 501–510 (2008)
3. Boneh, D., Shaw, J.: Collusion-Secure Fingerprinting for Digital Data. IEEE Transactions on Information Theory 44(5), 1897–1905 (1998)
4. Boneh, D., Sahai, A., Waters, B.: Fully Collusion Resistant Traitor Tracing with Short Ciphertexts and Private Keys. In: Vaudenay, S. (ed.) EUROCRYPT 2006. LNCS, vol. 4004, pp. 573–592. Springer, Heidelberg (2006)
5. Boneh, D., Waters, B.: A fully collusion resistant broadcast, trace, and revoke system. In: ACM Conference on Computer and Communications Security 2006, pp. 211–220 (2006)
6. Chor, B., Fiat, A., Naor, M.: Tracing Traitors. In: Desmedt, Y.G. (ed.) CRYPTO 1994. LNCS, vol. 839, pp. 257–270. Springer, Heidelberg (1994)
7. Cox, I.J., Kilian, J., Leighton, F.T., Shamoon, T.: Secure spread spectrum watermarking for multimedia. IEEE Transactions on Image Processing 6(12), 1673–1687 (1997)
8. Chabanne, H., Hieu Phan, D., Pointcheval, D.: Public Traceability in Traitor Tracing Schemes. In: Cramer, R. (ed.) EUROCRYPT 2005. LNCS, vol. 3494, pp. 542–558. Springer, Heidelberg (2005)
9. Cramer, R., Shoup, V.: Universal Hash Proofs and a Paradigm for Adaptive Chosen Ciphertext Secure Public-Key Encryption. In: Knudsen, L.R. (ed.) EUROCRYPT 2002. LNCS, vol. 2332, pp. 45–64. Springer, Heidelberg (2002)
10. Dodis, Y., Fazio, N., Kiayias, A., Yung, M.: Scalable public-key tracing and revoking. In: Proceedings of the Twenty-Second ACM Symposium on Principles of Distributed Computing (PODC 2003), Boston, Massachusetts, July 13-16 (2003)

11. Furukawa, J., Attrapadung, N.: Fully Collusion Resistant Black-Box Traitor Revocable Broadcast Encryption with Short Private Keys. In: Arge, L., Cachin, C., Jurdziński, T., Tarlecki, A. (eds.) ICALP 2007. LNCS, vol. 4596, pp. 496–508. Springer, Heidelberg (2007)
12. Fiat, A., Naor, M.: Broadcast Encryption. In: Stinson, D.R. (ed.) CRYPTO 1993. LNCS, vol. 773, pp. 480–491. Springer, Heidelberg (1994)
13. Fiat, A., Tassa, T.: Dynamic Traitor Tracing. Journal of Cryptology 4(3), 211–223 (2001)
14. Gafni, E., Staddon, J., Lisa Yin, Y.: Efficient Methods for Integrating Traceability and Broadcast Encryption. In: Wiener, M. (ed.) CRYPTO 1999. LNCS, vol. 1666, pp. 372–387. Springer, Heidelberg (1999)
15. Kurosawa, K., Desmedt, Y.G.: Optimum Traitor Tracing and Asymmetric Schemes. In: Nyberg, K. (ed.) EUROCRYPT 1998. LNCS, vol. 1403, pp. 145–157. Springer, Heidelberg (1998)
16. Kiayias, A., Pehlivanoglu, S.: On the security of a public-key traitor tracing scheme with sublinear ciphertext size. In: Digital Rights Management Workshop 2009, pp. 1–10 (2009)
17. Kiayias, A., Pehlivanoglu, S.: Improving the Round Complexity of Traitor Tracing Schemes. In: Zhou, J., Yung, M. (eds.) ACNS 2010. LNCS, vol. 6123, pp. 273–290. Springer, Heidelberg (2010)
18. Kiayias, A., Yung, M.: Self Protecting Pirates and Black-Box Traitor Tracing. In: Kilian, J. (ed.) CRYPTO 2001. LNCS, vol. 2139, pp. 63–79. Springer, Heidelberg (2001)
19. Kiayias, A., Yung, M.: On Crafty Pirates and Foxy Tracers. In: Sander, T. (ed.) DRM 2001. LNCS, vol. 2320, pp. 22–39. Springer, Heidelberg (2002)
20. Lee, M., Ma, D., Seo, M.: Breaking Two k-Resilient Traitor Tracing Schemes with Sublinear Ciphertext Size. In: Abdalla, M., Pointcheval, D., Fouque, P.-A., Vergnaud, D. (eds.) ACNS 2009. LNCS, vol. 5536, pp. 238–252. Springer, Heidelberg (2009)
21. Matsushita, T., Imai, H.: A Public-Key Black-Box Traitor Tracing Scheme with Sublinear Ciphertext Size Against Self-Defensive Pirates. In: Lee, P.J. (ed.) ASIACRYPT 2004. LNCS, vol. 3329, pp. 260–275. Springer, Heidelberg (2004)
22. Naor, D., Naor, M., Lotspiech, J.B.: Revocation and Tracing Schemes for Stateless Receivers. In: Kilian, J. (ed.) CRYPTO 2001. LNCS, vol. 2139, pp. 41–62. Springer, Heidelberg (2001)
23. Naor, D., Naor, M., Lotspiech, J.B.: Revocation and Tracing Schemes for Stateless Receivers, Electronic Colloquium on Computational Complexity (ECCC) 43 (2002)
24. Naor, M., Pinkas, B.: Threshold Traitor Tracing. In: Krawczyk, H. (ed.) CRYPTO 1998. LNCS, vol. 1462, pp. 502–517. Springer, Heidelberg (1998)
25. Hieu Phan, D., Safavi-Naini, R., Tonien, D.: Generic Construction of Hybrid Public Key Traitor Tracing with Full- Public-Traceability, pp. 264–275
26. Staddon, J.N., Stinson, D.R., Wei, R.: Combinatorial Properties of Frameproof and Traceability Codes. IEEE Transactions on Information Theory 47(3), 1042–1049 (2001)
27. Safavi-Naini, R., Wang, Y.: Sequential Traitor Tracing. In: Bellare, M. (ed.) CRYPTO 2000. LNCS, vol. 1880, pp. 316–332. Springer, Heidelberg (2000)
28. Safavi-Naini, R., Wang, Y.: Traitor Tracing for Shortened and Corrupted Fingerprints. In: Feigenbaum, J. (ed.) DRM 2002. LNCS, vol. 2696, pp. 81–100. Springer, Heidelberg (2003)

How to Find Short RC4 Colliding Key Pairs

Jiageng Chen* and Atsuko Miyaji**

School of Information Science,
Japan Advanced Institute of Science and Technology,
1-1 Asahidai, Nomi, Ishikawa 923-1292, Japan
{jg-chen,miyaji}@jaist.ac.jp

Abstract. The property that the stream cipher RC4 can generate the
same keystream outputs under two different secret keys has been discovered recently. The principle that how the two different keys can achieve
a collision is well known by investigating the key scheduling algorithm of
RC4. However, how to find those colliding key pairs is a different story,
which has been largely remained unexploited. Previous researches have
demonstrated that finding colliding key pairs becomes more difficult as
the key size decreases. The main contribution of this paper is proposing an efficient searching algorithm which can successfully find 22-byte
colliding key pairs, which are by far the shortest colliding key pairs ever
found.

1 Introduction

The stream cipher RC4 is one of the oldest and most wildly used stream ciphers in
the world. It has been deployed to innumerable real world applications including
Microsoft Office, Secure Socket Layer (SSL), Wired Equivalent Privacy (WEP),
etc. Since its debut in 1994 [1], many cryptanalysis works have been done on it,
and many weaknesses have been exploited, such as [5] [6] and [7]. However, if
RC4 is used in a proper way, it is still considered to be secure. Thus it is still
considered to be a high valuable cryptanalysis target both in the industrial and
academic world.

In this paper, we focus on exploiting the weakness that RC4 can generate
colliding key pairs, namely, two different keys will result in the same keystream
output. This weakness was first discovered by [2] and later generalized by [3].
For any ciphers, the first negative effects that this property could bring is the
reducing of the key space. It seems that it is not very dangerous if the colliding
key pairs are not so many. However, [4] demonstrated a key recovery attack by
making use of this weakness, and the complexity of the attack depends heavily
on how fast we can find those colliding key pairs. In [2], it has demonstrated that
the shorter the key is, the harder it is to find the colliding key pairs. A searching
algorithm was proposed in [2] and a 24-byte colliding key pair was the shortest
one that experimentally found. Finding short colliding key pairs has its practical

---

* This author is supported by the Graduate Research Program.
** This work is supported by Grant-in-Aid for Scientific Research (B), 20300003.

X. Lai, J. Zhou, and H. Li (Eds.): ISC 2011, LNCS 7001, pp. 32–46, 2011.

meaning mainly because that the key size deployed in most of the applications are short ones which are between 16 bytes to 32 bytes, and also the link between the attacks like [4].

Our main contribution is proposing a searching algorithm that can find short colliding key pairs efficiently. 22-byte colliding key pair is experimentally found by using our algorithm in about three days time while a 24-byte colliding key pair was found in about ten days time in [2]. We also analyze the complexity of both our algorithm and the one in [2] to support our experimental result from a theoretical point of view, so that we can understand the new searching techniques clearly.

This paper is organized as follows. In section 2, we give a short introduction on RC4 algorithm and the details on the key collisions. In section 3, we review the previous searching techniques including brute force searching and the one proposed in [2]. Section 4 covers the new techniques we propose to reduce the searching complexity followed by the new algorithm in section 5. Complexity evaluations are described in section 6.

2   RC4 Key Collision

First we shortly describe the RC4 algorithm. The internal state of RC4 consists of a permutation S of the numbers $0, ..., N-1$ and two indices $i, j \in \{0, ..., N-1\}$. The index i is determined and known to the public, while j and permutation S remain secret. RC4 consists of two algorithms: The Key Scheduling Algorithm (KSA) and the Pseudo Random Generator Algorithm (PRGA). The KSA generates an initial state from a random key K of k bytes as described in Algorithm 1. It starts with an array $\{0, 1, ..., N-1\}$ where $N = 256$ by default. At the end, we obtain the initial state S_{N-1}.

Once the initial state is created, it is used by PRGA. The purpose of PRGA is to generate a keystream of bytes which will be XORed with the plaintext to generate the ciphertext. PRGA is described in Algorithm 2. Since key collision is only related to KSA algorithm, we will ignore PRGA in the rest of the paper.

Algorithm 1. KSA	Algorithm 2. PRGA
1: **for** $i = 0$ **to** $N - 1$ **do**	1: $i \leftarrow 0$
2: $S[i] \leftarrow i$	2: $j \leftarrow 0$
3: **end for**	3: **loop**
4: $j \leftarrow 0$	4: $i \leftarrow i + 1$
5: **for** $i = 0$ **to** $N - 1$ **do**	5: $j \leftarrow j + S[i]$
6: $j \leftarrow j + S[i] + K[i \bmod l]$	6: swap$(S[i], S[j])$
7: swap$(S[i], S[j])$	7: keystream byte $z_i = S[S[i] + S[j]]$
8: **end for**	8: **end loop**

We focus on the key collision pattern discovered in [2], which can generate shorter colliding key pairs than other patterns discovered in [3]. In [2], it clearly described how two keys K_1 and K_2 with the only one difference $K_2[d] = K_1[d]+1$ can achieve a collision. It traced two KSA procedure and two S-Box states generated by the two keys, and pointed out how two S-Box states become equal to each other at the end of the KSA. Actually, the essence of the key collisions is only related to some j values at some specific locations. If these conditions once satisfied, a collision is expected. Thus we prefer to use another way to explain the collision by listing all the j conditions. In this way, we only need to exam the behavior of one key, since once the j values generated by this key satisfy all the conditions, then deterministically, there exists another key that they form a colliding key pair. To simplify, we check whether a given key K_1 has a related key K_2 such that $K_2[d] = K_1[d] + 1$ and K_1 and K_2 can achieve a collision. Then all we need is to confirm whether K_1's j behaviors satisfy the conditions in Table 1.

Table 1. j conditions required to achieve a collision

Round	Round Interval	Class 1	Class 2
1	$[0, d+1]$	$j_d = d,\ j_{d+1} = d+k$	$j_{0 \sim d-1} \neq d, d+1$
2	$[d+2, d+k]$	$j_{d+k} = d+2k$	$j_{d+2 \sim d+k-1} \neq d+k$
... ...	... ...	... ...	... ...
t	$[d+(t-2)k+1,$ $d+(t-1)k]$	$j_{d+(t-1)k} = d+tk$	$j_{d+(t-2)k+1 \sim d+(t-1)k-1} \neq$ $d+(t-1)k$
... ...	... ...	... ...	... ...
$n-1$	$[d+(n-3)k+1,$ $d+(n-2)k]$	$j_{d+(n-2)k} = (d-1)+(n-1)k$	$j_{d+(n-3)k+1 \sim d+(n-2)k-1} \neq$ $d+(n-2)k$
n	$[d+(n-2)k+1,$ $d+(n-1)k-1]$	$j_{d+(n-1)k-2} = S^{-1}_{d+(n-1)k-3}[d],$ $j_{d+(n-1)k-1} = d+(n-1)k-1$	$j_{d+(n-2)k+1 \sim d+(n-1)k-3} \neq$ $d+(n-1)k-1$

The Round column presents the round number in the KSA steps in the Round Interval column. There are $n = \lfloor \frac{256+k-1-d}{k} \rfloor$ rounds, which is also the times that the key difference repeats during KSA. We separate the conditions into two categories, Class 1 and Class 2. From Table 1, you see that the conditions in Class 1 column are computational dominant compared with Class 2. This is because for j at some time to be some exact value, probability will only be 2^{-8} assuming random distribution, while not equal to some exact value in Class 2 has a relatively much higher probability. Also the main point for finding a colliding key pair is how to meet those low probability conditions in Class 1 column. In the rest of the paper, we focus on the Class 1 conditions. When we say a KSA procedure (a trial) under some key K passes round i and fails at round $i+1$, we indicate that all the Class 1 j conditions are satisfied in the previous i rounds and fails at the $i+1$-th round.

3   Known Searching Techniques

3.1   Brute Force Search

The most trivial method is to do the brute force search. The attacker simply generates a random secret key K with length k, and runs the KSA to test its random variable j's behavior. If the trial fails, then repeat the procedure until one colliding key pair is found. In [2], it has been demonstrated that for each trial, the successful probability is around $(\frac{1}{256})^{n+2}$. Thus the complexity for the brute force searching is $2^{8(n+2)}$. For 24-byte keys, the complexity is around 2^{96}, and for 22-byte keys which is actually found by us, it is around 2^{104}.

3.2   Matsui's Searching Algorithm

A searching algorithm is proposed in [2]. Here we make a short introduction on his searching technique which is described in Table 2. It defines a search function with two related keys as input, and output a colliding key pair or fail. When some trial fails to find the colliding key pair, the algorithm does not restart by trying another random related key pair, instead, it modifies the keys as $K_1[x] = K_1[x] + y$, $K_1[x+1] = K_1[x+1] - y$ for every x and y. Since $j_x = j_x + y$ and $j_{x+1} = j_x + S_x[x+1] + K_1[x]$, thus j_{x+1} after the modification will not be changed. This means that by modifying in this way, the next trial will have a relatively close relation with the previous trial, in other words, if the previous trial before the modification tends to achieve a collision, then the next trial after the modification will also have the tendency. The algorithm recursively calls the function Search(K_1, K_2) until it return a colliding key or fail.

4   New Techniques to Reduce the Searching Complexity

In this section, we propose several techniques to reduce the searching complexity so that we can find short colliding keys in practical time.

4.1   Bypassing the First Round Deterministically

Our first observation is that we can pass the first round. Recall that in the first round, there are two j conditions in Class 1 that we need to satisfy, namely

$$j_d = d \quad \text{and} \quad j_{d+1} = k + d$$

As in [2], the setting of $K[d + 1] = k - d - 1$ always meets the condition $j_{d+1} = k + d$ since we have $j_{d+1} = j_d + d + 1 + K[d+1]$. But still we have another condition $j_d = d$ left in the first round. This condition can be easily satisfied by modifying

$$K[d] = 255 - j_{d-1}$$

at the time when KSA is proceeded at index $d - 1$ after the swap. Since $j_d = j_{d-1} + d + K[d] = d$, and by modifying $K[d]$ dynamically when the previous value j_{d-1} is known, $j_d = d$ will always be satisfied. Then we can bypass the first round and reduce the necessary number of rounds to $n - 1$.

Table 2. Matsui's Algorithm

Input: Key length k, $d = k - 1$

Output: colliding key pair K_1 and K_2 such that $K_2[d] = K_1[d] + 1$,
$K_1[i] = K_2[i]$ if $i \neq d$, $KSA(K_1) = KSA(K_2)$.

---

1. Generate a random key pair K_1 and K_2 which differs at position d by one.
Set $K_1[d + 1] = K_2[d + 1] = k - d - 1$.

2. Call function Search(K_1, K_2), if Search$(K_1, K_2)=1$, collision is found, else goto 1.

Search(K_1, K_2) :

    $s = MaxColStep(K_1, K_2)$

    If $s = 255$, then return 1.

    $MaxS = max_{x,y} MaxColStep(K_1\langle x, y \rangle, K_2 \langle x, y \rangle)$

    If $Maxs \leq s$, then return 0.

    C=0

    For all x and y, do the following:

        If $MaxColStep(K_1\langle x, y \rangle, K_2\langle x, y \rangle) = MaxS$, call $Search(K_1, K_2)$

        $C = C + 1$

        If $C = MaxC$, then return 0.

Notations:

$MaxColStep(K_1, K_2)$: The maximal number of S-Box elements that S_1 differs
                    from S_2.

$K\langle x, y \rangle : K[x] = K[x] + y, K[x + 1] = K[x + 1] - y, K[i] = K[i]$ if $i \neq x, x + 1$.

4.2 Bypassing the Second Round with High Probability

If we choose the differential key index carefully, we find that the second round
can also be skipped with very high probability compared with the uniform distribution. Generally speaking, we would like to choose $d = k - 1$ so that in the KSA
procedure, the key differential index will be repeated as few times as possible.
Actually choosing the d at the indices close to $k - 1$ will have the same affect as
the last index $k - 1$. For example, for key with length 20-24 bytes, setting the
key differential at indices $k - 1, k - 2, k - 3, k - 4$ will cause the key differential
index to be repeated the same times during the KSA. Thus instead of setting
$d = k - 1$, let's set

$$d = k - 3$$

so that after d, we have another two key bytes. For the first round and second
round, the following two j conditions are necessary to meet:

$$j_{d+1} = j_{k-2} = 2k - 3$$

$$j_{d+k} = j_{2k-3} = 3k - 3$$

and we have

$$j_{2k-3} = j_{k-2} + K[k-1] + \sum_{i=0}^{k-3} K[i] + \sum_{i=k-1}^{2k-3} S_{i-1}[i] \tag{1}$$

$$\stackrel{P_{2nd}}{=} j_{k-2} + K[k-1] + \sum_{i=0}^{k-3} K[i] + \sum_{i=k-1}^{2k-3} S_{k-2}[i] \tag{2}$$

Thus by modifying

$$K[d+2] = K[k-1] = j_{2k-3} - j_{k-2} - \sum_{i=0}^{k-3} K[i] - \sum_{i=k-1}^{2k-3} S_{k-2}[i]$$

at the time $i = k - 2$ after the swap, with probability

$$P_{2nd} = \frac{256 - (k-2)}{256} \times \frac{256 - (k-3)}{256} \times \cdots \times \frac{256 - 1}{256} = \prod_{i=1}^{k-2} \frac{256 - i}{256}$$

we can pass the second round.

This can be explained as follows. For two fixed j values j_{k-2} in the first round, and j_{2k-3} in the second round, we have equation (1). At the time $i = k - 2$ after the swap, we don't know $\sum_{i=k-1}^{2k-3} S_{i-1}[i]$, but we can approximate it by using $\sum_{i=k-1}^{2k-3} S_{k-2}[i]$. The conditions on this approximation is that for $i \in [k-1, 2k-4]$, j does not touch any indices $[i+1, 2k-3]$, which gives us the probability P_{2nd}. Then if we set the $K[d+2]$ as before, with P_{2nd} we can pass the second round. Notice that the reason why we can modify $K[d+2]$ is related to the choice of d. When modifying $K[d+2]$, we don't wish the modification will affect the previous execution, which has been successfully passed. When modifying $K[d+2]$ trying to meet the second round condition, this key byte is used for the first time during KSA, thus we won't have the previous concern. For short keys such as $k = 24$, $P_{2nd} = 0.36$, and for $k = 22$, $P_{2nd} = 0.43$. The successful probability is thus much bigger compared with the uniform probability $2^{-8} = 0.0039$.

4.3   Reducing the Complexity in the Last Round

In the last round, there are two j conditions need to be satisfied, namely,

$$j_{(n-1)k+d-2} = r \text{ such that } S_{(n-1)k+d-3}[r] = d$$

$$j_{(n-1)k+d-1} = d + (n-1)k - 1$$

And from $j_{(n-1)k+d-1} = j_{(n-1)k+d-2} + S_{(n-1)k+d-2}[(n-1)k+d-1] + K[d-1]$, $K[d-1]$ can be decided if $j_{(n-1)k+d-2}$ is fixed to some value. During the KSA procedure, $j_{(n-1)k+d-2}$ could be touching any indices, but with overwhelming probability, it will touch index d. This is because after step $i = d$, one of the two S-Box differentials will be staying at index d till step $i = (n-1)k + d - 2$ unless it is touched by any j during the steps $[d + 1, (n - 1)k + d - 3]$. Thus we can assume that

$$\dot{j}_{(n-1)k+d-2} = d$$

and we can thus modify $K[d - 1]$ at step $i = d - 1$ before the swap as follows:

$K[d - 1] = j_{(n-1)k+d-1} - j_{(n-1)k+d-2} - S_{(n-1)k+d-2}[(n - 1)k + d - 1] = (n - 1)k + d - 1 - d - (d + 1) = (n - 1)k - d - 2$

This modification indicates that if some trial meets the $\dot{j}_{(n-1)k+d-2} = d$ condition in the last round, then with probability 1, the other condition in this round on $j_{(n-1)k+d-1}$ will be satisfied. Simply speaking, 2^{16} computation cost is required to pass the final round, while we reduce it to

$$P_{last} = 2^8 \times (\frac{255}{256})^{-((n-1)k-3)}$$

For a 24-byte key, the computation cost can be reduced to around $2^{9.2}$, which is a significant improvement. The overall cost will be covered in the next section, here we just demonstrate to give a intuition.

4.4 Multi-key Modification

In the area of finding hash collisions, multi-message modification is a widely used technique that first proposed by [8]. MD5 and some other hash functions are broken by using this technique. The idea is that when modifying the message block at some later round i to satisfy the i-th round conditions, leaving the previous rounds conditions satisfied (In hash functions, a message block is usually processed for many rounds in different orders). Since finding the key collision of RC4, to some degree, is related to finding hash collisions, we are motivated by the multi-message modification technique and find that we can also do such efficient modifications in finding RC4 colliding key pairs. Thus we call it multi-key modification.

After adapting previous proposed techniques, we may easily bypass the previous two rounds. Start from the third round, however, all the key bytes have been used more than once. This means that modifying any key bytes will definitely affect the previous rounds, which could make the previous round conditions become unsatisfied. In case of RC4, due to its property, we can to some degree maintain the previous round conditions while modifying the key in any later round. Let's assume for some round $2 < t < n - 1$ for the easy demonstration, the t-th round conditions are not satisfied, namely, $j_{(t-1)k+d} \neq tk + d$, and all the previous rounds conditions are satisfied. The following equations should all be satisfied in order to pass the first t rounds.

$$j_{2k+d} = j_{k+d} + \sum_{j=0}^{k-1} K[j] + \sum_{j=k+d+1}^{2k+d} S_{j-1}[j] \qquad (3)$$

$$j_{3k+d} = j_{2k+d} + \sum_{j=0}^{k-1} K[j] + \sum_{j=2k+d+1}^{3k+d} S_{j-1}[j] \qquad (4)$$

....

$$j_{(t-1)k+d} = j_{(t-2)k+d} + \sum_{j=0}^{k-1} K[j] + \sum_{j=(t-2)k+d+1}^{(t-1)k+d} S_{j-1}[j] \qquad (5)$$

There are four parts in each of these equations, and when the trial fails to pass the round t, (5) does not hold while all the previous equations hold. From the satisfied equations, the sum of the secret key is fixed, and when modifying the secret key in round t, we should not change the sum $\sum_{j=0}^{k-1} K[j]$, otherwise the previous equations will not be satisfied anymore. Then our problem now becomes how to modify K to satisfy condition on $j_{(t-1)k+d}$. There are many ways to modify the secret key without changing the sum. Matsui's algorithm actually uses one of the ways, namely, $K[x] = K[x] + y$ and $K[x+1] = K[x+1] - y$. Setting the modification targets next to each other reduce the steps that different j values will change the previous correct S-Box sum. Matsui's algorithm tries this modification for every x and y ($x \in [0, k-2]$, $y \in [0, 255]$) one by one, hoping that for some x and y, the S-Box sum $\sum_{j=(t-2)k+d+1}^{(t-1)k+d} S_{j-1}[j]$ will be the correct one so that condition on $j_{(t-1)k+d}$ is satisfied, while leaving the all the previous S-Box sum satisfied. We point out that modifying the secret key in this way have some drawbacks. First, only some specific x and y values will satisfy the condition on $j_{(t-1)k+d}$ leaving the previous conditions satisfied, while most of the other modifications will fail. In other words, for passing round t, this modification can be seen as brute force search (but its effect on the previous rounds is less than brute force search, we will cover it in the complexity evaluation). Second, as also mentioned in [2], such modification will generate many duplicated searching paths. Especially, since it is a recursive algorithm, one duplication in the small depth of the tree will cause a considerable amount of computation waste.

We discover that by adding a strategy on x and y in the key modification instead of brute force search, we could overcome the previous two drawbacks. Let's again consider the trial that passes all the previous $t-1$ rounds and fails to pass the t-th round, where we assume $2 < t < n-2$. Let's run the KSA until step $i = (t-1)k+d-1$ after the swap, then we check if the Class 1 j conditions on round t is satisfied or not, namely whether

$$S_{(t-1)k+d-1}[(t-1)k+d] = j_{(t-1)k+d} - j_{(t-2)k+d} - \sum_{j=(t-2)k+d+1}^{(t-1)k+d-1} S_{j-1}[j] - \sum_{j=0}^{k-1} K[j]$$

If the equation holds, we pass the t-th round and proceed the next round. Otherwise, let's denote

$$\Delta_{(t-1)k+d} = j_{(t-1)k+d} - j_{(t-2)k+d} - \sum_{j=(t-2)k+d+1}^{(t-1)k+d-1} S_{j-1}[j] - \sum_{j=0}^{k-1} K[j]$$

And we wish the value $\Delta_{(t-1)k+d}$ could be at index $(t-1)k+d$ before i touches it. We check if $\Delta_{(t-1)k+d} \leq (t-2)k+d$. If this is the case, it means that we have available S-Box value that can be swapped here. In other words, modify the key as follows:

$$K[\Delta_{(t-1)k+d}] = K[\Delta_{(t-1)k+d}] + (t-1)k + d - j_{\Delta_{(t-1)k+d}}$$
$$K[\Delta_{(t-1)k+d} + 1] = K[\Delta_{(t-1)k+d} + 1] - (t-1)k - d + j_{\Delta_{(t-1)k+d}}$$

We can store all the previous j values so that $j_{\Delta_{(t-1)k+d}}$ is available when we need it for the key modification. If $\Delta_{(t-1)k+d} > (t-2)k+d$, it means that no matter how we modify the key, we can not pass the i-th round by changing $S[(t-1)k+d]$. In this case, we go back one step to test if $S_{(t-1)k+d-2}[(t-1)k+d-1]$ is the correct one assuming $S_{(t-1)k+d-2}[(t-1)k+d] = S_{(t-1)k+d-1}[(t-1)k+d]$. Keep testing until $i = (t-2)k+d+1$. Now modifying the key becomes target oriented instead of brute searching all x and y, and thus duplicated searches can be greatly reduced. And another big advantage is that once the modification succeeds, we pass the t-th round, while in [2], after the modification assures the passing of the previous $t-1$ rounds, we need to pass the t-th round in a random way.

4.5   New Searching Algorithm

All the techniques described previously compose our new searching algorithm, which is summarized in Table 3. It is a recursive algorithm with recursive depth set to be n, which is the maximum rounds. If the newsearch function returns the maximum rounds, then it indicates that a collision is found. Note that when implementing, it can be further optimized by combining Matsui's algorithm and our new proposed one to proceed part of the rounds accordingly, so that a better performance could be achieved. For the simplicity, we just describe the most straightforward way in Table 3.

5   Complexity Evaluation

5.1   Complexity for Our Proposed Algorithm

We will see from a theoretical point of view, how efficiently our proposed algorithm can perform. We start by giving the following theorem which is important to compute the complexity, and show the proof.

Table 3. Proposed Searching Algorithm

Input: Key length k, different index $d = k - 3$, $n = \lfloor \frac{256+k-1-d}{k} \rfloor$

Output: K_1 and K_2 such that $K_2[d] = K_1[d] + 1$, $K_1[i] = K_2[i]$ if $i \neq d$,

$\quad KSA(K_1) = KSA(K_2)$

---

1. Store the following j^* values in the table, which are the conditions needed to be satisfied. $j_d^* = d, j_{d+1}^* = k + d, j_i^* = i + k$ for $i \in \{d + k, ..., d + k(n - 2)\}$,

$j_{d-2+k(n-1)}^* = d$, $j_{d-1+k(n-1)}^* = d - 1 + k(n - 1)$. (Class 1 j conditions)

2. Randomly generate a key K_1 with key length k. Modify

$K_1[d - 1] = (n - 1)k - d - 2$, $K_1[d + 1] = k - d - 1$.

Set $K_2 = K_1$ and $K_2[d] = K_1[d] + 1$.

3. Run the KSA until $i = d - 1$ after the swap. Modify $K_1[d] = 256 - j_{d-1}$, and

$K_2[d] = K_1[d] + 1$.

4. Keep running the KSA until $i = d + 1$ after the swap. Modify

$K_1[d + 2] = j_{2k-3}^* - j_{k-2}^* - \sum_{i=0}^{k-3} K_1[i] - \sum_{i=k-1}^{2k-3} S_{1,k-2}[i]$

$K_2[d + 2] = j_{2k-3}^* - j_{k-2}^* - \sum_{i=0}^{k-3} K_2[i] - \sum_{i=k-1}^{2k-3} S_{2,k-2}[i]$

5. Set the recursive depth variable $R = 0$.

6. If newsearch(K_1, K_2)=n

$\quad$ Colliding key pair found. Output K_1 and K_2.

$\quad$ else goto 2.

---

newsearch(K_1, K_2):

If $Round(K_1, K_2) = n$

$\quad$ then return n.

$MaxR = Round(K_1, K_2) = t - 1$, set $r = (t - 1)k + d$

while $r > (t - 2)k + d$

$\quad$ set $\Delta_r = j_{(t-1)k+d} - j_{(t-2)k+d} - \sum_{j=r-k+1}^{r-1} S_{j-1}[j] - \sum_{j=0}^{k-1} K[j]$.

$\quad$ If $\Delta_r \leq (t - 2)k + d$

$\quad\quad$ modify the key as follows:

$\quad\quad$ $K_1[\Delta_r] = K_1[\Delta_r] + r - j_{\Delta_r}$, $K_1[\Delta_r + 1] = K_1[\Delta_r + 1] - r + j_{\Delta_r}$

$\quad\quad$ $K_2[\Delta_r] = K_2[\Delta_r] + r - j_{\Delta_r}$, $K_2[\Delta_r + 1] = K_2[\Delta_r + 1] - r + j_{\Delta_r}$

$\quad\quad$ If $Round(K_1, K_2) \leq MaxR$ or $R = n$

$\quad\quad\quad$ return $Round(K_1, K_2)$.

$\quad\quad$ Else $R = R + 1$, **newsearch(K_1, K_2)**

$\quad$ $r = r - 1$

Notation

$Round(K_1, K_2)$: The number of rounds that a key pair K_1, K_2 can pass. In other words, key pair K_1 and K_2 satisfy all the j conditions in the first $Round(K_1, K_2)$ rounds.

Theorem 1. *Define $Pr_{t,(x,y)}$ be the probability for a trial that passes round t $(t > 2)$ by modifying the secret key as $K[x] = K[x] + y, K[x+1] = K[x+1] - y$ according to the multi-key modification given the previous trial fails to pass the t-th round. Then*

$$Pr_{t,(x,y)} \approx \sum_{i=1}^{t} \left(\left(\frac{(t-1)k-2}{256} \right) \times \prod_{j=0}^{i-2} \left(\frac{256-(t-j)k+x+3}{256} \right)^4 \right.$$

$$\left. \times \frac{(t-i+1)k-x-3}{256} \times \sum_{j=0}^{3} \left(\frac{256-(t-i+1)k+x+3}{256} \right) \right)$$

$$+ \frac{256-(t-1)k+2}{256}$$

Proof. Now let's consider some trial that passes all the first $t-1$ rounds and fails to pass the t-th round. Then we modify the secret key at indices x and $x+1$ with value difference y so that $K[x] = K[x] + y, K[x+1] = K[x+1] - y$. Let's denote $j'_{s,x}, j'_{s,x+1}$ and $j_{s,x}, j_{s,x+1}$ be the j values for the current trial and the trial after the key modification at the modified key indices at round s. It is easy to see that for each such key modification, the change of the 4 j values at each round will cause 4 S-Box values to be changed.

For the trial before the key modification, the successful pass of the first $t-1$ rounds indicates the correct S-Box sum for some fixed key sum $\sum_{i=0}^{k-1} K[i]$. Since our modification doesn't change the key sum, thus, after the key modification, the previous correct S-Box sum should still be satisfied in order to have a chance to pass the t-th round. Otherwise, the key modification will only cause a failure at an rather early round. For example if the previous trial passes the first $t-1$ rounds, for the key modification in round $s \leq t-1$ (assuming this key modification passes all the previous s-1 rounds), the S-Box sum $\sum_{i=x+(s-1)k}^{(t-1)k-1} S_{i-1}[i]$ should not be violated by the 4 changed j values $j'_{s,x}, j'_{s,x+1}$ and $j_{s,x}, j_{s,x+1}$.

First let's consider the probability that due to the key modification that the previous correct S-Box sum is violated. The modification is processed in the same order as the KSA procedure. And notice that in each round, due to the key modification, we have 4 changed j values, and they are checked in the sequence $j'_{s,x}, j_{s,x}, j'_{s,x+1}, j_{s,x+1}$ whether the failure conditions are satisfied. Notice that due to the use of the multi-key modification technique, the S-Box sum in the t-th round can not be touched since we have already precomputed the sum and are expecting the corresponding swap. The following events define the the the S-Box intervals that once touched, the previous correct S-Box sums will be violated due to the modification in round s.

- $A_s : j'_{s,x} \in [x + (s-1)k, tk-3]$ (the original $j'_{s,x}$ violates the S-Box sum $\sum_{i=x+(s-1)k}^{tk-3} S_{i-1}[i]$)
- $B_s : j_{s,x} \in [x + (s-1)k, tk-3]$ (the newly modified $j'_{s,x}$ violates the S-Box sum $\sum_{i=x+(s-1)k}^{tk-3} S_{i-1}[i]$))

- $C_s : j'_{s,x+1} \in [x + (s-1)k + 1, tk - 3]$ (the original $j'_{s,x+1}$ violates the S-Box sum $\sum_{i=x+(s-1)k+1}^{tk-3} S_{i-1}[i]$)
- $D_s : j_{s,x+1} \in [x + (s-1)k + 1, tk - 3]$ (the newly modified $j_{s,x+1}$ violates the S-Box sum $\sum_{i=x+(s-1)k+1}^{tk-3} S_{i-1}[i]$)

Denote $Pr(S_s)$ to be the probability that the modification in round s will not break the Class 1 j conditions that have been satisfied in the previous trial.

$$Pr(S_s) = (1 - Pr(A_s)) \cdot (1 - Pr(B_s)) \cdot (1 - Pr(C_s)) \cdot (1 - Pr(D_s))$$
$$= Pr(\bar{A}_s) \cdot Pr(\bar{B}_s) \cdot Pr(\bar{C}_s) \cdot Pr(\bar{D}_s)$$

Denote $Pr(F_s)$ to be the probability that the modification in round s will break the Class 1 j conditions that have been satisfied in the previous trial so that the current trial fails to pass round t.

$$Pr(F_s) = 1 - Pr(S_s)$$

The exact values for the four events can be computed as follows for $s > 2$:

$$Pr(A_s) = Pr(B_s) = \frac{(t - s + 1) * k - x - 2}{256}$$

$$Pr(C_s) = Pr(D_s) = \frac{(t - s + 1) * k - x - 3}{256}$$

Recall that the multi-key modification may fail because no available S-Box element can be swapped to the corresponding location in the t-th round. We approximate this probability to be

$$Pr(F_{multi}) \approx \frac{256 - (t - 1)k + 2}{256}$$

And the probability that we successfully find a candidate for the multi-key modification is

$$Pr(S_{multi}) \approx \frac{(t - 1)k - 2}{256}$$

Then the total probability that after the key modification the trial fails to pass the t rounds can be computed as follows:

$$Pr(F) = Pr(F_{multi}) + Pr(S_{multi}) \cdot Pr(F_1) + Pr(S_{multi}) \cdot Pr(S_1) \cdot Pr(F_2) +$$
$$\cdots + Pr(S_{multi}) \prod_{i=1}^{t-1} Pr(S_i) \cdot Pr(F_t)$$

Thus the probability that for some key modification succeeds to pass the t-th round while the trial before the modification passes the previous $t - 1$ rounds is

$$Pr_{t,(x,y)} = 1 - Pr(F)$$

After replacing with detailed parameters we complete our proof.

Then the complexity can be derived by the Theorem 2.

Theorem 2. *The complexity to find a colliding key pair for secret key with key length k is*

$$Comp_{new} \approx Pr^{-1}_{n,(\bar{x},\bar{y})}$$

where $Pr_{n,(\bar{x},\bar{y})}$ is the average case on all possible x and y, and $n = \lfloor \frac{256+k-1-d}{k} \rfloor$, $d = k - 3$.

To find 22-byte and 24-byte colliding key pairs, the complexity is around 2^{45} and 2^{40}.

5.2   Complexity for Matsui's Algorithm

In [2], a searching algorithm was proposed without giving the complexity evaluation. In order to compare the efficiency, we also give the complexity evaluation for algorithm proposed in [2]. Since Matsui's algorithm is also a recursive based algorithm, we can use a similar way as previous to analyze. We point out the different points here.

Without using the multi-key modification technique that chooses the target position to modify the key, it tries all the values for x and y, thus the S-Box in the t-th round can be touched. Also they set key difference at $d = k - 1$. We can redefine the following events that for the changed j value violating the S-Box sum.

- $A_s^M : j'_{s,x} \in [x + (s-1)k, (t-1)k - 1]$
- $B_s^M : j_{s,x} \in [x + (s-1)k, (t-1)k - 1]$
- $C_s^M : j'_{s,x+1} \in [x + (s-1)k + 1, (t-1)k - 1]$
- $D_s^M : j_{s,x+1} \in [x + (s-1)k + 1, (t-1)k - 1]$

Since there is no concern for the multi-key modification failure , $Pr(F^M)$ can be denoted as

$$Pr(F^M) = Pr(F_1^M) + Pr(S_1^M) \cdot Pr(F_2^M) + \cdots + \prod_{i=1}^{t-2} Pr(S_i^M)Pr(F_{t-1}^M)$$

where

$$Pr(S_s^M) = (1 - Pr(A_s^M)) \cdot (1 - Pr(B_s^M)) \cdot (1 - Pr(C_s^M)) \cdot (1 - Pr(D_s^M))$$
$$= Pr(\bar{A}_s^M) \cdot Pr(\bar{B}_s^M) \cdot Pr(\bar{C}_s^M) \cdot Pr(\bar{D}_s^M)$$

and

$$Pr(F_s^M) = 1 - Pr(S_s^M)$$

Also another big difference is that the modification of the key cannot guarantee the passing of the t-th round. Thus we have to assume the t-th round Class 1 j conditions will be satisfied randomly, namely,

$$Pr^M_{t,(x,y)} = (1 - Pr(F^M)) \times 2^{-8 \cdot (1 + \lfloor \frac{t}{n} \rfloor)}$$

This is because for any rounds except the last round, we have one j condition to satisfy, and we have two in the last round. Then we have the following theorems.

Theorem 3. *Define $Pr^M_{t,(x,y)}$ be the probability for a trial that passes round t by modifying the secret key as $K[x] = K[x] + y, K[x+1] = K[x+1] - y$ according to the Matsui's algorithm given the previous trial fails to pass the t-th round. Then we have*

$$Pr^M_{t,(x,y)} \approx \left(1 - \frac{(t-1)k - x}{256} \times \sum_{i=0}^{3} (\frac{256 - (t-1)k + x}{256})^i - \sum_{i=2}^{t-1} (\frac{(t-j)k - x}{256} \times \right.$$
$$\left. \prod_{j=1}^{i-1} (\frac{256 - (t-j)k + x}{256})^4 \times \sum_{j=0}^{3} (\frac{256 - (t-i)k + x}{256})^j) \right) \times 2^{-8(1 + \lfloor \frac{t}{n} \rfloor)}$$

Theorem 4. *The complexity of Matsui's algorithm to find a colliding key pair for secret key with key length k is*

$$Comp_{matsui} = (Pr^M_{n,(\bar{x},\bar{y})})^{-1}$$

where $Pr_{n,(\bar{x},\bar{y})}$ is the average case on all possible x and y, and $n = \lfloor \frac{256 + k - 1 - d}{k} \rfloor$.

As a result, the complexity for finding 24-byte colliding key pair is around 2^{48} and 2^{53} for 22-byte keys. The following figure shows the complexity to search for different colliding key pairs using two different algorithms.

We run the experiment under our proposed algorithm and successfully find by far the shortest 22-byte colliding key pair in about three days computational time by using parallel computer Cray XT5 (Quad-Core AMD Opteron 2.4GHz, 10 cores are used). In case of [2], around 10 days computational time and multiple cpus were used (the detailed information was not published) to find a 24-byte colliding key pair. Also, our proposed algorithm has a better efficiency searching for other short colliding keys which seems difficult to find by using the algorithm

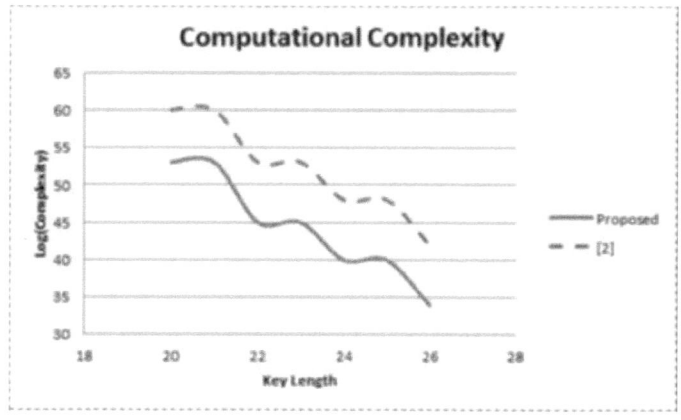

Fig. 1. Computational Complexity

in [2]. Here is the concrete 22-byte colliding key pair found by us in hexadecimal form:

$\mathbf{K_1}(\mathbf{K_2})$: A2 27 43 A7 03 94 2F 17 75 BB A7 27 8F DD 3E 7B C6 A1 C7 **81(82)**
    02 5A

6 Conclusion

In this paper, we investigate how to find RC4 colliding key pairs efficiently. We propose several techniques that can be used to bypass several rounds faster than brute force search, and the multi-key modification technique allows us to further increase the searching efficiency without drawback of duplicate searching which is the problem in [2]. We also give the complexity evaluation for both our proposed algorithm as well as the one in [2]. And finally by showing by far the shortest 22-byte colliding key pair ever found, we confirm that our algorithm does work efficiently as expected.

Acknowledgements. This work is inspired by the previous work of Mitsuru Matsui, and the authors wish to thank him for his invaluable comments. Also, the authors wish to thank all the anonymous reviewers for their useful suggestions to help to improve this paper.

References

1. Anonymous: RC4 Source Code. CypherPunks mailing list (September 9, 1994),
   http://cypherpunks.venona.com/date/1994/09/msg00304.html,
   http://groups.google.com/group/sci.crypt/msg/10a300c9d21afca0
2. Matsui, M.: Key Collisions of the RC4 Stream Cipher. In: Dunkelman, O., Preneel, B. (eds.) FSE 2009. LNCS, vol. 5665, pp. 38–50. Springer, Heidelberg (2009)
3. Chen, J., Miyaji, A.: Generalized RC4 Key Collisions and Hash Collisions. In: Garay, J.A., De Prisco, R. (eds.) SCN 2010. LNCS, vol. 6280, pp. 73–87. Springer, Heidelberg (2010)
4. Chen, J., Miyaji, A.: A New Practical Key Recovery Attack on the Stream Cipher RC4 Under Related-Key Model. In: Lai, X., Yung, M., Lin, D. (eds.) Inscrypt 2010. LNCS, vol. 6584, pp. 62–76. Springer, Heidelberg (2011)
5. Sepehrdad, P., Vaudenay, S., Vuagnoux, M.: Statistical Attack on RC4. In: Paterson, K. (ed.) EUROCRYPT 2011. LNCS, vol. 6632, pp. 343–363. Springer, Heidelberg (2011)
6. Sepehrdad, P., Vaudenay, S., Vuagnoux, M.: Discovery and Exploitation of New Biases in RC4. In: Biryukov, A., Gong, G., Stinson, D. (eds.) SAC 2010. LNCS, vol. 6544, pp. 74–91. Springer, Heidelberg (2011)
7. Maitra, S., Paul, G., Sen Gupta, S.: Attack on Broadcast RC4 Revisited. In: Joux, A. (ed.) FSE 2011. LNCS, vol. 6733, pp. 199–217. Springer, Heidelberg (2011)
8. Wang, X., Yu, H.: How to break MD5 and other hash functions. In: Cramer, R. (ed.) EUROCRYPT 2005. LNCS, vol. 3494, pp. 19–35. Springer, Heidelberg (2005)

A Formal Approach to
Distance-Bounding RFID Protocols

Ulrich Dürholz[2], Marc Fischlin[1], Michael Kasper[2], and Cristina Onete[1]

[1] Darmstadt University of Technology & CASED, Germany
www.cryptoplexity.de
[2] Fraunhofer Institute for Secure Information Technology (SIT) and CASED,
Germany

Abstract. Distance-bounding protocols aim at impeding man-in-the-middle(MITM) attacks by measuring response times. Three kinds of attacks are usually addressed: (1) Mafia attacks where adversaries relay communication between honest prover and honest verifier in different sessions; (2) Terrorist attacks where adversaries gets limited active support from the prover to impersonate; (3) Distance attacks where a malicious prover claims to be closer to the verifier than it really is. Many protocols in the literature address one or two such threats, but no rigorous security models —nor clean proofs— exist so far. For resource-constrained RFID tags, distance-bounding is more difficult to achieve. Our contribution here is to formally define security against the above-mentioned attacks and to relate the properties. We thus refute previous beliefs about relations between the notions, showing instead that they are independent. Finally we assess the security of the RFID distance-bounding scheme due to Kim and Avoine in our model, and enhance it to include impersonation security and allow for errors due to noisy channel transmissions.

1 Introduction

Man-in-the-middle attacks (MITM) are powerful strategies against identification schemes: the adversary relays communication between prover and verifier, making the latter accept. Following [16], pure relaying is called Mafia fraud. Environments with no central authority and certificates, like RFID identification, are especially subject to Mafia fraud, as indicated in [23,15,17,22,19]; several works also show attacks on the HB protocol [27,20,18,9,32,30], which is designed for low-power devices e.g. RFIDs. For an overview of RFID security see [28].

1.1 Distance Bounding Protocols

Distance bounding protocols were proposed by Brands and Chaum [8] as a countermeasure against MITM attacks. The basic idea is that pure relay takes longer than genuine responses; thus, if verifiers measure the time elapsed between sending a value and receiving the reply, MITM attacks should be infeasible. In practice, verifiers check round-times for many *fast* or *time-critical* communication phases (as opposed to *slow* or *lazy* phases, where round times do not matter).

X. Lai, J. Zhou, and H. Li (Eds.): ISC 2011, LNCS 7001, pp. 47–62, 2011.
© Springer-Verlag Berlin Heidelberg 2011

We mainly address RFID authentication, but our new framework applies to general distance bounding. Here verifiers (in RFID, readers) indicate by a bit if provers (in RFID, tags) are authenticated or not after some interaction. We often use the terms readers/tags for provers/verifiers. RFID distance bounding is extensively analysed in $[1, 2, 3, 4, 8, 10, 12, 13, 15, 16, 17, 22, 23, 26, 29, 34, 36, 37]$. See also [24] for a comprehensive overview. The three main attacks are: (1) Mafia fraud, where adversaries try to impersonate to the reader while communicating with an honest tag (the timing prevents pure message relay though); (2) Terrorist fraud, where tags may help the adversary by leaking information in offline phases (tags should reveal no trivial information though, like the secret key); (3) Distance fraud, where tags claim to be closer than it actually is. We also consider the basic (often neglected) requirement for identification, i.e. slow-round impersonation resistance, independent of the limited number of fast phases.

We exemplify the three attacks as follows: consider a gym locker with an inbuilt RFID reader, for which Alice holds the unique pass key (an RFID tag). One evening, Alice is not at the gym, but at a party. In the *Mafia fraud* scenario, Bob *is* at the gym; his accomplice, Bobette, is at the party with Alice. Bob wants to open the locker (without Alice's consent for Mafia fraud). In this attack, Bob and Bobette relay messages between the locker and Alice's tag. If, on the contrary, Alice *wants* Bob to use her locker (for this night only) we have *Terrorist fraud*. Alice may now give Bob information to help him use her locker, but she doesn't want Bob to abuse her kindness and open the locker on his own, this or any other time. For Terrorist attacks thus, Alice helps Bob herself: Bobette is not needed. Finally, if Alice parked her car in a bad spot, she might want to "prove" that she was at the gym instead (this is *distance fraud*) by opening the locker, which can be opened only if the unique key is in direct proximity.

Several existing protocols implement resistance against one (or more) of the above threats. A selection of such protocols is compared in Figure 1. The values mentioned for $[8, 26, 4, 34, 29]$ are those claimed by the respective papers (despite a lack of formal approaches). We note that public-key constructions, as opposed to private-key ones, are unsuitable for low-power devices like RFID. Also, most existing work permits adversaries to impersonate the reader to the tag, thus leaking information about fast-phase response times. If only bits are transmitted in fast phases, the ideal impersonation bound would be 2^{N_c} for N_c critical rounds; however, most protocols allow impersonation and thus reach a lower than ideal bound. To account for this Mafia fraud attack, under "Rounds", we give the number N_c of time-critical rounds required for a Mafia resistance of about 2^{-k}. We round down the number of rounds in $[26, 34]$ to $2k$. Note that [4] shows a construction with reduced complexity, at the expense of security.

We lastly outline some related literature. Further related work follows in Section 1.2. Recently, position-based cryptography [14] shows a model for determining if a prover is (exactly) at a claimed position — but in a single protocol run, with many verifiers, unlike in Mafia and terrorist fraud. Adversaries in [14] *all* have the same knowledge as the prover (knowing also the secret key). This notion is closer to distance fraud, where tags must prove they are closer to the

	[8]	[26]	[4]	[34]	[29]
Mafia	✓	✓	✓	✓	✓
Terror	×	×	×	$(\checkmark)^1$	×
Distance	✓	✓	✓	✓	✓
Impersonation	×	×	✓	×	×
Rounds N_c	k	$>2k$	k	$>2k$	k
Storage	N_c	$2N_c$	$O(2^{N_c})$	$2\ N_c$	$4N_c$
Private-key	×	✓	✓	✓	✓

Fig. 1. Claimed Security and Actual Efficiency of Distance Bounding Protocols at a glance ([1]only special terrorists, no formal proof)

reader than they really are. However, exact positioning is unachievable for RFID, requiring too many readers to deal with the high variance in response time. In fact, recent work due to Hancke [25] suggests designing a distance bounding channel limiting channel-specific variations of response times.

By contrast, self-delegation as in [21] and [11] resembles terrorist fraud. In [21], self-delegated secondary keys are used to authenticate. Losing too many such keys compromises the long-term key, like in terrorist fraud, where tags partly reveal the key when helping the MITM. The model in [21] mainly differs from ours in the following: [21] considers a public-key setting only (with server certification of secondary keys); they analyse only signature-leakage; and no online help is available. Also, the public-key cryptography and non-interactive zero-knowledge proofs used in [21] are unsuitable for RFID. Finally the "all-or-nothing" approach of transferable anonymous credentials [11] associates (with no formal security model) sharing secret pseudonyms or credentials with recovering a user's full secret, as in terrorist fraud. The use of public-key infrastructures also makes the idea inapplicable to RFID.

1.2   Our Contributions

Our contributions are threefold: (1) We give rigorous models for Mafia, terrorist, and distance fraud, thus (2) formally relating these notions (we also refute claims in [2,34] that terrorist fraud resistance implies distance fraud resistance). Finally, we (3) assess the security of the prominent scheme in [29] in our model, making it impersonation resistant and allowing for noisy channels.

The Practice behind the Theory. Practical RFID investigations [13,15,34,26,33, 25] show some design issues applicable for *all* low-power devices. We provide for them in our framework. Firstly, due to the unreliable and noisy transmissions, readers and tags should only exchange bits in time-critical phases [15, 25, 34]. Also, time-critical computations must be simple, taking consistent time so as not to bias round-trip time and threshold values. Low-power devices like RFID must use little storage; noise in both transmissions and time measurement must also be provided for [15, 25, 34]. Our thresholds for failures during timed steps add depth to the model, allowing pure relay in some phases.

Lastly, implementations may allow adversaries to predict a bit "halfway into the signal" [17]. Also, as computation complexity may vary with received input, adversaries can get information from the reader or tag faster than expected. In our model, adversaries may relay data as long as it is not purely duplicated. We also suggest, as in [4], to add offline authentication (preferably) before the necessarily-few fast phases: the often neglected, but essential requirement of impersonation resistance is strongly defined in our model. Though schemes like [8, 26] lack this enhancement, we use it for the scheme in [29].

The Models. A sound modeling of the above attacks is crucial to assessing protocol security. Many confusions regarding attack modes and successful MITM attacks appear e.g. for the HB protocol [27, 20, 18, 9, 32, 30]. As another example, the allegedly secure Hitomi and NUS protocols were recently proved insecure [1]. We formalize game-based models while also considering practical conditions, and can thus formally prove that, contrary to statements in [34] and [2], terrorist fraud resistance does not imply distance fraud resistance. In fact, we show that Mafia, terrorist, and distance fraud resistance are all independent (concretely, we show protocols that are vulnerable to one attack, but resistant to all others, including basic impersonation resistance for authentication). Thus terrorist fraud resistance does not imply Mafia fraud resistance, nor vice versa.

Some groundwork has already been laid in this field by Avoine et al. [2], who model Mafia, distance, and terrorist fraud in both a black-box and a white-box sense i.e. giving adversaries access to the implementation of the primitive or not. Distance-bounding protocols here have two main goals: authentication and distance checking; each type of fraud is also more formally defined. Adversaries may choose from three main strategies: pre-ask (query prover before being queried by verifier), post-ask (query prover after being queried by verifier), and early-reply (respond before verifier sends query, without querying the prover). In the black box model, Mafia and terrorist fraud are proved equivalent, whereas terrorist fraud resistance is said to imply distance fraud resistance. In the white-box model, terrorist fraud resistance implies both Mafia and distance fraud. Mafia fraud resistance is equivalent in the black and white box models, and white-box terrorist and distance fraud are strictly stronger than the black-box notions.

By contrast, our definitions are much more concrete and formal. Protocols have many rounds (lazy or time-critical), and adversaries choose (possibly different) strategies at each round, unlike [2]. Our Mafia adversaries may relay *parts of the communication*, e.g. flip bits, or purely relay (taint) some rounds. Our Mafia and terrorist provers may be anywhere, unlike [2], where provers are outside the target distance from the verifier. By using a simulator, we concretely define "advantage for future attacks" [2] for terrorist fraud. Hence, we prove that all security notions are independent. We also extend impersonation resistance to lazy-phase authentication, thus preventing information leaks to fake provers.

In other related work, [31] considers honest provers only, and states no security goals, while the formal methods approach in thoroughly models distance bounding, but treats wireless networks in general for provers and verifiers with equal capacities (unlike for RFID, where tags are computationally weaker). Also, some physical properties of RF communication, e.g. the unreliability of tags' backscattering and colliding signals, are unaccounted for. Cryptographically speaking, [35] provide no reliable definitions for the attacks above.

Using our Framework. We assess the security of the Kim and Avoine protocol [29] in our model. Here, mutual tag-to-reader and reader-to-tag fast-phase authentication yield good Mafia and distance fraud (but not impersonation-) resistance. If reader authentication fails, tags respond randomly every round. We first make the scheme in [29] impersonation resistant, then formally assess its security in our model, proving also that it is not terrorist fraud resistant.

2 Preliminaries

We consider a single reader $\mathcal{R}$ and a single tag $\mathcal{T}$, sharing a secret key generated in Kg. We associate to $\mathcal{R}$ a clock and a database entry storing $\mathcal{T}$'s secret key. Identification schemes $\mathcal{ID} = (\mathsf{Kg}, \mathcal{T}, \mathcal{R})$ mark (consecutive) steps of the protocol as *lazy* or *time-critical*. In time-critical steps, one party —usually the reader— compares measured round-times Δt to a predetermined threshold $t_{\max}$. Else the phase is *lazy*. Protocols consist of arbitrary *non-overlapping* sequences of lazy and time-critical phases, with possibly many consecutive time-critical phases. Denote by N_c the number of time-critical phases. At most $T_{\max}$-many round-times may exceed $t_{\max}$, to account for transmission-time lags. Similarly, up to $E_{\max}$-many time-critical responses may be erroneous.

Definition 1. *An* identification scheme for timing parameters $(t_{\max}, T_{\max}, E_{\max}, N_c)$ *is a triplet of efficient algorithms* $\mathcal{ID} = (\mathsf{Kg}, \mathcal{T}, \mathcal{R})$ *with:*

KEY GENERATION. *For parameter $n \in \mathbb{N}$,* Kg *generates a secret key sk.*
IDENTIFICATION. *The joint execution of algorithms $\mathcal{T}(sk)$ and $\mathcal{R}(sk)$ generates, depending on $t_{\max}, T_{\max}, E_{\max}, N_c$, a verifier output $b \in \{0, 1\}$.*

We assume that the scheme is complete: for any $n \in \mathbb{N}$ and any key $sk \leftarrow \mathsf{Kg}(1^n)$, the decision bit b produced by honest party $\mathcal{R}(sk)$ interacting with honest party $\mathcal{T}(sk)$ under the requirements following from the timing parameters, is 1 with probability (negligibly close to) 1.

As noted also in [2], distance bounding enhances authentication with distance seeking. Thus, $t_{\max}$ is a crucial parameter here. The parameters $E_{\max}$ and $T_{\max}$ are intrinsic to communication over noisy channels (e.g. RF channels between readers and passive and semi-passive RFID tags[1]). In distance bounding, it is unreasonable to separate transmission *reliability* from *security*, as round-time measurements are crucial towards acceptance or rejection. Bit errors are unavoidable in RF communication, as stated in point 4 of Clulow et al.'s principles for secure time-of-flight distance-bounding [15]. As shown in section 1, communication noise makes transmissions unreliable and introduce possible lags. We can, however, set $T_{\max} = E_{\max} = 0$ for extremely reliable scenarios.

---

[1] Passive RFID tags have no power source and are very sensitive to metals and liquids in particular. Semi-passive tags use their own power source for computation, but rely on readers for communication, and are also affected by metals and liquids.

3   Security Model

3.1   Communication Model

The adversary can access: a reader instance to which it impersonates the tag (a *reader-adversary session*), a tag instance to which it impersonates the reader (*adversary-tag session*), and an interface observing a genuine reader-tag protocol for which the adversary cannot change transmissions (*reader-tag session*). The adversary can access all interfaces concurrently and in many sessions (sessions share a secret key, but have different random tapes). Each session has an identifier sid (given to the adversary, but not to protocol participants). We assume that the adversary knows if an authentication attempt succeeded or not.[2]

In our concurrent single-reader-single-tag scenario (as opposed to a single reader and multiple tags), many instances of the single tag may exist in parallel, sharing the secret key, but not the random tape. The key is *static*, i.e., not updated after executions. For many independent keys (multiple tags), adversaries can always pick a tag to attack in our model. Three factors are crucial to multiple-tag scenarios: the interdependency of the keys; the noise in the communication due to tag-to-reader collisions (a factor modeled by $E_{\max}$); and key management. A formal approach for key update is, however, beyond the scope of this paper.

We assume message-driven attacks, i.e., honest parties reply as soon as they receive a (protocol) message. The adversary schedules message delivery to honest parties. We assume a global clock, assigning an integer $\mathsf{clock}(\mathsf{sid}, k)$ to the k-th protocol message, delivered in session sid to an honest party. The honest party's reply is assigned $\mathsf{clock}(\mathsf{sid}, k+1) = \mathsf{clock}(\mathsf{sid}, k)+1$.[3] Furthermore, $\mathsf{clock}(\mathsf{sid}, k) < \mathsf{clock}(\mathsf{sid}^*, k)$ if the adversary delivers the k-th message in session sid^* after the k-th message in session sid. Denote by $\Pi_{\mathsf{sid}}[i \ldots j]$ messages i to j exchanged in session sid and by $\Pi_{\mathsf{sid}}[1 \ldots]$ all messages exchanged in sid. Let $\mathsf{view}_{\mathcal{A}}$ denote the adversary's view in an attack, containing its internal randomness and all the transcripts (of communication with and among other parties).

Let t denote the adversary's running time, including steps of honest parties. Denote by $q_{\mathcal{R}}$ (resp. $q_{\mathcal{T}}$ and q_{OBS}) the maximal number of reader-adversary (resp. adversary-tag and reader-tag) sessions. Below we refine the attacks and define winning conditions for the adversary (who must non-trivially impersonate the tag in a reader-adversary session). For an attack att we write $\mathbf{Adv}_{\mathcal{ID}}^{\mathrm{att}}(\mathcal{A})$ for the probability that the $(t, q_{\mathcal{R}}, q_{\mathcal{T}}, q_{\mathrm{OBS}})$-adversary $\mathcal{A}$ wins.

3.2   Mafia Fraud Detection Model

Mafia fraud adversaries can communicate arbitrarily with tag and reader, *except for purely relaying time-critical transmissions*. We exclude only attacks where the adversary relays *exact* transmissions, calling such time-critical phases *tainted*:

---

[2] This is not a strong requirement. In practice the success of an authentication attempt is marked by a physical event: a beep, the opening of a door, a green light etc.

[3] We could also allow adversaries to delay message delivery *from* honest parties. Our model and results are robust with respect to this idea, but this contradicts the implementation of reliable time measurements and enable denial-of-service attacks.

Definition 2 (Tainted Time-Critical Phase (Mafia)). *A time-critical phase* $\Pi_{\mathsf{sid}}[k \ldots k+2\ell-1] = (m_k, \ldots, m_{k+2\ell-1})$ *for* $k, \ell \geq 1$ *of a reader-adversary session* sid, *with the* k-*th message being received by the adversary, is* tainted *by the phase* $\Pi_{\mathsf{sid}^*}[k \ldots k+2\ell-1] = (m_k^*, \ldots, m_{k+2\ell-1}^*)$ *of an adversary-tag session* sid^* *if for all* $i = 0, 1, \ldots, \ell-1$ *we have:*

$$(m_k, \ldots, m_{k+2\ell-1}) = (m_k^*, \ldots, m_{k+2\ell-1}^*),$$
$$clock(\mathsf{sid}, k+2i) < clock(\mathsf{sid}^*, k+2i),$$
$$and \quad clock(\mathsf{sid}, k+2i+1) > clock(\mathsf{sid}^*, k+2i+1).$$

As shown in Figure 2, our notion is slightly conservative. We account for computation complexity depending on input values, allowing adversaries to receive one reply, change the response, and relay it in time. But now adversaries could flip redundant bits and relay crucial ones without tainting a phase. We nonetheless prefer to err on the safe side and give adversary more freedom, as obvious redundancy is easily modified as shown for key exchange protocols [7, 6]. Secondly, time-critical phases are tainted if *all* transmitted messages are relayed in two sessions. However, if *a single* transmission is relayed, the phase is untainted. Here we give adversaries more freedom and get a stronger notion.

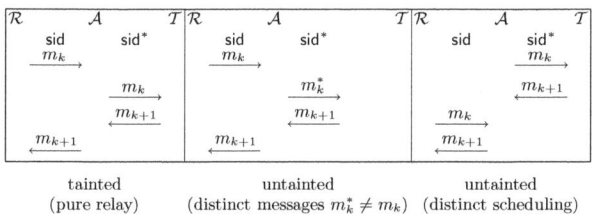

tainted   untainted   untainted

(pure relay)   (distinct messages $m_k^* \neq m_k$)   (distinct scheduling)

Fig. 2. Examples of Tainted and Untainted Time-Critical Phases

The adversary must now make the reader accept in session sid such that for each adversary-tag session sid^* at most $T_{\max}$ phases of sid are tainted by sid^*:

Definition 3 (Mafia Fraud Resistance). *For a distance-bounding identification scheme* $\mathcal{ID}$ *with parameters* $(t_{\max}, T_{\max}, E_{\max}, N_c)$, *a* $(t, q_{\mathcal{R}}, q_{\mathcal{T}}, q_{\mathrm{OBS}})$-*Mafia-fraud adversary* $\mathcal{A}$ *wins against* $\mathcal{ID}$ *if the reader accepts in a reader-adversary session* sid *such that any adversary-tag session* sid^* *taints at most* $T_{\max}$ *time-critical phases of* sid. *Let* $\mathbf{Adv}_{\mathcal{ID}}^{mafia}(\mathcal{A})$ *denote the probability that* $\mathcal{A}$ *wins.*

Different adversary-tag sessions may taint different rounds of reader-adversary session sid. As we count $T_{\max}$ over all adversary-tag sessions the adversary wins if it taints at most $T_{\max}$ *distinct* phases. Protocols must prevent such attacks to be Mafia fraud secure in concurrent settings. Further session interdependencies should also be avoided so that messages from another session do not taint sid.

3.3   Terrorist Attack Model

In a terrorist attack the tag aids the adversary in all short of revealing its secret key, in fact wanting to ensure that the adversary only wins with the tag's aid (the dishonest prover controls the adversary's access). Desmedt [16] concretely describes the tag's involvement as offline help in a single impersonation attempt. The adversary now wins if the reader accepts, but the adversary cannot use the help given by tag $\mathcal{T}'$ to impersonate further.

We formalize the idea by using ideas from proofs of computational ability [38, 5], which exactly capture the intuition of terrorist attacks: given support from a prover e.g. $\mathcal{T}'$, one can solve a hard problem e.g. identifying to the reader. This is independent of how the prover gives support. We are not, however, interested in the cases where $\mathcal{T}'$ yields the entire key (or large parts of it) and mark certain auxiliary data given by $\mathcal{T}'$ as trivial, i.e. the data is trivial if it allows one to successfully complete a "fresh" identification attempt *without help from $\mathcal{T}'$*. This includes the case when $\mathcal{T}'$ gives the secret key, but circumvents the problem of determining which parts of the key are helpful. Data is trivial if it aids identification beyond the dedicated help in the session where $\mathcal{T}'$ helps.

We formalize the latter by demanding that no algorithm $\mathcal{S}$, called simulator, can use the data passed by $\mathcal{T}'$ to $\mathcal{A}$ to authenticate without the help of $\mathcal{T}'$ (to be fair, we allow $\mathcal{S}$ the same number $q_{\mathcal{R}}$ of attempts as $\mathcal{A}$). This is in line with well-known simulation paradigms, and allows to compare the respective success probabilities of the adversary $\mathcal{A}$ aided by $\mathcal{T}'$, and the simulator $\mathcal{S}$ using $\mathcal{A}$'s information to authenticate. If $\mathcal{A}$ is significantly more successful than $\mathcal{S}$, the attack is non-trivial and the protocol is insecure against terrorist attacks. Note that "unsophisticated" adversaries may do worse than simulators for secure schemes, thus yielding negative advantages.

For terrorist fraud, $\mathcal{A}$ acts as for Mafia fraud, but may query the "malicious" interface $\mathcal{T}'$ in lazy phases. Sessions sid′ with $\mathcal{T}'$ are arbitrary, not following protocol.In fact we may consider only one session sid′ when $\mathcal{T}'$ helps $\mathcal{A}$. The tag may *not* aid $\mathcal{A}$ in time-critical phases, a fact which we model by defining tainted time-critical phases as pure-relay phases or rounds where $\mathcal{A}$ queries $\mathcal{T}'$.

Definition 4 (Tainted Time-Critical Phase (Terror)). *A time-critical phase $\Pi_{\mathsf{sid}}[k \ldots k+2\ell-1] = (m_k, \ldots, m_{k+2\ell-1})$ for $k, \ell \geq 1$ of a reader-adversary session* sid, *with the k-th message being received by the adversary, is* tainted *if there is a session* sid′ *between the adversary and $\mathcal{T}'$ such that, for some i,*

$$clock(\mathsf{sid}, k) < clock(\mathsf{sid}', i) < clock(\mathsf{sid}, k + 2\ell - 1).$$

For the new definition of tainted phases, terrorist fraud resistance demands that for any terrorist fraud attacker $\mathcal{A}$ there exists a simulator $\mathcal{S}$ such that for any supporting $\mathcal{T}'$, $\mathcal{S}$ is essentially as successful as $\mathcal{A}$. We use concrete security statements and omit quantification over $\mathcal{A}$, $\mathcal{S}$, and $\mathcal{T}'$ algorithms; this quantification is included in subsequent security claims in the usual form (i.e., for any adversary there exists a simulator such that for all tags the advantage is small).

Definition 5 (Terrorist Fraud Resistance). *Let* $\mathcal{ID}$ *be a distance-bounding identification scheme with parameters* $(t_{\max}, T_{\max}, E_{\max}, N_c)$. *Let* $\mathcal{A}$ *be a* $(t, q_{\mathcal{R}}, q'_T)$-*terrorist-fraud adversary*, $\mathcal{S}$ *be an algorithm running in time* $t_{\mathcal{S}}$, *and* T' *be an algorithm running in time* t'. *Denote*

$$\boldsymbol{Adv}_{\mathcal{ID}}^{terror}(\mathcal{A}, \mathcal{S}, T') = p_{\mathcal{A}} - p_{\mathcal{S}}$$

where $p_{\mathcal{A}}$ *is the probability that the reader accepts in one of the* $q_{\mathcal{R}}$ *reader-adversary sessions* sid *such that at most* $T_{\max}$ *time-critical phases of* sid *are tainted, and* $p_{\mathcal{S}}$ *is the probability that, given* view$_{\mathcal{A}}$ *in an attack of* $\mathcal{A}$, $\mathcal{S}$ *makes the reader accept in one of* $q_{\mathcal{R}}$ *subsequent executions.*

Again, if the advantage is negative, $\mathcal{A}$ performs worse than $\mathcal{S}$. Our notion is quite strong: the simulator only gets to see $\mathcal{A}$'s transcript in an offline phase, instead of communicating with T' online. This guarantees stronger security and saves us from dealing with issues related to the number of queries and successful attacks (adversary vs. simulator).

How does our definition fit into previous efforts? Previous protocols [34, 4] claim to achieve a security of $(1/2)^{-N_c}$. This, however, corresponds to a tailor-made strategy of T'; other strategies may still exist. Proving that the advantage in Definition 5 is negligible, then we *prove* that T' can only help trivially.

3.4 Distance-Fraud Model

For distance fraud an adversary must reply ahead of a time-critical phase or it cannot respond in time. In practice this is enforced by a tight value of $t_{\max}$. For any time-critical phase, with possibly many communication rounds, the adversary must commit to the *first* message to be sent. For any later rounds in the phase, the adversary has time to reply even from farther away.

The order of committed and sent values is determined by on oracle CommitTo with a single session sid$_{\text{CommitTo}}$, taking tuples (sid, i, m_i) from the adversary and giving empty responses. The adversary commits to the first message of time-critical phase i of session sid (message j in sid) at time clock(sid$_{\text{CommitTo}}, j$). As the adversary may repeatedly commit to this message, we take the last commitment before phase i begins. A time-critical phase is tainted if the adversary returns an answer it has not committed to.

Definition 6 (Tainted Time-Critical Phase (Distance)). *A time-critical phase* $\Pi_{\text{sid}}[k \ldots k+2\ell-1] = (m_k, \ldots, m_{k+2\ell-1})$ *for* $k, \ell \geq 1$ *of a reader-adversary session* sid, *with the* k-*th message being received by the adversary, is tainted if the maximal* j *with* $\Pi_{\text{sid}_{\text{CommitTo}}}[j] = (\text{sid}, k+1, m^*_{k+1})$ *for some* m^*_{k+1} *and* clock(sid, k) > clock(sid$_{\text{CommitTo}}, j$) *satisfies* $m^*_{k+1} \neq m_{k+1}$ *(or no such* j *exists).*

Definition 7 (Distance Fraud Resistance). *For an identification scheme* $\mathcal{ID}$ *with parameters* $(t_{\max}, T_{\max}, E_{\max}, N_c)$, *a* $(t, q_{\mathcal{R}}, q_T, q_{\text{OBS}})$-*distance-fraud adversary* $\mathcal{A}$ *wins against* $\mathcal{ID}$ *if the reader accepts in one of* $q_{\mathcal{R}}$ *reader-adversary sessions* sid *with at most* $T_{\max}$ *tainted time-critical phases. Let* $\boldsymbol{Adv}_{\mathcal{ID}}^{dist}(\mathcal{A})$ *be the probability of* $\mathcal{A}$ *winning.*

3.5   Impersonation Resistance

We suggest a simple, but very strong definition of impersonation security as a basic requirement of identification in our concurrent setting. Thus even adversaries who actively take part in intertwined prover and verifier runs cannot impersonate the prover. Whereas the previous properties concern time-critical phases, impersonation security requires that an adversary cannot impersonate a tag in *lazy* phases. This ensures that the reader leaks no time-critical information to an invalid tag. Following the idea that parties should authenticate even if the time-critical phases are not executed, we consider projections $\Pi_{\mathsf{sid}}^{\mathrm{lazy}}[1\ldots]$ of $\Pi_{\mathsf{sid}}[1\ldots]$ containing lazy phases transmissions only, and (not necessarily consecutive) indices $\iota_{\mathsf{sid}}^{\mathrm{lazy}} = (i_1, i_2, \ldots)$ of lazy phase messages. The adversary wins if a reader-adversary session succeeds and no adversary-tag session has the same "lazy transcript", created via pure relaying.

Definition 8 (Impersonation Resistance). *In a distance-bounding identification scheme $\mathcal{ID}$ with parameters $(t_{\max}, T_{\max}, E_{\max}, N_c)$ where $\mathcal{R}$ always go first, a $(t, q_{\mathcal{R}}, q_{\mathcal{T}}, q_{\mathrm{OBS}})$-impersonation adversary $\mathcal{A}$ wins against $\mathcal{ID}$ if $\mathcal{R}$ accepts in a reader-adversary session sid such that no adversary-tag session sid^* has*

$$\Pi_{\mathsf{sid}}^{lazy}[1\ldots] = \Pi_{\mathsf{sid}^*}^{lazy}[1\ldots],$$

and

$$clock(\mathsf{sid}, i) < clock(\mathsf{sid}^*, i)$$

for any $i \in \iota_{\mathsf{sid}}^{lazy} \cap \iota_{\mathsf{sid}^*}^{lazy}$ s.t. $\mathcal{R}$ has sent the i-th message to $\mathcal{A}$ in sid, and

$$clock(\mathsf{sid}, j) > clock(\mathsf{sid}^*, j)$$

for any $j \in \iota_{\mathsf{sid}}^{lazy} \cap \iota_{\mathsf{sid}^*}^{lazy}$ such that the adversary has sent the j-th message to the reader in sid. Let $\mathbf{Adv}_{\mathcal{ID}}^{imp}(\mathcal{A})$ be the probability that $\mathcal{A}$ wins.

4   Relationship between Fraud Types

Impersonation security concerns lazy protocol phases, while Terrorist, Mafia, and distance fraud attack time-critical phases. In our framework we refute the idea in [34] that terrorist fraud resistance implies distance fraud resistance and show that all properties are independent. Due to limited space, we leave the formal proofs for the full version and give only an intuition below.

Theorem 1 (Security Diagram — Informal). *If pseudorandom functions exist, the following holds:*

1. *There exists a distance-bounding identification scheme that is impersonation-secure, Mafia and distance fraud resistant, but not terrorist fraud resistant.*
2. *There exists a distance-bounding identification scheme that is impersonation-secure, Terrorist and Mafia fraud resistant, but not distance fraud resistant. Thus, terrorist fraud resistance does not imply distance fraud resistance.*
3. *There exists a distance-bounding identification scheme that is impersonation-secure, Terrorist and distance fraud resistant, but not Mafia fraud resistant. Thus, terrorist fraud resistance does not imply Mafia fraud resistance.*

Terrorist-Fraud Resistance. The enhanced Kim-Avoine scheme in Section 5 has all properties except for terrorist-fraud resistance. The reason it fails against terrorist attacks is that time-critical messages are predetermined by the lazy phase and can be revealed without disclosing the secret key (thus providing sufficient, but non-trivial offline help). In general, terrorist attacks are thwarted by interlinking authentication sessions, such that malicious tags (partially) reveal long-term secrets if they help the adversary. The difficulty in designing terrorist-fraud resistant schemes is formally ensuring that the simulator can extract the secret from the adversary and thus authenticate. The simulator's only advantage is that it can rewind executions and get responses for different challenges.

Distance-Fraud Resistance. We separate distance-fraud resistance from the other properties by giving the tag a special key which makes time-critical responses predictable. Honest parties never use this key, but malicious tags may use it to commit distance fraud. Other security properties are unaffected, as the special key is never used by honest parties. Distance-fraud resistance depends on the unpredictability of each round's answer. This is easily achieved by adding some time-critical rounds where tags echo random bits.

Mafia-Fraud Resistance. We show Mafia fraud resistance independence by starting with a protocol having all other security properties; the tag may use a bit to indicate that time-critical bits are flipped. Then a man-in-the-middle adversary can flip replies from an adversary-tag session and authenticate to the reader without tainting the phases. There are two options to prevent Mafia fraud attacks. Assume that in each fast phase the reader sends a random challenge. If the adversary correctly predicts the challenge in a reader impersonation, it can use the reply in the reader-adversary session without tainting the phase; for a wrong prediction, the adversary guesses the answer instead. The overall success is $\frac{3}{4}$ per round as in, e.g., the Hancke and Kuhn protocol [26]. The other option is to authenticate the reader by the fast phase challenges. Now the adversary-tag session in the above attack aborts for a wrong prediction, dropping the adversary's success probability in the reader-adversary execution to $\frac{1}{2}$ for subsequent rounds. This is the strategy of the Kim-Avoine as discussed next.

5   Case Study: The Construction Due to Kim and Avoine

The scheme in [29] is Mafia and Distance fraud resistant. We tweak it to add impersonation security, provide for noisy channels as in Section 2, then prove it secure in our framework. The proof relies on the fact that the nonce pairs exchanged in each run are quasi unique; also for any efficient adversary $\mathcal{A}'$ the advantage $\mathbf{Adv}_{\mathsf{PRF}}^{\mathsf{dist}}(\mathcal{A}')$ of distinguishing a pseudorandom function from a truly random one is small (see Appendix A for a formal proof).

Theorem 2 (Security Properties). *The distance bounding protocol $\mathcal{ID}$ in Fig. 3 with parameters $(T_{\max}, t_{\max}, E_{\max}, N_c)$ has the following properties:*

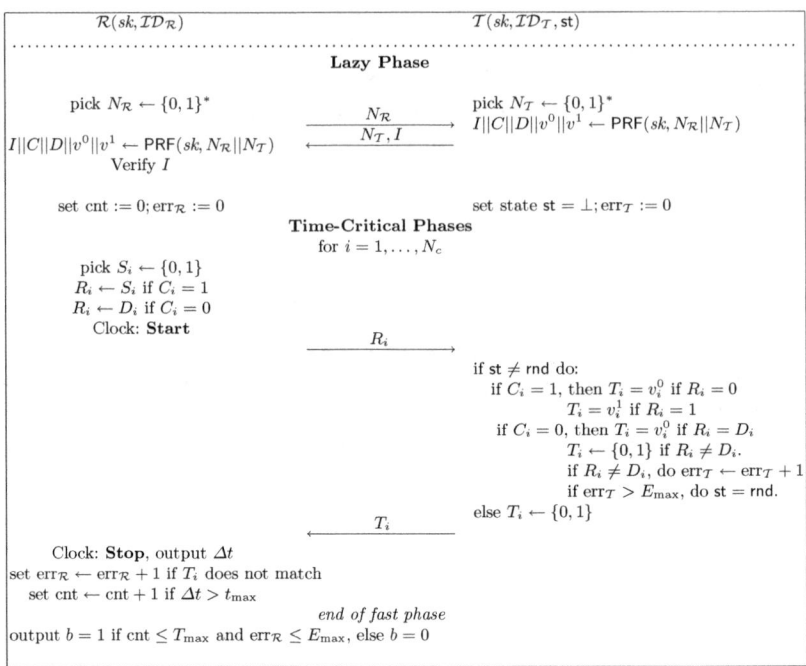

Fig. 3. Enhanced Kim/Avoine protocol

- *It is not terrorist-fraud resistant.*
- *For any $(t, q_\mathcal{R}, q_\mathcal{T}, q_{\text{OBS}})$-impersonation adversary $\mathcal{A}$ against $\mathcal{ID}$ there exists a (t', q')-distinguisher $\mathcal{A}'$ against PRF (with $t' = t + O(n)$ and $q' = q_\mathcal{R} + q_\mathcal{T} + q_{\text{OBS}}$) such that,*

$$\boldsymbol{Adv}_{\mathcal{ID}}^{imp}(\mathcal{A}) \leq q_\mathcal{R} \cdot 2^{-|I|} + \boldsymbol{Adv}_{\mathsf{PRF}}^{dist}(\mathcal{A}') + \binom{q_\mathcal{T}}{2} \cdot 2^{-|N_\mathcal{T}|}$$
$$+ \binom{q_\mathcal{R} + q_{\text{OBS}}}{2} \cdot 2^{-|N_\mathcal{R}|}.$$

- *For any $(t, q_\mathcal{R}, q_\mathcal{T}, q_{\text{OBS}})$-distance-fraud adversary $\mathcal{A}$ against $\mathcal{ID}$ there is a (t', q')-distinguisher $\mathcal{A}'$ against PRF (where $t' = t + O(n)$ and $q' = q_\mathcal{R} + q_\mathcal{T} + q_{\text{OBS}}$) such that, for $N_t = T_{\max} + E_{\max}$*

$$\boldsymbol{Adv}_{\mathcal{ID}}^{dist}(\mathcal{A}) \leq q_\mathcal{R} \cdot \binom{N_c}{N_t} \left(\frac{7}{8}\right)^{N_c - N_t} + \boldsymbol{Adv}_{\mathsf{PRF}}^{dist}(\mathcal{A}') + \binom{q_\mathcal{R} + q_{\text{OBS}}}{2} \cdot 2^{-|N_\mathcal{R}|}$$

- *For any $(t, q_\mathcal{R}, q_\mathcal{T}, q_{\text{OBS}})$-Mafia-fraud adversary $\mathcal{A}$ against $\mathcal{ID}$ there exists a (t', q')-distinguisher $\mathcal{A}'$ against PRF (where $t' = t + O(n)$ and $q' = q_\mathcal{R} + q_\mathcal{T} + q_{\text{OBS}}$) such that, for $N_t = T_{\max} + 2E_{\max}$*

$$Adv_{\mathcal{ID}}^{mafia}(\mathcal{A}) \leq \tfrac{5}{8} \cdot q_{\mathcal{R}} \binom{N_c}{N_t} \cdot (N_c - N_t + 2) \cdot 2^{-(N_c - N_t)} + Adv_{\mathsf{PRF}}^{dist}(\mathcal{A}')$$

$$+ \binom{q_{\mathcal{R}} + q_{\mathrm{OBS}}}{2} \cdot 2^{-|N_{\mathcal{R}}|} + \binom{q_{\mathcal{T}}}{2} \cdot 2^{-|N_{\mathcal{T}}|}$$

For a single impersonation attempt and $T_{\max} = E_{\max} = 0$ we have up to small terms the (almost optimal) bound $\tfrac{1}{2}(N_c + 2) \cdot 2^{-N_c}$ for Mafia-Fraud resistance. The distance fraud resistance of $\tfrac{7}{8}$ per round is tight, corresponding to an adversary who sends v_i^0 in round i (v^0 is precomputed in the lazy phase).

References

1. Abyneh, M.R.S.: Security analysis of two distance-bounding protocols. In: Proceedings of RFIDSec 2011. LNCS. Springer, Heidelberg (2011)
2. Avoine, G., Bingol, M.A., Karda, S., Lauradoux, C., Martin, B.: A formal framework for analyzing RFID distance bounding protocols. Journal of Computer Security - Special Issue on RFID System Security (2010)
3. Kara, O., Kardaş, S., Bingöl, M.A., Avoine, G.: Optimal security limits of RFID distance bounding protocols. In: Ors Yalcin, S.B. (ed.) RFIDSec 2010. LNCS, vol. 6370, pp. 220–238. Springer, Heidelberg (2010)
4. Avoine, G., Tchamkerten, A.: An efficient distance bounding RFID authentication protocol: Balancing false-acceptance rate and memory requirement. In: Samarati, P., Yung, M., Martinelli, F., Ardagna, C.A. (eds.) ISC 2009. LNCS, vol. 5735, pp. 250–261. Springer, Heidelberg (2009)
5. Bellare, M., Goldreich, O.: Proving computational ability (1992), http://www.wisdom.weizmann.ac.il/~oded/PS/poa.ps
6. Bellare, M., Pointcheval, D., Rogaway, P.: Authenticated key exchange secure against dictionary attacks. In: Preneel, B. (ed.) EUROCRYPT 2000. LNCS, vol. 1807, pp. 139–155. Springer, Heidelberg (2000)
7. Bellare, M., Rogaway, P.: Entity authentication and key distribution. In: Stinson, D.R. (ed.) CRYPTO 1993. LNCS, vol. 773, pp. 232–249. Springer, Heidelberg (1994)
8. Brands, S., Chaum, D.: Distance-bounding protocols. In: Helleseth, T. (ed.) EUROCRYPT 1993. LNCS, vol. 765, pp. 344–359. Springer, Heidelberg (1994)
9. Bringer, J., Chabanne, H.: Trusted-hb: A low-cost version of hb $^+$ secure against man-in-the-middle attacks. Transactions on Information Theory 54(9), 4339–4342 (2008)
10. Bussard, L., Bagga, W.: Distance-bounding proof of knowledge to avoid real-time attacks. Security and Privacy in the Age of Ubiquitous Computing 181, 222–238 (2005)
11. Camenisch, J., Lysyanskaya, A.: An efficient system for non-transferable anonymous credentials with optional anonymity revocation. In: Pfitzmann, B. (ed.) EUROCRYPT 2001. LNCS, vol. 2045, pp. 93–118. Springer, Heidelberg (2001)
12. Capkun, S., Butty'an, L., Hubaux, J.P.: Sector: Secure tracking of node encounters in multi-hop wireless networks. In: ACM Workshop on Security of Ad Hoc and Sensor Networks - SASN, pp. 21–32. ACM Press, New York (2003)
13. Carluccio, D., Kasper, T., Paar, C.: Implementation details of a multi purpose ISO 14443 RFIDtool. Printed handout of Workshop on RFID Security - RFIDSec 2006 (July 2006)

14. Chandran, N., Goyal, V., Moriarty, R., Ostrovsky, R.: Position based cryptography. In: Halevi, S. (ed.) CRYPTO 2009. LNCS, vol. 5677, pp. 391–407. Springer, Heidelberg (2009)

15. Clulow, J., Hancke, G.P., Kuhn, M.G., Moore, T.: So near and yet so far: Distance-bounding attacks in wireless networks. In: Buttyán, L., Gligor, V.D., Westhoff, D. (eds.) ESAS 2006. LNCS, vol. 4357, pp. 83–97. Springer, Heidelberg (2006)

16. Desmedt, Y.: Major security problems with the 'unforgeable' (feige)-fiat-shamir proofs of identity and how to overcome them. In: SecuriCom, pp. 15–17. SEDEP, Paris (1988)

17. Drimer, S., Murdoch, S.J.: Keep your enemies close: distance bounding against smartcard relay attacks. In: Proc. of the 16-th USENIX Security Symposium on USENIX Security Symposium, article no. 7. ACM Press, New York (2007)

18. Duc, D., Kim, K.: Securing HB+ against GRS man-in-the-middle attack. In: Symposium on Cryptography and Information Security (SCIS). The Institute of Electronics, Information and Communication Engineers (2007)

19. Francillon, A., Danev, B., Capkun, S.: Relay attacks on passive keyless entry and start systems in modern cars. Cryptology ePrint Archive, Report 2010/332 (2010), ePRINTURL

20. Gilbert, H., Robshaw, M., Sibert, H.: An active attack against hb+ - a provably secure lightweight authentication protocol. Cryptology ePrint Archive, Report 2005/237 (2005), ePRINTURL

21. Goldreich, O., Pfitzmann, B., Rivest, R.L.: Self-delegation with controlled propagation - or - what if you lose your laptop. In: Krawczyk, H. (ed.) CRYPTO 1998. LNCS, vol. 1462, pp. 153–168. Springer, Heidelberg (1998)

22. Haataja, K., Toivanen, P.: Two practical man-in-the-middle attacks on bluetooth secure simple pairing and countermeasures. Transactions on Wireless Communications 9(1), 384–392 (2010)

23. Hancke, G.P.: A practical relay attack on ISO 14443 proximity cards (2005), http://www.cl.cam.ac.uk/gh275/relay.pdf

24. Hancke, G.: Distance bounding publication database (2010), http://www.rfidblog.org.uk/db.html

25. Hancke, G.P.: Design of a secure distance-bounding channel for RFID. Journal of Network and Computer Applications (2010)

26. Hancke, G.P., Kuhn, M.G.: An RFID distance bounding protocol. In: SECURECOMM, pp. 67–73. ACM Press, New York (2005)

27. Hopper, N.J., Blum, M.: Secure human identification protocols. In: Boyd, C. (ed.) ASIACRYPT 2001. LNCS, vol. 2248, pp. 52–66. Springer, Heidelberg (2001)

28. Juels, A.: RFID security and privacy: a research survey. IEEE Journal on Selected Areas in Communications 24(2), 381–394 (2006)

29. Kim, C.H., Avoine, G.: RFID distance bounding protocol with mixed challenges to prevent relay attacks. In: Garay, J.A., Miyaji, A., Otsuka, A. (eds.) CANS 2009. LNCS, vol. 5888, pp. 119–133. Springer, Heidelberg (2009)

30. Leng, X., Mayes, K., Markantonakis, K.: HB-MP+ protocol: An improvement on the HB-MP protocol. In: International Conference on RFID, pp. 118–124. IEEE Computer Society Press, Los Alamitos (2008)

31. Meadows, C., Poovendran, R., Pavlovic, D., Chang, L., Syverson, P.: Distance bounding protocols: Authentication logic analysis and collusion attacks. In: Proceedings of Secure Localization and Time Synchronization for Wireless Sensor and Ad Hoc Networks. Springer, Heidelberg (2007)

32. Ouafi, K., Overbeck, R., Vaudenay, S.: On the security of hb# against a man-in-the-middle attack. In: Pieprzyk, J. (ed.) ASIACRYPT 2008. LNCS, vol. 5350, pp. 108–124. Springer, Heidelberg (2008)
33. Rasmussen, K.B., Čapkun, S.: Realization of RF distance bounding. In: USENIX Security Symposium (2010)
34. Reid, J., Nieto, J.M.G., Tang, T., Senadji, B.: Detecting relay attacks with timing-based protocols. In: ASIACCS, pp. 204–213. ACM Press, New York (2007)
35. Schaller, P., Schmidt, B., Basin, D., Capkun, S.: Modeling and verifying physical properties of security protocols for wireless networks. In: Proceedings of the 22nd IEEE Computer Security Foundations Symposium 2009, pp. 109–123. ACM, New York (2009)
36. Singelée, D., Preneel, B.: Distance bounding in noisy environments. In: Stajano, F., Meadows, C., Capkun, S., Moore, T. (eds.) ESAS 2007. LNCS, vol. 4572, pp. 101–115. Springer, Heidelberg (2007)
37. Trujillo-Rasua, R., Martin, B., Avoine, G.: The poulidor distance-bounding protocol. In: Ors Yalcin, S.B. (ed.) RFIDSec 2010. LNCS, vol. 6370, pp. 239–257. Springer, Heidelberg (2010)
38. Yung, M.: Zero-knowledge proofs of computational power. In: Quisquater, J.-J., Vandewalle, J. (eds.) EUROCRYPT 1989. LNCS, vol. 434, pp. 196–207. Springer, Heidelberg (1990)

A   Security Proof of the Protocol of Kim and Avoine

Proof. The protocol is not terrorist-fraud resistant: $\mathcal{T}'$ can forward adversary $\mathcal{A}$ the value $I||C||S||v^0||v^1$. Now $\mathcal{A}$ authenticates successfully; a simulator can't authenticate, however, as a fresh session has new nonces in the lazy phase.

We prove Mafia-fraud resistance as follows: (1) replace honest parties' PRF output by independent random values $I||C||D||v^0||v^1$ for new nonces $(N_\mathcal{R}, N_\mathcal{T})$; (2) show quasi-uniqueness of nonce pairs except in 1 adversary-tag session and 1 reader-adversary session s.t. $\mathcal{A}$ relays the nonces; (3) bound $\mathcal{A}$'s winning probability in time-critical phases for at most one adversary-tag interaction.

Due to space limits, we only sketch steps (1) and (2). In (1), replacing PRF-values by random (but consistent) values decreases $\mathcal{A}$'s success probability by at most the distinguishing advantage for PRF (else we use $\mathcal{A}$ to distinguish PRF). For (2), if $\mathcal{A}$ does *not* relay nonces, the probability of colliding nonces is exactly $\binom{q_\mathcal{R} + q_{\text{obs}}}{2} \cdot 2^{-|N_\mathcal{R}|} + \binom{q_\mathcal{T}}{2} \cdot 2^{-|N_\mathcal{T}|}$.

Now let $\mathcal{A}$ lose if the nonces match apart from the case above. Thus the values $I||C||D||v^0||v^1$ are independent. Let sid be a reader-adversary session where $\mathcal{A}$ successfully impersonates to $\mathcal{R}$. By assumption at most one other adversary-tag session sid* has the same nonce pair. If sid* exists, it taints sid with high probability (if sid* doesn't exist, $\mathcal{A}$ can't benefit from it). Suppose that sid* taints at most $T_{\max}$ time-critical phases of sid. For the moment let $E_{\max} = 0$; we allow for $E_{\max} > 0$ later.

Consider an untainted time-critical phase of sid where $\mathcal{R}$ sends R_i and expects T_i, i.e. assume $\mathcal{A}$ successfully passed the first $i - 1$ time-critical phases. There are four strategies for the adversary in this i-th phase:

GO-EARLY. In session sid^* $\mathcal{A}$ sends bit R_i^* to $\mathcal{T}$ before receiving R_i (i.e., $\mathsf{clock}(\mathsf{sid}, i + 2) > \mathsf{clock}(\mathsf{sid}^*, i + 2)$). As R_i is random and independently chosen, $R_i^* \neq R_i$ w.p. $\frac{1}{2}$ – then $\mathcal{A}$ doesn't receive T_i in sid^* and must guess T_i in sid. Also, session sid^* becomes invalid with probability $\frac{1}{4}$.

GO-LATE. In session sid, $\mathcal{A}$ replies to R_i with T_i before receiving T_i^* in session sid^* ($\mathsf{clock}(\mathsf{sid}, i + 3) < \mathsf{clock}(\mathsf{sid}^*, i + 3)$). Now $\mathcal{A}$ wins the phase w.p. $\frac{1}{2}$.

MODIFY-IT. $\mathcal{A}$ receives R_i in sid, sends R_i^* in sid^*, gets T_i^* in sid^*, and forwards T_i in sid. This scheduling is pure relay, but $R_i \neq R_i^*$ or $T_i \neq T_i^*$. If R_i^* is wrong then T_i^* was never sent by $\mathcal{T}$ in sid^* and $\mathcal{A}$ can only guess T_i w.p. $\frac{1}{2}$; if $R_i = R_i^*$ then $T_i \neq T_i^*$ makes the reader reject.

TAINT-IT. The adversary taints this phase of sid through sid^*.

Tainting the phase makes $\mathcal{R}$ accept with probability 1, deducting 1 from the remaining taintable phases. The Go-Late and Modify-it Strategy both succeed w.p. at most $\frac{1}{2}$. Go-Early succeeds w.p. $\frac{3}{4}$, inactivating sid^* w.p. $\frac{1}{2}$. Assume that $\mathcal{A}$ taints the last $T_{\max}$ time-critical phases (else we renumber the phases). For the other $P := N_c - T_{\max}$ phases let pass_i denote the event that $\mathcal{A}$ passes phase i of sid. We have

$$\mathrm{Prob}\left[\bigwedge_{j=i}^{P} \mathsf{pass}_j \,\middle|\, \bigwedge_{j=1}^{i-1} \mathsf{pass}_j\right] \leq \frac{5}{8} \cdot \mathrm{Prob}\left[\bigwedge_{j=i+1}^{P} \mathsf{pass}_j \,\middle|\, \bigwedge_{j=1}^{i} \mathsf{pass}_j\right] + \frac{1}{2} \cdot \frac{1}{2} \cdot 2^{-P+i+1}.$$

The first term captures the success of Go-Late, Modify-It, and correct Go-Early-prediction. The second term covers incorrect Go-Early prediction (w.p. $\frac{1}{4}$); now sid^* is inactivated, and $\mathcal{A}$ must guess T_i for this and the next $P - i - 1$ rounds (the responses are independent). Expanding the probabilities we obtain

$$\mathrm{Prob}\left[\bigwedge_{j=1}^{P} \mathsf{pass}_j\right] \leq 2^{-P} + \sum_{j=0}^{P-1} \frac{5}{8} \cdot 2^{-j} \cdot 2^{-P+j} = \tfrac{5}{8} \cdot (P + 2) \cdot 2^{-P}.$$

We sum over $q_{\mathcal{R}}$ reader-adversary sessions, distribute $T_{\max} + E_{\max}$ "jokers" on the reader side and $E_{\max}$ on the tag side, and obtain the claimed bound.

For impersonation security, the only way to generate colliding nonce pairs (and produce authentication string I) is by lazy phase relay, which is an invalid impersonation attack. For distinct nonce pairs, the probability that $\mathcal{A}$ sends a correct I in a reader-adversary session is: $q_{\mathcal{R}} \cdot 2^{-|I|}$ plus the distinguishing advantage for the PRF plus the probability of colliding nonces.

Distance-bounding (the third statement) is proved as above: once the pseudorandom values are replaced by truly random ones, the probability that $C_i = 1$ and $v_i^0 \neq v_i^1$ is at least $\frac{1}{4}$ for round i. Since $\mathcal{A}$ can commit only then, $\mathcal{A}$ fails with probability at least $\frac{1}{8}$. Overall, $\mathcal{A}$ succeeds only w.p. $\frac{7}{8}$ per round, except for a number $T_{\max} + E_{\max}$ of phases.

$\square$

MASHA – Low Cost Authentication with a New Stream Cipher

Shinsaku Kiyomoto[1], Matt Henricksen[2], Wun-She Yap[2],
Yuto Nakano[1], and Kazuhide Fukushima[1]

[1] KDDI R & D Laboratories Inc.
2-1-15 Ohara, Fujimino-shi, Saitama, 356-8502, Japan
`kiyomoto@kddilabs.jp`
[2] Institute for Infocomm Research
1 Fusionopolis Way #21-01 Connexis (South Tower) Singapore 138632
`mhenricksen@i2r.a-star.edu.sg`

Abstract. In this paper, we propose a new high-speed stream cipher
called MASHA (Message Authenticated Streaming-encryption Heterogeneous Algorithm) with integrated MAC functionality. It simultaneously
encrypts plaintext and produces an authentication tag that assures data
and origin integrity. On the Intel Core 2, its speed is 11.92 cycles/byte,
which is faster than the time it takes to encrypt and authenticate using
well-known primitives SNOW 2.0 and SHA-256 in conjunction. We show
that MASHA is secure against all known attacks.

1 Introduction

Encryption and message authentication are necessary for secure communication
on a public network. These functions are frequently implemented separately
using a stream or block cipher to provide confidentiality, and an authentication
tag algorithm in order to provide data and origin integrity.

For example, AES-CBC can be run in two passes to generate ciphertext and an
authentication tag respectively. The mode of operation necessarily incorporates
a state, since in order to to provide integrity, the authentication tag must change
if encrypted blocks are swapped. There are modes of operation that encrypt and
authenticate n-block messages in $O(n \cdot log(n))$ time [21]. However, it is more
natural to use stream ciphers, which incorporate the state, to perform both
tasks. Stream ciphers are both faster than block ciphers in software, and more
compact in hardware [20].

Stream ciphers that provide both encryption and authentication functionality
can be categorized on whether the authentication is 'integrated' or 'external'. In
ciphers with integrated authentication, the keystream generation and authentication algorithms share at least some of their state. The advantage of this is
efficiency, but the design of such algorithms must be carefully considered as the
attacker not only influences the key and IV during state initialization, but also
the plaintext during keystream generation. He may use carefully chosen plaintext to commit message forgery, or to derive information about the shared state

X. Lai, J. Zhou, and H. Li (Eds.): ISC 2011, LNCS 7001, pp. 63–78, 2011.

leading to key recovery. Stream ciphers using external authentication dedicate extra state to the authentication task. The plaintext does not affect state used by the keystream generation algorithm, limiting the attacker's ability to recover the key. Because the cipher has to manage the sufficient mixing of the dedicated authentication state, external authentication algorithms tend to be slower, reducing the advantage of using a stream rather than block cipher.

We propose an efficient stream cipher algorithm with integrated message authentication called Message Authenticated Streaming-encryption Heterogeneous Algorithm (MASHA). The cipher uses a 128-bit key in conjunction with a 192-bit Initialization Vector (IV) to provide a 128-bit security level for both encryption and message authentication. This is commensurate with a tag size of 256 bits, providing much better security than the smaller tags produced by other stream ciphers (see Section 2). We specify MASHA in Section 3. In Sections 4 and 5, we outline design decisions made during its development, and show how it resists common attacks. In Section 6, we show metrics for MASHA implemented on common software platforms. In Section 7, we give concluding notes.

2   Related Work

The recent ECRYPT eSTREAM [9] call for ciphers requested ciphers that included an associated authentication component. Of the thirty-four candidates submitted to eSTREAM, six included message authentication mechanisms. Three of these were archived at the end of the first phase due to security or efficiency issues. Of the remaining three ciphers, Phelix [22] offers integrated authentication. The ability of the attacker to control the plaintext, which is injected directly into the Phelix state, leads to an attack [23] with complexity $O(2^{37})$ and with 2^{34} chosen IVs. The attack is attributed to the passage of the plaintext word through the state to keystream without passing through sufficient confusion and diffusion layers (we note that Phelix contains no s-boxes). It also requires that the attacker can reuse key-IV pairs, which is in contradiction to the commonly assumed model of usage for stream ciphers (ie. the cipher can only be attacked by this method when it is used incorrectly). However, it was disqualified from round 3 of the eSTREAM process due to the attack [2], which means that vulnerabilities due to reuse of key-IV pairs should be taken seriously.

Of the other two ciphers, VEST [18] was discarded in eSTREAM round 2, due to poor efficiency. NLS [11] uses an external authentication algorithm, called Mundja [10], which uses modified SHA-256 in conjunction with a Cyclic Redundancy Checksum (CRC). The use of an external algorithm means that the attacks that apply to the authentication algorithm are unlikely to apply to the keystream generation algorithm. However, Mundja was discarded at the end of round two of the competition due to poor performance. This highlights the advantage of integrated authentication.

Grain [12] was one of the 'winners' of eSTREAMs hardware portfolio, which requires ciphers to use 80-bit keys for the sake of minimizing hardware gate counts. Agren, Hell, Johansson and Meier [1] revised Grain by upsizing the key to

128 bits, using a slightly different keystream generation mechanism, and adding an external authentication mechanism that uses two thirty-two bit registers. These registers are updated using a combination of plaintext and Grain state. The plaintext never enters into the Grain state. The tag size of 32 bits is restrictive and small relative to the key size. No third-party analysis has followed, but the algorithm is not very efficient in software.

Dragon-MAC [15] is an external message authentication algorithm based on the Dragon stream cipher, a phase three candidate in eSTREAM. The MAC is external, using the the Dragon F function but not its state, to provide data integrity to the keystream. The MAC's application is limited: being designed for wireless sensor networks, it cannot be used with messages of longer than 256 bits. It also suffers from security problems due to the non-bijectivity of parts of the F function, so that the attacker can commit message forgery in less than $O(2^{32})$ time without changing multiple words of the message.

Hummingbird-1 and Hummingbird-2 [8] are lightweight stream ciphers with integrated authentication. Hummingbird-2 uses a 128-bit key and 64-bit IV to initialize a 128-bit state. Later on, the state is used to produce a tag of up to 128 bits. However, it is clear that the small state size relative to key size allows the cipher to be attacked generically, so the security of the cipher and the tag are unable to reach the security of MASHA: ie. 128 bits. Hummingbird is optimized for hardware.

As can be seen above, none of the algorithms achieved a good balance between utility, security and performance. In this paper, we provide the specification of MASHA, which we believe achieves a good balance. MASHA is based on the K2 stream cipher [14]. K2 has survived for four years without any successful attacks on the full version. It is currently being standardized by the ISO/IEC JTC 1/SC 27 committee, and has excellent performance in hardware and software. It seems to be an ideal base for adding authentication.

3   MASHA

In this section, we describe the specification of MASHA. MASHA consists of two feedback shift registers, FSR-A and FSR-B, comprising five and eleven 32-bit stages respectively, and a Finite State Machine (FSM) with four internal 32-bit registers $R1$, $R2$, $L1$, and $L2$. The total size of the internal state is 640 bits. MASHA is shown in Figure 1, and its FSM in Figure 2.

3.1   Feedback Shift Registers

Let A_t^i and B_t^i respectively denote the ith register of FSR-A and FSR-B at time t where A_t^0 and B_t^0 are produced as output at time t. Let $A_t[p] = \{0,1\}$ denote the pth bit of A_t, where $A_t[31]$ is the most significant bit of A_t. The symbol $\oplus$ denotes bitwise exclusive-or operation and the symbol $\boxplus$ denotes 32-bit addition.

A byte string x denotes $(x_7, x_6, ..., x_1, x_0)$, where x_7 is the most significant bit and x_0 is the least significant bit. Let γ and δ respectively be the roots of the primitive polynomials

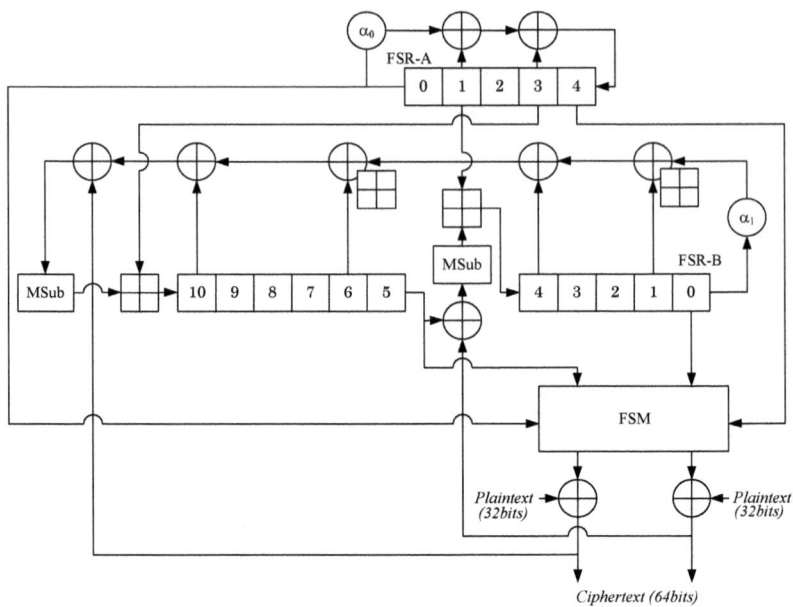

Fig. 1. MASHA

$$x^8 + x^7 + x^5 + x^3 + 1 \in GF(2)[x]$$
$$x^8 + x^6 + x^3 + x^2 + 1 \in GF(2)[x]$$

Let α_0 and α_1 be the roots of the irreducible polynomials of degree four

$$x^4 + \gamma^{173}x^3 + \gamma^{248}x^2 + \gamma^{38}x + \gamma^{121} \in GF(2^8)[x]$$
$$x^4 + \delta^8 x^3 + \delta^{254}x^2 + \delta^{123}x + \delta^{77} \in GF(2^8)[x]$$

respectively. A 32-bit string Y denotes (Y_3, Y_2, Y_1, Y_0), where Y_i is a byte string and Y_3 is the most significant byte. Y is represented by $Y = Y_3\alpha_i^3 + Y_2\alpha_i^2 + Y_1\alpha_i + Y_0$ $(i = 1, 2, 3)$.

The feedback polynomial for FSR-A is described as follows:

$$f(x) = \alpha_0 x^5 + x^4 + x^2 + 1$$

The feedback function of FSR-A is defined as follows:

$$A_{t+1}^4 = \alpha_0 A_t^0 \oplus A_t^1 \oplus A_t^3$$
$$A_{t+1}^i = A_t^{i+1} \quad (0 \leq i \leq 3)$$

The feedback polynomial for FSR-B is described as follows:

$$f(x) = \alpha_1 x^{11} + x^{10} + x^7 + x^5 + x + 1$$

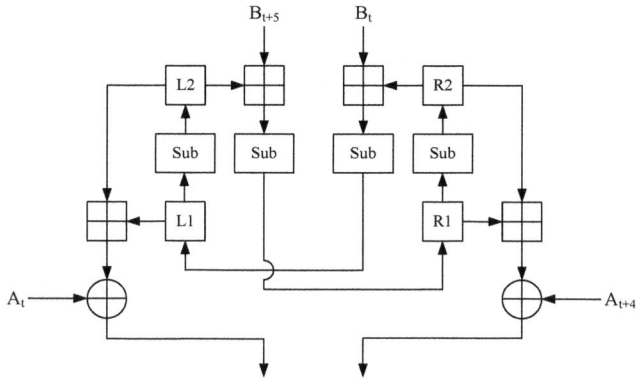

Fig. 2. FSM of MASHA

The feedback function of FSR-B is based upon this polynomial, but during each clock of the cycle, some of the operations in the FSR-B feedback polynomial are modified by the values of 'control bits' in FSR-A. Control bits $cl0_t$ and $cl1_t$ are binary variables defined as $cl0_t = A_{t+2}[31]$, $cl1_t = A_{t+2}[30]$. The feedback function is also modified by ciphertext components C_t^H and C_t^L, which respectively represent the higher and lower 32-bits of ciphertext C_t. The feedback function is defined as follows:

$$B_{t+1}^{10} = A_t^3 \boxplus MSub(\alpha_1 B_t^0[\oplus, \boxplus]_{cl0_t} B_t^1 \oplus B_t^4[\oplus, \boxplus]_{cl1_t} B_t^6 \oplus B_t^{10} \oplus C_t^H)$$
$$B_{t+1}^4 = A_t^1 \boxplus MSub(B_t^5 \oplus C_t^L)$$
$$B_{t+1}^i = B_t^{i+1} \quad (0 \le i \le 3, 5 \le i \le 9)$$

3.2 Nonlinear Function

The non-linear function of MASHA is fed the values of four registers of *FSR-B* and that of internal registers $R1$, $R2$, $L1$, $L2$, and outputs 64 bits of the keystream every clock. The primary source of non-linearity is the *Sub* function, which is invoked four times per cycle.

The *Sub* step divides the 32-bit input into four bytes, applying a bijective substitution to each, then uses the four bytes as input to a 32-to-32 bit linear permutation. The substitution uses the AES s-box [6], and the permutation the AES *Mix Column* operation. The substitution consists of two functions: g and f. The g calculates the multiplicative inverse modulo the irreducible polynomial $m(x) = x^8 + x^4 + x^3 + x + 1$ without 0x00, and 0x00 is transformed to itself (0x00). f is an affine transformation defined by;

$$\begin{bmatrix} b_7 \\ b_6 \\ b_5 \\ b_4 \\ b_3 \\ b_2 \\ b_1 \\ b_0 \end{bmatrix} = \begin{bmatrix} 11111000 \\ 01111100 \\ 00111110 \\ 00011111 \\ 10001111 \\ 11000111 \\ 11100011 \\ 11110001 \end{bmatrix} \times \begin{bmatrix} a_7 \\ a_6 \\ a_5 \\ a_4 \\ a_3 \\ a_2 \\ a_1 \\ a_0 \end{bmatrix} \oplus \begin{bmatrix} 0 \\ 1 \\ 1 \\ 0 \\ 0 \\ 0 \\ 1 \\ 1 \end{bmatrix}$$

where $a = (a_7, ..., a_0)$ is the input and $b = (b_7, ..., b_0)$ is an output, and a_0 and b_0 are the least significant bit (LSB). Let C be (c_3, c_2, c_1, c_0) and output D be (d_3, d_2, d_1, d_0), where c_i, d_i are 8-bit values. The linear permutation $D = p(C)$ is described as follows;

$$\begin{pmatrix} d_0 \\ d_1 \\ d_2 \\ d_3 \end{pmatrix} = \begin{pmatrix} 02 \ 03 \ 01 \ 01 \\ 01 \ 02 \ 03 \ 01 \\ 01 \ 01 \ 02 \ 03 \\ 03 \ 01 \ 01 \ 02 \end{pmatrix} \begin{pmatrix} c_0 \\ c_1 \\ c_2 \\ c_3 \end{pmatrix}$$

in $GF(2^8)$ of the irreducible polynomial $m(x) = x^8 + x^4 + x^3 + x + 1$.

The points at which the ciphertext is prepared to enter FSR-B are protected by a modified Sub, called $MSub$, which consists of the Sub prepended by the AES $MixColumn$ operations:

$$MSub(x) = Sub(p(x))$$

The permutation $p(x)$ is the same as AES $Mix \ Column$ operation.

3.3 Encryption and Decryption

Let keystream at time t be $Z_t = (z_t^H, z_t^L)$ (each z_t^x is a 32-bit value, and z_t^H is a higher string). The keystream z_t^H, z_t^L is calculated as follows:

$$z_t^L = A_t^4 \oplus (R2_t \boxplus R1_t)$$
$$z_t^H = A_t^0 \oplus (L2_t \boxplus L1_t)$$

A 64-bit plaintext message (P_t^H, P_t^L) is encrypted to the ciphertext (C_t^H, C_t^L) as;

$$C_t^L = z_t^L \oplus P_t^L$$
$$C_t^H = z_t^H \oplus P_t^H$$

During decryption, each 64-bit plaintext message (P_t^H, P_t^L) is calculated from the the ciphertext (C_t^H, C_t^L) as;

$$M_t^L = z_t^L \oplus C_t^L$$
$$M_t^H = z_t^H \oplus C_t^H$$

The plaintext message is padded by adding the byte $0x80$ followed by zero bits until the padding reaches a multiple of 64 bits.

3.4   State Update

After the encryption, the internal registers are updated as follows;

$$R1_{t+1} = Sub(L2_t \boxplus B_t^5), \quad R2_{t+1} = Sub(R1_t)$$
$$L1_{t+1} = Sub(R2_t \boxplus B_t^0), \quad L2_{t+1} = Sub(L1_t)$$

where $Sub(X)$ is an output of the Sub step for X.

The registers of FSRs are also updated by the feedback functions which are described in 3.1.

3.5   Streaming MAC Generation

After encryption of a plaintext message, constants $Const1$, $Const2$, $Const3$, and $Const4$ are added to $L1$, $L2$, $R1$, and $R2$.

$$L1 = L1 \oplus Const1, L2 = L2 + Const2$$
$$R1 = R1 \oplus Const3, R2 = R2 + Const4$$

The constants $Const1$, $Const2$, $Const3$, and $Const4$ are 0xE10C33E5, 0x4C7 A8DCB, 0xC0EAD75E, and 0x394E6808, respectively. The algorithm then executes additional 22 steps ('blanking rounds', using plaintext 0x0000) as a finalization process. A 256-bit authentication tag is produced by clocking the cipher a further four steps. The concatenated keystream from this phase forms the tag.

3.6   Initialization Process

The internal state is initialized with a 128-bit initial key $IK = (IK_0, IK_1, IK_2, IK_3)$ and a 192-bit initial vector $IV = (IV_0, IV_1, IV_2, IV_3, IV_4, IV_5)$ as follows:

$$A^0 = IK_0, A^1 = IK_1, A^2 = 0\text{x}00, A^3 = IK_2, A^4 = IK_3,$$
$$B^0 = IV_0, B^1 = IV_1, B^2 = IV_2, B^3 = 0\text{x}00, B^4 = 0\text{x}00,$$
$$B^5 = IV_3, B^6 = IV_4, B^7 = IV_5, B^8 = 0\text{x}00, B^9 = 0\text{x}00,$$
$$B^{10} = 0\text{x}00$$

The remaining internal registers, $R1$, $R2$, $L1$ and $L2$ are set to 0x00. After the above processes, the cipher clocks 28 times (t= 1, ..., 28), updating the internal states both of FSR and the internal registers. The internal states are updated as follows:

$$s_t^L = A_t^4 \boxplus (R2_t \boxplus R1_t)$$
$$s_t^H = A_t^0 \oplus (L2_t \boxplus L1_t)$$
$$A_{t+1}^4 = \alpha_0 A_t^0 \oplus A_t^1 \oplus A_t^3 \oplus s_t^L$$

$$A_{t+1}^i = A_t^{i+1} \ (0 \leq i \leq 3)$$
$$B_{t+1}^{10} = \alpha_1 B_t^0[\oplus, \boxplus]_{cl0_t} B_t^1 \oplus B_t^4[\oplus, \boxplus]_{cl1_t} B_t^6 \boxplus B_t^{10} \oplus s_t^H$$
$$B_{t+1}^i = B_t^{i+1} \ (0 \leq i \leq 9)$$
$$R1_{t+1} = Sub(L2_t \boxplus B_t^5), \quad R2_{t+1} = Sub(R1_t)$$
$$L1_{t+1} = Sub(R2_t \boxplus B_t^0), \quad L2_{t+1} = Sub(L1_t)$$

The recommended maximum number of cycles for MASHA without re-initializing is 2^{58} cycles (2^{64} keystream bits). MASHA cannot be used with repeated key-IV pairs, due to the presence of generic attacks against all stream ciphers in this model.

4   Design Decisions

The main challenge in designing a stream cipher that offers integrated authentication is that this mode opens up an extra avenue of attack. For encryption-only ciphers, the attacker is active only during the state initialization phase, when it can orchestrate related-key attacks or manipulate the IV, to see how changes influence the keystream. During keystream generation, the attacker reverts to the passive role of monitoring the keystream.

For ciphers that offers authentication, the attacker has an active role during keystream generation, by manipulating plaintext and/or ciphertext. For ciphers with integrated authentication, choosing differences in the ciphertext can produce differences in keystream relatively quickly compared to during state initialization, where the state is mixed and keystream suppressed for some non-trivial number of cycles. In a chosen ciphertext attack, the attacker can perform message modification at any time, to try to cancel differences in the internal state. If the path between the ciphertext input position and the keystream is too short, or does not contain sufficient non-linear components, it will be easy to learn information about the internal state.

This kind of attack obtains maximum power when the attacker can replay key-IVs, thereby generating the same keystream prefixes multiple times. Applying ciphertexts with controlled differences when the differences in keystream is uncontrolled isn't very effective, so the attacker needs to ensure that the keystreams have no differences up to the first point of ciphertext injection. During our design and analysis phases, we assumed that the attacker can replay key-IVs a limited number of times to produce identical keystreams. We designed MASHA to provide maximum defence for this usage model. We should clearly state: we do not permit MASHA to be reused with the same key-IV pairs.

4.1   Basing MASHA on the K2

The K2 cipher is based on the SNOW 2.0 cipher, which is one of the most successful word-based stream ciphers, being faster than the Advanced Encryption

Standard, and at least as secure for a 128-bit key. K2 is secure relative to SNOW 2.0 while retaining efficiency. Our design decision means that most of the analysis on SNOW 2.0 and K2 can be leveraged to show the security of MASHA and improve confidence in the design.

4.2 Effectiveness of FSR-A

MASHA contains a 160 bit autonomous FSR-A. The decision to maintain its autonomy both guarantees a long period in the keystream, and ensures that an attacker who manipulates the ciphertext, has no way to directly measure or influence the contents of FSR-A. The output of the register is combined using the unbiased exclusive-or operation to produce the keystream. It is analogous to key whitening in block ciphers.

For many types of attacks, such as guess-and-determine and divide-and-conquer attacks, the attacker must guess the contents of the relevant parts of FSR-A in order to get the contributions of FSR-B and the FSM to the keystream. Since FSR-A is larger than the 128-bit key, and its autonomy and means of combinations are unbiased, many statistical attacks immediately infeasible.

4.3 Using MSub to Protect against Differential Plaintext Attacks

The operation for each tap of the feedback polynomial of FSR-B depends upon the value of particular bits in the FSM, but with probability $p = \frac{1}{2}^t$, t of the taps are combined using exclusive-or. This means that without the $MSubs$ and subsequent addition, the attacker can use message modification to cancel differences at B_{10} and B_4. For example, if the attacker injects Δ at time t, $t+1$, $t+5$ and $t+7$, with probability $p = \frac{1}{8}$ the attacker is able to isolate a single difference in the FSM and analyse its affect in the keystream, uncomplicated by other differences (see Table 1). The propagation of a difference through a single Sub operation allows two candidates for the real values to be deduced with complexity $O(2^{10})$, after which a guess-and-determine attack becomes more feasible.

With $MSubs$, the combination of B_{10} (respectively B_5) with the ciphertext means that a single byte difference Δ injected by the attacker is translated by the

Table 1. Propagation of Δ through reduced MASHA with no pre-B_{10} additions) to keystream z_H

t	C_H	C_L	B_{10}	B_9	B_8	B_7	B_6	B_5	B_4	z_H	p
0	Δ										p
1	Δ		Δ								p^{-1}
2				Δ							
3					Δ						
4						Δ					
5	Δ					Δ					p^{-2}
6							Δ				
7		Δ									p^{-3}
8										$Sub(\Delta + c) + d$	

first permutation of $MSub$ into four byte differences, each of which pass through an s-box to emerge to a known output difference with maximum probability 2^{-24}. This is then combined with unknown material from FSR-A (if the combination with B_{10} uses exclusive-or, the output of $MSub$ is combined with A_3 using the incompatible modular addition operation). The opportunity for the attacker to use message modification in FSR-B is virtually non-existent. With differential $\Delta \rightarrow \Delta$ across $MSub$, p (of Table 1) has maximum 2^{-25} rather than 2^{-1}. For many differences, this differential will have probability 0. The non-linear addition will further decrease the probability, although this is not shown in the calculation.

4.4   Separating MAC Generation from Encryption

Assume that we do not add constants before the pre-authentication blanking rounds. The attacker acquire suppressed keystream words during blanking by encrypting $(P^A: x)$ and $(P^B: x \parallel 0_{22})$ under the same key-IV.

MASHA's twenty-two blanking rounds use an all-zero ciphertext and do not emit ciphertext. The twenty-two 'all-zero' words appended to x in the P^B message simulate the blanking rounds during the encryption of P^A, but emit keystream - the same keystream as that suppressed during encryption of P^A, since the key and IV are the same for both invocations. This leakage of suppressed words can be used to attack the authentication process.

By adding constants to the internal state before the blanking rounds, that are subsequently confounded by non-linear operations, we remove the ability of the attacker to simulate the blanking rounds without keystream suppression, for the cost of a few simple operations.

5   Analysis of MASHA

In this section, we discuss MASHA's statistical properties and resistance against common attacks.

5.1   Statistical Properties

The statistical properties of MASHA depend largely on those of output sequences of FSR-A and FSR-B. FSR-A is unbiased and has a period of $2^{160} - 1$ as expected for an LFSR of length 160 bits using an irreducible primitive feedback polynomial.

We evaluated the statistical properties for the register outputs, and the cipher keystream using the NIST Test Suite [17] and confirmed that they passed all the relevant statistical tests. We did not find any short periods.

5.2   Resistance against Attacks

Time-Memory-Data Trade-Off attacks. MASHA was designed with two Time-Memory Trade-off attacks in mind. The first, which inverts keystream into internal state, demands that the state size be at least twice the length of the combined key and IV. The second, which inverts keystream to the key and IV, requires that the IV be at least 1.5 times the length of the key. According to Hong and Sarker [13], a TMD attack requires $O(2^{\frac{3(k+v)}{4}})$ pre-computation, $O(2^{\frac{k+v}{2}})$ memory, and $O(2^{\frac{k+v}{2}})$ available data, enabling an online attack with time complexity of $O(2^{\frac{k+v}{2}})$, where the lengths of the secret key and IV are k and v, respectively. The IV, the secret key, and the internal state are sufficiently large such that the cipher is immune to time-memory-data trade-off attacks.

Guess-and-Determine Attacks. The most efficient approach to launching a guess-and-determine attack on MASHA is to completely guess the contents of FSR-A, and virutally remove it from the cipher. Then by guessing five components of the remaining cipher, the keystream and feedback polynomial of FSR-B can be used to determine the remaining unknown components. Several words of keystream can be used for consistency checking. The complexity of the attack is $O(2^{160+160} = 2^{320})$, far above that of brute force.

Linear Masking. Distinguishing attacks require a probabilistic linear relation between keystream bits. It is easy to leverage the distinguishing attack by Nyberg and Wallen [16] on SNOW 2.0 to analyze MASHA with respect to distinguishing attacks, since both ciphers share the same non-linear components.

Their attack had a computation complexity of $O(2^{174})$, which far exceeds the complexity of brute-forcing MASHA's 128-bit key.

SNOW 2.0 uses one Sub operation per generation of keystream word, whereas MASHA uses four in the generation of keystream words. The attacker can construct a linear recurrence from output keystream bits with fixed clock control bits $cl0_t = cl1_t$ for each cycle, in which two Sub operations are affected by the linearization.

The number of involved substitutions is the same number of attacks on SNOW 2.0. Thus, we expect that the security of the cipher is the same level of SNOW 2.0 against distinguishing attacks, particularly as we cannot find a linear approximation with the probability less than 2^{64}.

Algebraic Attacks. We investigated the possibility of algebraic attacks [5] under the assumption that FSR has regular clocking and the addition modulo 2^{32} operation is replaced by the XOR operation. This attack is effective for stream cipher algorithms that have a non-linear function with internal memory. An algebraic attack on SNOW 2.0 was proposed by O. Billet and H. Gilbert[3], which can also be applied to MASHA. The attacker determines internal registers at time t, defined by linear equations that consist of initial values of internal state variables, and constructs relationships between the input values of a non-linear substitution and the corresponding output values, which are low degree algebraic expressions. First, we obtain the following equation from the assumption:

$$R1_t = Sub(A_{t-1} \oplus L1_{t-1} \oplus z_{t-1}^H), \quad L1_{t-1} = Sub(R2_{t-2} \oplus B_t),$$
$$R2_t = A_{t+4} \oplus R1_t \oplus z_t^L, = A_{t+4} \oplus Sub(A_{t-1} \oplus Sub(R2_{t-2} \oplus B_t) \oplus z_{t-1}^H) \oplus z_t^L$$

Table 2. Diffusion Process of Initial Key and IV

Register	0	2	8	12	16
A^0	*	*	K	K	K, IV
A^1	*	*	K	K, IV	K, IV
A^2	*	*	K	K, IV	K, IV
A^3	*	K	K	K, IV	K, IV
A^4	*	K	K	K, IV	K, IV
B^0	*	*	*	IV	K, IV
B^1	*	*	*	IV	K, IV
B^2	*	*	*	IV	K, IV
B^3	*	*	*	IV	K, IV
B^4	*	*	IV	K, IV	K, IV
B^5	*	*	IV	K, IV	K, IV
B^6	*	*	IV	K, IV	K, IV
B^7	*	*	IV	K, IV	K, IV
B^8	*	*	K, IV	K, IV	K, IV
B^9	*	*	K, IV	K, IV	K, IV
B^{10}	*	IV	K, IV	K, IV	K, IV
$R1$	*	*	IV	K, IV	K, IV
$R2$	*	*	*	IV	K, IV
$L1$	*	*	*	IV	K, IV
$L2$	*	*	*	IV	K, IV

We cannot obtain the linear recurrence as per the algebraic attack on SNOW 2.0. Furthermore, the linear approximation of the additions increases the computational complexity of the algebraic attack [4]. Thus it is impossible to apply this attack for the full-version of the cipher.

Analysis of the Initialization Process. The state initialization process generates the internal state of the cipher by thoroughly mixing the key and IV using the keystream generation function (suppressing keystream, and feeding it back into the state). All registers in the internal state depend on the initial key and IV after sixteen cycles, as shown in Table 2. After this, the initialization process still continues twelve cycles to further refine the mixing. This means that the attacker who launches related key-IV attacks in this phase has no control over differences in the state by the time keystream is generated.

Another attack on the initialization process is to generate collisions between key-IV pairs such that the keystream produced is identical. The large state size relative to combined key-IV size, and the combination of high non-linearity in the FSM and its extensive diffusion process both mean it is infeasible to generate collisions.

Attacks on MAC. Table 3 and Table 4 show diffusion of each difference in messages P^H and P^L. After ten steps, the difference has diffused into all of internal registers: $R1$, $R2$, $L1$, $L2$, and registers of FSR-B.

Table 3. Diffusion Process of Difference in M^H

Register	0	1	2	3	4	5	6	7	8	9	10
B^0	*	*	*	*	*	*	*	*	*	*	Δ
B^1	*	*	*	*	*	*	*	*	*	Δ	Δ
B^2	*	*	*	*	*	*	*	*	Δ	Δ	Δ
B^3	*	*	*	*	*	*	*	Δ	Δ	Δ	Δ
B^4	*	*	*	*	*	*	Δ	Δ	Δ	Δ	Δ
B^5	*	*	*	*	*	Δ	Δ	Δ	Δ	Δ	Δ
B^6	*	*	*	*	Δ	Δ	Δ	Δ	Δ	Δ	Δ
B^7	*	*	*	Δ	Δ	Δ	Δ	Δ	Δ	Δ	Δ
B^8	*	*	Δ	Δ	Δ	Δ	Δ	Δ	Δ	Δ	Δ
B^9	*	Δ	Δ	Δ	Δ	Δ	Δ	Δ	Δ	Δ	Δ
B^{10}	Δ	Δ	Δ	Δ	Δ	Δ	Δ	Δ	Δ	Δ	Δ
$R1$	*	*	*	*	*	*	Δ	Δ	Δ	Δ	Δ
$R2$	*	*	*	*	*	*	*	Δ	Δ	Δ	Δ
$L1$	*	*	*	*	*	*	*	*	Δ	Δ	Δ
$L2$	*	*	*	*	*	*	*	*	*	Δ	Δ

Table 4. Diffusion Process of Difference in M^L

Register	0	1	2	3	4	5	6	7	8	9	10	11
B^0	*	*	*	*	Δ	*	*	*	*	*	*	Δ
B^1	*	*	*	Δ	*	*	*	*	*	*	Δ	Δ
B^2	*	*	Δ	*	*	*	*	*	*	Δ	Δ	Δ
B^3	*	Δ	*	*	*	*	*	*	Δ	Δ	Δ	Δ
B^4	Δ	*	*	*	*	*	*	Δ	Δ	Δ	Δ	Δ
B^5	*	*	*	*	*	*	Δ	Δ	Δ	Δ	Δ	Δ
B^6	*	*	*	*	*	Δ	Δ	Δ	Δ	Δ	Δ	Δ
B^7	*	*	*	*	Δ	Δ	Δ	Δ	Δ	Δ	Δ	Δ
B^8	*	*	*	Δ	Δ	Δ	Δ	Δ	Δ	Δ	Δ	Δ
B^9	*	*	Δ	Δ	Δ	Δ	Δ	Δ	Δ	Δ	Δ	Δ
B^{10}	*	Δ	Δ	Δ	Δ	Δ	Δ	Δ	Δ	Δ	Δ	Δ
$R1$	*	*	*	*	*	*	*	Δ	Δ	Δ	Δ	Δ
$R2$	*	*	*	*	*	*	*	*	Δ	Δ	Δ	Δ
$L1$	*	*	*	*	*	Δ	*	*	*	Δ	Δ	Δ
$L2$	*	*	*	*	*	*	Δ	*	*	*	Δ	Δ

Table 5. Performance Evaluation

MASHA	11.92 cycle/byte + 2016 cycle
AES-CCM	60.40 cycle/byte
SHA-256	18.06 cycle/byte [19]
SNOW 2.0	4.50 cycle/byte [7]

Thus, the number of steps for finalizing MAC generation is sufficient. To forge a MAC, the adversary need to know all internal state bits, which requires more efficient attack techniques than those that we have analysed (and we are not aware of the existence of any).

The adversary can strive to accomplish message forgery, whereby different plaintexts collide. As we showed in Section 4, MASHA is designed to resist this attack, partly due to the use of $MSub$ at the points at which ciphertexts enters into the internal state. A single byte difference in a ciphertext propagates to another with probability at most 2^{-24} before it enters the state, and faces several non-linear operations before it emerges as keystream. It is difficult for the attacker to cancel differences using message modification, and the system quickly escalates out of his control.

6   Performance Analysis

Evaluation results on Core 2 E8600 3.33GHz (in 32-bit mode) are shown in Table 4. The encryption/decryption speed of MASHA is 11.92 cycle/byte, and additional 2016 cycles are required for the streaming MAC generation. The speed of MASHA is much faster than that of AES-CCM. If we use SNOW 2.0 and SHA-256 for encryption and message authentication, the throughput is totally 22.56 cycle/byte. Thus, MASHA is competitive against existing algorithms. The initialization process of MASHA requires 1633 cycles.

The Sub operation can be implemented by instructions in the Intel AES-NI (Advanced Encryption Standard) set. This is likely to result in a significant speed-up that can't be duplicated in external authenticators such as SHA-256 or OCB, or using block and stream ciphers that don't incorporate AES components.

7   Conclusion

The history of stream ciphers with authentication is not long, and can be summed up by ciphers that use external authentication (slow), or integrated authentication (vulnerable to chosen ciphertext attacks under reuse of key-IV pairs).

It is likely that the NLS external MAC is more secure than its own keystream generation algorithm, but its uncompetitive performance meant that it attracted little interest. The Phelix cipher did attract a lot of interest, as one of few ciphers to use integrated authentication, and with efficiency far surpassing block ciphers. Its speed was in part due to a novel design style that eschewed s-boxes, and relied only on repeated use of modular addition for non-linearity. It is consequently not very surprising that it was broken by a differential-linear attack, albeit under a forbidden usage model.

MASHA is based upon a conservative design style - a Linear Feedback Shift Register, with non-linear filter, but specified with intelligence to ensure peak performance and security. Core to its security is the repeated use of the Sub function, which makes use of very well analysed AES confusion and diffusion components. The feedback taps in one of the shift registers are designed to remove attacker's control of differences entering in the state as quickly as possible. The path between the entry into the state of a difference as ciphertext, and its exit as keystream is as long as possible, and traverses several highly non-linear components. The other shift register acts as whitening to defeat a whole range of statistical attacks.

Consequently, we believe that MASHA offers excellent security for a very low price, providing 128-bit security for both encryption and generation of its 256-bit authentication tag.

References

1. Agren, M., Hell, M., Johansson, T., Meier, W.: A new version of grain-128 with authentication. In: Proc. of SKEW 2011 (2011)
2. Babbage, S., de Canniere, C., Canteaut, A., Cid, C., Gilbert, H., Johansson, T., Paar, C., Parker, M., Preneel, B., Rijmen, V., Robshaw, M., Wu, H.: Short report on the end of the second phase. In: ECRYPT (2007), eStream Project
   http://www.ecrypt.eu.org/stream/PhaseIIreport.pdf
3. Billet, O., Gilbert, H.: Resistance of SNOW 2.0 against algebraic attacks. In: Menezes, A. (ed.) CT-RSA 2005. LNCS, vol. 3376, pp. 19–28. Springer, Heidelberg (2005)
4. Courtois, N., Debraize, B.: Algebraic description and simultaneous linear approximations of addition modulo 2^n. In: SASC 2008, pp. 69–86 (2008)
5. Coutois, N.: Algebraic attacks on combiners with memory and several outputs. In: Park, C.-s., Chee, S. (eds.) ICISC 2004. LNCS, vol. 3506, pp. 3–20. Springer, Heidelberg (2005)
6. Daemen, J., Rijmen, V.: The Design of Rijndael: AES - The Advanced Encryption Standard. Information Security and Cryptography, Texts and Monographs (2002)
7. Ekdahl, P., Johansson, T.: A new version of the stream cipher SNOW. In: Nyberg, K., Heys, H.M. (eds.) SAC 2002. LNCS, vol. 2595, pp. 47–61. Springer, Heidelberg (2003)
8. Engels, D., Saarinen, M.-J.O., Smith, E.M.: The Hummingbird-2 Lightweight Authenticated Encryption Algorithm (2011),
   http://eprint.iacr.org/2011/126.pdf
9. ECRYPT eSTREAM. the ECRYPT stream cipher project,
   http://www.ecrypt.eu.org/stream/
10. Hawkes, P., Paddon, M., Rose, G.: The Mundja streaming MAC. IACR ePrint Archive, 2004/271 (2004)
11. Hawkes, P., Paddon, M., Rose, G., de Vries, M.W.: Primitive specification for NLSv2. eSTREAM report, 2006/036 (2006)
12. Hell, M., Johansson, T., Maximov, A., Meier, W.: The Grain family of stream ciphers. In: Robshaw, M.J.B., Billet, O. (eds.) New Stream Cipher Designs. LNCS, vol. 4986, pp. 179–190. Springer, Heidelberg (2008)
13. Hong, J., Sarkar, P.: Rediscovery of time memory tradeoffs. IACR ePrint Archive, Report 2005/090 (2005)
14. Kiyomoto, S., Tanaka, T., Sakurai, K.: K2: A stream cipher algorithm using dynamic feedback control. In: SECRYPT 2007, International Conference on Security and Cryptography, Barcelona, Spain (July 28-31, 2007)

15. Lim, S.Y., Pu, C.C., Lim, H.T., Lee, H.J.: Dragon-MAC: Securing wireless sensor networks with authenticated encryption. IACR ePrint Archive, 2007/024 (2007)
16. Nyberg, K., Wallén, J.: Improved linear distinguishers for SNOW 2.0. In: Robshaw, M.J.B. (ed.) FSE 2006. LNCS, vol. 4047, pp. 144–162. Springer, Heidelberg (2006)
17. National Institute of Standards and Technology. Random number generation and testing, NIST Test Suite (2000),
    http://csrc.nist.gov/groups/ST/toolkit/rng/index.html
18. O'Neil, S., Gittins, B., Landman, H.: VEST - hardware-dedicated stream ciphers. eSTREAM report, 2005/032 (2005)
19. Pornin, T.: Comparative performance review of most of the sha-3 second-round candidates. In: Proc. of The Second SHA-3 Candidate Conference (2010)
20. Robshaw, M., Billet, O. (eds.): New Stream Cipher Designs: The eSTREAM Finalists. LNCS, vol. 4986. Springer, Heidelberg (2008)
21. Rogaway, P., Bellare, M., Black, J.: OCB: A block-cipher mode of operation for efficient authenticated encryption. ACM Transactions on Information and System Security (TISSEC) 6(3), 365–403 (2003)
22. Whiting, D., Schneier, B., Lucks, S., Muller, F.: Phelix - fast encryption and authentication in a single cryptographic primitive. eSTREAM report, 2005/020 (2005)
23. Wu, H., Preneel, B.: Differential-linear attacks against the stream cipher Phelix. In: Biryukov, A. (ed.) FSE 2007. LNCS, vol. 4593, pp. 87–100. Springer, Heidelberg (2007)

A   Test Vector

A test vector for MASHA is as follows;

Initial Key (128 bits):
0x00000000000000000000000000000000

Initial Vector (192 bits):
0x000000000000000000000000000000000000000000000000

Plaintext (256 bits):
0x0000000000000000000000000000000000000000000000000000000000000000

Ciphertext (256 bits):
0x2ab0811fe3f52101f78054c236992469d4bd4f907e42210a85cf6431b41ee4b5

Message Authentication Code (256 bits):
0x8aed4124ae5c1e830e48469f04a15265f4dcdbc2eb7240a3252cb71beff0df4a

Toward Pairing-Free Certificateless Authenticated Key Exchanges

Hu Xiong[1], Qianhong Wu[2,3], and Zhong Chen[1]

[1] Key Laboratory of Network and Software Security Assurance of the Ministry of Education, Institute of Software, School of Electronics Engineering and Computer Science, Peking University, Beijing 100871, China
[2] Key Lab. of Aerospace Information Security and Trusted Computing Ministry of Education, School of Computer, Wuhan University, China
[3] Universitat Rovira i Virgili, Department of Computer Engineering and Mathematics, Av. Països Catalans 26, E-43007 Tarragona, Spain

Abstract. Certificateless authenticated key exchange (CL-AKE) protocols do not suffer from intricate certificate management or heavy trust reliance on a third party. Unfortunately, these advantages are partially counteracted in most CL-AKE protocols which require expensive pairing operations. This paper proposes a new CL-AKE protocol without requiring any pairing operation during the protocol execution, although a pairing map may be required to realize a Decisional Diffie-Hellman (DDH) oracle in the security proof. With implicit authentication, we illustrate modular proofs in a security model incorporating standard definitions of AKE protocols and certificateless cryptography. Analysis shows that our protocol is also efficient.

1 Introduction

Since the seminal inception by Diffie and Hellman [9], key exchange (KE) protocols have attracted much attention. Various approaches have been developed to improve security and efficiency of KE protocols [10,20]. Among them, authenticated key exchange (AKE) is one of the most attracting concepts. It allows two or more parties to exchange a secret session key over an open network fully controlled by an attacker. As a fundamental primitive, secure AKE protocols can then serve as basic building blocks for constructing advanced protocols.

Most AKE protocols have been realized in the traditional public-key infrastructure (PKI) or Identity-based (ID-based) cryptosystem. However, PKI-based protocols suffer from complicated certificate management while ID-based systems are subject to the so-called key escrow problem. One may note the key generation center (KGC) as a passive attacker in ID-based AKE protocols cannot learn the negotiated session keys among the users. However, there is no way to guarantee that a malicious KGC can only launch passive attacks, and indeed, an actively attacking KGC can impersonate any user to accomplish the AKE protocol and then eavesdrop the communications without being detected. Thus, in addition to avoiding the heavy certificate management burden in PKI based

X. Lai, J. Zhou, and H. Li (Eds.): ISC 2011, LNCS 7001, pp. 79–94, 2011.
© Springer-Verlag Berlin Heidelberg 2011

AKEs, it is also of interest to relieve the heavy trust reliance on KGC in ID-based AKEs.

Recent efforts have been devoted to eliminate these limitations in AKE protocols. The most promising approach is the Certificateless cryptography (CLC) [1]. In a CLC setting, a semi-trusted Key Generation Center (KGC) helps each user to generate his/her private key but cannot access the full private key of any user. Thus, the key escrow problem is removed in CLC. Further, the public key of a user is computed from KGC's public parameters and a secret value chosen by the user in CLC. Hence, CLC does not need an additional certificate to show the authorship of a public key. Pioneered by the work of Al-Riyami and Paterson who presented the first certificateless AKE (CL-AKE) protocol from pairings [1], quite a number of bilinear pairings based CL-AKE protocols have been proposed [1,18,19,21,22,23,25]. Most of these protocols make the efforts to reduce the number of pairing operations required in their protocols. The up-to-date protocol due to *Zhang et al.* [25] requires only one pairing operation.

In view of both security and efficiency, it is of great interest to build CL-AKE protocols without bilinear pairings. In 2009, Catalano *et al.* [6] proposed a CL-AKE protocol based on Fiore-Gennaro's ID-based key exchange [11]. Almost the same time, two independent CL-AKE protocols without bilinear pairings were proposed in [12,15]. However, Yang and Tan [24] showed that both protocols are insecure and proposed a provably secure pairing-free CL-AKE protocol along with a new model for CL-AKE. Their CL-AKE model integrates the typical two types of attacks in CLC, and also incorporates the underlying idea of PKI based AKE protocols. To this goal, a signature is included in the public key of each user. Independent of Yang-Tan's work, He *et al.* [14] also proposed a CL-AKE protocol without pairings recently. Unfortunately, Han [13] later showed that He *et al.*'s protocol can not resist the Type I adversary in CLC. These studies reveal that it is far from trivial to construct secure pairing-free CL-AKE protocols, although they are desirable in practice.

In this paper, we make further efforts toward secure pairing-free CL-AKE protocols. We propose a new CL-AKE protocol without requiring any pairing operation during the protocol execution. To achieve active security, we employ implicit authentication. We define a formal security model for CL-AKE protocols incorporating standard definitions of AKE protocols and CLC. We then prove the security of the proposed protocol with Kulda-Paterson's modular approach [17] in the mBR model. One may note that the proof of one of our main security results relies on the GDH assumption which requires a DDH oracle. This implies that our protocol can only be implemented in GDH groups. So far, the only known GDH groups are constructed from symmetric pairing groups. Indeed, in all existing secure paring-free CL-AKE protocols [6,24], oracles similar to the DDH oracle are required. With available implementation technologies, a pairing map is necessary for their security proofs to be valid in these protocols. However, since the goal of pairing-free CL-AKE protocols is to eliminate expensive pairing operations during the run of the AKE protocol, the requirement of pairings merely in security proofs does not deviate this goal.

The rest of this paper is organized as follows. A brief review of some basic concepts and security notions used in our scheme is described in Section 2. In Section 3, we propose a new pairing-free CL-AKE protocol based on the GDH assumption. In Section 4, the efficiency and security comparison between the our proposal and up-to-date protocols is conducted. Finally, the conclusions are given in Section 5.

2 Preliminaries

2.1 Computational Assumptions

Our protocol relies on some computational assumptions in a finite cyclic group G of a prime order p. Specifically, the proposal is built on the GDH assumption which states that the CDH problem is hard even if a probabilistic polynomial-time (PPT) attacker can query a DDH oracle whether $\{P, aP, bP, cP\} \in G$ for randomly chosen $a, b, c \in \mathbb{Z}_p$ is a DDH tuple. Here, $\{P, aP, bP, cP\}$ is called a DDH tuple if $c = ab \bmod p$.

Definition 1 (CDH Problem). *Given $\{P, aP, bP\} \in G$ for a randomly sampled element $P \in G$ and random values $a, b \in \mathbb{Z}_p$, compute $abP \in G$.*

Definition 2 (GDH Problem). *Given $\{P, aP, bP\} \in G$ for some random values $a, b \in \mathbb{Z}_p$, compute the element abP with the help of a DDH oracle.*

Since the only known DDH oracle is realized in symmetric pairing groups over elliptical curves [8], this implies that our protocol can only be realized in symmetric pairing groups. However, this does not violate our goal to eliminate time-consuming pairing operations in our protocol because the DDH oracle is only used in the proof while we do not require any pairing computation when running the protocol.

Also, for clarity, we employ the equivalent variant of the CDH assumption, i.e., the Divisible CDH (DCDH) assumption [2] reviewed as follows.

Definition 3 (DCDH Assumption). *Given $\{P, aP, bP\} \in G$ for some random values $a, b \in \mathbb{Z}_p$, for any PPT algorithm, it is hard to compute $ab^{-1}P$.*

2.2 Modeling CL-AKE Protocols

A CL-AKE protocol consists of four PPT algorithms: Setup, Partial-Private-Key-Generate, User-key-Generate and Key-Agreement. These algorithms are defined as follows.

Setup. It is run by KGC. It takes as input a security parameter k and returns a master-key and a list of system parameters *params*.

Partial-Private-Key-Generate. It is also run by KGC. It takes as input the parameter list *params*, master-key and an entity's identity ID_i, to produce the entity's partial private key D_{ID_i}.

User-key-Generate. It takes as input a parameter list *params*, an entity's identity ID_i, to produce a secret value usk_{ID_i} and public key upk_{ID_i} for this entity.

Key-Agreement. This is a PPT interactive algorithm involving two entities A and B. The inputs are the system parameters *params* for both A and B, plus $\{D_{ID_A}, usk_{ID_A}, upk_{ID_A}, ID_A\}$ for A and $\{D_{ID_B}, usk_{ID_B}, upk_{ID_B}, ID_B\}$ for B. Here, D_{ID_A} and D_{ID_B} are the respective partial private keys of A and B; $\{usk_{ID_A}, upk_{ID_A}\}$ and $\{usk_{ID_B}, upk_{ID_B}\}$ are the user secret value and public key of A and B respectively; ID_A is the identity of A and ID_B is the identity of B. Eventually, if the protocol does not fail, A and B obtain a secret session key $K_{AB} = K_{BA} = K$.

2.3   Security Definitions of CL-AKE Protocol

Motivated by the model of Zhang *et al.* [25] and the modified Bellare-Rogaway model (mBR model) [3], we present a security model for CL-AKE protocols. The security of our protocol Π is defined by the following game between a challenger $\mathcal{CH}$ and an adversary $\mathcal{A} \in \{\mathcal{A}_1, \mathcal{A}_2\}$. Here, Type I adversary $\mathcal{A}_1$ models an adversary who does not know the master-key of KGC, but has the ability to replace the public key of any entity with a value of her choice. Type II adversary $\mathcal{A}_2$ models a malicious KGC who knows the master-key, but cannot replace the target user's public key. We use the oracle $\Pi_{i,j}^s$ to represent the s-th instance between participant i and partner participant j in a session. At the beginning of the game, $\mathcal{CH}$ runs the Setup algorithm, takes as input a security parameter k to obtain the master-key and the system parameters *params*. If $\mathcal{A}$ is a Type I adversary $\mathcal{A}_1$, $\mathcal{CH}$ sends *params* to $\mathcal{A}$ and keeps the master-key secret; otherwise, $\mathcal{A}$ is a Type II adversary $\mathcal{A}_2$, and $\mathcal{CH}$ sends *params* with the master key to $\mathcal{A}$.

$\mathcal{A}$ is modeled by a PPT Turing machine. All communications go through the adversary $\mathcal{A}$. Participants only respond to the queries by $\mathcal{A}$ and do not communicate directly among themselves. Assume that participant i has identity i. $\mathcal{A}$ can relay, delete, modify, interleave or delete all the message flows in the system. Note that $\mathcal{A}$ is allowed to make a polynomial number of queries, including one Test query defined as follows.

- Create(ID_i): This allows $\mathcal{A}$ to ask $\mathcal{CH}$ to set up a new participant i. On receiving such a query, $\mathcal{CH}$ generates the public/private key pair for i.
- Public-Key(ID_i): $\mathcal{A}$ can request the public key of a participant i. To respond, $\mathcal{CH}$ outputs the public key upk_{ID_i} of participant i.
- Partial-Private-Key(ID_i): On input an identity ID_i, $\mathcal{CH}$ outputs the corresponding partial private key D_{ID_i} of participant i.
- Corrupt(ID_i): On input an identity ID_i, $\mathcal{CH}$ outputs the partial private key D_{ID_i} and secret value usk_{ID_i} of participant i.
- Public-Key-Replacement(ID_i, upk'_{ID_i}): For a participant i, $\mathcal{A}$ can choose a new public key upk'_{ID_i} and then set upk'_{ID_i} as the new public key of this participant. $\mathcal{CH}$ will record these replacements which will be used later.

- Send($\Pi_{j,i}^t$,M): $\mathcal{A}$ can send a message M of her choice to an oracle, say $\Pi_{j,i}^t$, in which case participant i assumes that the message has been sent by participant j. $\mathcal{A}$ may also make a special Send query with $M = \lambda$ to the oracle $\Pi_{j,i}^t$, which demonstrates i to initiate a protocol run with j. An oracle is an initiator if the first message it has received is λ. If an oracle does not receive a message λ as its first message, then it is a responder oracle.
- Reveal($\Pi_{j,i}^t$): $\mathcal{A}$ can ask a particular oracle to reveal the session key (if any) it currently holds to $\mathcal{A}$.
- Test($\Pi_{I,J}^T$): At some point, $\mathcal{A}$ has to make a Test query to a *fresh* oracle $\Pi_{I,J}^T$ (see Definition 4). To answer the query, $\mathcal{CH}$ flips a fair coin $b \in \{0,1\}$, and returns the session key held by $\Pi_{I,J}^T$ if $b = 0$, or a random sample from the distribution of the session key if $b = 1$.

Definition 4. *(Fresh oracle). Here, $\Pi_{i,j}^s$ is fresh if (1) $\Pi_{i,j}^s$ has accepted the request to establish a session key; (2) $\Pi_{i,j}^s$ has not been revealed; (3) there is no matching conversation[1] of oracle $\Pi_{i,j}^s$ has been revealed; (4) participant $j \neq i$ has not been corrupted; (5) if $\mathcal{A}$ is a Type I adversary, $\mathcal{A}$ has never requested the partial private key of participant j; and, if $\mathcal{A}$ is a Type II adversary, $\mathcal{A}$ has never replaced the public key of participant j.*

Note that definition 4 allows participant i to be corrupted, and thus can be used to address the key-compromise impersonation property as well as the partial forward secrecy property. After a Test query, $\mathcal{A}$ can continue to query the oracles except that it cannot make a Reveal query to the test oracle $\Pi_{I,J}^T$ or to $\Pi_{J,I}^S$ who has a matching conversation with $\Pi_{I,J}^T$ (if it exists). In addition, if $\mathcal{A}$ is a Type I adversary, $\mathcal{A}$ cannot request the partial private key of the participant J; and if $\mathcal{A}$ is a Type II adversary, $\mathcal{A}$ cannot replace the public key of the participant J. Finally, $\mathcal{A}$ outputs its guess b' for b. $\mathcal{A}$'s advantage $Advantage^{\mathcal{A}}(k)$ is defined as the probability $|\Pr[b = b'] - \frac{1}{2}|$.

The security of CL-AKE protocol can be defined using the concept of $\mathcal{A}$'s advantage as follows:

Definition 5. *A certificateless two-party AKE protocol is said to be secure if:*

1. *In the presence of a benign adversary, two oracles, $\Pi_{i,j}^s$ and $\Pi_{j,i}^t$, running the protocol both accept holding the same session key, and the session key is distributed uniformly at random on $\{0,1\}^k$; and*
2. *For any adversary $\mathcal{A}$, $Advantage^{\mathcal{A}}(k)$ is negligible.*

Similarly to [25], since $\mathcal{A}$ is formalized in a way that it can perform all kinds of known attacks in the real world, then a protocol provides desirable security attributes including *known session key security, forward secrecy, key compromise impersonation resilience* and *unknown key-share resilience* when it satisfies Definition 5.

---

[1] Let the session ID be the concatenation of the messages in a session. Two oracles $\Pi_{i,j}^s$ and $\Pi_{j,i}^t$ are said to have a matching conversation with each other if they have the same session ID.

2.4   Kudla and Paterson's Modular Approach

To provide a concise but precise security proof for AKE protocols in mBR model, Kudla and Paterson proposed a reduced game called Computational No Reveal-mBR game (cNR-mBR game) [17], which is regarded as one of the best solutions to prove AKE protocols [5,7]. In this subsection, we explore this approach to simplify our security model. The simplified cNR-mBR game is identical to the security game described in section 2.3 except that $\mathcal{A}$ is not allowed to ask Reveal queries and $\mathcal{A}$ no longer makes a Test query. Instead, an adversary must choose a fresh oracle $\Pi_{i,j}^{s}$ at the end of the game, and it must compute the session key instead of deciding between a session key and a random value to win the game. In such a game, the security of the protocol is defined as the probability that $\mathcal{A}$ outputs a session key K such that $K = K_{\Pi_{i,j}^{s}}$. Thus, the simplified cNR-mBR security model can be defined as follows:

Definition 6. *A CL-AKE protocol is said to be cNR-mBR-secure if:*

1. *In the presence of a benign adversary, two oracles running the protocol both accept holding the same session key, and the session key is distributed uniformly at random on $\{0,1\}^{k}$; and*
2. *For any adversary $\mathcal{A}$, Advantage$^{\mathcal{A}}(k)$ in the cNR-mBR game is negligible.*

To employ the modular approach, we need first transform the target protocol Π into a corresponding protocol π which is identical to Π except that Π produces a hashed session key while π utilizes the input string of the hash function as the session key. Then prove the security of π in the simplified cNR-mBR game using Definition 6. Finally, the security of π in cNR-mBR game can be associated with that of Π in the mBR game according to the following theorem.

Theorem 1. *[17] Suppose that a key exchange protocol Π produces a hashed session key on completion of the protocol (via hash function H) and that Π has strong partnering[2]. If the cNR-mBR security of the corresponding π is probabilistic polynomial time reducible to the hardness of the computational problem of some relation f, and the session string decisional problem for Π is polynomial time reducible to the decisional problem of f, then the mBR security of Π is probabilistic polynomial time reducible to the hardness of the Gap problem of f, assuming that H is a random oracle.*

2.5   Preciseness of Modular Approach

Kudla and Paterson's modular proof removes the requirement for *Reveal* queries, and simplifies the proof while still maintaining the precision and correctness of the proof result. In this subsection, we analyze the reason.

---

[2] If there exists an adversary $\mathcal{A}$, which when attacking Π in an mBR game with non-negligible probability in the security parameter k, can make any two oracles $\Pi_{a,b}^{u}$ and $\Pi_{b,a}^{v}$ accept holding the same session key when they are not partners, then we say that Π has weak partnering. If Π does not have weak partnering, then we say that Π has strong partnering.

In the security proof of the mBR model, the challenger $\mathcal{CH}$ has to answer the adversary's *Reveal* queries. For queries $Reveal(\Pi_{I,j}^s)$ where I denotes the attacked party, $\mathcal{CH}$ is not able to answer since $\mathcal{CH}$ usually has to solve a difficult computational problem which $\mathcal{CH}$ doesn't know how to solve. Otherwise, the session key could be computed with the oracle's ephemeral random value and the partner's secret value and partial private key, which violates key compromise impersonation resilience. Regarding the other possible *Reveal* queries, $\mathcal{CH}$ can answer them easily because $\mathcal{CH}$ knows the corresponding secret value and partial private key. However, $\mathcal{CH}$ has to answer both kinds of *Reveal* queries because this simulates the known session key attack. In order to respond the queries to maintain the indistinguishability with the real world, various methods have been proposed, and a survey of these methods can be found in [7].

The modular approach addresses this problem with the cNR-mBR game which removes the *Reveal* query. This removal will not affect the correctness of the security proof since the modular approach assumes that there exists a certain method for $\mathcal{CH}$ to answer the impossible $Reveal(\Pi_{I,j}^s)$ queries, and this method is given in the proof process of Theorem 1. The method which employs a random oracle and a decisional oracle is identical to all AKE protocols Π of the similar category. A direct proof for any correct Π can be given by combining the method and the proof in the cNR-mBR game (see an example of our protocol in Appendix A). However, instead of doing so, one can use the modular approach to simply obtain the cNR-mBR security and then apply Theorem 1 to obtain the security of Π. Since the cNR-mBR game is conceptually simple, it is easy to maintain the indistinguishability and to obtain precisely the cNR-mBR security. By employing Theorem 1, the final security conclusion for Π is also precise. For a comprehensive study of Kudla and Paterson's modular proof we refer the readers to [17].

3   Proposed CL-AKE Protocol without Pairings

3.1   Protocol Description

Our proposed protocol is inspired by the work due to Cao *et al.* [4] and Zhang *et al.* [25]. Similarly to other CL-AKE protocols, the proposed one requires a key generator center (KGC) and consists of four phases: system setup, partial key extraction, user key generation and key exchange phase.

Setup: Given a security parameter $k \in \mathbb{Z}$, the algorithm works as follows:

1. Run the parameter generator on input k to generate a prime p and determine the tuple $\{\mathbb{F}_p, E/\mathbb{F}_p, G, P\}$ as defined in Sect. 2.
2. Choose a master-key $x \in_R \mathbb{Z}_p^*$, and compute $P_{pub} = xP$.
3. Choose cryptographic hash functions $H_1 : \{0,1\}^* \times G \rightarrow \mathbb{Z}_p^*$ and $H_2 : \{0,1\}^{*2} \times G^9 \rightarrow \{0,1\}^k$. Finally the PKG's master-key x is kept secret and the system parameters $\{\mathbb{F}_p, E/\mathbb{F}_p, G, P, P_{pub}, H_1, H_2\}$ are published.

Partial-Private-Key-Generate: Given a user's identity $ID_U \in \{0,1\}^*$, KGC first chooses at random $r_U \in_R \mathbb{Z}_p^*$, and computes $R_U = r_U P$, $h = H_1(ID_U \| R_U)$

and $s_U = (r_U + hx)^{-1}$. It then sets this user's partial private key (s_U, R_U) and transmits it to user ID_i secretly.

It is easy to see that user ID_U can validate her partial private key by checking whether the equation $s_U(R_U + H_1(ID_U \| R_U)P_{pub}) = P$ holds. The partial private key is valid if the equation holds and vice versa.

User-key-Generate: The user ID_U selects a secret value $x_U \in_R \mathbb{Z}_p^*$ as his user secret key usk_U, and computes his public key as $upk_U = x_U P$.

key exchange: Assume that an entity A with identity ID_A has full private key (s_A, R_A, x_A) and public key upk_A, and an entity B with identity ID_B has full private key (s_B, R_B, x_B) and public key upk_B. The message flows and computations of a protocol run are described below.

1. To start an AKE session with the intended responder B, the initiator A will send $\{ID_A, upk_A, R_A\}$ to B. On receiving the initiation message from A, B does the following.
   (a) Choose at random the ephemeral key $b \in_R \mathbb{Z}_p^*$ and compute the key token $T_B = b(R_A + H_1(ID_A \| R_A)P_{pub})$;
   (b) Send $\{ID_B, upk_B, R_B\}$ and the key token T_B to A.
2. On receiving $\{ID_B, upk_B, R_B, T_B\}$, A does the following.
   (a) Choose at random the ephemeral key $a \in_R \mathbb{Z}_p^*$ and compute the key token $T_A = a(R_B + H_1(ID_B \| R_B)P_{pub})$;
   (b) Send the key token T_A to B.
3. Then both A and B can compute the shared secrets as follows. Participant A computes $s_A T_B = bP$ and $K_{AB}^1 = bP + aP$, $K_{AB}^2 = a \cdot bP$ and $K_{AB}^3 = a \cdot upk_B + usk_A \cdot bP$. Participant B computes $s_B T_A = aP$ and $K_{BA}^1 = aP + bP$, $K_{BA}^2 = b \cdot aP$ and $K_{BA}^3 = b \cdot upk_A + usk_B \cdot aP$. The shared secrets are consistent because $K_{AB}^1 = bP + aP = K_{BA}^1$, $K_{AB}^2 = abP = K_{BA}^2$, $K_{AB}^3 = a \cdot upk_B + usk_A \cdot bP = usk_B \cdot aP + b \cdot upk_A = K_{BA}^3$. Thus the agreed session key for A and B can be computed as $K = H_2(ID_A, ID_B, upk_A, upk_B, R_A, R_B, T_A, T_B, K_{AB}^1, K_{AB}^2, K_{AB}^3)$

3.2   Security Analysis

Next we prove the security of the new protocol using Kudla and Paterson's model. We first turn our new protocol Π into a related protocol π, which is similar to the former except that π uses the string $(ID_A, ID_B, upk_A, upk_B, R_A, R_B, T_A, T_B, K_{AB}^1, K_{AB}^2, K_{AB}^3)$ as the session key while Π uses $H_2(ID_A, ID_B, upk_A, upk_B, R_A, R_B, T_A, T_B, K_{AB}^1, K_{AB}^2, K_{AB}^3)$. Then we prove the cNR-mBR security of π.

Lemma 1. *Suppose that if for protocol π there is a Type I adversary $\mathcal{A}_1$ who can win in the cNR-mBR game with advantage at least ε, then the CDH problem can be solved with non-negligible advantage by an algorithm $\mathcal{CH}$.*

Proof. Suppose $\mathcal{CH}$ is given an instance $(aP, bP) \in G$ of the CDH problem, and is tasked to compute $cP \in G$ with $c = ab \bmod p$. To do this, $\mathcal{CH}$ simulates a challenger with $\mathcal{A}_1$. $\mathcal{CH}$ stipulates the hash function H_1 and maintains an H_1-list which is initialized empty. The number of participants in the game is denoted by $n_p(k)$ and the number of sessions each participant may be involved in is denoted by $n_s(k)$. The full private key for the i-th participant ID_i is (s_i, R_i, x_i) and ID_i is the corresponding identifier. $\mathcal{CH}$ generates ID_i's partial private key as follows:

$\mathcal{CH}$ first chooses at random $I \in \{1, \cdots, n_p(k)\}$, then chooses $R_I \in_R G$ and sets $\{\perp, R_I\}$ as ID_I's partial private key. The system public key can be denoted as $P_{pub} = H_1(ID_I, R_I)^{-1}(bP - R_I)$ which implicitly means that $s_I^{-1}P = bP$. For ID_i with $i \in \{1, \cdots, n_p(k)\}$ and $i \neq I$, $\mathcal{CH}$ sets the partial private key by first choosing at random $(s_i, h_i) \in_R \mathbb{Z}_p^*$. Then $\mathcal{CH}$ computes $R_i = s_i^{-1}P - h_i P_{pub}$ and sets (s_i, R_i) as ID_i's partial private key. After that, $\mathcal{CH}$ passes R_i and ID_i to $\mathcal{A}_1$ and adds $\{ID_i, R_i, s_i, h_i\}$ to the H_1-list for $i \in \{1, \cdots, n_p(k)\}$.

Then $\mathcal{CH}$ picks at random $J \in \{1, \cdots, n_p(k)\} \neq I$, $v \in \{1, \cdots, n_s(k)\}$, and $\mathcal{CH}$ starts $\mathcal{A}_1$ by answering $\mathcal{A}_1$'s queries as follows:

$H_1(ID_i, R_i)$ query: If the tuple $\{ID_i, R_i, s_i, h_i\}$ is already in the H_1-list, $\mathcal{CH}$ responds with h_i, otherwise, $\mathcal{CH}$ chooses $h_i \in_R \mathbb{Z}_p^*$, adds $\{ID_i, R_i, s_i, h_i\}$ to the H_1-list and returns h_i to $\mathcal{A}_1$.

Create(ID_i): $\mathcal{CH}$ maintains an initially empty list $\mathbf{C}$ consisting of tuples of the form $(ID_i, R_i, s_i, x_i, upk_i)$. For simplicity, we assume that all the Create queries are distinct. On receiving a Create query on ID_i, $\mathcal{CH}$ first makes an H_1 query to obtain a tuple $\{ID_i, R_i, s_i, h_i\}$, then chooses a random $x_i \in_R \mathbb{Z}_p^*$ and computes the public key $upk_i = x_i P$ for ID_i with $i \in \{1, \cdots, n_p(k)\}$. Finally the tuple $(ID_i, R_i, s_i, x_i, upk_i)$ is added to $\mathbf{C}$. Without loss of generality, we assume that, before asking the following queries, $\mathcal{A}_1$ has already asked some Create queries on the related queries.

Public-key(ID_i): On receiving this query, $\mathcal{CH}$ first searches for a tuple $(ID_i, R_i, s_i, x_i, upk_i)$ in $\mathbf{C}$ which is indexed by ID_i, then returns upk_i as the answer.

Partial-Private-Key(ID_i): Whenever $\mathcal{CH}$ receives this query, if $ID_i = ID_I$, $\mathcal{CH}$ aborts; else, $\mathcal{CH}$ searches for a tuple $(ID_i, R_i, s_i, x_i, upk_i)$ in $\mathbf{C}$ which is indexed by ID_i and returns (R_i, s_i) as the answer.

Corrupt(ID_i): Whenever $\mathcal{CH}$ receives this query, if $ID_i = ID_I$, $\mathcal{CH}$ aborts; else, $\mathcal{CH}$ searches for a tuple $(ID_i, R_i, s_i, x_i, upk_i)$ in $\mathbf{C}$ which is indexed by ID_i and if $x_i = null$, $\mathcal{CH}$ returns $null$; otherwise $\mathcal{CH}$ returns (R_i, s_i, x_i) as the answer.

Public-Key-Replacement(ID_i, upk_i'): On receiving this query, $\mathcal{CH}$ searches for a tuple $(ID_i, R_i, s_i, x_i, upk_i)$ in $\mathbf{C}$ which is indexed by ID_i, then updates upk_i to upk_i' and sets $x_i = null$.

Send($\Pi_{i,j}^s, M$): If $\Pi_{i,j}^s \neq \Pi_{J,j}^v$, then $\mathcal{CH}$ acts according to the protocol specification. Otherwise, $\mathcal{CH}$ responds with the tuple (ID_j, upk_j, R_j, aP).

The probability that $\mathcal{A}_1$ chooses $\Pi_{J,j}^v$ as the Test oracle and that $ID_j = ID_I$ is $\frac{1}{n_p^2(k)n_s(k)}$. In this case, $\mathcal{A}_1$ would not have corrupted ID_I, and so $\mathcal{CH}$ would not have aborted. If $\mathcal{A}_1$ can win in such a game, then at the end of this game, $\mathcal{A}_1$ will output its guess of the session key of the form $\{0,1\}^* \times \{0,1\}^* \times A \times$

$B \times C \times D \times E \times F \times G \times H \times I$, and $\mathcal{CH}$ can output $G - s_J M$ where M is the input message of $\mathsf{Send}(\Pi_{J,j}^v, M)$ query. Thus $\mathcal{CH}$ can solve the DCDH problem with non-negligible probability $\frac{c}{n_p^2(k)n_s(k)}$ within $t(k)$ where c is a constant. Then according to the equivalence between the DCDH and the CDH problems, the CDH problem can be solved with advantage at least $(\frac{c}{n_p^2(k)n_s(k)})^2$. □

Lemma 2. *Suppose that if for protocol π there is a Type II adversary $\mathcal{A}_2$ who can win in the cNR-mBR game with advantage at least ε, then the CDH problem can be solved with non-negligible advantage by an algorithm $\mathcal{CH}$.*

Proof. Suppose that there exists a Type II adversary $\mathcal{A}_2$ who can win the game with a non-negligible advantage in polynomial-time. Then we show that there is an algorithm $\mathcal{CH}$ that solves the CDH problem with non-negligible probability.

Suppose $\mathcal{CH}$ is given an arbitrary input $(P, aP, bP) \in G$ of the CDH problem. We show how $\mathcal{CH}$ can use $\mathcal{A}_2$ to solve the CDH problem, i.e., to compute abP. $\mathcal{CH}$ first chooses $x \in_R \mathbb{Z}_p^*$ at random and sets $P_{pub} = xP$ as the system public key. After that, x is sent to $\mathcal{A}_2$ by $\mathcal{CH}$. $\mathcal{CH}$ stipulates the hash function H_1 and maintains an H_1-list which is initialized empty. The number of participants in the game is denoted by $n_p(k)$ and the number of sessions each participant may be involved in is denoted by $n_s(k)$. The full private key for the i-th participant ID_i is (s_i, R_i, x_i) and ID_i is the corresponding identifier. $\mathcal{CH}$ chooses at random $r_i \in_R \mathbb{Z}_p^*$, computes $R_i = r_i P$ and $s_i = (r_i + H_1(ID_i \| R_i)x)^{-1}$, and sets (s_i, R_i) as ID_i's partial private key.

Then $\mathcal{CH}$ picks at random $I, J \in \{1, \cdots, n_p(k)\}$, $v \in \{1, \cdots, n_s(k)\}$, and $\mathcal{CH}$ starts $\mathcal{A}_2$ by answering $\mathcal{A}_2$'s queries as follows:

$H_1(ID_i, R_i)$ query: If the tuple $\{ID_i, R_i, s_i, h_i\}$ is already in the H_1-list, $\mathcal{CH}$ responds with h_i; otherwise, $\mathcal{CH}$ chooses $h_i \in_R \mathbb{Z}_p^*$, adds $\{ID_i, R_i, s_i, h_i\}$ to the H_1-list and returns h_i to $\mathcal{A}_2$.

$\mathsf{Create}(ID_i)$: $\mathcal{CH}$ maintains an initially empty list $\mathbf{C}$ consisting of tuples of the form $(ID_i, R_i, s_i, x_i, upk_i)$. For simplicity, we assume that all the Create queries are distinct. On receiving a Create query on ID_i, $\mathcal{CH}$ sets the public key as bP and the user secret key as $\perp$ for ID_i with $i = I$; otherwise, $\mathcal{CH}$ chooses a random $x_i \in_R \mathbb{Z}_p^*$ and computes the public key as $x_i P$ and user secret key as x_i for ID_i with $i \in \{1, \cdots, n_p(k)\}$ and $i \neq I$. After that, $\mathcal{CH}$ makes an H_1 query to obtain a tuple $\{ID_i, R_i, s_i, h_i\}$. Finally the tuple $(ID_i, R_i, s_i, x_i, upk_i)$ is added to $\mathbf{C}$. Without loss of generality, we assume that, before asking the following queries, $\mathcal{A}_2$ has already asked some Create queries on the related queries.

$\mathsf{Public\text{-}key}(ID_i)$: On receiving this query, $\mathcal{CH}$ first searches for a tuple $(ID_i, R_i, s_i, x_i, upk_i)$ in $\mathbf{C}$ which is indexed by ID_i, then returns upk_i as the answer.

$\mathsf{Corrupt}(ID_i)$: Whenever $\mathcal{CH}$ receives this query, if $ID_i = ID_I$, $\mathcal{CH}$ aborts; else $\mathcal{CH}$ searches for a tuple $(ID_i, R_i, s_i, x_i, upk_i)$ in $\mathbf{C}$ which is indexed by ID_i and returns (R_i, s_i, x_i) as the answer.

$\mathsf{Public\text{-}Key\text{-}Replacement}(ID_i, upk_i')$: On receiving this query, if $ID_i = ID_I$, $\mathcal{CH}$ aborts; otherwise, $\mathcal{CH}$ searches for a tuple $(ID_i, R_i, s_i, x_i, upk_i)$ in $\mathbf{C}$ which is indexed by ID_i, then updates upk_i to upk_i' and sets $x_i = null$.

Send($\Pi_{i,j}^s, M$): If $\Pi_{i,j}^s \neq \Pi_{J,j}^v$, then $\mathcal{CH}$ acts according to the protocol specification. Otherwise, $\mathcal{CH}$ responds with the tuple (ID_j, upk_j, R_j, aP).

The probability that $\mathcal{A}_2$ chooses $\Pi_{J,j}^v$ as the Test oracle and that $ID_j = ID_I$ is $\frac{1}{n_p^2(k)n_s(k)}$. In this case, $\mathcal{A}_2$ would not have corrupted ID_I, and so $\mathcal{CH}$ would not have aborted. If $\mathcal{A}_2$ can win in such a game, then at the end of this game, $\mathcal{A}_2$ will output its guess of the session key of the form $\{0,1\}^* \times \{0,1\}^* \times A \times B \times C \times D \times E \times F \times G \times H \times I$, and $\mathcal{CH}$ can output $I - s_J M$ where M is the input message of Send($\Pi_{J,j}^v, M$) query. Thus $\mathcal{CH}$ can solve the DCDH problem with non-negligible probability $\frac{c}{n_p^2(k)n_s(k)}$ within $t(k)$ where c is a constant. Then according to the equivalence between the DCDH and the CDH problems, the CDH problem can be solved with advantage at least $(\frac{c}{n_p^2(k)n_s(k)})^2$.   □

Before applying Theorem 1 to the above result, we have to prove that protocol Π satisfies the property of strong partnering.

Lemma 3. *Protocol Π has strong partnering in the random oracle model.*

Proof. Let pa denote the partner of user a. Suppose an adversary $\mathcal{A}$ can have two oracles $\Pi_{i,pi}^s$ and $\Pi_{j,pj}^t$ accept holding the same session key when $pi \neq j$ and $pj \neq i$.

If $\Pi_{i,pi}^s$ is an initiator for the session key, it has to make a query of the form $\{ID_i, ID_{pi}, A, B, C, D, E, F, G, H, I\}$ to the random oracle H_2, and to receive h_2^i as the session key. For $\Pi_{j,pj}^t$ to have the same session key, it must have made the H_2 query of the form $\{ID_{pj}, ID_j, A, B, C, D, E, F, G, H, I\}$ since $ID_j \neq ID_i$. Then it must have $ID_{pj} = ID_i$ and vice versa. Thus $\Pi_{i,pi}^s$ and $\Pi_{j,pj}^t$ are partners, which contradicts the assumption. Thus it is impossible for the adversary $\mathcal{A}$ to obtain a qualified $\Pi_{j,pj}^t$ when $\Pi_{i,pi}^s$ is an initiator; the same can be proved when $\Pi_{i,pi}^s$ is a responder. This completes the proof.   □

Theorem 2. *Our protocol Π is secure in the random oracle model assuming the hardness of Gap Diffie-Hellman problem.*

Proof. The theorem follows directly from Theorem 1 and Lemmas 1-3. Thus, the protocol provides known session key security, key compromise impersonation resilience and unknown key share resilience, which is satisfied even in face of Kaliski's UKS attack, as shown in [16].   □

Theorem 3. *Our protocol has the perfect forward secrecy property if the CDH problem in G is hard.*

Proof. Suppose that A and B established a session key K using our CL-AKE protocol, and later, their full private key (s_A, R_A, x_A) and (s_B, R_B, x_B) were compromised. Let a and b be the secret random numbers chosen by A and B, respectively, during the establishment of their session key. It is easy to see that, to compute the established session key K, the adversary who owns (s_A, R_A, x_A), (s_B, R_B, x_B), $s_B T_a = aP$ and $s_A T_b = bP$ for unknown a, b must know the value of abP. However, to compute the value of abP without the knowledge of either a

or b, the adversary must have the ability to solve the CDH problem in G. Under the CDH assumption, this probability is negligible. Hence, our protocol has the perfect forward secrecy property.   □

4   Comparison

In this section we compare the efficiency of the proposed protocols, Zhang *et al.* [25]'s protocol which is the most efficient existing CL-AKE protocol based on pairings, and Yang-Tan [24]'s protocol which is the up-to-date provably secure pairing-free CL-AKE protocol. Here we only consider the costly operations and we omit the computation efforts which can be pre-computed. We denote by P a pairing operation, by E an exponentiation, by ME a multi-exponentiation ($\approx 1.5E$), by V a signature verification and by σ a signature.

Table 1. Performance comparison of different protocols

Protocol	Computation	Bandwidth	Assumption
Yang-Tan's [24]	$5E + 3ME + 1V$	$6|G| + 2|\sigma|$	Gap-DH
Zhang *et al.*'s [25]	$5E + 1P$	$4|G|$	CDH,BDH
Our protocol	$7E$	$6|G|$	Gap-DH

According to the benchmark of pairing [7], 1 pairing operation is roughly equivalent to 3 exponentiation in computation. Thus, comparing with these two protocols, our protocol is the most efficient one in computation, while Zhang *et al.*'s protocol requires less communication. Note that the security proofs of existing pairing-free CL-AKEs require the GDH assumption. With available implementation technologies, a pairing map is needed to realize the DDH oracle in the GDH assumption. However, since the goal of pairing-free CL-AKE protocols is to eliminate expensive pairing operations during the run of the AKE protocol, the requirement of pairings merely in security proofs does not deviate this goal.

5   Conclusions

We have proposed CL-AKE protocol without any pairing operation during the protocol execution. The proposed protocol realizes a pairing-free CL-AKE protocol with implicit authentication, and its security has been proved in Kudla and Paterson's security model. Since the existing pairing-free CL-AKE protocols rely on the GDH assumption to complete the proofs, this implies that the CL-AKE protocols so far are not fully pairing-free. Hence, it remains an interesting problem to construct efficient fully pairing-free CL-AKE protocols without relying on GDH-like assumptions.

Acknowledgements. This work is partially supported by National Natural Science Foundation of China under Grant No. 61003230, State Key Laboratory of Information Security under Grant No.01-02-4 and National Research Foundation for the Doctoral Program of Higher Education of China under Grant No. 200806140010. The second author is partly supported by the

EU 7FP through project "DwB", the Spanish Government through projects CTV-09-634, PTA2009-2738-E, TSI-020302-2010-153, PT-430000- 2010-31, TIN2009-11689, CONSOLIDER INGENIO 2010 "ARES" CSD2007-0004 and TSI2007-65406-C03-01, by the Government of Catalonia under grant SGR2009-1135, and by the Chinese NSF projects 60970115, 60970116, 61003214 and 61021004. The second author also acknowledges support by the Fundamental Research Funds for the Central Universities of China to Project 3103004, Beijing Municipal Natural Science Foundation to Project 4112052 and Shaanxi Provincial Education Department through Scientific Research Program 2010JK727.

References

1. Al-Riyami, S., Paterson, K.: Certificateless Public Key Cryptography. In: Laih, C.S. (ed.) ASIACRYPT 2003. LNCS, vol. 2894, pp. 452–473. Springer, Heidelberg (2003)
2. Bao, F., Deng, R., Zhu, H.: Variations of Diffie-Hellman Problem. In: Qing, S., Gollmann, D., Zhou, J. (eds.) ICICS 2003. LNCS, vol. 2836, pp. 301–312. Springer, Heidelberg (2003)
3. Bellare, M., Rogaway, P.: Random oracles are practical: a paradigm for designing efficient protocols. In: Proc. 1st ACM CCS, pp. 62–73 (1993)
4. Cao, X., Kou, W., Yu, Y., Sun, Y.: Identity-based authenticated key agreement protocols without bilinear pairings. IEICE Transactions on Fundamentals of Electronics, Communications and Computer Sciences E91.A(12), 3833–3836 (2009)
5. Cao, X., Kou, W., Du, X.: A pairing-free identity-based authenticated key agreement protocol with minimal message exchanges. Information Sciences 180(15), 2895–2903 (2010)
6. Catalano, D., Fiore, D., Gennaro, R.: Certificateless onion routing. In: Proc. 16th ACM CCS, pp. 151–160 (2009)
7. Chen, L., Cheng, Z., Smart, N.: Identity-based key agreement protocols from pairings. International Journal of Information Security 6(4), 213–241 (2007)
8. Cilardo, A., Coppolino, L., Mazzocca, N., Romano, L.: Elliptic curve cryptography engineering. Proceedings of the IEEE 94(2), 395–406 (2006)
9. Diffie, W., Hellman, M.: New Directions in Cryptography. IEEE Transactions on Information Theory 22(6), 644–654 (1976)
10. Dutta, R., Barua, R.: Overview of Key Agreement Protocols. Cryptology ePrint Archive, Report 2005/289 (2005), http://eprint.iacr.org/
11. Fiore, D., Gennaro, R.: Making the Diffie-Hellman Protocol Identity-Based. In: Pieprzyk, J. (ed.) CT-RSA 2010. LNCS, vol. 5985, pp. 165–178. Springer, Heidelberg (2010)
12. Geng, M., Zhang, F.: Provably secure certificateless two-party authenticated key agreement protocol without pairing. In: IEEE CIS 2009, pp. 208–212 (2009)
13. Han, W.: Breaking a certificateless key agreement protocol withour bilinear pairing. Cryptology ePrint Archive, Report 11/249 (2011), http://eprint.iacr.org/
14. He, D., Chen, J., Hu, J.: A pairing-free certificateless authenticated key agreement protocol. International Journal of Communication Systems (2011), doi:10.1002/dac.1265
15. Hou, M., Xu, Q.: A two-party certificateless authenticated key agreement protocol without pairing. In: 2nd IEEE ICCSIT, pp. 412–416 (2009)

16. Kaliski Jr., B.S.: An unknown key-share attack on the MQV key agreement proto-col. ACM Transactions on Information and System Security 4(3), 275–288 (2001)
17. Kudla, C., Paterson, K.G.: Modular security proofs for key agreement protocols. In: Roy, B. (ed.) ASIACRYPT 2005. LNCS, vol. 3788, pp. 549–565. Springer, Heidelberg (2005)
18. Luo, M., Wen, Y., Zhao, H.: An Enhanced Authentication and Key Agreement Mechanism for SIP Using Certificateless Public-key Cryptography. In: 9th ICYCS 2008, pp. 1577–1582 (2008)
19. Mandt, T.K., Tan, C.H.: Certificateless authenticated two-party key agreement protocols. In: Okada, M., Satoh, I. (eds.) ASIAN 2006. LNCS, vol. 4435, pp. 37–44. Springer, Heidelberg (2008)
20. Menezes, A.J., Van Oorschot, P.C., Vanstone, S.A.: Handbook of applied cryptography. CRC Press, USA (1997)
21. Swanson, C., Jao, D.: A Study of Two-Party Certificateless Authenticated Key-Agreement Protocols. In: Roy, B., Sendrier, N. (eds.) INDOCRYPT 2009. LNCS, vol. 5922, pp. 57–71. Springer, Heidelberg (2009)
22. Wang, F., Zhang, Y.: A new provably secure authentication and key agreement mechanism for SIP using certificateless public-key cryptography. Computer Communications 31(10), 2142–2149 (2008)
23. Wang, S., Cao, Z., Wang, L.: Efficient certificateless authenticated key agreement protocol from pairings. Wuhan University Journal of Natural Sciences 11(5), 1278–1282 (2006)
24. Yang, G., Tan, C.-H.: Strongly secure certificateless key exchange without pairing. In: 6th ACM ASIACCS, pp. 71–79 (2011)
25. Zhang, L., Zhang, F., Wu, Q., Domingo-Ferrer, J.: Simulatable certificateless two-party authenticated key agreement protocol. Information Sciences 180(6), 1020–1030 (2010)

A   A Direct Proof for the Proposed Protocol

In this section, we present a direct proof in the original security model for our CL-AKE protocol to show the preciseness of the modular approach. The proof is based on a combination of the method proposed in the modular approach and the proof in the cNR-mBR game.

Theorem 4. *Suppose that if for our protocol Π there is a Type I adversary $\mathcal{A}_1$ who can win in the security game described in Definition 5 with advantage at least ε, then the CDH problem can be solved with non-negligible advantage.*

Proof. Suppose $\mathcal{CH}$ is given an instance $(aP, bP) \in G$ of the CDH problem, and is tasked to compute $cP \in G$ with $c = ab \bmod p$. To do this, $\mathcal{CH}$ simulates a challenger with $\mathcal{A}_1$. $\mathcal{CH}$ stipulates the hash function $H_i (i = 1, 2)$ and maintains an H_i-list which is initialized empty; $\mathcal{CH}$ also maintains for Send queries an initially empty Ω-list and for Reveal queries an initially empty Λ-list. The number of participants in the game is denoted by $n_p(k)$ and the number of sessions each participant may be involved in is denoted by $n_s(k)$. The full private key for the i-th participant ID_i is (s_i, R_i, x_i) and ID_i is the corresponding identifier. $\mathcal{CH}$ generates ID_i's partial private key as follows:

$\mathcal{CH}$ first chooses $P_0 \in G$ at random and sets P_0 as the system public key P_{pub}. Then $\mathcal{CH}$ chooses at random $I \in \{1, \cdots, n_p(k)\}$ and generates the partial private key for ID_I. $\mathcal{CH}$ chooses $h_I \in_R \mathbb{Z}_p^*$ and computes $R_I = bP - h_I P_0$ which implicitly means that $s_I^{-1} P = bP$. Then $(\perp, R_I)$ is set as ID_I's partial private key. For ID_i with $i \in \{1, \cdots, n_p(k)\}$ and $i \neq I$, $\mathcal{CH}$ sets the partial private key by first choosing at random $(s_i, h_i) \in_R \mathbb{Z}_p^*$. Then $\mathcal{CH}$ computes $R_i = s_i^{-1} P - h_i P_{pub}$ and sets (s_i, R_i) as ID_i's partial private key. After that, $\mathcal{CH}$ passes (s_i, R_i) and ID_i to $\mathcal{A}_1$ and adds $\{ID_i, R_i, s_i, h_i\}$ to the H_1-list for $i \in \{1, \cdots, n_p(k)\}$.

Then $\mathcal{CH}$ picks at random $J \in \{1, \cdots, n_p(k)\} \neq I$, $v \in \{1, \cdots, n_s(k)\}$, and $\mathcal{CH}$ starts $\mathcal{A}_1$ by answering $\mathcal{A}_1$'s queries as follows:

$H_1(ID_i, R_i)$ query: If the tuple $\{ID_i, R_i, s_i, h_i\}$ is already in the H_1-list, $\mathcal{CH}$ responds with h_i, otherwise, $\mathcal{CH}$ chooses $h_i \in_R \mathbb{Z}_p^*$, adds $\{ID_i, R_i, s_i, h_i\}$ to the H_1-list and returns h_i to $\mathcal{A}_1$.

Create(ID_i): $\mathcal{CH}$ maintains an initially empty list $\mathbf{C}$ consisting of tuples of the form $(ID_i, R_i, s_i, usk_i, upk_i)$. On receiving a Create query on ID_i, $\mathcal{CH}$ first makes an H_1 query to obtain a tuple $\{ID_i, R_i, s_i, h_i\}$, then chooses a random $x_i \in_R \mathbb{Z}_p^*$ as the secret value usk_i and computes the public key $upk_i = x_i P$ for ID_i with $i \in \{1, \cdots, n_p(k)\}$. Finally the tuple $(ID_i, R_i, s_i, usk_i, upk_i)$ is added to $\mathbf{C}$.

Corrupt(ID_i): Whenever $\mathcal{CH}$ receives this query, if $ID_i = ID_I$, $\mathcal{CH}$ aborts; else, $\mathcal{CH}$ searches for a tuple $(ID_i, R_i, s_i, usk_i, upk_i)$ in $\mathbf{C}$ which is indexed by ID_i and returns (R_i, s_i, usk_i) as the answer.

Send($\Pi_{i,j}^s, M$): $\mathcal{CH}$ maintains an Ω-list of the form $\{\Pi_{i,j}^s, trans_{i,j}^s, t_{i,j}^s\}$ where $trans_{i,j}^s$ is the transcript of $\Pi_{i,j}^s$ so far and $t_{i,j}^s$ will be described later.

- If $\Pi_{i,j}^s \neq \Pi_{i,J}^v$, then $\mathcal{CH}$ acts according to the specification of protocol Π. Note that when M is not the second message to $\Pi_{i,j}^s$, $\mathcal{CH}$ chooses at random $t_{i,j}^s \in_R \mathbb{Z}_p^*$ and computes the reply as $t_{i,j}^s P$. Then $\mathcal{CH}$ updates the tuple indexed by $\Pi_{i,j}^s$ in Ω-list.
- Otherwise, $\mathcal{CH}$ responds with aP and update the tuple $\{\Pi_{i,j}^s, trans_{i,j}^s, t_{i,j}^s\}$ with $t_{i,j}^s = \perp$.

Reveal($\Pi_{i,j}^s$): $\mathcal{CH}$ maintains a Λ-list of the form $\{ID_{ini}^s, ID_{resp}^s, upk_{ini}^s, upk_{resp}^s, X_{ini}^s, Y_{resp}^s, \Pi_{i,j}^s, K_{i,j}^s\}$ where ID_{ini}^s and upk_{ini}^s is the identification and public key of the initiator, and ID_{resp}^s and upk_{resp}^s is the identification and public key of the responder in the session which $\Pi_{i,j}^s$ engages in. The description of the other items will be given below.

- If $\Pi_{i,j}^s = \Pi_{i,J}^v$, abort.
- Else if $i \neq I$, $\mathcal{CH}$ extracts (s_i, R_i, usk_i) from $\mathbf{C}$, and goes through Ω-list for corresponding (X_i, Y_j) and $t_{i,j}^s$. According to the specification of security game in Definition 5, there must be such an item in Ω-list. So $\mathcal{CH}$ computes $Z_s^1 = s_i Y_j + t_{i,j}^s P$, $Z_s^2 = t_{i,j}^s s_i Y_j$ and $Z_s^3 = t_{i,j}^s upk_{resp}^s + usk_i \cdot s_i Y_j$. After that, $\mathcal{CH}$ makes an H_2 query. If $\Pi_{i,j}^s$ is the initiator oracle, then the query is of the form $\{ID_i, ID_j, upk_i, upk_j, R_i, R_j, X_i, Y_j, Z_s^1, Z_s^2, Z_s^3\}$ or else of the form $\{ID_j, ID_i, upk_j, upk_i, R_j, R_i, Y_j, X_i, Z_s^1, Z_s^2, Z_s^3\}$. At last, $\mathcal{CH}$ sets H_2 responses h_s^2 as $K_{i,j}^s$.

- Else if $i = I$, $\mathcal{CH}$ goes through Ω-list for corresponding (X_i, Y_j) and $t_{i,j}^s$. $\mathcal{CH}$ go through the H_2-list to see if there exists a tuple indexed by $\{ID_i, ID_j, upk_i, upk_j, X_i, Y_j\}$ if $\Pi_{i,j}^s$ is a initiator; otherwise indexed by $\{ID_j, ID_i, upk_j, upk_i, Y_j, X_i\}$. If there exists such a tuple, and the corresponding Z_u^1, Z_u^2 and Z_u^3 satisfies the equations $\hat{e}(Z_u^1 - t_{i,j}^s P, R_i + h_i P_{pub}) = \hat{e}(P, Y_j)$, $\hat{e}(Z_u^2, P) = \hat{e}(X_i, Y_j)$ and $\hat{e}(Z_u^3 - t_{i,j}^s upk_j, R_i + h_i P_{pub}) = \hat{e}(upk_i, Y_j)$ given a proper bilinear map $\hat{e}$ for group G, then obtain the corresponding h_u^2 as $K_{i,j}^s$. Else $\mathcal{CH}$ chooses at random $K_{i,j}^s \in \{0,1\}^k$.
- Insert the tuple $\{ID_{ini}^s, ID_{resp}^s, upk_{ini}^s, upk_{resp}^s, X_{ini}^s, Y_{resp}^s, \Pi_{i,j}^s, K_{i,j}^s\}$ into the Λ-list and return $K_{i,j}^s$.

$H_2(ID_i, ID_j, upk_i, upk_j, R_i, R_j, X_i, Y_j, Z_u^1, Z_u^2, Z_u^3)$ query: H_2-list is of the form $\{ID_i, ID_j, upk_i, upk_j, R_i, R_j, X_i, Y_j, Z_s^1, Z_s^2, Z_s^3, h_u^1\}$, and $\mathcal{CH}$ responds with H_2 queries $(ID_i, ID_j, upk_i, upk_j, R_i, R_j, X_i, Y_j, Z_u^1, Z_u^2, Z_u^3)$ as follows:

- If a tuple indexed by $(ID_i, ID_j, upk_i, upk_j, R_i, R_j, X_i, Y_j, Z_u^1, Z_u^2, Z_u^3)$ is already in H_2-list, reply with the corresponding h_u^1.
- If there is no such a tuple,

    • If the equations $\hat{e}(Z_u^1 - t_{i,j}^s P, R_i + h_i P_{pub}) = \hat{e}(P, Y_j)$, $\hat{e}(Z_u^2, P) = \hat{e}(X_i, Y_j)$ and $\hat{e}(Z_u^3 - t_{i,j}^s upk_j, R_i + h_i P_{pub}) = \hat{e}(upk_i, Y_j)$ hold given a proper bilinear map $\hat{e}$ for group G, $\mathcal{CH}$ goes through the Λ-list. If there is such a tuple indexed by $\{ID_u^i, ID_u^j, upk_u^i, upk_u^j, X_u, Y_u\}$ in Λ-list, then obtain the corresponding $K_{i,j}^s$ as h_u^2. Else $\mathcal{CH}$ chooses at random $h_u^2 \in \{0,1\}^k$.
    • Else if the equations do not hold, $\mathcal{CH}$ chooses at random $h_u^2 \in \{0,1\}^k$.

- Insert the tuple $\{ID_i, ID_j, upk_i, upk_j, R_i, R_j, X_i, Y_j, Z_s^1, Z_s^2, Z_s^3, h_u^1\}$ into the H_2-list and return h_u^2.

The probability that $\mathcal{A}_1$ chooses $\Pi_{J,j}^v$ as the Test oracle and that $ID_j = ID_I$ is $\frac{1}{n_p^2(k)n_s(k)}$. In this case, $\mathcal{A}_1$ would not have corrupted ID_I or reveal $\Pi_{J,j}^v$, and so $\mathcal{CH}$ would not have aborted. If $\mathcal{A}_1$ can win in such game, then at the end of this game, $\mathcal{A}_1$ will output its guess of the session key of the form $\{0,1\}^* \times \{0,1\}^* \times A \times B \times C \times D \times E \times F \times G \times H \times I$, and $\mathcal{CH}$ can output $G - s_J M$ where M is the input message of $\mathsf{Send}(\Pi_{J,j}^v, M)$ query. Thus $\mathcal{CH}$ can solve the DCDH problem with non-negligible probability $\frac{c}{n_p^2(k)n_s(k)}$ within $t(k)$ where c is a constant. Then according to the equivalence of the DCDH and the CDH assumptions, the CDH problem can be solved with advantage at least $(\frac{c}{n_p^2(k)n_s(k)})^2$.

As can be seen, in the above proof, the Reveal queries are answered using the method proposed by the modular proof. The rest of the proof is identical to the proof in the cNR-mBR game. Since the method to answer the Reveal queries is common to every protocol, it's safe to omit the proof for the Type II adversary $\mathcal{A}_2$. Thus by providing the cNR-mBR security of a protocol, one can obtain the corresponding ordinary security claim. Therefore, the modular approach can provide a correct security proof for a protocol.

Security Analysis of an RSA Key Generation Algorithm with a Large Private Key

Fanyu Kong[1,2], Jia Yu[3], and Lei Wu[4]

[1] Institute of Network Security, Shandong University, Jinan 250100, China
fanyukong@sdu.edu.cn
[2] Key Laboratory of Cryptologic Technology and Information Security,
Ministry of Education, Jinan 250100, China
[3] College of Information Engineering, Qingdao University, Qingdao 266071, China
yujia@qdu.edu.cn
[4] School of Information Science and Engineering, Shandong Normal University,
Jinan 250014, China

Abstract. In 2003, L. H. Encinas, J. M. Masqué and A. Q. Dios proposed an algorithm for generating the RSA modulus N with a large private key d, which was claimed secure. In this paper, we propose an attack on Encinas-Masqué-Dios algorithm and find its security flaw. Firstly, we prove that Encinas-Masqué-Dios algorithm is totally insecure when the public exponent e is larger than the sum of the two primes p and q. Secondly, we show that when e is larger than $N^{\frac{1}{4}}$, Encinas-Masqué-Dios algorithm leaks sufficient secret information and then everyone can recover the factorization of the RSA modulus N in polynomial time.

Keywords: Cryptanalysis, RSA, Key generation algorithm, Lattice basis reduction, Partial key exposure attack.

1 Introduction

The RSA cryptosystem [13], which is one of the most important public key cryptosystems, has been widely used in electronic commerce, secure communication, digital signature and so on. Let $N = pq$ be an n-bit RSA modulus, where p and q are two distinct large primes. Let e and d be the public and private exponents satisfying $ed \equiv 1 \pmod{\phi(N)}$, where $\phi(N)$ is the Euler totient function. First of all, the parameters d, p, q and $\phi(N)$ must be kept secret from the attackers.

In order to speed up the decryption or signing process in the RSA cryptosystem, a short private exponent d may be used to accelerate the computation $m = c^d \pmod{N}$. However, the breakthrough idea of M. Wiener [15] introduced a short private key attack with the continued fraction algorithm. M. Wiener proved that every public exponent $e < N^{1.5}$ corresponding to a private exponent $d < \frac{1}{3}N^{0.25}$ yields the factorization of the RSA modulus $N = pq$ in polynomial time. E. Verheul and H. Tilborg [14] extended the attack to $d > \frac{1}{3}N^{0.25}$, which required an exhaustive search of $2t + 8$ bits with $t = \log_2(d/n^{0.25})$. At EUROCRYPT 1999, D. Boneh and G. Durfee [2] improved the attack bound

X. Lai, J. Zhou, and H. Li (Eds.): ISC 2011, LNCS 7001, pp. 95–101, 2011.

to $d < N^{0.292}$ by using LLL algorithm [10] and Coppersmith's method for finding small roots of modular polynomial equations [4]. At ASIACRYPT 1998, D. Boneh, G. Durfee and Y. Frankel [3] proposed the partial key exposure attacks on RSA cryptosystem and showed that a quarter of the leaked least significant bits of d are sufficient for recover all of d efficiently. Many other lattice-based attacks on RSA and other cryptosystems can be found in the literature [1,5,8,9,11,12].

In spite of choosing a short private key, L. H. Encinas, J. M. Masqué and A. Q. Dios [6,7] proposed an algorithm for generating the RSA modulus $N = pq$ with a large private key d. Note that the public and private keys satisfy $ed - k\phi(N) = 1$ in RSA cryptosystem. In Encinas-Masqué-Dios algorithm, the private key is generated with the form of $d = \frac{1+k\phi(N)}{e}$ where $k = e - 1$. In [6,7], L. H. Encinas et al. claimed that the RSA key generated by their algorithm was secure against the aforementioned attacks.

In this paper, we propose an attack on Encinas-Masqué-Dios algorithm and show that it is insecure when $e > N^{\frac{1}{4}}$. Firstly, we prove that Encinas-Masqué-Dios algorithm can be broken when $e > p + q$. Secondly, it is showed that when $e > N^{\frac{1}{4}}$, everyone can also break Encinas-Masqué-Dios algorithm and compute the private exponent d in polynomial time.

The rest of this paper is organized as follows. In Section 2, we review Encinas-Masqué-Dios RSA key generation algorithm. In Section 3, we propose the attack on Encinas-Masqué-Dios algorithm when $e > N^{\frac{1}{4}}$. Finally, in Section 4 we conclude the paper.

2   Preliminaries

2.1   Notations and Definitions

The RSA cryptosystem [13] can be described as follows. Let $N = pq$ be an n-bit RSA modulus, where p and q are two distinct primes. Let e, d be the public and private exponents, i.e. $ed \equiv 1 \pmod{\phi(N)}$.

For convenience, we set $s = p + q$ and denote by k the integer such that

$$ed - k\phi(N) = ed - k(N - s + 1) = 1.$$

Note that $\phi(N) = N - p - q + 1 < N$. When $d < \phi(N)$, we obtain that $k < e$. The public key $\langle N, e \rangle$ is published to all while the private key $\langle p, q, d \rangle$ must be kept secret. The encryption process (or verification of a signature) is to calculate $c = m^e \pmod{N}$ while the decryption process (or signing of a signature) is to compute $m = c^d \pmod{N}$.

2.2   Review of Encinas-Masqué-Dios RSA Key Generation Algorithm

The fundamental idea of Encinas-Masqué-Dios algorithm [6,7] is to generate the special RSA key satisfying $ed - k\phi(N) = 1$ where $k = e - 1$.

Encinas-Masqué-Dios algorithm is shown in Algorithm 1 and its properties are shown in Proposition 1 and 2, which are described in detail in [6,7].

Algorithm 1. Encinas-Masqué-Dios algorithm [6,7].

Output: The RSA modulus $N = pq$ and the public and private keys e and d.
1. Choose a public exponent $e > 2$.
2. Generate a large prime p such that $r_p(r_p - 1) \in Z_e^*$, where $r_p = p \pmod{e}$.
3. Compute the prime $q = r_p(r_p - 1)^{-1} \pmod{e} + ke$, where k is an integer.
4. Compute $N = pq$ and $\phi(N) = (p - 1)(q - 1)$, and verify that $1 < e < \phi(N)$ and $\gcd(e, \phi(N)) = 1$.
5. Finally, compute the private exponent d, where $1 < d < \phi(N)$ and $ed \equiv 1 \pmod{\phi(N)}$. It can be obtained by $d = \frac{1+(e-1)\phi(N)}{e}$.

Proposition 1. *[6,7] Let e and d be the public and private keys with the RSA modulus $N = pq$, where $ed = 1 + k\phi(N)$. If $k = e - 1$, it follows that $d > \frac{2}{3}\phi(N)$ and $bitlength(d) > bitlength(N) - 2$, where $bitlength(a)$ means the length of binary representation of the integer a.*

Proposition 2. *[6,7] Let e and d be the public and private keys with the RSA modulus $N = pq$, where $ed = 1 + k\phi(N)$ with $k = e - 1$. Let $r_p = p \pmod{e}$ and $r_q = q \pmod{e}$ be the residues of p and q modulo e respectively, and let $S_e = \{r \in Z_e^* | r(r - 1) \text{ is invertible modulo } e\}$. Then we have*
(1) It holds that $r_p, r_q \in S_e$ and $r_q = \frac{r_p}{r_p - 1}$.
(2) If $e = p_1^{m_1} \cdots p_t^{m_t}$ is the prime factorization of e, we have

$$\#S_e = \prod_{i=1}^{t}(p_i - 2)p_i^{m_i - 1}.$$

Conversely, if p and q are arbitrary primes satisfying (1), we have $k = e - 1$.

3   The Proposed Attack on Encinas-Masqué-Dios RSA Key Generation Algorithm

The basic idea of our attack is described as follows. In Encinas-Masqué-Dios algorithm [6,7], the private key d and the public key e satisfy the following equation

$$ed - (e - 1)\phi(N) = ed - (e - 1)(N - p - q + 1) = 1.$$

Let $\phi(N)_e = \phi(N) \pmod{e}$ be the residue of $\phi(N)$ modulo e, i.e. $0 \leq \phi(N)_e < e$. We can compute $\phi(N)_e = 1 = -(e - 1)^{-1} \pmod{e}$ by using the extended Euclidean algorithm. It follows that

$$\phi(N) = N - p - q + 1 \equiv 1 \pmod{e}.$$

Thus we have $N \equiv p + q \pmod{e}$, which means that $p + q \pmod{e}$ is revealed to everyone.

In this section, we first give a simple attack on Encinas-Masqué-Dios algorithm when $e > p + q$. Secondly, we prove that when $e > N^{\frac{1}{4}}$ holds, the leakage of $\phi(N)_e = \phi(N) \pmod{e}$ suffices to break the whole RSA cryptosystem since we can recover the factorization of $N = pq$ and compute the private exponent d.

3.1   Encinas-Masqué-Dios Algorithm Is Insecure When $e > p + q$

When $e > p + q$, we can obtain the large integer $p + q$ directly by computing N mod e since that $N \equiv p + q \pmod{e}$. The revealed value $p + q$ is sufficient for recovering the factorization of the RSA modulus N.

The secret primes p and q are computed easily by solving the following equation, where $s = N \pmod{e}$ is known,

$$\begin{cases} N = pq, \\ s = p + q. \end{cases}$$

When the primes p and q are known, one can compute the Euler function $\phi(N) = (p-1)(q-1) = N + 1 - (p+q)$ and the private key d by $d = e^{-1} \pmod{\phi(N)}$ with the extended Euclidean algorithm. Hence Encinas-Masqué-Dios algorithm is insecure when $e > p + q$.

3.2   Encinas-Masqué-Dios Algorithm Is Insecure When $e > N^{\frac{1}{4}}$

When $e > N^{\frac{1}{4}}$, we can also compute $p + q \pmod{e}$ by using a similar method. However, since it holds that $p + q > N^{\frac{1}{2}}$, we can only obtain $p + q \pmod{e}$, which may be only a part of the integer $p + q$. Thus the method in Subsection 3.1 can not work.

Now we propose the attack on Encinas-Masqué-Dios algorithm, which is based on lattice-based method [3,4,10], when it holds that $e > N^{\frac{1}{4}}$. The method is described in Attack 1. Note that the attack in Subsection 3.1 can be included in Attack 1 since the assumption $e > p + q > N^{\frac{1}{2}}$ is a part of the assumption $e > N^{\frac{1}{4}}$.

Attack 1. Attack on Encinas-Masqué-Dios algorithm when $e > N^{\frac{1}{4}}$.

Input: The RSA modulus N and the public key e, where $e > N^{\frac{1}{4}}$ and
$\quad\quad ed - (e-1)\phi(N) = 1$.
Output: The primes p and q, which satisfy $N = pq$, and the private key d.
1.   We compute $s_e = N \pmod{e}$. Let $t = s_e$ or $t = s_e + e$.
2.   By solving the equations in two variables r_p and r_q, $t = r_p + r_q$ and $r_q = \frac{r_p}{r_p - 1}$,
     we obtain r_p and r_q, where $r_p = p \pmod{e}$ and $r_q = q \pmod{e}$.
3.   Using Coppersmith's lattice-based method, we calculate the primes p and q
     by finding the small solution (x_0, y_0) of the bivariate equation
     $\quad\quad f(x, y) = (ex + r_p)(ey + r_q) - N = 0$.
4.   Finally, we compute the RSA private key $d = e^{-1} \pmod{\phi(N)}$.

Before analyzing the correctness and efficiency of Attack 1, we first review the results of Coppersmith [3,4] and Boneh-Durfee-Frankel method [3] as follows.

Theorem 1. [3,4] Let $p(x, y)$ be an irreducible polynomial in two variables over Z, of maximum degree δ in each variable separately. Let X and Y be upper bounds on the desired integer solution (x_0, y_0), and let $W = max_{i,j}|p_{ij}|X^i Y^j$. If for some $\varepsilon > 0$, $XY < W^{2/(3\delta)-\varepsilon}$ then in time polynomial in $(\log W, 2^\delta)$, one can find all integer pairs (x_0, y_0) such that $p(x_0, y_0) = 0$, $|x_0| \leq X$ and $|y_0| \leq Y$.

Theorem 2. *[3] Let $N = pq$ be an n-bit RSA modulus. Let $r \geq 2^{\frac{n}{4}}$ be given and suppose $p_0 = p \pmod{r}$ is known. Then it is possible to factor N in time polynomial in n. (We denote the running time $T_c(n)$).*

Now we give the analysis of Attack 1 in Proposition 3.

Proposition 3. *Let $N = pq$ be an n-bit RSA modulus. Let $1 < e, d < \phi(N)$ satisfy $ed - (e-1)\phi(N) = 1$, where e is the public exponent. If $e > N^{\frac{1}{4}}$ holds, then there is a $O(T_c(\log N) + \log^3 N)$ polynomial time algorithm that can recover the private exponent d and the factorization of $N = pq$.*

Proof. (*Correctness*) Let $r_p = p \pmod{e}$ and $r_q = q \pmod{e}$, where $0 < r_p < e$ and $0 < r_q < e$, be the residues of p and q modulo e respectively. Let $s = p + q$ be the sum of the primes p and q, and $s_e = p + q \pmod{e}$ be the residue of $p + q$ modulo e, where $0 < s_e < e$.

Then it follows that $r_p + r_q = s_e$ or $r_p + r_q = s_e + e$ holds.

In Encinas-Masqué-Dios algorithm [6,7], the private key d and the public key e satisfy the equation $ed - (e-1)\phi(N) = 1$. Let $\phi(N)_e = \phi(N) \pmod{e}$ be the residue of $\phi(N)$ modulo e. We can compute $\phi(N)_e = 1 = -(e-1)^{-1} \pmod{e}$ by using the extended Euclidean algorithm. It follows that

$$\phi(N) = N - p - q + 1 \equiv 1 \pmod{e}.$$

Thus we have $N \equiv p + q \pmod{e}$ and compute the value $s_e = N \pmod{e}$ directly.

According to Proposition 2, we have that $r_q = \frac{r_p}{r_p - 1}$. Thus the following quadratic equations in two variables r_p and r_q are obtained, one of which is correct.

$$\begin{cases} r_p + r_q = s_e, \\ r_q = \frac{r_p}{r_p - 1}. \end{cases}$$

or

$$\begin{cases} r_p + r_q = s_e + e, \\ r_q = \frac{r_p}{r_p - 1}. \end{cases}$$

By solving the above equations, we can obtain r_p and r_q. According to Theorem 2 [3], if $r_p = p \pmod{e}$ is known with $e > N^{\frac{1}{4}}$, we can factor $N = pq$ in polynomial time in $\log N$.

Finally, when the primes p and q are known, we can compute the Euler function $\phi(N) = (p-1)(q-1) = N + 1 - (p+q)$. Then it follows that the private key d can be obtained by $d = e^{-1} \pmod{\phi(N)}$ with the extended Euclidean algorithm. Hence Encinas-Masqué-Dios algorithm can be broken completely when $e > N^{\frac{1}{4}}$.

(*Efficiency*) In Step 4 of Attack 1, the main operation is the extended Euclidean algorithm with the computational complexity $O(\log^3 N)$. Step 1 consists of arithmetic operations such as modular additions and reductions modulo e, which have the computational complexity $O(\log N)$. Step 2 of Attack 1 includes the computation of finding the square roots of an integer, which have the computational complexity $O(\log^3 N)$. In Step 3 of Attack 1, the LLL algorithm is applied to finding the small solution of a bivariate equation, which can be completed in polynomial time in $\log N$ (We denote $T_c(\log N)$).

Thus, Attack 1 can also compute the private key d and recover the factorization of the RSA modulus N in polynomial time $O(T_c(\log N) + \log^3 N)$.   □

4   Conclusion

In this paper, we propose an attack on Encinas-Masqué-Dios algorithm and show that Encinas-Masqué-Dios algorithm is insecure when $e > N^{\frac{1}{4}}$, in which case one can compute the private exponent d in polynomial time. Therefore, it is still an interesting problem to choose secure and efficient keys in the RSA cryptosystem.

Acknowledgments. The authors would like to thank the anonymous reviewers for their valuable comments and suggestions. This work is partially supported by the Shandong Provincial Natural Science Foundation (No. ZR2010FQ015, ZR2010FQ019).

References

1. Bleichenbacher, D., May, A.: New Attacks on RSA with Small Secret CRT-Exponents. In: Yung, M., Dodis, Y., Kiayias, A., Malkin, T. (eds.) PKC 2006. LNCS, vol. 3958, pp. 1–13. Springer, Heidelberg (2006)
2. Boneh, D., Durfee, G.: Cryptanalysis of RSA with Private Key d Less Than $N^{0.292}$. IEEE Transactions on Information Theory 46, 1339–1349 (2000)
3. Boneh, D., Durfee, G., Frankel, Y.: An Attack on RSA given a Small Fraction of the Private Key Bits. In: Ohta, K., Pei, D. (eds.) ASIACRYPT 1998. LNCS, vol. 1514, pp. 25–34. Springer, Heidelberg (1998)
4. Coppersmith, D.: Small solutions to polynomial equations and low exponent vulnerabilities. Journal of Cryptology 10(4), 223–260 (1997)
5. Coron, J.-S., May, A.: Deterministic Polynomial-Time Equivalence of Computing the RSA Secret Key and Factoring. Journal of Cryptology 20(1), 39–50 (2007)
6. Encinas, L.H., Masqué, J.M., Dios, A.Q.: Large decryption exponents in RSA. Applied Mathematics Letters 16, 293–295 (2003)
7. Encinas, L.H., Masqué, J.M., Dios, A.Q.: An algorithm to obtain an RSA modulus with a large private key. Cryptology ePrint Archive: Report 2003/045 (2003)
8. Jochemsz, E., May, A.: A Strategy for Finding Roots of Multivariate Polynomials with New Applications in Attacking RSA Variants. In: Lai, X., Chen, K. (eds.) ASIACRYPT 2006. LNCS, vol. 4284, pp. 267–282. Springer, Heidelberg (2006)
9. Jochemsz, E., May, A.: A Polynomial Time Attack on RSA with Private CRT-Exponents Smaller Than $N^{0.073}$. In: Menezes, A. (ed.) CRYPTO 2007. LNCS, vol. 4622, pp. 395–411. Springer, Heidelberg (2007)

10. Lenstra, A.K., Lenstra, H.W., Lovász, L.: Factoring polynomials with rational co-efficients. Mathematische Annalen. 261, 513–534 (1982)
11. May, A., Ritzenhofen, M.: Solving Systems of Modular Equations in One Variable: How Many RSA-Encrypted Messages Does Eve Need to Know? In: Cramer, R. (ed.) PKC 2008. LNCS, vol. 4939, pp. 37–46. Springer, Heidelberg (2008)
12. Nguyen, P.Q., Shparlinski, I.E.: The insecurity of the elliptic curve digital signature algorithm with partially known nonces. Designs, Codes and Cryptography 30(2), 201–217 (2003)
13. Rivest, R.L., Shamir, A., Adleman, L.: A method for obtaining digital signatures and public-key cryptosystems. Communications of the ACM 21(2), 120–126 (1978)
14. Verheul, E., Tilborg, H.: Cryptanalysis of less short RSA secret exponents. Applicable Algebra in Engineering, Communication and Computing 8(5), 425–435 (1997)
15. Wiener, M.: Cryptanalysis of Short RSA Secret Exponents. IEEE Transactions on Information Theory 36(3), 553–558 (1990)

Adaptive Secure-Channel Free Public-Key Encryption with Keyword Search Implies Timed Release Encryption

Keita Emura[1], Atsuko Miyaji[2], and Kazumasa Omote[2]

[1] Center for Highly Dependable Embedded Systems Technology
[2] School of Information Science
Japan Advanced Institute of Science and Technology, 1-1, Asahidai, Nomi, Ishikawa,
923-1292, Japan
{k-emura,miyaji,omote}@jaist.ac.jp

Abstract. As well-known results, timed-release encryption (TRE) and public key encryption scheme with keyword search (PEKS) are very close to identity-based encryption (IBE), respectively. It seems natural that there is a close relationship between TRE and PEKS. However, no explicit bridge has been shown between TRE and PEKS so far. In this paper, we show that TRE can be generically constructed by PEKS with extended functionalities, called secure-channel free PEKS (SCF-PEKS) with adaptive security, and discuss the reason why PEKS and (non-adaptive) SCF-PEKS are not suitable for constructing TRE. In addition to this result, we also show that adaptive SCF-PEKS can be generically constructed by anonymous IBE only. That is, for constructing adaptive SCF-PEKS we do not have to require any additional cryptographic primitive compared to the Abdalla et al. PEKS construction (J. Cryptology 2008), even though adaptive SCF-PEKS requires additional functionalities. This result seems also independently interesting.

1 Introduction

Timed-Release Encryption (TRE): Timed-release encryption (TRE) was proposed by May [22], where even a legitimate recipient cannot decrypt a ciphertext before a semi-trusted time server (TS) sends (or broadcasts) a time-release key s_T assigned with a release time T of the encryptor's choice. As a well-known result, (public key based) TRE is very close to identity-based encryption (IBE). More precisely, generic constructions of TRE based on IBE, public key encryption (PKE), and one-time signature (OTS) have been proposed [8,21,24]. Since PKE can be constructed by IBE (and OTS) [6], and digital signature can be constructed by the extraction algorithm of IBE [9], we can say that TRE can be generically constructed by IBE. As an intuition, TRE might be close to other cryptographic primitives which are also close to IBE. So, next we introduce public key encryption scheme with keyword search (PEKS) as such a primitive.

Public key Encryption scheme with Keyword Search (PEKS): PEKS was proposed by Boneh et al. [5]. This scheme considers searching keywords

X. Lai, J. Zhou, and H. Li (Eds.): ISC 2011, LNCS 7001, pp. 102–118, 2011.

from encrypted data. Briefly, the flow of PEKS is as follows: A receiver makes a trapdoor t_ω for a keyword ω, and uploads it on an e-mail server. A sender makes an encrypted keyword (which is encrypted by using a keyword ω' and the receiver's public key), and sends it to the server. The server outputs 1 if $\omega = \omega'$, by using t_ω, and 0 otherwise. As a way to construct PEKS, Abdalla et al. [1] showed that a generic construction of PEKS based on anonymous IBE is sufficient. Next, we discuss the relationships among IBE, TRE, and PEKS.

The Relationships among IBE, TRE, and PEKS: One may think that TRE can be generically constructed by PEKS, since IBE (with 1-bit plaintext space) can be constructed by PEKS [5], and TRE can be generically constructed by IBE [8,24]. However, since generic constructions of TRE based on IBE [8,24] implicitly[1] require multi-bit plaintext space, we cannot conclude that TRE can be generically constructed by PEKS (so we denote it with "?" in Fig 1 later). There are two easy-to-find ways for showing a relationship between TRE and PEKS: (1) for IBE, show that 1-bit plaintext space is enough to make multi-bit plaintext space (as in the PKE case shown by Myers and Shelat [23]), or (2) propose a generic construction of TRE based on IBE with just 1-bit plaintext space. We do not conclude that these are possible or impossible, but try to establish a bridge between TRE and PEKS from another perspective in this paper. To do so, we revisit PEKS with extended functionalities, called secure-channel free PEKS (SCF-PEKS).

Security Conditions of Previous Secure Channel Free PEKS (SCF-PEKS) Schemes and its Theoretic Extension: PEKS schemes ensure that the server (or an outsider) does not learn anything about keywords chosen by the sender *without trapdoor information*. If trapdoors are revealed, then anyone can execute the test procedure. Therefore, trapdoors cannot be sent via public (i.e., insecure) channels. So, in PEKS schemes, a secure channel (such as secure socket layer (SSL) and transport layer security (TLS)) between a receiver and a server is required, and establishing secure channel requires additional setup costs. To solve this problem, secure channel-free PEKS (SCF-PEKS) have been proposed [2,15,16,19], where the server has a public/secret key pair, and the sender makes an encrypted keyword (which is encrypted by using a keyword ω' and both the server's public key and the receiver's public key), and sends it to the server. The server outputs 1 if $\omega = \omega'$ by using the trapdoor t_ω and its own secret key, and 0 otherwise. Even if t_ω is sent via an insecure channel, no entity (except the server) can run the test procedure.

Next, we discuss the security conditions of the previous SCF-PEKS. The security model considered in [2,15,16,19] does not capture the test queries (i.e., "CPA-like" security). As an exception, Rhee et al. have considered test queries [26]. However, this definition is still weak (i.e., "Unquoted CCA-like" security [23]), where an adversary is not allowed to issue the test queries

---

[1] That is, a plaintext of IBE has the form $K_v \| (M \oplus r)$, where K_v is a verification key of OTS, M is a plaintext of TRE, and r is a random number.

adaptively. By considering the CCA2 security, SCF-PEKS must be secure even if a "malicious-but-legitimate" receiver can be admitted to issue test queries *adaptively*. We insist that this adaptive (i.e., "CCA2-like") security is the natural extension of the SCF-PEKS security theoretically[2] , which is called adaptive SCF-PEKS.

Our Contribution: In this paper, we show the relationships of IBE, TRE, and adaptive SCF-PEKS (dashed arrows in Fig 1).

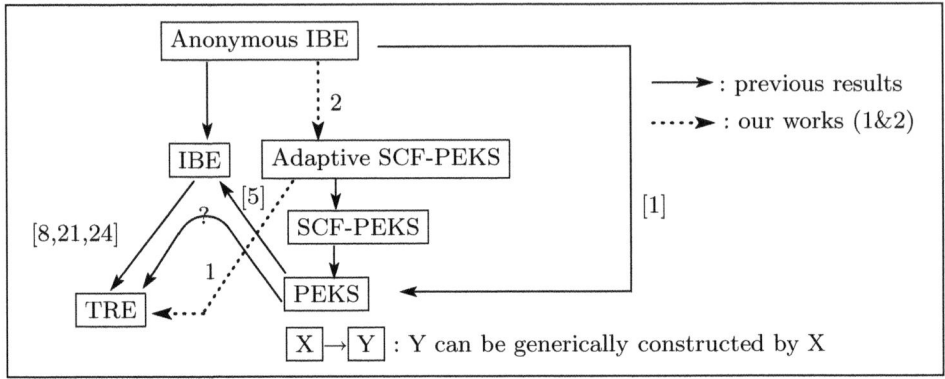

Fig. 1. Relationships of IBE, TRE, and adaptive SCF-PEKS

1. We show that TRE (with 1-bit plaintext space) can be constructed generically from adaptive SCF-PEKS.

   – We discuss the detailed reason why PEKS and (non-adaptive) SCF-PEKS are not suitable for constructing TRE in Section 4.3.

2. We propose a generic construction of adaptive SCF-PEKS based on anonymous IBE, selective-tag chosen-ciphertext (IND-stag-CCA) secure tag-based encryption (TBE), and strongly existentially unforgeable (sUF) OTS. This is the first generic construction of SCF-PEKS.

   – IND-stag-CCA-secure TBE can be constructed by selective-ID chosen plaintext (sID-CPA) secure IBE [20], and digital signature can be constructed by IBE [9]. So, our result shows that adaptive SCF-PEKS can be constructed by anonymous IBE only.

   – No additional cryptographic primitive is required from a generic construction of PEKS [1], even though adaptive SCF-PEKS requires additional functionalities.

---

[2] The word "theoretically" means that here we do not discuss the necessity and practicality of adaptive SCF-PEKS. However, since malicious receivers can use the server as the test oracle in the SCF-PEKS usage, our adaptive security notion might be useful in practice.

2 Preliminaries

In this section, we define the building tools for our generic TRE and adaptive SCF-PEKS construction. $x \xleftarrow{\$} S$ means that x is chosen uniformly from a set S. $y \leftarrow A(x)$ means that y is an output of an algorithm A under an input x. We denote $State$ as the state information transmitted by the adversary to himself across stages of the attack in experiments.

2.1 Definition of TRE

We refer the Dent et al. TRE definition [10] (which is used by Nakai et al. [24] and Matsuda et al. [21]). As an exception, we exclude pre-open capability from the Dent et al. definition. In the following, $\mathcal{T}$ and $\mathcal{M}_{TRE}$ are a release-time space and a plaintext space, respectively.

TRE scheme Π consists of four algorithms, TRE.Setup, TRE.UKG, TRE.Ext, TRE.Enc, and TRE.Dec. A global parameter prm and a master secret key msk are given by executing TRE.Setup(1^κ). A user's public key pk_u and a user's secret key sk_u are given by executing TRE.UKG(1^κ). For a release-time $T \in \mathcal{T}$, a time-release key s_T corresponding to release-time T is given by executing TRE.Ext(prm, msk, T). For a message $M \in \mathcal{M}_{TRE}$ and $T \in \mathcal{T}$, where $\mathcal{M}_{TRE}$ is the message space of TRE, an encryptor runs TRE.Enc(prm, pk_u, T, M), and obtains a ciphertext C_{TRE}. The message M is computed by executing TRE.Dec(prm, sk_u, s_T, C_{TRE}). Correctness is defined as follows: For all (prm, msk) $\leftarrow$ TRE.Setup(1^κ), all $(pk_u, sk_u) \leftarrow$ TRE.UKG(1^κ), all $M \in \mathcal{M}_{TRE}$, and all $T \in \mathcal{T}$, TRE.Dec(prm, sk_u, s_T, C) $= M$ holds, where $C \leftarrow$ TRE.Enc(prm, pk_u, T, M) and $s_T \leftarrow$ TRE.Ext(prm, msk, T).

Next, we define time-server security, called IND-TR-CCA$_{TS}$. It guarantees that no TS can decrypt a ciphertext.

Definition 1 (Time-server Security). *A TRE scheme Π is said to be IND-TR-CCA$_{TS}$ secure if the advantage is negligible for any PPT adversary $\mathcal{A}$, where*

$$Adv_{\Pi,\mathcal{A}}^{IND\text{-}TR\text{-}CCA_{TS}}(1^\kappa) = \big| \Pr \big[(\text{prm}, \text{msk}) \leftarrow \text{TRE.Setup}(1^\kappa);$$

$$(pk_u, sk_u) \leftarrow \text{TRE.UKG}(1^\kappa); \; (M_0^*, M_1^*, T^*, State) \leftarrow \mathcal{A}^{\mathcal{DEC}}(\text{find}, \text{prm}, \text{msk}, pk_u);$$

$$\mu \xleftarrow{\$} \{0,1\}; \; C_{TRE}^* \leftarrow \text{TRE.Enc}(\text{prm}, pk_u, T^*, M_\mu^*); \; \mu' \leftarrow \mathcal{A}^{\mathcal{DEC}}(\text{guess}, C_{TRE}^*, State);$$

$$\mu = \mu' \big] - 1/2 \big|$$

that $\mathcal{DEC}$ is a decryption oracle, where, for input of a ciphertext C_{TRE} and T, it returns the corresponding plaintext M. Note that (C_{TRE}^*, T^*) are not allowed as input to $\mathcal{DEC}$.

Next, we define insider security, called IND-TR-CPA$_{IS}$. It guarantees that no receiver can decrypt a ciphertext before the corresponding time-release key is published.

Definition 2 (Insider Security). *A TRE scheme Π is said to be IND-TR-CPA$_{IS}$ secure if the advantage is negligible for any PPT adversary $\mathcal{A}$, where*

$$Adv_{\Pi,\mathcal{A}}^{IND\text{-}TR\text{-}CPA_{IS}}(1^\kappa) =$$
$$\big| \Pr \big[(\mathsf{prm}, \mathsf{msk}) \leftarrow \mathsf{TRE.Setup}(1^\kappa);\ (pk_u, sk_u) \leftarrow \mathsf{TRE.UKG}(1^\kappa);$$
$$(M_0^*, M_1^*, T^*, State) \leftarrow \mathcal{A}^{\mathcal{EXTRACT}}(\mathsf{find}, \mathsf{prm}, pk_u, sk_u);\ \mu \xleftarrow{\$} \{0,1\};$$
$$C_{TRE}^* \leftarrow \mathsf{TRE.Enc}(\mathsf{prm}, pk_u, T^*, M_\mu^*);\ \mu' \leftarrow \mathcal{A}^{\mathcal{EXTRACT}}(\mathsf{guess}, C_{TRE}^*, State);$$
$$\mu = \mu'\big] - 1/2 \big|$$

that $\mathcal{EXTRACT}$ is an extract oracle, where, for input of T, it returns the corresponding time-release key s_T. $T \geq T^*$ is not allowed as input to $\mathcal{EXTRACT}$.

One may think that the notion "CPA" is weak, and a stronger notion can be defined. Actually, the TRE definition [7,14] achieves such strong security, where $\mathcal{A}$ can access the decryption oracle. However, if no other public key ($pk \neq pk_u$) is considered, such decryption oracle is redundant, since $\mathcal{A}$ has sk_u. Therefore, as in [10,21,24], we adopt the CPA notion in this paper.

2.2 Definitions of sUF OTS

A strongly existentially unforgeable (sUF) OTS against adaptively chosen message attack (CMA) [4] Π consists of three algorithms, Sig.KeyGen, Sign and Verify. Sig.KeyGen is a probabilistic algorithm which outputs a signing/verification key pair (K_s, K_v) from the security parameter 1^κ. Sign is a probabilistic algorithm which outputs a signature σ from K_s, and a message $M \in \mathcal{M}_{Sig}$, where $\mathcal{M}_{Sig}$ is the message space of a signature scheme. Verify is a deterministic algorithm which outputs a bit (1 means that σ is a valid signature, and 0 otherwise) from $\sigma \in \mathcal{S}_{sig}$, K_v and M, where $\mathcal{S}_{sig}$ is the signature space. Correctness is defined as follows: For all $(K_s, K_v) \leftarrow \mathsf{Sig.KeyGen}(1^\kappa)$ and all $M \in \mathcal{M}_{Sig}$, $\mathsf{Verify}(K_v, \sigma, M) = 1$ holds, where $\sigma \leftarrow \mathsf{Sign}(K_s, M)$.

Definition 3 (one-time sUF-CMA). *A signature scheme is said to be one-time sUF-CMA secure if the advantage $Adv_{\Pi,\mathcal{A}}^{one\text{-}time\ sUF\text{-}CMA}(1^\kappa)$ is negligible for any probabilistic polynomial-time (PPT) adversary $\mathcal{A}$ in the following experiment.*

$$Adv_{\Pi,\mathcal{A}}^{one\text{-}time\ sUF\text{-}CMA}(1^\kappa) := \Pr \big[(K_s, K_v) \leftarrow \mathsf{Sig.KeyGen}(1^\kappa);$$
$$(M, State) \leftarrow \mathcal{A}(K_v);\ \sigma \leftarrow \mathsf{Sign}(K_s, M);\ (M^*, \sigma^*) \leftarrow \mathcal{A}(State, \sigma);$$
$$(M^*, \sigma^*) \neq (M, \sigma);\ \mathsf{Verify}(K_v, \sigma^*, M^*) = 1\big]$$

2.3 Definitions of IND-Stag-CCA Secure TBE

A TBE scheme [20] Π consists of three algorithms, TBE.KeyGen, TBE.Enc and TBE.Dec. The public key pk and the secret key sk are given by executing TBE.KeyGen(1^κ), where $\kappa \in \mathbb{N}$ is the security parameter. For a message

$M \in \mathcal{M}_{TBE}$ with a tag $t \in \mathcal{TAG}$, where $\mathcal{M}_{TBE}$ is the message space and $\mathcal{TAG}$ is the tag space of TBE, an encryptor runs TBE.Enc(pk, t, M), and obtains a ciphertext C_{TBE}. The message M is computed by executing TBE.Dec(sk, t, C_{TBE}). Correctness is defined as follows: For all $(pk, sk) \leftarrow$ TBE.KeyGen(1^κ), all $M \in \mathcal{M}_{TBE}$, and all $t \in \mathcal{TAG}$, TBE.Dec$(sk, t, C_{TBE}) = M$ holds, where $C_{TBE} \leftarrow$ TBE.Enc(pk, t, M).

The security experiment of TBE under selective-tag CCA (IND-stag-CCA) is defined as follows.

Definition 4 (IND-stag-CCA). *A TBE scheme is said to be IND-stag-CCA secure if the advantage is negligible for any PPT adversary $\mathcal{A}$ in the following experiment.*

$$Adv_{\Pi,\mathcal{A}}^{IND\text{-}stag\text{-}CCA}(1^\kappa) = \Big| \Pr\big[(t^*, State) \leftarrow \mathcal{A}(1^\kappa); \; (pk, sk) \leftarrow \text{TBE.KeyGen}(1^\kappa);$$

$$(M_0^*, M_1^*, State) \leftarrow \mathcal{A}^{\mathcal{DEC}}(\text{find}, pk, State); \; \mu \xleftarrow{\$} \{0,1\};$$

$$C_{TBE}^* \leftarrow \text{TBE.Enc}(pk, t^*, M_\mu^*); \; \mu' \leftarrow \mathcal{A}^{\mathcal{DEC}}(\text{guess}, C^*, State); \mu = \mu'\big] - 1/2\Big|$$

that $\mathcal{DEC}$ is a decryption oracle for any tag $t \neq t^*$, where for input of a ciphertext $(C_{TBE}, t) \neq (C_{TBE}^*, t^*)$, it returns the corresponding plaintext M. Note that (C_{TBE}^*, t^*) is not allowed as input to $\mathcal{DEC}$.

2.4   Definitions of Anonymous IBE

IBE scheme Π consists of four algorithms, IBE.Setup, IBE.Extract, IBE.Enc and IBE.Dec. The public key pk and the master key mk are given by executing IBE.Setup(1^κ). For an identity $ID \in \mathcal{ID}$, where $\mathcal{ID}$ is the identity space, a secret key corresponding to ID sk_{ID} is given by executing IBE.Extract(pk, mk, ID). For a message $M \in \mathcal{M}_{IBE}$ and $ID \in \mathcal{ID}$, where $\mathcal{M}_{IBE}$ is the message space of IBE, an encryptor runs IBE.Enc(pk, ID, M), and obtains a ciphertext C_{IBE}. The message M is computed by executing IBE.Dec(sk_{ID}, C_{IBE}). Correctness is defined as follows: For all $(pk, mk) \leftarrow$ IBE.Setup(1^κ), all $M \in \mathcal{M}_{IBE}$, and all $ID \in \mathcal{ID}$, IBE.Dec$(sk_{ID}, C_{IBE}) = M$ holds, where $C_{IBE} \leftarrow$ IBE.Enc(pk, ID, M) and $sk_{ID} \leftarrow$ IBE.Extract(pk, mk, ID).

Next, we define the security experiment of IBE under chosen ciphertext attack (IBE-IND-CCA) as follows.

Definition 5 (IBE-IND-CCA). *An IBE scheme is said to be IBE-IND-CCA secure if the advantage is negligible for any PPT adversary $\mathcal{A}$ in the following experiment.*

$$Adv_{\Pi,\mathcal{A}}^{IBE\text{-}IND\text{-}CCA}(1^\kappa) = \Big| \Pr\big[(pk, mk) \leftarrow \text{IBE.Setup}(1^\kappa);$$

$$(M_0^*, M_1^*, ID^*, State) \leftarrow \mathcal{A}^{\mathcal{EXTRACT}, \mathcal{DEC}}(\text{find}, pk); \; \mu \xleftarrow{\$} \{0,1\};$$

$$C_{IBE}^* \leftarrow \text{IBE.Enc}(pk, ID^*, M_\mu^*); \; \mu' \leftarrow \mathcal{A}^{\mathcal{EXTRACT}, \mathcal{DEC}}(\text{guess}, C_{IBE}^*, State);$$

$$\mu = \mu'\big] - 1/2\Big|$$

that $\mathcal{EXTRACT}$ is an extract oracle, where, for input of an identity ID, it returns the corresponding secret key sk_{ID}. ID^* is not allowed as input to $\mathcal{EXTRACT}$. $\mathcal{DEC}$ is a decryption oracle, where, for input of a ciphertext C and an identity ID, it returns the corresponding plaintext M. (ID^*, C^*_{IBE}) is not allowed as input to $\mathcal{DEC}$. Chosen plaintext security (IBE-IND-CPA) is simply defined by removing $\mathcal{DEC}$ from the IBE-IND-CCA experiment.

Next, we define anonymity experiment of IBE under CPA (IBE-ANO-CPA).

Definition 6 (IBE-ANO-CPA). *An IBE scheme is said to be IBE-ANO-CPA secure if the advantage is negligible for any PPT adversary $\mathcal{A}$, where*

$$Adv^{IBE\text{-}ANO\text{-}CPA}_{\Pi,\mathcal{A}}(1^\kappa) = \Big| \Pr\big[(pk, mk) \leftarrow \mathsf{IBE.Setup}(1^\kappa);$$

$$(ID^*_0, ID^*_1, M^*, State) \leftarrow \mathcal{A}^{\mathcal{EXTRACT}}(\mathsf{find}, pk); \ \mu \xleftarrow{\$} \{0,1\};$$

$$C^*_{IBE} \leftarrow \mathsf{IBE.Enc}(pk, ID^*_\mu, M^*); \ \mu' \leftarrow \mathcal{A}^{\mathcal{EXTRACT}}(\mathsf{guess}, C^*_{IBE}, State);$$

$$\mu = \mu'\big] - 1/2 \Big|$$

Note that ID^*_0 and ID^*_1 are not allowed as input to $\mathcal{EXTRACT}$.

Definition 7 (Anonymous IBE). *An IBE scheme is said to be anonymous IBE if the IBE scheme is both IBE-IND-CPA secure and IBE-ANO-CPA secure.*

3   Definitions of Adaptive SCF-PEKS

In this section, we define security requirements of SCF-PEKS. An SCF-PEKS scheme Π consists of five algorithms, SCF-PEKS.KeyGen$_S$, SCF-PEKS.KeyGen$_R$, SCF-PEKS.Trapdoor, SCF-PEKS.Enc and SCF-PEKS.Test. The server public key pk_S and the server secret key sk_S are given by executing SCF-PEKS.KeyGen$_S(1^\kappa)$, and the receiver public key pk_R and the receiver secret key sk_R are given by executing SCF-PEKS.KeyGen$_R(1^\kappa)$, where $\kappa \in \mathbb{N}$ is the security parameter. For a keyword ω, a trapdoor t_ω is given by executing SCF-PEKS.Trapdoor(sk_R, ω), and a ciphertext λ is given by executing SCF-PEKS.Enc(pk_S, pk_R, ω). The server has a public/secret key pair (pk_S, sk_S), and a sender makes a ciphertext λ (which is encrypted by using a keyword ω', pk_S, and pk_R), and sends λ to the server. The server runs SCF-PEKS.Test$(\lambda, sk_S, t_\omega)$, whose output is 1 if $\omega = \omega'$, and 0 otherwise. Note that obviously SCF-PEKS implies PEKS (i.e., if sk_S is publicly opened and (t_ω, sk_S) is regarded as a trapdoor of PEKS, then such a function-downgraded SCF-PEKS is PEKS). Correctness is defined as follows: For all $(pk_S, sk_S) \leftarrow$ SCF-PEKS.KeyGen$_S(1^\kappa)$, all $(pk_R, sk_R) \leftarrow$ SCF-PEKS.KeyGen$_R(1^\kappa)$, and all $\omega \in \mathcal{K}$, SCF-PEKS.Test$(\lambda, sk_S, t_\omega) = 1$ holds, where $\lambda \leftarrow$ SCF-PEKS.Enc(pk_R, pk_S, ω), $t_\omega \leftarrow$ SCF-PEKS.Trapdoor(sk_R, ω), and $\mathcal{K}$ is a keyword space.

Next, we consider two security requirements "consistency" and "keyword privacy".

Definition 8 (Consistency). *The SCF-PEKS scheme is said to be computationally consistent if the advantage is negligible for any PPT adversary $\mathcal{A}$ in the following experiment.*

$$Adv_{\Pi,\mathcal{A}}^{SCF\text{-}PEKS\text{-}CONSIST}(1^\kappa) = \Pr\big[(pk_S, sk_S) \leftarrow \text{SCF-PEKS.KeyGen}_S(1^\kappa);$$
$$(pk_R, sk_R) \leftarrow \text{SCF-PEKS.KeyGen}_R(1^\kappa);\ (\omega, \omega') \leftarrow \mathcal{A}(pk_S, pk_R); \omega \neq \omega';$$
$$\lambda \leftarrow \text{SCF-PEKS.Enc}(pk_S, pk_R, \omega);\ t_{\omega'} \leftarrow \text{SCF-PEKS.Trapdoor}(sk_R, \omega');$$
$$\text{SCF-PEKS.Test}(\lambda, sk_S, t_{\omega'}) = 1\big]$$

Next, we define two security notions for keyword privacy, "indistinguishability against chosen keyword attack with the server's secret key" (IND-CKA-SSK for short) and "indistinguishability against chosen keyword attack with all trapdoors" (IND-CKA-AT for short). In the IND-CKA-SSK experiment, an adversary $\mathcal{A}$ is assumed to be a malicious server. Therefore, $\mathcal{A}$ is given the server's secret key sk_S, whereas $\mathcal{A}$ cannot obtain the receiver's secret key sk_R. Instead of obtaining sk_R, $\mathcal{A}$ can issue a query to a trapdoor oracle $\mathcal{TRAP}$, which for an input keyword ω, returns a trapdoor t_ω. Note that $\mathcal{A}$ cannot query the challenge keywords ω_0^* and ω_1^* to $\mathcal{TRAP}$. As in the definition of [26], $\mathcal{A}$ computes (pk_S, sk_S), and gives pk_S to the challenger. So, we omit sk_S in the following experiment.

Definition 9 (IND-CKA-SSK). *An SCF-PEKS scheme is said to be IND-CKA-SSK-secure if the advantage is negligible for any PPT adversary $\mathcal{A}$ in the following experiment.*

$$Adv_{\Pi,\mathcal{A}}^{IND\text{-}CKA\text{-}SSK}(1^\kappa) =$$
$$\big| \Pr\big[(pk_S, State) \leftarrow \mathcal{A}(1^\kappa);\ (pk_R, sk_R) \leftarrow \text{SCF-PEKS.KeyGen}_R(1^\kappa);$$
$$(\omega_0^*, \omega_1^*, State) \leftarrow \mathcal{A}^{\mathcal{TRAP}}(\text{find}, pk_R, State);\ \mu \xleftarrow{\$} \{0,1\};$$
$$\lambda^* \leftarrow \text{SCF-PEKS.Enc}(pk_S, pk_R, \omega_\mu^*); \mu' \leftarrow \mathcal{A}^{\mathcal{TRAP}}(\text{guess}, \lambda^*, State);$$
$$\mu = \mu'\big] - 1/2\big|$$

Remark: Note that, for our TRE construction, the adversarial server's key generation above is not required. That is, the weaker definition can be used, where $\mathcal{C}$ can run $(pk_S, sk_S) \leftarrow \text{SCF-PEKS.KeyGen}_S(1^\kappa)$, and sends (pk_S, sk_S) to $\mathcal{A}$ in our proof of Theorem 2.

Next, we define the adaptive-IND-CKA-AT experiment. In this experiment, an adversary $\mathcal{A}$ is assumed to be a malicious-but-legitimate receiver or outsider. Therefore, $\mathcal{A}$ is given the receiver's secret key sk_R, whereas $\mathcal{A}$ cannot obtain the server's secret key sk_S. This means that $\mathcal{A}$ knows *all* trapdoors. $\mathcal{A}$ can issue a query to a test oracle $\mathcal{TEST}$, which for an input (λ, t_ω) which satisfies $(\lambda, t_\omega) \notin \{(\lambda^*, t_{\omega_0^*}), (\lambda^*, t_{\omega_1^*})\}$, returns the result of the test algorithm. As in the definition of [26], $\mathcal{A}$ computes (pk_R, sk_R), and gives pk_R to the challenger. So, we omit sk_R in the following experiment.

Definition 10 (Adaptive-IND-CKA-AT). *An SCF-PEKS scheme is said to be adaptive-IND-CKA-AT-secure if the advantage is negligible for any PPT adversary $\mathcal{A}$ in the following experiment.*

$$Adv_{\Pi,\mathcal{A}}^{Adaptive\text{-}IND\text{-}CKA\text{-}AT}(1^\kappa) =$$
$$\big| \Pr\big[(pk_S, sk_S) \leftarrow \text{SCF-PEKS.KeyGen}_S(1^\kappa);\ (pk_R, State) \leftarrow \mathcal{A}(1^\kappa);$$
$$(\omega_0^*, \omega_1^*, State) \leftarrow \mathcal{A}^{\mathcal{TEST}}(\text{find}, pk_S, State);\ \mu \xleftarrow{\$} \{0,1\};$$
$$\lambda^* \leftarrow \text{SCF-PEKS.Enc}(pk_S, pk_R, \omega_\mu^*); \mu' \leftarrow \mathcal{A}^{\mathcal{TEST}}(\text{guess}, \lambda^*, State);$$
$$\mu = \mu'\big] - 1/2 \big|$$

Remark: As in the IND-CKA-SSK, for TRE construction, the adversarial receiver's key generation above is not required. That is, we use the weaker definition, where $\mathcal{C}$ can run $(pk_R, sk_R) \leftarrow \text{SCF-PEKS.KeyGen}_R(1^\kappa)$, and sends (pk_R, sk_R) to $\mathcal{A}$ in our proof of Theorem 1.

4   Adaptive SCF-PEKS Implies TRE

4.1   Proposed TRE Construction Based on Adaptive SCF-PEKS

In this section, we propose a generic construction of TRE (with 1-bit plaintext space) based on adaptive SCF-PEKS. Our construction adopts the Boneh et al. IBE construction from PEKS [5], namely, for a plaintext 0 (resp. 1) and a release-time T, the time-release key is a trapdoor of the keyword $T\|0$ (resp. $T\|1$). In the following construction, a SCF-PEKS receiver is regarded as a TS, and a SCF-PEKS server is regarded as a TRE receiver. We set $\mathcal{T} = \mathcal{K}$ and $\mathcal{M}_{TRE} = \{0,1\}$.

Protocol 1 (TRE based on adaptive SCF-PEKS)

TRE.Setup(1^κ) : *Run* $(pk_R, sk_R) \leftarrow \text{SCF-PEKS.KeyGen}_R(1^\kappa)$, *set* prm $= pk_R$ *and* msk $= sk_R$, *and return* prm *and* msk.

TRE.UKG(1^κ) : *Run* $(pk_S, sk_S) \leftarrow \text{SCF-PEKS.KeyGen}_S(1^\kappa)$, *set* $pk_u = pk_S$ *and* $sk_u = sk_S$, *and return* pk_u *and* sk_u.

TRE.Ext(prm, msk, T) : *Run* $t_{T0} \leftarrow \text{SCF-PEKS.Trapdoor}(\text{msk}, T\|0)$ *and* $t_{T1} \leftarrow \text{SCF-PEKS.Trapdoor}(\text{msk}, T\|1)$, *set* $s_T = (t_{T0}, t_{T1})$, *and return* s_T.

TRE.Enc(prm, pk_u, T, M) : *For* $M \in \{0,1\}$, *run* $\lambda \leftarrow \text{SCF-PEKS.Enc}(\text{prm}, pk_u, T\|M)$, *set* $C = \lambda$, *and return* C.

TRE.Dec(prm, sk_u, s_T, C) :   *Parse* s_T $=$ (t_{T0}, t_{T1}).   *If* SCF-PEKS.Test(C, sk_u, t_{T0}) $= 1$ *holds, then output* $M = 0$. *Else if* SCF-PEKS.Test(C, sk_u, t_{T1}) $= 1$ *holds, then output* $M = 1$. *Otherwise, output* $\perp$.

Obviously, correctness holds if the underlying SCF-PEKS satisfies correctness.

4.2   Security Analysis of Our TRE Construction

Theorem 1. *Our TRE construction satisfies IND-TR-CCA$_{TS}$ if the underlying SCF-PEKS satisfies adaptive IND-CKA-AT and consistency.*

Proof: Let $\mathcal{A}$ be an adversary who can break the IND-TR-CCA$_{TS}$ security of our TRE construction, and $\mathcal{C}$ be the challenger of the adaptive IND-CKA-AT game. Then we construct an algorithm $\mathcal{B}$ which can break the adaptive IND-CKA-AT security (or consistency) of the underlying SCF-PEKS. First, $\mathcal{C}$ runs $(pk_R, sk_R) \leftarrow$ SCF-PEKS.KeyGen$_R(1^\kappa)$ and $(pk_S, sk_S) \leftarrow$ SCF-PEKS.KeyGen$_S(1^\kappa)$, and sends (pk_R, sk_R, pk_S) to $\mathcal{B}$. $\mathcal{B}$ sets prm $= pk_R$, msk $= sk_R$, and $pk_u = pk_S$, and sends (prm, msk, pk_u) to $\mathcal{A}$.

Phase 1: For a decryption query (C, T), $\mathcal{B}$ issues two test queries $(C, T\|0)$ and $(C, T\|1)$ to $\mathcal{C}$. If $\mathcal{C}$ answers 0 for both queries, then $\mathcal{B}$ answers $\perp$. If $\mathcal{C}$ answers 1 for both queries, $\mathcal{B}$ can break consistency and aborts. Else, $\mathcal{C}$ answers 1 for the query $(C, T\|M)$ $(M \in \{0, 1\})$. Then $\mathcal{B}$ answers M.

Challenge: $\mathcal{A}$ sends (M_0^*, M_1^*, T^*) to $\mathcal{B}$. W.l.o.g, we set $M_0^* = 0$ and $M_1^* = 1$. $\mathcal{B}$ sends $(T^*\|M_0^*, T^*\|M_1^*) = (T^*\|0, T^*\|1)$ to $\mathcal{C}$ as the challenge keywords. $\mathcal{C}$ sends λ^* to $\mathcal{B}$. $\mathcal{B}$ sets $C^* = \lambda^*$, and sends C^* to $\mathcal{A}$. Note that C^* is a TRE ciphertext against either M_0^* or M_1^*.

Phase 2: For a decryption query $(C, T) \neq (C^*, T^*)$, $\mathcal{B}$ issues two test queries $(C, T\|0)$ and $(C, T\|1)$ to $\mathcal{C}$. If $\mathcal{C}$ answers 0 for both queries, then $\mathcal{B}$ answers $\perp$. If $\mathcal{C}$ answers 1 for both queries, $\mathcal{B}$ can break consistency and aborts. Else, $\mathcal{C}$ answers 1 for the query $(C, T\|M)$ $(M \in \{0, 1\})$. Then $\mathcal{B}$ answers M.

Guess: Finally, $\mathcal{A}$ outputs the guessing bit $\mu' \in \{0, 1\}$. $\mathcal{B}$ outputs μ' as the guessing bit of the adaptive IND-CKA-AT game.   □

Theorem 2. *Our TRE construction satisfies IND-TR-CPA$_{IS}$ if the underlying SCF-PEKS satisfies IND-CKA-SSK.*

Proof: Let $\mathcal{A}$ be an adversary who can break the IND-TR-CPA$_{IS}$ security of our TRE construction, and $\mathcal{C}$ be the challenger of the IND-CKA-SSK game. Then we construct an algorithm $\mathcal{B}$ which can break the IND-CKA-SSK security of the underlying SCF-PEKS. First, $\mathcal{C}$ runs $(pk_R, sk_R) \leftarrow$ SCF-PEKS.KeyGen$_R(1^\kappa)$ and $(pk_S, sk_S) \leftarrow$ SCF-PEKS.KeyGen$_S(1^\kappa)$, and sends (pk_R, pk_S, sk_S) to $\mathcal{B}$. $\mathcal{B}$ sets prm $= pk_R$, $pk_u = pk_S$, and $sk_u = sk_S$, and sends (prm, pk_u, sk_u) to $\mathcal{A}$.

Phase 1: For an extraction query T, $\mathcal{B}$ issues two trapdoor queries $T\|0$ and $T\|1$. $\mathcal{C}$ sends t_{T0} and t_{T1} to $\mathcal{B}$. $\mathcal{B}$ sets $s_T = (t_{T0}, t_{T1})$, and sends s_T to $\mathcal{A}$.

Challenge: $\mathcal{A}$ sends (M_0^*, M_1^*, T^*) to $\mathcal{B}$. $\mathcal{B}$ sends $(T^*\|0, T^*\|1)$ to $\mathcal{C}$ as the challenge keywords. $\mathcal{C}$ sends λ^* to $\mathcal{B}$. $\mathcal{B}$ sets $C^* = \lambda^*$, and sends C^* to $\mathcal{A}$. Note that C^* is a TRE ciphertext against either M_0^* or M_1^*.

Phase 2: For an extraction query $T \neq T^*$, $\mathcal{B}$ issues two trapdoor queries $T\|0$ and $T\|1$. $\mathcal{C}$ sends t_{T0} and t_{T1} to $\mathcal{B}$. $\mathcal{B}$ sets $s_T = (t_{T0}, t_{T1})$, and sends s_T to $\mathcal{A}$.

Guess: Finally, $\mathcal{A}$ outputs the guessing bit $\mu' \in \{0, 1\}$. $\mathcal{B}$ outputs μ' as the guessing bit of the IND-CKA-SSK game.   □

4.3 Discussion: The Reason Why PEKS and Non-adaptive SCF-PEKS Are Not Suitable for Constructing TRE

First, we make it clear that we do not deny the possibility of TRE construction based on either PEKS or non-adaptive SCF-PEKS in the following discussion. But we observe that TRE requires two entities, called TS and receiver, and these entities have their public/secret key pair, respectively. So, it is hard to directly implement TRE from PEKS since PEKS requires just one entity (i.e., receiver). From the above considerations, SCF-PEKS is suitable for constructing TRE, since SCF-PEKS requires two entities, called server and receiver. Next, we need to implement the oracles defined in TRE security requirements in the SCF-PEKS context. The extraction query (in the IND-TR-CPA$_{IS}$ experiment) can be implemented by the trapdoor oracle (IND-CKA-SSK) in the non-adaptive SCF-PEKS context. However, the decryption query (in the IND-TR-CCA$_{TS}$ experiment) is hard to be implemented in the "non-adaptive" SCF-PEKS context, since no test query is considered in the IND-CKA-AT experiment. On the contrary, the decryption query can be handled by the test oracle in the adaptive SCF-PEKS context. This is the reason why we apply SCF-PEKS with adaptive security for constructing TRE. Note that although decryptable PEKS [12,13,17] might handle the decryption query, it requires just one entity, and therefore it is hard to directly implement TRE from decryptable PEKS. As a remark, IND-TR-CPA$_{TS}$ secure TRE can be constructed from non-adaptive SCF-PEKS from the above considerations.

5   Anonymous IBE Implies Adaptive SCF-PEKS

5.1 Proposed Adaptive SCF-PEKS Construction

In this section, we give a generic construction of adaptive SCF-PEKS based on anonymous IBE, IND-stag-CCA TBE, and sUF OTS. In our construction, a ciphertext of an anonymous IBE scheme (say C_{IBE}) is used as a "plaintext" of a TBE scheme to hide keyword information from an adversary. From the result of the decryption of the TBE scheme, the ciphertext C_{IBE} must be obtained. In addition, usually, $C_{IBE} \notin \mathcal{M}_{TBE}$. To handle this condition, we apply the KEM/DEM framework [28] (a.k.a. hybrid encryption), where KEM stands for key encapsulation mechanism, and DEM stands for data encapsulation mechanism. By using TBE KEM (see Section 6 of [20]), compute $(K, C_{TBE}) \leftarrow \mathsf{TBE.Enc}(pk, t)$, and encrypt C_{IBE} as a plaintext of the CCA secure DEM such that $e = \mathsf{E}_K(C_{IBE})$. Note that a CCA-secure DEM can be generically constructed from any pseudorandom functions without redundancy. So, even if we assume that a CCA secure DEM exists, we do not need any additional cryptographic primitive, except anonymous IBE, for constructing adaptive SCF-PEKS. From now on, we assume that $C_{IBE} \in \mathcal{M}_{TBE}$ and $e = \mathsf{E}_K(C_{IBE})$ is implicitly included in C_{TBE} (i.e., C_{IBE} is obtained from the decryption of C_{TBE}).

In the following construction, we use a target collision resistant (TCR) hash function [3] $H_{tag} : \{0, 1\}^* \rightarrow \mathcal{TAG}$.

Protocol 2 (Our Construction of Adaptive SCF-PEKS)

SCF-PEKS.KeyGen$_S(1^\kappa)$: *Run* $(pk_S, sk_S) \leftarrow$ TBE.KeyGen(1^κ), *and output* $(pk_S,$ $sk_S)$.

SCF-PEKS.KeyGen$_R(1^\kappa)$: *Run* $(pk_R, sk_R) \leftarrow$ IBE.KeyGen(1^κ), *and output* $(pk_R,$ $sk_R)$.

SCF-PEKS.Trapdoor(sk_R, ω): *Run* $t_\omega \leftarrow$ IBE.Extract(sk_R, ω), *and output* t_ω.

SCF-PEKS.Enc(pk_S, pk_R, ω): *Generate* $(K_s, K_v) \xleftarrow{} $ Sig.KeyGen, *compute* $t =$ $H_{tag}(K_v)$, *choose* $R \xleftarrow{\$} \mathcal{M}_{IBE}$, *run* $C_{IBE} \leftarrow$ IBE.Enc(pk_R, ω, R), $C_{TBE} \leftarrow$ TBE.Enc(pk_S, t, C_{IBE}), *and* $\sigma \leftarrow$ Sign$(K_s, (C_{TBE}, R))$, *and output* $\lambda =$ (C_{TBE}, K_v, σ).

SCF-PEKS.Test$(\lambda, sk_S, t_\omega)$: *Let* $\lambda = (C_{TBE}, K_v, \sigma)$. *Compute* $t = H_{tag}(K_v)$, *run* $C'_{IBE} \leftarrow$ TBE.Dec(sk_S, t, C_{TBE}) *and* $R' \leftarrow$ IBE.Dec(t_ω, C'_{IBE}). *Output 1 if* $1 =$Verify$(K_v, \sigma, (C_{TBE}, R'))$, *and 0 otherwise.*

Obviously, correctness holds if the underlying TBE, IBE, and OTS satisfy correctness.

By observing our construction, non-adaptive SCF-PEKS (i.e., IND-CKA-AT without test queries, which has the same security requirement with Fang et al. [11]) can be constructed by reducing the one-time signature part and replacing the TBE part with CPA-secure PKE (i.e., chosen plaintext security is enough). A ciphertext is (C_{PKE}, R), where $C_{IBE} \leftarrow$ IBE.Enc(pk_R, ω, R) and $C_{PKE} \leftarrow$ PKE.Enc(pk_S, C_{IBE}). As in our adaptive SCF-PEKS construction, we assume that $C_{IBE} \in \mathcal{M}_{PKE}$, where $\mathcal{M}_{PKE}$ is the message space of the underlying PKE scheme. The test procedure is described as follows. Compute $C'_{IBE} \leftarrow$ PKE.Dec(sk_S, C_{PKE}) and $R' \leftarrow$ IBE.Dec(t_ω, C'_{IBE}). Output 1 if $R' = R$, and 0 otherwise.

5.2   Security Analysis of Our Adaptive SCF-PEKS Construction

Theorem 3. *The SCF-PEKS scheme constructed by our method is computationally consistent if the underlying IBE scheme is IBE-IND-CPA secure.*

Proof: Let $\mathcal{A}$ be an adversary who breaks the computational consistency of SCF-PEKS constructed by the protocol 1, and $\mathcal{C}$ be the challenger of the IBE-IND-CPA experiment. Then, we can construct an algorithm $\mathcal{B}$ that breaks the IBE-IND-CPA security of the IBE scheme. First, $\mathcal{C}$ runs IBE.Setup(1^κ), and gives pk to $\mathcal{B}$. $\mathcal{B}$ sets pk as pk_R, runs $(pk_S, sk_S) \leftarrow$ TBE.KeyGen(1^κ), and gives (pk_R, pk_S) to $\mathcal{A}$. $\mathcal{B}$ obtains keywords ω and ω' from $\mathcal{A}$. $\mathcal{B}$ chooses $R_0, R_1 \xleftarrow{\$} \mathcal{M}_{IBE}$ as the challenge messages, and sends (ω, R_0, R_1) to $\mathcal{C}$. $\mathcal{C}$ gives $C^*_{IBE} \leftarrow$ IBE.Enc(pk_R, ω, R_μ) to $\mathcal{B}$, where $\mu \in \{0, 1\}$. $\mathcal{B}$ gets a trapdoor $t_{\omega'}$ by issuing an $\mathcal{EXTRACT}$ query. If IBE.Dec$(t_{\omega'}, C^*_{IBE}) = R_1$, then $\mathcal{B}$ outputs 1, otherwise $\mathcal{B}$ outputs 0. $\qquad\square$

Theorem 4. *The SCF-PEKS scheme constructed by our method is IND-CKA-SSK secure if the underlying IBE scheme is IBE-ANO-CPA secure.*

Proof: Let $\mathcal{A}$ be an adversary who breaks the IND-CKA-SSK security of SCF-PEKS constructed by the protocol 1, and $\mathcal{C}$ be the challenger of the IBE-ANO-CPA experiment. Then we can construct an algorithm $\mathcal{B}$ that breaks the IBE-ANO-CPA security of the underlying IBE scheme. First, $\mathcal{C}$ runs IBE.Setup(1^κ), and gives pk to $\mathcal{B}$. $\mathcal{B}$ sets pk as pk_R. $\mathcal{A}$ runs $(pk_S, sk_S) \leftarrow$ TBE.KeyGen(1^κ), and gives pk_S to $\mathcal{B}$. For a $\mathcal{TRAP}$ query ω_i, $\mathcal{B}$ forwards ω_i to $\mathcal{C}$ as an $\mathcal{EXTRACT}$ query of the IBE scheme, gets t_{ω_i}, and answers t_{ω_i} to $\mathcal{A}$.

In the Challenge phase, $\mathcal{A}$ sends the challenge keywords ω_0^* and ω_1^* to $\mathcal{B}$. $\mathcal{B}$ chooses $R^* \xleftarrow{\$} \mathcal{M}_{IBE}$, and computes the challenge ciphertext as follows:

1. $\mathcal{B}$ sends $(R^*, \omega_0^*, \omega_1^*)$ to $\mathcal{C}$.
2. $\mathcal{C}$ gives $C_{IBE}^* \leftarrow$ IBE.Enc(pk_R, ω_μ^*, R^*) to $\mathcal{B}$, where $\mu \in \{0,1\}$.
3. $\mathcal{B}$ generates $(K_s^*, K_v^*) \xleftarrow{\$}$ Sig.KeyGen, and computes $t^* = H_{tag}(K_v^*)$, $C_3^* \leftarrow$ TBE.Enc(pk_S, t^*, C_{IBE}^*), and $\sigma^* \leftarrow$ Sign($K_s^*, (C_{TBE}^*, R^*)$).
4. $\mathcal{B}$ sends $\lambda^* = (C_{TBE}^*, K_v^*, \sigma^*)$ to $\mathcal{A}$.

Note that $\mathcal{A}$ can compute $C_{IBE}^* \leftarrow$ TBE.Dec($sk_S, H_{tag}(K_v^*), C_{TBE}^*$). In addition, R^* may be revealed from σ^* without contradicting unforgeability property. However, this situation is the same as in the IBE-ANO-CPA experiment, where $\mathcal{A}$ inputs $ID_0^* := \omega_0^*$, $ID_1^* := \omega_1^*$, and $M^* := R^*$, and gets the challenge ciphertext C_{IBE}^*. Finally, $\mathcal{B}$ outputs μ', where $\mu' \in \{0,1\}$ is the output of $\mathcal{A}$.   □

Theorem 5. *The SCF-PEKS scheme constructed by our method is adaptive-IND-CKA-AT secure if the underlying TBE scheme is IND-stag-CCA secure, the underlying signature is one-time sUF-CMA secure, and H_{tag} is a TCR hash function.*

Proof: Let $\mathcal{A}$ be an adversary who breaks the adaptive-IND-CKA-AT security of SCF-PEKS constructed by the protocol 1, and $\mathcal{C}$ be the challenger of the IND-stag-CCA experiment. Then, we can construct an algorithm $\mathcal{B}$ that breaks the IND-stag-CCA security of the underlying TBE scheme. First, $\mathcal{B}$ runs $(K_s^*, K_v^*) \leftarrow$ Sig.KeyGen(1^κ), and sends $t^* := H_{tag}(K_v^*)$ to $\mathcal{C}$ as the challenge tag. $\mathcal{C}$ runs TBE.KeyGen(1^κ), and gives pk to $\mathcal{B}$. $\mathcal{B}$ sets pk as pk_S. $\mathcal{A}$ runs $(pk_R, sk_R) \leftarrow$ IBE.Setup(1^κ), and gives pk_R to $\mathcal{B}$. Let (SCF-PEKS.Enc(pk_S, pk_R, ω_j) := (C_{TBE}, K_v, σ), t_{ω_j}) be a $\mathcal{TEST}$ query, where $\omega_j \in \mathcal{ID}$. $\mathcal{B}$ computes $t = H_{tag}(K_v)$, and answers as follows:

$t \neq t^*$: $\mathcal{B}$ can use the $\mathcal{DEC}$ oracle of the underlying TBE scheme as follows.
1. $\mathcal{B}$ forwards (C_{TBE}, t) to $\mathcal{C}$ as a $\mathcal{DEC}$ query of the TBE scheme.
2. $\mathcal{C}$ answers $C_{IBE}' \leftarrow$ TBE.Dec(sk, t, C_{TBE}).
   - Note that if t is not the legitimate tag of C_{TBE}, then $\mathcal{C}$ answers $\bot$. In this case, $\mathcal{B}$ answers 0.
3. $\mathcal{B}$ computes $R' \leftarrow$ IBE.Dec(t_{ω_j}, C_{IBE}').
4. If Verify($K_v, \sigma, (C_{TBE}, R')$) = 1, then $\mathcal{B}$ returns 1, and 0 otherwise.

$t = t^*$: If $K_v \neq K_v^*$, then $\mathcal{B}$ breaks the TCR property of H_{tag}. If $K_v = K_v^*$ (we call this a forge$_1$ event), then $\mathcal{B}$ gives a random answer in $\mathcal{C}$, and aborts.

In the Challenge phase, $\mathcal{A}$ sends the challenge keywords ω_0^* and ω_1^* to $\mathcal{B}$. $\mathcal{B}$ chooses $R^* \xleftarrow{\$} \mathcal{M}_{IBE}$, and computes the challenge ciphertext as follows:

1. $\mathcal{B}$ computes $C_{IBE,0} \leftarrow$ IBE.Enc(pk_R, ω_0^*, R^*) and $C_{IBE,1} \leftarrow$ IBE.Enc(pk_R, ω_1^*, R^*).
2. $\mathcal{B}$ sends $(M_0^*, M_1^*) := (C_{IBE,0}, C_{IBE,1})$ to $\mathcal{C}$ as the challenge messages of the IND-stag-CCA experiment of the TBE scheme.
3. $\mathcal{C}$ gives $C_{TBE}^* \leftarrow$ TBE.Enc(pk_S, t^*, M_μ^*) to $\mathcal{B}$.
4. $\mathcal{B}$ computes $\sigma^* \leftarrow$ Sign$(K_s^*, (C_{TBE}^*, R^*))$, and sends $\lambda^* = (C_{TBE}^*, K_v^*, \sigma^*)$ to $\mathcal{A}$.

Again, let (SCF-PEKS.Enc$(pk_S, pk_R, \omega_j) := (C_{TBE}, K_v, \sigma), t_{\omega_j})$ be a $\mathcal{TEST}$ query, where $\omega_j \in \mathcal{ID}$. $\mathcal{B}$ computes $t = H_{tag}(K_v)$, and answers as follows:

In the case $t_{\omega_j} \in \{t_{\omega_0^*}, t_{\omega_1^*}\}$:

$t = t^*$: If $K_v \neq K_v^*$, then $\mathcal{B}$ breaks the TCR property of H_{tag}. If $K_v = K_v^*$ (we call this a forge$_2$ event), then $\mathcal{B}$ gives a random answer in $\mathcal{C}$, and aborts.

$t \neq t^*$: Then $\mathcal{B}$ can use the $\mathcal{DEC}$ oracle of the underlying TBE scheme as follows. .

  1. $\mathcal{B}$ forwards (C_{TBE}, t) to $\mathcal{C}$ as a $\mathcal{DEC}$ query of the TBE scheme.
  2. $\mathcal{C}$ answers $C'_{IBE} \leftarrow$ TBE.Dec(sk, t, C_{TBE}).
     – Note that if t is not the legitimate tag of C_{TBE}, then $\mathcal{C}$ answers $\bot$. In this case, $\mathcal{B}$ answers 0.
  3. $\mathcal{B}$ computes $R' \leftarrow$ IBE.Dec(t_{ω_j}, C'_{IBE}).
  4. If Verify$(K_v, \sigma, (C_{TBE}, R')) = 1$, then $\mathcal{B}$ returns 1, and 0 otherwise.

In the case $t_{\omega_j} \notin \{t_{\omega_0^*}, t_{\omega_1^*}\}$:

$(C_{TBE}, K_v, \sigma) = (C_{TBE}^*, K_v^*, \sigma^*)$: $\mathcal{B}$ returns 0, since $(C_{TBE}^*, K_v^*, \sigma^*)$ is an SCF-PEKS ciphertext of either ω_0^* or ω_1^*.

$(C_{TBE}, K_v, \sigma) \neq (C_{TBE}^*, K_v^*, \sigma^*)$: $\mathcal{B}$ runs the same simulation as in the find stage.

If $\mathcal{B}$ does not abort, then our simulation is perfect. Finally, $\mathcal{B}$ outputs μ', where $\mu' \in \{0, 1\}$ is the output of $\mathcal{A}$.

Next, we prove that $\Pr[\text{forge}] := \Pr[\text{forge}_1 \vee \text{forge}_2]$ is negligible. We construct an algorithm $\mathcal{B}'$ which can win the sUF game with probability at least $\Pr[\text{forge}]$. $\mathcal{B}'$ obtains K_v^* from the sUF challenger, instead of executing Sig.KeyGen(1^κ). $\mathcal{B}'$ runs $(pk_S, sk_S) \leftarrow$ TBE.KeyGen(1^κ), and gives pk_S to $\mathcal{A}$. $\mathcal{A}$ runs $(pk_R, sk_R) \leftarrow$ IBE.Setup(1^κ), and gives pk_R to $\mathcal{B}$. Since $\mathcal{B}'$ has sk_S, $\mathcal{B}'$ can answer any $\mathcal{TEST}$ queries. In the challenge phase of the adaptive-IND-CKA-AT experiment, $\mathcal{B}'$ computes $t^* = H_{tag}(K_v^*)$, chooses $R^* \xleftarrow{\$} \mathcal{M}_{IBE}$, runs $C_{IBE}^* \leftarrow$ IBE.Enc(pk_R, ω_μ, R), and $C_{TBE}^* \leftarrow$ TBE.Enc(pk_S, t^*, C_{IBE}^*), sets $M^* := (C_{TBE}^*, R^*)$, sends M^* to the sUF challenger, and obtains σ^* from the sUF challenger. Therefore, $\mathcal{B}'$ makes at most one signature query. Note that we do not have to care about the value $\mu \in \{0, 1\}$, since we only have to guarantee that $\lambda^* = (C_{TBE}^*, K_v^*, \sigma^*)$ is a valid SCF-PEKS ciphertext. In the forge events, $\mathcal{A}$ sends a $\mathcal{TEST}$ query $((C_{TBE}, K_v, \sigma), t_{\omega_j})$ with $K_v = K_v^*$.

forge$_1$: In this case, $\mathcal{B}'$ can obtain a signature without issuing the signature query. $\mathcal{B}'$ computes $C_{IBE} \leftarrow$ TBE.Dec$(sk_S, H_{tag}(K_v), C_{TBE})$ and $R' \leftarrow$ IBE.Dec(t_{ω_j}, C_{IBE}). If $((C_{TBE}, R'), \sigma)$ is not a valid signature pair, then $\mathcal{B}'$ returns 0 as the answer of this $\mathcal{TEST}$ query. Otherwise, if $((C_{TBE}, R'), \sigma)$ is a valid signature pair, then $\mathcal{B}'$ submits a forged pair $((C_{TBE}, R'), \sigma)$ to the sUF challenger and wins.

forge$_2$: Now $t_{\omega_j} \in \{t_{\omega_0^*}, t_{\omega_1^*}\}$. Then $(C_{TBE}, \sigma) \neq (C_{TBE}^*, \sigma^*)$. $\mathcal{B}'$ computes $C_{IBE} \leftarrow$ TBE.Dec$(sk_S, H_{tag}(K_v), C_{TBE})$ and $R' \leftarrow$ IBE.Dec(t_{ω_j}, C_{IBE}). If $((C_{TBE}, R'), \sigma)$ is not a valid signature pair, then $\mathcal{B}'$ returns 0 as the answer of this $\mathcal{TEST}$ query. Otherwise, if $((C_{TBE}, R'), \sigma)$ is a valid signature pair, then $\mathcal{B}'$ submits a forged pair $((C_{TBE}, R'), \sigma)$ to the sUF challenger and wins.

Therefore, $\Pr[\text{forge}] := \Pr[\text{forge}_1 \vee \text{forge}_2]$ is negligible, since the underlying signature is sUF. □

6 Conclusion

In this paper, to show the relationships of IBE, TRE, and adaptive SCF-PEKS, we propose a generic construction of TRE with 1-bit plaintext space (resp. adaptive SCF-PEKS) from adaptive SCF-PEKS (resp. anonymous IBE). Our first result seems interesting since no bridge between TRE and PEKS primitive has been known before. In addition, no generic construction of SCF-PEKS has been proposed so far. That is, our second construction also seems independently interesting.

As future works, it is interesting to consider the keyword guessing attacks [18,29], namely, if adaptive SCF-PEKS can handle keyword guessing attack, then what happens in the TRE context. In addition, we expect that the wildcard searching capability [27] might lead to a construction of time-specific encryption [25], where the time "interval" can be specified. Finally, a construction of TRE with multi-bit plaintext space from adaptive SCF-PEKS needs to be revisited.

References

1. Abdalla, M., Bellare, M., Catalano, D., Kiltz, E., Kohno, T., Lange, T., Malone-Lee, J., Neven, G., Paillier, P., Shi, H.: Searchable encryption revisited: Consistency properties, relation to anonymous IBE, and extensions. J. Cryptology 21(3), 350–391 (2008)
2. Baek, J., Safavi-Naini, R., Susilo, W.: Public key encryption with keyword search revisited. In: Gervasi, O., Murgante, B., Laganà, A., Taniar, D., Mun, Y., Gavrilova, M.L. (eds.) ICCSA 2008, Part I. LNCS, vol. 5072, pp. 1249–1259. Springer, Heidelberg (2008)
3. Bellare, M., Rogaway, P.: Collision-resistant hashing: Towards making UOWHFs practical. In: Kaliski Jr., B.S. (ed.) CRYPTO 1997. LNCS, vol. 1294, pp. 470–484. Springer, Heidelberg (1997)

4. Bellare, M., Shoup, S.: Two-tier signatures, strongly unforgeable signatures, and Fiat-Shamir without random oracles. In: Okamoto, T., Wang, X. (eds.) PKC 2007. LNCS, vol. 4450, pp. 201–216. Springer, Heidelberg (2007)

5. Boneh, D., Di Crescenzo, G., Ostrovsky, R., Persiano, G.: Public key encryption with keyword search. In: Cachin, C., Camenisch, J.L. (eds.) EUROCRYPT 2004. LNCS, vol. 3027, pp. 506–522. Springer, Heidelberg (2004)

6. Canetti, R., Halevi, S., Katz, J.: Chosen-ciphertext security from identity-based encryption. In: Cachin, C., Camenisch, J.L. (eds.) EUROCRYPT 2004. LNCS, vol. 3027, pp. 207–222. Springer, Heidelberg (2004)

7. Cathalo, J., Libert, B., Quisquater, J.J.: Efficient and non-interactive timed-release encryption. In: Qing, S., Mao, W., López, J., Wang, G. (eds.) ICICS 2005. LNCS, vol. 3783, pp. 291–303. Springer, Heidelberg (2005)

8. Cheon, J.H., Hopper, N., Kim, Y., Osipkov, I.: Provably secure timed-release public key encryption. ACM Trans. Inf. Syst. Secur. 11(2) (2008)

9. Cui, Y., Fujisaki, E., Hanaoka, G., Imai, H., Zhang, R.: Formal security treatments for IBE-to-signature transformation: Relations among security notions. IEICE Transactions 92-A(1), 53–66 (2009)

10. Dent, A.W., Tang, Q.: Revisiting the security model for timed-release encryption with pre-open capability. In: Garay, J.A., Lenstra, A.K., Mambo, M., Peralta, R. (eds.) ISC 2007. LNCS, vol. 4779, pp. 158–174. Springer, Heidelberg (2007)

11. Fang, L., Susilo, W., Ge, C., Wang, J.: A secure channel free public key encryption with keyword search scheme without random oracles. In: Garay, J.A., Miyaji, A., Otsuka, A. (eds.) CANS 2009. LNCS, vol. 5888, pp. 248–258. Springer, Heidelberg (2009)

12. Fang, L., Wang, J., Ge, C., Ren, Y.: Decryptable public key encryption with keyword search schemes. JDCTA 4(9), 141–150 (2010)

13. Fuhr, T., Paillier, P.: Decryptable searchable encryption. In: Susilo, W., Liu, J.K., Mu, Y. (eds.) ProvSec 2007. LNCS, vol. 4784, pp. 228–236. Springer, Heidelberg (2007)

14. Fujioka, A., Okamoto, Y., Saito, T.: Generic construction of strongly secure timed-release public-key encryption. In: Parampalli, U., Hawkes, P. (eds.) ACISP 2011. LNCS, vol. 6812, pp. 319–336. Springer, Heidelberg (2011)

15. Gu, C., Zhu, Y.: New efficient searchable encryption schemes from bilinear pairings. International Journal of Network Security 10(1), 25–31 (2010)

16. Gu, C., Zhu, Y., Pan, H.: Efficient public key encryption with keyword search schemes from pairings. In: Pei, D., Yung, M., Lin, D., Wu, C. (eds.) Inscrypt 2007. LNCS, vol. 4990, pp. 372–383. Springer, Heidelberg (2008)

17. Hofheinz, D., Weinreb, E.: Searchable encryption with decryption in the standard model. Cryptology ePrint Archive, Report 2008/423 (2008), http://eprint.iacr.org/

18. Jeong, I.R., Kwon, J.O., Hong, D., Lee, D.H.: Constructing PEKS schemes secure against keyword guessing attacks is possible? Computer Communications 32(2), 394–396 (2009)

19. Khader, D.: Public key encryption with keyword search based on k-resilient IBE. In: Gervasi, O., Gavrilova, M.L. (eds.) ICCSA 2007, Part III. LNCS, vol. 4707, pp. 1086–1095. Springer, Heidelberg (2007)

20. Kiltz, E.: Chosen-ciphertext security from tag-based encryption. In: Halevi, S., Rabin, T. (eds.) TCC 2006. LNCS, vol. 3876, pp. 581–600. Springer, Heidelberg (2006)

21. Matsuda, T., Nakai, Y., Matsuura, K.: Efficient generic constructions of timed-release encryption with pre-open capability. In: Joye, M., Miyaji, A., Otsuka, A. (eds.) Pairing 2010. LNCS, vol. 6487, pp. 225–245. Springer, Heidelberg (2010)
22. May, T.C.: Time-release crypto (1993) (unpublished manuscript)
23. Myers, S., Shelat, A.: Bit encryption is complete. In: FOCS, pp. 607–616 (2009)
24. Nakai, Y., Matsuda, T., Kitada, W., Matsuura, K.: A generic construction of timed-release encryption with pre-open capability. In: Takagi, T., Mambo, M. (eds.) IWSEC 2009. LNCS, vol. 5824, pp. 53–70. Springer, Heidelberg (2009)
25. Paterson, K.G., Quaglia, E.A.: Time-specific encryption. In: Garay, J.A., De Prisco, R. (eds.) SCN 2010. LNCS, vol. 6280, pp. 1–16. Springer, Heidelberg (2010)
26. Rhee, H.S., Park, J.H., Susilo, W., Lee, D.H.: Improved searchable public key encryption with designated tester. In: ASIACCS 2009, pp. 376–379 (2009)
27. Sedghi, S., van Liesdonk, P., Nikova, S., Hartel, P., Jonker, W.: Searching keywords with wildcards on encrypted data. In: Garay, J.A., De Prisco, R. (eds.) SCN 2010. LNCS, vol. 6280, pp. 138–153. Springer, Heidelberg (2010)
28. Shoup, V.: Using hash functions as a hedge against chosen ciphertext attack. In: Preneel, B. (ed.) EUROCRYPT 2000. LNCS, vol. 1807, pp. 275–288. Springer, Heidelberg (2000)
29. Yau, W.C., Heng, S.H., Goi, B.M.: Off-line keyword guessing attacks on recent public key encryption with keyword search schemes. In: Rong, C., Jaatun, M.G., Sandnes, F.E., Yang, L.T., Ma, J. (eds.) ATC 2008. LNCS, vol. 5060, pp. 100–105. Springer, Heidelberg (2008)

The n-Diffie-Hellman Problem and Its Applications

Liqun Chen[1] and Yu Chen[2,3]

[1] Hewlett-Packard Laboratories, Bristol, UK
liqun.chen@hp.com
[2] School of Computer Science, Peking University, Beijing, China
[3] Institute of Information Engineering, Chinese Academy of Sciences
cycosmic@gmail.com

Abstract. The main contributions of this paper are twofold. On the one hand, the twin Diffie-Hellman (twin DH) problem proposed by Cash, Kiltz and Shoup is extended to the n-Diffie-Hellman (n-DH) problem for an arbitrary integer n, and this new problem is shown to be at least as hard as the ordinary DH problem. Like the twin DH problem, the n-DH problem remains hard even in the presence of a decision oracle that recognizes solution to the problem. On the other hand, observe that the double-size key in the Cash et al. twin DH based encryption scheme can be replaced by two separated keys each for one entity, that results in a 2-party encryption scheme which holds the same security feature as the original scheme but removes the key redundancy. This idea is further extended to an n-party case, which is also known as n-out-of-n encryption. As examples, a variant of ElGamal encryption and a variant of Boneh-Franklin IBE have been presented; both of them have proved to be CCA secure under the computational DH assumption and the computational bilinear Diffie-Hellman (BDH) assumption respectively, in the random oracle model. The two schemes are efficient, due partially to the size of their ciphertext, which is independent to the value n.

Keywords: the (strong) n-DH assumption, the (strong) n-BDH assumption, multiple public key encryption, multiple identity-based encryption.

1 Introduction

In EUROCRYPT 2008 [6], Cash, Kiltz and Shoup proposed a new computational problem and named it the *twin Diffie-Hellman* (twin DH) problem with the meaning that given a random triple of the form $(X_1, X_2, Y) \in \mathbb{G}^3$ for a cyclic group $\mathbb{G}$, compute $\mathrm{dh}(X_1, Y)$ and $\mathrm{dh}(X_2, Y)$, where dh is the DH function. They also proposed the *strong twin DH* problem, which is the twin DH problem under the condition that an adversary is given access to a corresponding decision twin DH oracle. They proved that the strong twin DH problem is as hard as the (ordinary) DH problem, i.e., given a random pair of the form $(X, Y) \in \mathbb{G}^2$, compute $\mathrm{dh}(X, Y)$.

X. Lai, J. Zhou, and H. Li (Eds.): ISC 2011, LNCS 7001, pp. 119–134, 2011.
© Springer-Verlag Berlin Heidelberg 2011

The motivation of their introducing the (strong) twin DH problem is the following: it is well-known that there exist many cryptographic constructions (e.g., the Diffie-Hellman non-interactive key exchange protocol [17] and the Cramer-Shoup encryption scheme [13]) which are based on the DH problem, but security of these constructions can only be proved under the strong DH problem, i.e., the adversary is given access to a decision DH oracle. The reason is that in the security proof, the simulator need the help of the decision oracle to keep the simulation coherent throughout the game. By employing the strong twin DH problem in these constructions, they can successfully prove that the modified constructions are secure under the DH problem, since the strong twin DH problem implies the DH problem. This is a clever trick.

However, their method is not cost free. In order to employ the twin DH problem, their modified construction is "a bit less efficient" than the original one; specifically, the modified construction doubles the key of the original one. For example, in their twin Identity-Based Encryption (IBE) scheme, a master key of a Key Generation Center (KGC) is twin private/public key pairs, written as $((x_1, X_1), (x_2, X_2))$, instead of one (x, X) in the original IBE scheme, and accordingly, an user's secret key associated with this user's identity id (served as a public key of the user) is also two secret values written as (S_1, S_2), each of which is computed under one master key pair. Therefore, a key redundancy is the cost of tighter security reduction.

Can we use this key redundancy to achieve some extra useful function without imposing an efficiency penalty? Observe that in their twin IBE scheme, the identity value id in computing S_1 does not have to be the same as in computing S_2; the two private/public master key pairs (x_1, X_1) and (x_2, X_2) can each belong to an individual KGC. With this slight modification, a user can have two independent identities each associated with one KGC. For example, Alice has her working email address associated with her employer as one KGC and her passport number associated with the government of her country as another KGC. These two KGCs are independent authorities, and do not necessarily have any trust relation or communication between them. Furthermore, the number of the identities and KGCs in the IBE scheme may not be restricted to two[1].

This observation leads to the main contributions of our paper that the twin DH problem can be extended to the n-DH problem for an arbitrary number n, which enables us to build an efficient encryption scheme with multiple public keys and an efficient IBE scheme with multiple KGCs and identities. This type of encryption is also known as n-out-of-n encryption, in which a given message is encrypted under a set of n individual public keys, and the associated decryption operation makes use of the n corresponding secret keys. It is relevant to other well-known encryption primitives with multi-receivers, such as broadcast encryption [5, 16] (known as 1-out-of-n encryption) and threshold cryptosystem [15] (known as t-out-of-n encryption). The latter has an attractive

---

[1] The multi-KGC IBE is not an unsolved problem and could be implemented from extending an existing IBE scheme, but we want to show how we can do it *efficiently* using n-out-of-n encryption.

application, namely attribute-based encryption (ABE) [20, 3]. Compared with the well-explored t-out-of-n threshold encryption or ABE schemes, e.g. using a secret sharing technique [24], an n-out-of-n encryption scheme seems a naive solution. But we think it is worthy studying this solution properly since it has the advantage of simplicity in both algorithm implementation and security analysis.

More specifically, there are a number of contributions in this paper. Here we describe a brief overview of each contribution individually.

THE n-DH PROBLEM. We present a modification of the twin DH problem [6] by extending the number of the (ordinary) DH instances from 2 to an arbitrary integer n, and name it the n-DH problem. Intuitively, the n-DH problem is that given a random $n+1$ tuple of the form $(X_1, \ldots, X_n, Y) \in \mathbb{G}^{n+1}$ for a cyclic group $\mathbb{G}$, compute $(\mathrm{dh}(X_1, Y), \ldots, \mathrm{dh}(X_n, Y))$ where dh is the DH function. We also present the *strong n-DH problem* which is the n-DH problem under the condition that an adversary is given access to a corresponding decision n-DH oracle. We prove that the strong n-DH problem is just as hard as the DH problem.

THE n-BDH PROBLEM. We present a modification of the twin Bilinear-DH (twin BDH) problem [6, 12]. by extending the number of the (ordinary) BDH instances from 2 to an arbitrary integer n, and name it the n-BDH problem. Intuitively, the n-BDH problem is that given a random $2n + 1$ tuple of the form $(X_1, \ldots, X_n, Y, Z_1, \ldots, Z_n) \in \mathbb{G}^{2n+1}$ for a cyclic group $\mathbb{G}$, compute $(\mathrm{bdh}(X_1, Y, Z_1), \ldots, \mathrm{bdh}(X_n, Y, Z_n))$ where bdh is the BDH function. We also present the *strong n-BDH problem* which is the n-BDH problem under the condition that an adversary is given access to a corresponding decision n-BDH oracle. We prove that the strong n-BDH problem is just as hard as the BDH problem.

CONCEPT AND EXAMPLE OF AN MPKE SCHEME. We formalize the concept of an n-out-of-n public key encryption scheme and call it a Multiple Public Key Encryption (MPKE) scheme. MPKE schemes can be used in those applications, which requires that either a decryptor must be in the possession of n private keys (e.g., each can be bound with an particular attribute) or that n decryptors (each with an individual key) must work together, in order to decrypt a given ciphertext. As a concrete MPKE example, we present a new modification of the hashed ElGamal encryption scheme [1], and name it the n-ElGamal encryption scheme. Based on the strong n-DH assumption (that implies based on the ordinary DH assumption), we prove that the n-ElGamal encryption scheme has chosen ciphertext security in the random oracle [2].

CONCEPT AND EXAMPLE OF AN MIBE SCHEME. We formalize the concept of a Multiple Identity-Based Encryption (MIBE) scheme, which is an MPKE scheme with the identity-based key setting under the condition that the n KGCs, each generating a private key from an identity value, can be independent to each other. This type of IBE schemes has already been introduced in the literature, e.g. [7, 10, 11]. To the best of our knowledge, the security of the schemes in [7, 10, 11] have not been rigorously analyzed. As a concrete MIBE example, we present a new modification of the Boneh-Franklin IBE scheme [4] and name it the n-IBE scheme. Based on the strong n-BDH assumption (that implies based on the ordinary BDH assumption), we prove that the n-IBE scheme has chosen ciphertext security in the random oracle [2].

The rest of this paper is organized as follows. We describe definitions of the (strong) n-BDH assumption in Section 2 and of the (strong) n-BDH assumption in Section 3. After that, we present definitions of security models for MPKE schemes and MIBE schemes in Section 4, followed by a concrete MPKE scheme with a rigorous security analysis in Section 5, and a concrete MIBE scheme in Section 6 (due to the limited space, its rigorous security analysis is in the full paper [8]). We end the paper with conclusions and some open questions for future work in Section 7.

2   The n-DH Assumption

Let $\mathbb{G}$ be a cyclic group of prime order p and with generator g, and let dh be the DH function defined as

$$\mathrm{dh}(X, Y) := Z, \text{ where } X = g^x, Y = g^y \text{ and } Z = g^{xy}.$$

Recall that the DH assumption states it is hard to compute $\mathrm{dh}(X, Y)$ given random $X, Y \in \mathbb{G}$. We define the n-DH function function by

$$\mathrm{ndh} : \mathbb{G}^{n+1} \to \mathbb{G}^n, (X_1, \ldots, X_n, Y) \mapsto (\mathrm{dh}(X_1, Y), \ldots, \mathrm{dh}(X_n, Y)).$$

We also define a corresponding n-DH predicate by

$$\mathrm{ndhp}(X_1, \ldots, X_n, \hat{Y}, \hat{Z}_1, \ldots, \hat{Z}_n) := \mathrm{ndh}(X_1, \ldots, X_n, \hat{Y}) \stackrel{?}{=} (\hat{Z}_1, \ldots, \hat{Z}_n).$$

The n-DH assumption states that it is hard to compute $\mathrm{ndh}(X_1, \ldots, X_n, Y)$ given random $X_1, \ldots, X_n, Y \in \mathbb{G}$. Accordingly, the *strong n-DH assumption* states that it is hard to compute $\mathrm{ndh}(X_1, \ldots, X_n, Y)$ given random $X_1, \ldots, X_n, Y \in \mathbb{G}$ along with access to the predicate $\mathrm{ndhp}(X_1, \ldots, X_n, \cdot, \cdot, \cdot, \ldots, \cdot)$, which returns $\mathrm{ndhp}(X_1, \ldots, X_n, \hat{Y}, \hat{Z}_1, \ldots, \hat{Z}_n)$ on input $(\hat{Y}, \hat{Z}_1, \ldots, \hat{Z}_n)$. We have the following theorem to address the relation between the DH assumption and the (strong) n-DH assumption:

Theorem 2.1 (DH via strong n-DH). *The (ordinary) DH assumption holds if and only if the strong n-DH assumption holds.*

It is clear that the DH assumption implies the n-DH assumption. We now prove that the DH assumption implies the strong n-DH assumption. To do this, by following the trapdoor test technique of [6], we first create a trapdoor test.

Theorem 2.2 (Trapdoor Test for n-DH). *Let $\mathbb{G}$ be a cyclic group of prime order p with generator g. Let $I = \{2, \ldots, n\}$, and suppose X_1, r_i, s_i for all $i \in I$ are mutually independent random variables, where X_1 is randomly taken in $\mathbb{G}$, and each of r_i and s_i is uniformly distributed over $\mathbb{Z}_p$, and define the random variables $X_i := g^{s_i}/X_1^{r_i}$. Further suppose that $\hat{Y}, \hat{Z}_1, \cdots, \hat{Z}_n$ are random variables taking values in $\mathbb{G}$, each of which is defined as some function of X_i for all $i \in \{1\} \cup I$. Then we have:*

1. Each X_i for $i \in I$ is uniformly distributed over $\mathbb{G}$;
2. All X_i for $i \in \{1\} \cup I$ are mutually independent;
3. If $X_i = g^{x_i}$ for $i \in \{1\} \cup I$, then the probability that the truth value of

$$\hat{Z}_1^{r_2} \hat{Z}_2 = \hat{Y}^{s_2} \wedge \cdots \wedge \hat{Z}_1^{r_i} \hat{Z}_i = \hat{Y}^{s_i} \wedge \cdots \wedge \hat{Z}_1^{r_n} \hat{Z}_n = \hat{Y}^{s_n} \qquad (1)$$

does not agree with the truth value of

$$\hat{Z}_1 = \hat{Y}^{x_1} \wedge \cdots \wedge \hat{Z}_i = \hat{Y}^{x_i} \wedge \cdots \wedge \hat{Z}_n = \hat{Y}^{x_n} \qquad (2)$$

is at most $(1/p)^{n-1}$; moreover if (2) holds, then (1) certainly holds.

Proof. Observe that $s_i = r_i x_1 + x_i$ for $i \in I$ where $I = \{2, \ldots, n\}$. It is not difficult to verify that each X_i for $i \in I$ is uniformly distributed over $\mathbb{G}$, and that all X_i for $i \in \{1\} \cup I$ and r_i for $i \in I$ are mutually independent, from which the items 1 and 2 follow. To prove the item 3, condition on fixed values of X_i for $i \in \{1\} \cup I$. In the resulting conditional probability space, each r_i for $i \in I$ is uniformly distributed over $\mathbb{Z}_p$, while all x_i, $\hat{Y}$, $\hat{Z}_i$ for $i \in \{1\} \cup I$ are fixed. If (2) holds, (1) certainly holds, because $s_i = r_i x_1 + x_i$ for $i \in I$. Conversely, if (2) does not hold, we show that (1) holds with probability at most $(1/p)^{n-1}$. We take the $n-1$ equations of (1) separately. Each of them uses the same argument as in the proof of the trapdoor test of [6]. Observe that (1) is equivalent to

$$(\hat{Z}_1/\hat{Y}^{x_1})^{r_2} = \hat{Y}^{x_2}/\hat{Z}_2 \wedge \cdots \wedge (\hat{Z}_1/\hat{Y}^{x_1})^{r_i} = \hat{Y}^{x_i}/\hat{Z}_i \wedge \cdots \wedge (\hat{Z}_1/\hat{Y}^{x_1})^{r_n} = \hat{Y}^{x_n}/\hat{Z}_n. \qquad (3)$$

Let us take a look at the $(i-1)^{th}$ equation of (3). We can see that if $\hat{Z}_1 = \hat{Y}^{x_1}$ and $\hat{Z}_i \neq \hat{Y}^{x_i}$ no matter whether the other equations of (2) holds or not, then this equation certainly does not hold. This leaves us with the case $\hat{Z}_1 \neq \hat{Y}^{x_1}$. In this case, the left hand side of the equation is a random element of $\mathbb{G}$ (since r_i is uniformly distributed over $\mathbb{Z}_p$), but the right hand side is a fixed element of $\mathbb{G}$. So this equation holds with probability $1/p$. (3) holds if and only if $n-1$ different equations all hold. Now, we argue that these $n-1$ equations are mutually independent, because each r_i for $i \in I$ is uniformly distributed over $\mathbb{Z}_p$, therefore, the probability that (3) holds is at most $(1/p)^{n-1}$. $\qquad \square$

Using this trapdoor test as a tool, we can prove Theorem 2.1. Let $\mathcal{B}$ be a DH adversary. Denote its advantage by $\mathsf{AdvDH}_{\mathcal{B},\mathbb{G}}$ with the meaning of the probability that $\mathcal{B}$ computes $\mathrm{dh}(X, Y)$, given random $X, Y \in \mathbb{G}$. Let $\mathcal{A}$ be a strong n-DH adversary. Denote its advantage by $\mathsf{AdvnDH}_{\mathcal{A},\mathbb{G}}$ with the meaning of the probability that $\mathcal{A}$ computes $\mathrm{ndh}(X_1, \ldots, X_n, Y)$, given random $X_i, Y \in \mathbb{G}$ for $i \in \{1, \ldots, n\}$, along with access to the predicate $\mathrm{ndhp}(X_1, \ldots, X_n, \cdot, \cdot, \ldots, \cdot)$, which on input $(\hat{Y}, \hat{Z}_1, \ldots, \hat{Z}_n)$, returns $\mathrm{ndhp}(X_1, \ldots, X_n, \hat{Y}, \hat{Z}_1, \ldots, \hat{Z}_n)$. Theorem 2.1 is a special case of the following:

Theorem 2.3. *Suppose $\mathcal{A}$ is a strong n-DH adversary that makes at most Q_d queries to its decision oracle, and runs in time at most τ. Then there exists a DH adversary $\mathcal{B}$ with the following properties: $\mathcal{B}$ runs in time at most τ, plus* *the time to perform $O(Q_d \log q)$ group operations and some minor bookkeeping; moreover,*

$$\left(1 - \frac{Q_d}{p^{n-1}}\right) \mathsf{AdvnDH}_{\mathcal{A},\mathbb{G}} \leq \mathsf{AdvDH}_{\mathcal{B},\mathbb{G}}.$$

In addition, if $\mathcal{B}$ does not output "failure", then its output is correct with probability at least $1 - Q_d/p^{n-1}$.

Proof. The DH adversary $\mathcal{B}$ works as follows, given a challenge instance (X, Y) of the DH problem. First, $\mathcal{B}$ chooses $r_i, s_i \in \mathbb{Z}_p$ for $i \in I$ and $I = \{2, ..., n\}$ at random, sets $X_1 := X$ and $X_i := g^s/X_1^{r_i}$, and gives $\mathcal{A}$ the challenge instance $(X_1, \ldots, X_n, Y)$. Second, $\mathcal{B}$ processes each decision query $(\hat{Y}, \hat{Z}_1, \ldots, \hat{Z}_n)$ by testing if

$$\hat{Z}_1^{r_2} \hat{Z}_2 = \hat{Y}^{s_2} \wedge \cdots \wedge \hat{Z}_1^{r_i} \hat{Z}_i = \hat{Y}^{s_i} \wedge \cdots \wedge \hat{Z}_1^{r_n} \hat{Z}_n = \hat{Y}^{s_n}$$

holds. Finally, if and when $\mathcal{A}$ outputs $(Z_1, \ldots, Z_n)$, $\mathcal{B}$ tests if this output is correct by testing whether

$$Z_1^{r_2} Z_2 = Y^{s_2} \wedge \cdots \wedge Z_1^{r_i} Z_i = Y^{s_i} \wedge \cdots \wedge Z_1^{r_n} Z_n = Y^{s_n}$$

holds; if this does not hold, $\mathcal{B}$ outputs "failure", and otherwise, $\mathcal{B}$ outputs Z_1.

Provide the oracle simulation is perfect, adversary $\mathcal{A}$'s view is identical to its view in the real environment. It remains to calculate the accuracy of the trapdoor test. Note that the probability of the trapdoor test returning a wrong decision result for a query is at most $(1/p)^{n-1}$, and this happens at most Q_d times. Therefore the trapdoor test can simulate the decision oracle perfectly with probability at least $1 - Q_d/p^{n-1}$. Theorem 2.3 follows immediately. □

3   The n-BDH Assumption

In groups equipped with a pairing $e : \mathbb{G} \times \mathbb{G} \to \mathbb{G}_T$ where $\mathbb{G}$ and $\mathbb{G}_T$ are cyclic groups of prime order p and $\mathbb{G}$ is with generate g, we recall that the BDH function is defined as

$$\mathrm{bdh}(X, Y, Z) := W, \text{ where } X = g^x, Y = g^y, Z = g^z, \text{ and } W = e(g, g)^{xyz}.$$

The BDH assumption states that computing $\mathrm{bdh}(X, Y, Z)$ for random $X, Y, Z \in \mathbb{G}$ is a hard problem. The strong BDH assumption [21] states that the BDH problem remains hard even with the help of a corresponding decision oracle.

Note that for the purpose of describing our main results as simply as possible, without loss of the generality, we make use of symmetric pairings (also called Type-1 pairings). It does not mean that our proposed assumptions and schemes only work with symmetric pairings. Without changing the main results of this paper, this symmetric pairing representation can be modified to the asymmetric pairing one (i.e., $e : \mathbb{G}_1 \times \mathbb{G}_2 \to \mathbb{G}_T$ where $\mathbb{G}_1$, $\mathbb{G}_2$ and $\mathbb{G}_T$ are cyclic groups of prime order p). More specifically, one may use Type-2 pairings, where there is an efficiently computable group isomorphism $\psi : \mathbb{G}_2 \to \mathbb{G}_1$ mapping $g_2 \in \mathbb{G}_2$ to $g_1 \in \mathbb{G}_1$, or Type-3 pairings, where there is no known efficiently computable group isomorphism $\psi : \mathbb{G}_2 \to \mathbb{G}_2$ mapping g_2 to g_1. We refer readers to [19] for the details of these three types of pairings.

We define the n-*BDH* function by

$$\mathrm{nbdh} : \mathbb{G}_n \to \mathbb{G}_T^n,$$

$$(X_1, \ldots, X_n, Y, Z_1, \ldots, Z_n) \mapsto (\mathrm{bdh}(X_1, Y, Z_1), \ldots, \mathrm{bdh}(X_n, Y, Z_n)).$$

We also define a corresponding n-*BDH* predicate by

$$\mathrm{nbdhp}(X_1, \ldots, X_n, \hat{Y}, \hat{Z}_1, \ldots, \hat{Z}_n, \hat{W}_1, \ldots, \hat{W}_n) :=$$

$$\mathrm{nbdh}(X_1, \ldots, X_n, \hat{Y}, \hat{Z}_1, \ldots, \hat{Z}_n) \overset{?}{=} (\hat{W}_1, \ldots, \hat{W}_n).$$

The n-*BDH assumption* states that it is hard to compute $\mathrm{nbdh}(X_1, \ldots, X_n, Y, Z_1, \ldots, Z_n)$ given random $X_1, \ldots, X_n, Y, Z_1, \ldots, Z_n \in \mathbb{G}$. The *strong n-BDH assumption* states that it is hard to compute $\mathrm{nbdh}(X_1, \ldots, X_n, Y, Z_1, \ldots, Z_n)$, given random $X_1, \ldots, X_n, Y, Z_1, \ldots, Z_n \in \mathbb{G}$, along with the access to the predicate $\mathrm{nbdh}(X_1, \ldots, X_n, \cdot, \cdot, \ldots, \cdot, \cdot, \ldots, \cdot)$, which on input $(\hat{Y}, \hat{Z}_1, \ldots, \hat{Z}_n, \hat{W}_1, \ldots, \hat{W}_n)$, returns $\mathrm{nbdhp}(X_1, \ldots, X_n, \hat{Y}, \hat{Z}_1, \ldots, \hat{Z}_n, \hat{W}_1, \ldots, \hat{W}_n)$.

We have the following result to address the relation between the BDH assumption and the (strong) n-BDH assumption:

Theorem 3.1 (BDH via strong n-BDH). *The (ordinary) BDH assumption holds if and only if the strong n-BDH assumption holds.*

It is clear that the BDH assumption implies the n-BDH assumption. We prove that the BDH assumption implies the strong n-BDH assumption. Again, by following the technique developed in [6], we first create a trapdoor test.

Theorem 3.2 (Trapdoor Test for n-BDH). *Let $\mathbb{G}$ be a cyclic group of prime order p with a generator g and a pairing $e : \mathbb{G} \times \mathbb{G} \to \mathbb{G}_T$, where $\mathbb{G}_T$ is another cyclic group of order p. Let $I = \{2, \ldots, n\}$, and suppose X_1, r_i, s_i for $i \in I$ are all mutually independent random variables, where X_1 is randomly taken in $\mathbb{G}$, and each of r_i and s_i is uniformly distributed over $\mathbb{Z}_p$, and define the random variables $X_i := g^{s_i}/X_1^{r_i}$ for $i \in I$. Further suppose that $(\hat{Y}_1, \ldots, \hat{Y}_n, \hat{Z}, \hat{W}_i, \ldots, \hat{W}_n)$ are random variables taking values in $\mathbb{G}$, each of which is defined as some function of X_i for all $i \in \{1\} \cup I$. Then we have:*

1. *Each X_i for $i \in I$ is uniformly distributed over $\mathbb{G}$;*
2. *All X_i for $i \in \{1\} \cup I$ are mutually independent;*
3. *If $X_i = g^{x_i}$ for $i \in \{1\} \cup I$, the probability that the truth value of*

$$\hat{W}_1^{r_2}\hat{W}_2 = e(\hat{Y}_2, \hat{Z})^{s_2} \wedge \cdots \wedge \hat{W}_1^{r_i}\hat{W}_i = e(\hat{Y}_i, \hat{Z})^{s_i} \wedge \cdots \wedge \hat{W}_1^{r_n}\hat{W}_n = e(\hat{Y}_n, \hat{Z})^{s_n} \tag{4}$$

does not agree with the truth value of

$$\hat{W}_1 = e(\hat{Y}_1, \hat{Z})^{x_1} \wedge \cdots \wedge \hat{W}_i = e(\hat{Y}_i, \hat{Z})^{x_i} \wedge \cdots \wedge \hat{W}_n = e(\hat{Y}_n, \hat{Z})^{x_n} \tag{5}$$

is at most $(1/p)^{n-1}$; moreover if (5) holds, then (4) certainly holds.

The proof of this theorem is similar to the proof of Theorem 2.2. Due to the limited space, we have put this proof in the full paper [8].

Using this trapdoor test as a tool, we can prove Theorem 3.1. Let $\mathcal{B}$ be a BDH adversary. Denote its BDH advantage by $\mathsf{AdvBDH}_{\mathcal{B},\mathbb{G}}$ with the meaning of the probability that $\mathcal{B}$ computes $\mathrm{bdh}(X,Y,Z)$, given random $X, Y, Z \in \mathbb{G}$. Let $\mathcal{A}$ be a strong nbdh adversary. Denote its advantage by $\mathsf{AdvnBDH}_{\mathcal{A},\mathbb{G}}$ with the meaning of the probability that $\mathcal{A}$ computes $\mathrm{ndh}(X_1, \ldots, X_n, Y, Z_1, \ldots, Z_n)$, given random $X_i, Y, Z_i \in \mathbb{G}$ for $i \in \{1, \ldots, n\}$, along with access to a decision oracle for the predicate $\mathrm{nbdhp}(X_1, \ldots, X_n, \cdot, \cdot, \ldots, \cdot, \cdot, \ldots, \cdot)$, which on input $(\hat{Y}, \hat{Z}_1, \ldots, \hat{Z}_n, \hat{W}_1, \ldots, \hat{W}_n)$, returns $\mathrm{nbdhp}(X_1, \ldots, X_n, \hat{Y}, \hat{Z}_1, \ldots, \hat{Z}_n, \hat{W}_1, \ldots, \hat{W}_n)$. Theorem 3.1 is a special case of the following:

Theorem 3.3. *Suppose $\mathcal{A}$ is a strong n-BDH adversary that makes at most Q_d queries to its decision oracle, and runs in time at most τ. Then there exists a BDH adversary $\mathcal{B}$ with the following properties: $\mathcal{B}$ runs in time at most τ, plus the time to perform $O(Q_d \log q)$ group operations and some minor bookkeeping; moreover,*

$$\left(1 - \frac{Q_d}{p^{n-1}}\right) \mathsf{AdvnBDH}_{\mathcal{A},\mathbb{G}} \leq \mathsf{AdvBDH}_{\mathcal{B},\mathbb{G}}.$$

In addition, if $\mathcal{B}$ does not output "failure", then its output is correct with probability at least $1 - Q_d/p^{n-1}$.

The proof of this theorem is similar to the proof of Theorem 2.3. Again, due to the limited space, we have put this proof in the full paper [8].

4   Definitions of MPKE and MIBE

In this section we present formal definitions of a Multiple Public Key Encryption (MPKE) scheme and of a Multiple Indentity-Based Encryption (MIBE) scheme, including their security notion: chosen ciphertext security, which are based on the usual definitions of chosen ciphertext security for a public key encryption scheme [22] and an identity-based encryption scheme [4]. Recall that these two types of encryption schemes are n-out-of-n encryption schemes. In the security model an adversary is not allowed to corrupt any decryption key from the entirely n set of the keys.

4.1   Multiple Public Key Encryption

A Multiple Public Key Encryption scheme (say MPKE), with a security parameter 1^κ and associated system parameters params (include descriptions of a finite key space $\mathcal{K}$, a finite message space $\mathcal{M}$, and a finite ciphertext space $\mathcal{C}$), is specified by three algorithms: KeyGen, Encrypt, and Decrypt:

KeyGen: takes 1^κ and params as input, and generates a set n of public and secret key pairs, written as $(pk_i, sk_i) \in \mathcal{K}$ for $i = 1, \ldots, n$. We also denote the n public keys by $\mathbf{pk} = (pk_1, \ldots, pk_n)$ and the n secret keys by $\mathbf{sk} = (sk_1, \ldots, sk_n)$.

Encrypt: takes as input params, $\mathbf{pk}$, and a message $M \in \mathcal{M}$. It returns a cipher-text $C \in \mathcal{C}$.

Decrypt: takes as input params, a ciphertext $C \in \mathcal{C}$ and $\mathbf{sk}$, and returns M.

These algorithms must satisfy the standard consistency constraint, namely when $(\mathbf{pk}, \mathbf{sk}) \leftarrow \mathsf{KeyGen}(1^\kappa, \mathsf{params})$, then

$$\forall M \in \mathcal{M} : \mathsf{Decrypt}(\mathsf{params}, C, \mathbf{sk}) = M \text{ where } C = \mathsf{Encrypt}(\mathsf{params}, \mathbf{pk}, M).$$

Chosen ciphertext security of the scheme MPKE is defined by the following attack game, played between a challenger $\mathcal{CH}$ and an adversary $\mathcal{A}$:

Setup. The challenger takes a security parameter 1^κ and associated params, and runs the KeyGen algorithm. It gives the resulting $\mathbf{pk}$ together with params to $\mathcal{A}$, and keeps the corresponding $\mathbf{sk}$ to itself.

Phase 1. $\mathcal{A}$ makes a number of decryption queries to the challenger, where the input to each query is a ciphertext, say $\hat{C}$. To answer such a query, the challenger decrypts $\hat{C}$ and sends the result to $\mathcal{A}$. These queries may be asked adaptively, that is, each query may depend on the replies to previous queries.

Challenge. Once the adversary decides that Phase 1 is over, it outputs two equal length plaintexts $M_0, M_1 \in \mathcal{M}$ on which it wishes to be challenged. The challenger picks a random bit $\beta \in \{0, 1\}$, encrypts M_β, and sends the resulting ciphertext C^* as the challenge to $\mathcal{A}$.

Phase 2. $\mathcal{A}$ issues more decryption queries as in Phase 1, but with the restriction that $\hat{C} \neq C^*$. These queries may be asked adaptively as in Phase 1.

Guess. Finally, $\mathcal{A}$ outputs a guess $\beta' \in \{0, 1\}$ and wins the game if $\beta = \beta'$.

We refer to such an adversary $\mathcal{A}$ as an IND-CCA adversary. We define adversary $\mathcal{A}$'s advantage over the scheme MPKE by $\mathsf{AdvCCA}_{\mathcal{A}, \mathsf{MPKE}}(\kappa) = \left| \Pr[\beta = \beta'] - \frac{1}{2} \right|$. The probability is over the random bits used by the challenger and the adversary.

Definition 4.1. *We say that a multiple public key encryption scheme* MPKE *is* IND-CCA *secure if for any probabilistic polynomial time* IND-CCA *adversary* $\mathcal{A}$ *the advantage* $\mathsf{AdvCCA}_{\mathcal{A}, \mathsf{MPKE}}(\kappa)$ *is negligible*[2].

When we analyze the scheme MPKE in the random oracle model, then hash functions are modeled as random oracles, and both the challenger and adversary are given access to the random oracles in the above attack game. In that case, we write $\mathsf{AdvCCA}^{\mathrm{ro}}_{\mathcal{A}, \mathsf{MPKE}}(\kappa)$ for the corresponding advantage.

4.2 Multiple Identity-Based Encryption

A Multiple Identity-Based Encryption scheme, denoted by MIBE, is specified by four algorithms: Setup, Extract, Encrypt and Decrypt:

---

[2] We say that a function $f(\kappa)$ is negligible if for every $c > 0$ there exists a value κ_c such that $f(\kappa) < 1/\kappa^c$ for all $\kappa < \kappa_c$.

Setup: takes a security parameter 1^{κ}, and returns system parameters params and a set n of master public and secret key pairs, written as (mpk_i, msk_i) for $i = 1, \ldots, n$; without loss of generality, each key pair (mpk_i, msk_i) is associated with the i-th of a set n KGCs. We denote the n master public keys by $\mathbf{mpk} = (mpk_1, \ldots, mpk_n)$ and the n master secret keys by $\mathbf{msk} = (msk_1, \ldots, msk_n)$. The parameters params include a description of a finite message space $\mathcal{M}$, and a description of a finite ciphertext space $\mathcal{C}$.

Extract: takes as input params, a master key msk_i and an arbitrary identity $id_i \in \{0,1\}^*$ for $i \in \{1, \ldots, n\}$. It returns a secret key sk_i. By repeating the Extract algorithm n times with different i values, one can obtain $\mathbf{sk} = (sk_1, \ldots, sk_n)$ associated with $\mathbf{id} = (id_1, \ldots, id_n)$. Note that msk_i and id_i do not have to uniquely match to each other. Theoretically speaking, any arbitrary identity can bind with any master key, and therefore, the case $id_i = id_j$ for $i \neq j$ is allowed.

Encrypt: takes as input params, $\mathbf{pk}$, $\mathbf{id}$ and a message $M \in \mathcal{M}$. It returns a ciphertext $C \in \mathcal{C}$.

Decrypt: takes as input params, a ciphertext $C \in \mathcal{C}$ and $\mathbf{sk}$, and returns M.

These algorithms must satisfy the standard consistency constraint, namely when $(\mathbf{mpk}, \mathbf{msk}, \mathsf{params}) \leftarrow \mathsf{Setup}(1^{\kappa})$ and $\mathbf{sk} \leftarrow \mathsf{Extract}(\mathsf{params}, \mathbf{msk}, \mathbf{id})$, then

$$\forall m \in \mathcal{M} : \mathsf{Decrypt}(\mathsf{params}, C, \mathbf{sk}) = M \text{ where } C = \mathsf{Encrypt}(\mathsf{params}, \mathbf{mpk}, \mathbf{id}, M).$$

Chosen ciphertext security of scheme MIBE is defined by the following attack game, played between a challenger $\mathcal{CH}$ and an adversary $\mathcal{A}$:

Setup. The challenger runs the Setup algorithm. It gives the adversary the resulting params and $\mathbf{mpk}$, and keeps the associated $\mathbf{msk}$ to itself.

Phase 1. The adversary issues queries $q_1, \ldots, q_m$ where query q_i is one of:

- Extraction query $\langle i, \hat{id}_i \rangle$. The challenger responds by running algorithm Extract to generate the private key $\hat{sk}_i$ associated with $\hat{id}_i$ and msk_i. It sends $\hat{sk}_i$ to $\mathcal{A}$.
- Decryption query $\langle \hat{\mathbf{id}}, \hat{C} \rangle$. The challenger responds by running algorithm Extract n times to generate the private key $\hat{\mathbf{sk}}$ corresponding to $\hat{\mathbf{id}}$. It then runs algorithm Decrypt to decrypt the ciphertext $\hat{C}$. It sends the resulting plaintext to $\mathcal{A}$.

These queries may be asked adaptively, that is, each query q_i may depend on the replies to $q_1, \ldots, q_{i-1}$.

Challenge. Once the adversary decides that Phase 1 is over it outputs two equal length plaintexts $M_0, M_1 \in \mathcal{M}$ and a set of identities $\hat{\mathbf{id}}^*$ on which it wishes to be challenged. The only constraint is that each element id_i^* of $\hat{\mathbf{id}}^*$ did not appear in any private key extraction query associated with msk_i in Phase 1. The challenger picks a random bit $\beta \in \{0,1\}$ and set $C^* = \mathsf{Encrypt}(\mathsf{params}, \mathbf{mpk}, \hat{\mathbf{id}}^*, M_\beta)$. It sends C^* as the challenge to the adversary.

Phase 2. The adversary issues more queries $q_{m+1}, \ldots, q_r$ where q_i is one of:

- Extraction query $\langle i, \hat{id}_i \rangle$, where $\hat{id}_i \neq$ the i-th element of $\hat{\mathbf{id}}^*$. Challenger responds as in Phase 1.
- Decryption query $\langle \hat{\mathbf{id}}, \hat{C} \rangle \neq \langle \hat{\mathbf{id}}^*, C^* \rangle$. Challenger responds as in Phase 1.

These queries may be asked adaptively as in Phase 1.

Guess. The adversary outputs a guess $\beta' \in \{0, 1\}$ and wins the game if $\beta = \beta'$.

We refer to such an adversary $\mathcal{A}$ as an IND-ID-CCA adversary. We define $\mathcal{A}$'s advantage over the scheme MIBE by $\mathsf{AdvCCA}_{\mathcal{A}, \mathsf{MIBE}}(\kappa) = |\Pr[\beta = \beta'] - \frac{1}{2}|$. The probability is over the random bits used by the challenger and the adversary.

Definition 4.2. *We say that a Multiple IBE scheme* MIBE *is* IND-ID-CCA *secure if for any probabilistic polynomial time* IND-ID-CCA *adversary* $\mathcal{A}$ *the advantage* $\mathsf{AdvCCA}_{\mathcal{A}, \mathsf{MIBE}}(\kappa)$ *is negligible.*

When we analyze such a scheme MIBE in the random oracle model, we write $\mathsf{AdvCCA}^{\mathrm{ro}}_{\mathcal{A}, \mathsf{MIBE}}(\kappa)$ for the corresponding advantage.

5 The n-ElGamal Encryption Scheme

In this section, we present details of the n-ElGamal encryption scheme. The scheme makes use of a hash function H and a symmetric cipher $\mathsf{SE} = (\mathsf{E}, \mathsf{D})$. Let $\mathbb{G}$ be a cyclic group of prime order p and with generator g. A set of public keys for this scheme is denoted by a n-tuple of random group elements $\mathbf{pk} = (X_1, \ldots, X_n)$, with a set of corresponding secret keys denoted by $\mathbf{sk} = (x_1, \ldots, x_n)$, where $X_i = g^{x_i}$ for $i \in I$ and $I = (1, \ldots, n)$. To encrypt a message $m \in \mathcal{M}$, one chooses a random $y \in \mathbb{Z}_p$, and computes

$$Y := g^y, Z_i := X_i^y \text{ for } i \in I, k := \mathsf{H}(Y, Z_1, \ldots, Z_n), C := \mathsf{E}(k, M).$$

The ciphertext is (Y, c). Decryption works accordingly: given (Y, c) and secret key $\mathbf{sk}$, one computes

$$Z_i := Y^{x_i} \text{ for } i \in I, k := \mathsf{H}(Y, Z_1, \ldots, Z_n), M := \mathsf{D}(k, C).$$

As mentioned earlier, the size of the ciphertext in this scheme is independent to the number of public and secret keys n. Like the twin ElGamal encryption scheme [6], the scheme does not add redundancy in the ciphertext in order to achieve CCA security, as in the Fujisaki-Okamoto transformation [18]. Following the arguments in [1, 6, 14], we now show that the n-ElGamal encryption scheme is secure against chosen ciphertext attack, under the strong n-DH assumption. By Theorem 2.1, the same holds under the (ordinary) DH assumption. Formally speaking, we denote the n-ElGamal encryption scheme $\mathsf{MPKE}_{\mathrm{ndh}}$, and analyze security of this scheme with the following theorem, under the security model previously defined in Section 4.1.

Theorem 5.1. *Suppose* H *is modeled as a random oracle,* SE *is secure against chosen ciphertext attack, and the DH assumption holds in* $\mathbb{G}$. *The* $\mathsf{MPKE}_{\mathrm{ndh}}$ *is* secure against chosen ciphertext attack. In particular, suppose $\mathcal{A}$ is an adversary that carries out a chosen ciphertext attack against MPKE_{ndh} in the random oracle model, and $\mathcal{A}$ runs in time τ, and makes at most Q_h hash queries and Q_d decryption queries. Then there exists an adversary $\mathcal{B}_{dh}$ against the DH problem and an adversary $\mathcal{B}_{sym}$ against the chosen ciphertext security of SE, such that both $\mathcal{B}_{dh}$ and $\mathcal{B}_{sym}$ run in time at most τ, plus the time to perform $O((Q_h + Q_d)\log p)$ group operations; moreover,

$$\mathsf{AdvCCA}^{ro}_{\mathcal{A},\mathsf{MPKE}_{ndh}} \leq \left(\frac{p^{n-1}}{p^{n-1} - Q_h}\right) \mathsf{AdvDH}_{\mathcal{B}_{dh},\mathbb{G}} + \mathsf{AdvCCA}_{\mathcal{B}_{sym},\mathsf{SE}}.$$

Proof. We proceed with a sequence of games.

Game 0. This is the original chosen ciphertext attack game for a MPKE scheme as defined in Section 4.1. Let S_0 be the event that $\beta' = \beta$ in this game.

Setup: To start the game, the challenger generates the secret key set $\mathbf{sk} = (x_1, \ldots, x_n)$ and their corresponding public key set $\mathbf{pk} = (X_1, \ldots, X_n)$. The challenger gives $\mathbf{pk}$ to the adversary.

Hash oracle query $\langle \hat{Y}, \hat{Z}_1, \ldots, \hat{Z}_n \rangle$: The challenger maintains a list of tuples $(Y, Z_1, \ldots, Z_n, k)$ as explained below. We refer to this list as the L list, which is initially empty and indexed by elements of $\mathbb{G}^{n+1}$. Whenever the adversary makes a query $\langle \hat{Y}, \hat{Z}_1, \ldots, \hat{Z}_n \rangle$, if there is already a tuple on the L list indexed by it then the challenger responds with $L[\hat{Y}, \hat{Z}_1, \ldots, \hat{Z}_n] = \hat{k}$. Otherwise, the challenger picks a random symmetric key $\hat{k}$, adds the tuple $\langle \hat{Y}, \hat{Z}_1, \ldots, \hat{Z}_n, \hat{k} \rangle$ to the L list and responds the adversary with $\hat{k}$.

Phase 1 - Decryption query $\langle \hat{Y}, \hat{C} \rangle$: The challenger answers the decryption queries using $\mathbf{sk}$. The challenger need to call the H query in this operation.

Challenge: Once the adversary decides that Phase 1 is over it outputs two messages M_0, M_1 on which it wishes to be challenged. The challenger chooses a random $y \in \mathbb{Z}_p$, sets $Y := g^y$, $Z_i := X_i^y$ for $i = 1, \ldots, n$, then fetches the symmetric key k by querying H with $\langle Y, Z_1, \ldots Z_n \rangle$, and computes $c := \mathsf{E}_k(M_\beta)$, and returns the ciphertext (Y, C) to $\mathcal{A}$.

Phase 2. The decryption queries in Phase 2 are processed just as in Phase 1.

Guess: The adversary $\mathcal{A}$ outputs its guess β' for β.

That finishes the description of Game 0. Despite the syntactic difference, it is clear that

$$\mathsf{AdvCCA}^{ro}_{\mathcal{A},\mathsf{MPKE}_{ndh}} = |\Pr[S_0] - 1/2|. \tag{6}$$

Game 1. We now describe Game 1, which is the same as Game 0, but with the following difference: the challenger will abort the game if the adversary query H at $\langle Y, Z_1, \ldots, Z_n \rangle$ either in Phase 1 or Phase 2. Everything else remains exactly the same as Game 0. Let S_1 be the event that $\beta' = \beta$ in Game 1 and F be the event that the adversary queries the random oracle at $\langle Y, Z_1, \ldots Z_n \rangle$ in Game 1. Since Game 0 and Game 1 proceed identically unless F occurs, we have

$$|\Pr[S_1] - \Pr[S_0]| \leq \Pr[F]. \tag{7}$$

We claim that

$$\Pr[F] \leq \mathsf{AdvnDH}_{\mathcal{B}_{\mathrm{ndh}},\mathbb{G}}, \tag{8}$$

where $\mathcal{B}_{\mathrm{ndh}}$ is an efficient strong n-DH adversary that makes at most Q_h decison oracle queries. Next we detail how $\mathcal{B}_{\mathrm{ndh}}$ plays the role of the challenger in Game 1 to gain the advantage as claimed.

Setup: $\mathcal{B}_{\mathrm{ndh}}$ is given $(X_1, \ldots, X_n, Y)$ as the n-DH challenge instance. $\mathcal{B}_{\mathrm{ndh}}$ gives the adversary $\mathbf{pk} = (X_1, \ldots, X_n)$. Note that the only difference between $\mathcal{B}_{\mathrm{ndh}}$ and the challenger in Game 1 is that the former does not know the $\mathbf{sk} = (x_1, \ldots, x_n)$.

Hash oracle queries: Except processes the queries the same way as the challenger does in Game 1, for every random oracle query $(\hat{Y}, \hat{Z}_1, \ldots, \hat{Z}_n)$, $\mathcal{B}_{\mathrm{ndh}}$ sends this tuple to its own decision oracle, and marks it "good" or "bad" accordingly.

Phase 1 - Decryption queries: $\mathcal{B}_{\mathrm{ndh}}$ can process the decryption queries without using the secret key: given a ciphertext $(\hat{Y}, \hat{c})$, it checks if it has already seen a "good" tuple of the form $(\hat{Y}, \cdot, \ldots, \cdot)$ in L; if so, it uses the key associated with that tuple; if not, it generates a random key, and it will stay on the lookout for a "good" tuple of the form $(\hat{Y}, \cdot, \ldots, \cdot)$ in future random oracle queries, associating this key with that tuple to keep things consistent.

Challenge: Once the adversary decides that Phase 1 is over it outputs two messages M_0, M_1 on which it wishes to be challenged. $\mathcal{B}_{\mathrm{ndh}}$ checks if there is a "good" tuple of the form $(Y, \cdot, \ldots, \cdot)$, if so, it aborts; if not, it generates a random key k (it will stay on the lookout for a "good" tuple of the form $(\hat{Y}, \cdot, \ldots, \cdot)$ in future random oracle queries, associating this key with that tuple to keep things consistent), and computes $c := \mathsf{E}_k(M_\beta)$, and returns the ciphertext (Y, c) to $\mathcal{A}$.

Phase 2 - Decryption queries: The decryption queries in Phase 2 are processed just as in Phase 1. If the adversary issues a "good" tuple of the form $(Y, \cdot, \ldots, \cdot)$, $\mathcal{B}_{\mathrm{ndh}}$ aborts.

Guess: The adversary $\mathcal{A}$ outputs its guess β' for β.

At the end of the game, $\mathcal{B}_{\mathrm{ndh}}$ checks if it has seen a "good" tuple of the form $(Y, \cdot, \ldots, \cdot)$; if so, it outputs the last n components. According to the definition of event F, Equation (8) follows immediately. Theorem 2.3 gives us an efficient DH adversary $\mathcal{B}_{\mathrm{dh}}$ with

$$\mathsf{AdvnDH}_{\mathcal{B}_{\mathrm{ndh}},\mathbb{G}} \leq \frac{p^{n-1}}{p^{n-1} - Q_h} \mathsf{AdvDH}_{\mathcal{B}_{\mathrm{dh}},\mathbb{G}}.$$

Finally, it is easy to see that in Game 1, the adversary is essentially playing the chosen ciphertext attack game against SE. Thus, there is an efficient adversary $\mathcal{B}_{\mathrm{sym}}$ such that

$$|\Pr[S_1] - 1/2| = \mathsf{AdvCCA}_{\mathcal{B}_{\mathrm{sym}},\mathsf{SE}}. \tag{9}$$

Theorem 5.1 now follows by combining (6)-(9). $\square$

6   The n-IBE Scheme

We now present details of the n-IBE scheme. Let $\mathbb{G}$ and $\mathbb{G}_T$ be two cyclic groups of prime order p and $\mathbb{G}$ with generator g, and further let the two groups be equipped with a pairing $e : \mathbb{G} \times \mathbb{G} \to \mathbb{G}_T$. A master public key set is a tuple of n group elements $\mathbf{mpk} = (X_1, \ldots, X_n)$, where $X_i = g^{x_i}$ for $i \in I$ and $I = \{1, \ldots, n\}$. The corresponding master private key set is a tuple $\mathbf{msk} = (x_1, \ldots, x_n)$, which are selected at random from $\mathbb{Z}_p$. We treat the secret/public master key set $(\mathbf{msk}, \mathbf{mpk})$ as n separate key pairs $(x_1, X_1), \ldots, (x_n, X_n)$, which belong to n Key Generation Centers (KGCs) respectively. This scheme uses a symmetric cipher $\mathsf{SE} = (\mathsf{E}, \mathsf{D})$ and two hash functions H and G, where G is defined as $\mathbb{G} \times \{0, 1\}^* \to \mathbb{G}$, and H is defined as $(\{0, 1\}^*)^n \times \mathbb{G} \times \mathbb{G}_T^n \times \to \{0, 1\}^\lambda$ (λ is the length of a symmetric key in algorithm SE).

A private key set associated with n individual identities, denoted by $\mathbf{id} = (id_1, \ldots, id_n)$ for $id_i \in \{0, 1\}^*$ and $i \in I$, is a tuple of n group elements $\mathbf{sk} = (S_1, \ldots, S_n)$. The i-th element of $\mathbf{sk}$ is $S_i = \mathsf{G}(X_i, id_i)^{x_i}$. To encrypt a message $M \in \mathcal{M}$ for $\mathbf{id}$, one chooses $y \in \mathbb{Z}_p$ at random and sets

$$Y := g^y, W_i := e(\mathsf{G}(X_i, id_i), X_i)^y \text{ for } i \in I,$$

$$k := \mathsf{H}(id_1, \ldots, id_n, Y, W_1, \ldots, W_n), C := \mathsf{E}(k, M).$$

The ciphertext is (Y, C). To decrypt using $\mathbf{sk}$ for $\mathbf{id}$, one computes

$$W_i := e(S_i, Y) \text{ for } i \in I, k := \mathsf{H}(id_1, \ldots, id_n, Y, W_1, \ldots, W_n), M := \mathsf{D}(k, C).$$

Similar to the n-ElGamal encryption scheme in Section 5, the length of the ciphertext in the n-IBE scheme is independent to the number of KGCs and identities n. Like the twin IBE scheme of [6], the n-IBE scheme does not add redundancy to the ciphertext as in the Fujisaki-Okamoto transformation [18], which, e.g., is used in the Boneh-Franklin IBE scheme [4] and the Sakai-Kasahara IBE scheme [9, 23]. Now we denote our n-IBE scheme by $\mathsf{MIBE}_{\mathrm{nbdh}}$. It holds chosen ciphertext attack security under the strong n-BDH assumption, as shown in Theorem 6.1. By Theorem 3.1, it also means to be secure under the BDH assumption. The theorem can be proved by following the security analysis approach for the twin IBE scheme in [6] (the approach was originally proposed in [21]). Due to the limited space, we have put this proof in the full paper [8].

Theorem 6.1. *Suppose H and G are modeled as random oracles. Further, suppose the BDH assumption holds with $(\mathbb{G}, \mathbb{G}_T, e)$, and that the symmetric cipher $\mathsf{SE} = (\mathsf{E}, \mathsf{D})$ is secure against chosen ciphertext attack. Then $\mathsf{MIBE}_{\mathrm{nbdh}}$ is secure against the chosen ciphertext attack. In particular, suppose $\mathcal{A}$ is an adversary that carries out a chosen ciphertext attack against $\mathsf{MIBE}_{\mathrm{nbdh}}$, and that $\mathcal{A}$ runs in time τ, and makes at most Q_h hash H queries, Q_g hash G queries, Q_d decryption queries, and Q_e secret key sk_i extraction queries associated with id_i, where sk_i (id_i) is an element of $\mathbf{id}$ $(\mathbf{sk})$. Then there exist a BDH adversary $\mathcal{B}_{\mathrm{bdh}}$ and an adversary $\mathcal{B}_{\mathrm{sym}}$ against the chosen ciphertext security of SE,* *such that both $\mathcal{B}_{\mathrm{bdh}}$ and $\mathcal{B}_{\mathrm{sym}}$ run in time at most τ, plus that time to perform* $O((Q_e + Q_h + Q_g + Q_d) \log p)$ *group operations; moreover*[3]

$$\mathsf{AdvCCA}^{\mathrm{ro}}_{\mathcal{A},\mathrm{MIBE}_{\mathrm{nbdh}}} \leq \left(\frac{eQ_e}{n}\right)^n \cdot \left(\frac{q^{n-1}}{q^{n-1} - Q_h} \cdot \mathsf{AdvBDH}_{\mathcal{B}_{\mathrm{bdh}},\mathbb{G}} + \mathsf{AdvCCA}_{\mathcal{B}_{\mathrm{sym}},\mathrm{SE}}\right).$$

7 Conclusions

We have proposed a new computational problem called the n-DH problem, which is an extension of the twin DH problem of [6], and also proposed the associated strong n-DH problem and the (strong) n-BDH problem. We have shown that the strong n-DH (n-BDH) problem is as hard as the ordinary DH (BDH) problem. We have introduced a formal definition of n-out-of-n encryption which has two versions, namely MPKE and MIBE for the conventional public key setting and identity-based key setting respectively. We have also proposed an efficient MPKE (MIBE) scheme and proved it is CCA secure under the DH (BDH) assumption.

In our security model for an MPKE (MIBE) scheme, the adversary is not allowed to corrupt any individual key in the whole set of n keys, which is used in the challenge phase. This security model suits our target applications of multiple key encryption very well, where the decryption process requires that either a decryptor must holds n keys or that n decryptors much work together. However, whether this model can be strengthened and whether there is any practical motivation to any enhancement of the model might be an interesting topic for further investigation. Whether there are other applications which can benefit from the (strong) n-DH/n-BDH problem is another question which could lead to some future research.

References

1. Abdalla, M., Bellare, M., Rogaway, P.: The oracle Diffie-Hellman assumptions and an analysis of DHIES. In: Naccache, D. (ed.) CT-RSA 2001. LNCS, vol. 2020, pp. 143–158. Springer, Heidelberg (2001)
2. Bellare, M., Rogaway, P.: Random oracles are practical: A paradigm for designing efficient protocols. In: The 1st ACM Conference on Computer and Communications Security, pp. 62–73. ACM Press, New York (1993)
3. Bethencourt, J., Sahai, A., Waters, B.: Ciphertext-policy attribute-based encryption. In: IEEE Symposium on Security and Privacy (SP 2007), pp. 321–334 (2007)
4. Boneh, D., Franklin, M.: Identity-based encryption from the weil pairing. In: Kilian, J. (ed.) CRYPTO 2001. LNCS, vol. 2139, pp. 213–229. Springer, Heidelberg (2001)
5. Boneh, D., Gentry, C., Waters, B.: Collusion resistant broadcast encryption with short ciphertexts and private keys. In: Shoup, V. (ed.) CRYPTO 2005. LNCS, vol. 3621, pp. 258–275. Springer, Heidelberg (2005)
6. Cash, D.M., Kiltz, E., Shoup, V.: The twin Diffie-Hellman problem and applications. In: Smart, N.P. (ed.) EUROCRYPT 2008. LNCS, vol. 4965, pp. 127–145. Springer, Heidelberg (2008)

---

[3] Here $e \approx 2.71$ is the base of the natural logarithm.

7. Chen, L.: An interpretation of identity-based cryptography. In: Aldini, A., Gorrieri, R. (eds.) FOSAD 2007. LNCS, vol. 4677, pp. 183–208. Springer, Heidelberg (2007)
8. Chen, L., Chen, Y.: The n-Diffie-Hellman problem and its applications, Cryptology ePrint Archive, Report 2011/397 (2011)
9. Chen, L., Cheng, Z.: Security proof of sakai-kasahara's identity-based encryption scheme. In: Smart, N.P. (ed.) Cryptography and Coding 2005. LNCS, vol. 3796, pp. 442–459. Springer, Heidelberg (2005)
10. Chen, L., Harrison, K.: Multiple trusted authorities in identifier based cryptography from pairings on elliptic curves, HP Labs Technical Reports, HPL-2003-48
11. Chen, L., Harrison, K., Soldera, D., Smart, N.: Applications of multiple trust authorities in pairing based cryptosystems. In: Davida, G.I., Frankel, Y., Rees, O. (eds.) InfraSec 2002. LNCS, vol. 2437, pp. 260–275. Springer, Heidelberg (2002)
12. Chen, Y., Chen, L.: Twin bilinear Diffie-Hellman inversion problem and its application. To appear in the Proceedings of the 13th Annual International Conference on Information Security and Cryptology, ICISC 2010 (2010)
13. Cramer, R., Shoup, V.: A practical public key cryptosystem provably secure against adaptive chosen ciphertext attack. In: Krawczyk, H. (ed.) CRYPTO 1998. LNCS, vol. 1462, pp. 13–25. Springer, Heidelberg (1998)
14. Cramer, R., Shoup, V.: Design and analysis of practical public-key encryption schemes secure against adaptive chosen ciphertext attack. SIAM Journal on Computing 33, 167–226 (2001)
15. Damgård, I., Jurik, M.: A length-flexible threshold cryptosystem with applications. In: Safavi-Naini, R., Seberry, J. (eds.) ACISP 2003. LNCS, vol. 2727, pp. 350–364. Springer, Heidelberg (2003)
16. Delerablée, C., Paillier, P., Pointcheval, D.: Fully Collusion Secure Dynamic Broadcast Encryption with Constant-Size Ciphertexts or Decryption Keys. In: Takagi, T., Okamoto, T., Okamoto, E., Okamoto, T. (eds.) Pairing 2007. LNCS, vol. 4575, pp. 39–59. Springer, Heidelberg (2007)
17. Diffie, W., Hellman, M.E.: New directions in cryptograpgy. IEEE Transactions on Infomation Theory 22(6), 644–654 (1976)
18. Fujisaki, E., Okamoto, T.: Secure integration of asymmetric and symmetric encryption schemes. In: Wiener, M. (ed.) CRYPTO 1999. LNCS, vol. 1666, pp. 537–554. Springer, Heidelberg (1999)
19. Galbraith, S., Paterson, K., Smart, N.P.: Pairings for cryptographers. Discrete Applied Mathematics 156(16), 3113–3121 (2008)
20. Goyal, V., Pandey, O., Sahai, A., Waters, B.: Attribute-based encryption for fine-grained access control of encrypted data. In: ACM Conference on Computer and Communications Security, ACM CCS 2006, pp. 89–98. ACM, New York (2006)
21. Libert, B., Quisquater, J.-J.: Identity Based Encryption Without Redundancy. In: Ioannidis, J., Keromytis, A.D., Yung, M. (eds.) ACNS 2005. LNCS, vol. 3531, pp. 285–300. Springer, Heidelberg (2005)
22. Rackoff, C., Simon, D.R.: Non-interactive Zero-Knowledge Proof of Knowledge and Chosen Ciphertext Attack. In: Feigenbaum, J. (ed.) CRYPTO 1991. LNCS, vol. 576, pp. 433–444. Springer, Heidelberg (1992)
23. Sakai, R., Kasahara, M.: ID based cryptosystems with pairing on elliptic curve, Cryptology ePrint Archive, Report 2003/054 (2003)
24. Shamir, A.: How to share a secret. Commun. ACM 22(11), 612–613 (1979)

RatBot: Anti-enumeration Peer-to-Peer Botnets

Guanhua Yan[1], Songqing Chen[2,*], and Stephan Eidenbenz[1]

[1] Information Sciences (CCS-3)[**]
Los Alamos National Laboratory
[2] Department of Computer Science
George Mason University

Abstract. As evidenced by the recent botnet turf war between SpyEye and Zeus, the cyber space has been witnessing an increasing number of battles or wars involving botnets among different groups, organizations, or even countries. One important aspect of a cyber war is accurately estimating the attack capacity of the enemy. Particularly, each party in a botnet war would be interested in knowing how many compromised machines his adversaries possess. Towards this end, a technique often adopted is to infiltrate into an adversary's botnet and enumerate observed bots through active crawling or passive monitoring methods.

In this work, we study potential tactics that a botnet can deploy to protect itself from being enumerated. More specifically, we are interested in how a botnet owner can bluff the botnet size in order to intimidate the adversary, gain media attention, or win a contract. We introduce RatBot, a P2P botnet that is able to defeat existing botnet enumeration methods. The key idea of RatBot is the existence of a fraction of bots that are indistinguishable from their fake identities. RatBot prevents adversaries from inferring its size even after its executables are fully exposed. To study the practical feasibility of RatBot, we implement it based on KAD, and use large-scale high-fidelity simulation to quantify the estimation errors under diverse settings. The results show that a naive enumeration technique can significantly overestimate the sizes of P2P botnets. We further present a few countermeasures that can potentially defeat RatBot's anti-enumeration scheme.

1 Introduction

Due to its open nature, the cyber space has been witnessing a growing number of battles or wars among different groups, organizations, or even countries. The recent botnet turf war between SpyEye and Zeus fighting for bots [7] suggests that botnets can play an important role in cyber warfare. In a real battle or war, it is crucial for each party to know the attack capacities of his adversaries. Similarly, in a cyber war involving botnets, a party would be interested in estimating accurately how many compromised machines his opponents possess.

---

[*] Songqing Chen is partially supported by AFOSR grant FA9550-09-1-0071 and NSF grant CNS-0746649.

[**] Los Alamos National Laboratory Publication No. LA-UR 10-03929.

X. Lai, J. Zhou, and H. Li (Eds.): ISC 2011, LNCS 7001, pp. 135–151, 2011.

Currently, a commonly adopted approach to estimating botnet sizes is to infiltrate into an adversary's botnet and enumerate observed bots through either active crawling or passive monitoring methods [13,12]. Different techniques have been used to enumerate existing botnets, such as the Storm botnet, and they sometimes led to inconsistent results, spanning from 500,000 [13] to 50 million [25]. Despite technical challenges such as NAT and DHCP that render it difficult to estimate botnet sizes accurately, advanced techniques can be applied to sift out these effects. For instance, the passive enumeration approach proposed by Kang et al. can enumerate bots sitting behind a firewall or a NAT [13], and the UDmap algorithm developed by Xie et al. [29] helps mitigate the effects of dynamic IP addresses when enumerating bots based on their IP addresses.

In this work, however, we aim to address a more fundamental question: *can a botnet be intelligently designed so that accurately estimating its size is inherently difficult?* Particularly, we are interested in exploring potential tactics that a botmaster can use to *bluff* his botnet size. In a cyber battle, overestimating the size of the adversary's botnet can lead to the effect of intimidation: a party at a disadvantageous position can deploy this tactic to scare off a stronger opponent. In another example, a party can use this tactic to trick his adversary into using an overly high amount of resources to defend against an attack launched from one botnet so that he would hold advantage over his adversary in a different cyber battle that takes place simultaneously. Furthermore, when two botnet owners compete for the same customer who wants to use the larger botnet for, say, spamming or DDoS attacks, one botnet owner may apply the bluffing tactics to get the bid. Sometimes, a botnet owner may want his botnet size to be overestimated so that he can draw some media attention.

To study the power of such bluffing tactics, we design a hypothetical botnet called *RatBot*, which protects itself from being enumerated. RatBot employs the peer-to-peer (P2P) structure to improve its resilience against a single point of failure. The key idea of RatBot is the existence of a fraction of bots that are indistinguishable from their fake identities, which are spoofing IP addresses they use to hide themselves. RatBot prevents adversaries from inferring its size even after its executables are fully exposed. This is done with heavy-tailed distributions to generate the number of fake identities for each bot so that the sum of observed fake identities converges only slowly and thus has high variation.

Due to its anti-enumeration mechanism by design, RatBot distinguishes itself from those technical challenges (e.g., NAT and DHCP) making it difficult to enumerate bots accurately and is thus immune to existing solutions that aim to address these challenges. The wide deployment of NAT actually leads to underestimation of botnet sizes, which is contrary to the design goal of RatBot. Another distinguishing feature is that the degree to which RatBot can bluff about its size is controllable by the attacker. This is ideal in some situations (e.g., cyber war) where the attacker wants to adjust his bluffing tactics dynamically.

To study the practical feasibility of RatBot, we implement it using the actual development code of aMule, a P2P client software that uses KAD for its P2P communications [2]. We further develop a distributed simulation testbed to evaluate the effectiveness of RatBot in misleading botnet size estimation. We perform a variety of tests with different settings and the results show that a naive botnet enumeration approach by counting the IP addresses observed from the P2P botnets could significantly overestimate their sizes.

The remainder of this paper is organized as follows. Section 2 presents related work and Section 3 gives the threat model. In Section 4, we discuss the design of RatBot, and provide the rationale of such design in Section 5. We introduce the implementation of RatBot in Section 6 and use large-scale simulation to evaluate its performance in Section 7. In Section 8, we further discuss potential countermeasures against RatBot and draw concluding remarks in Section 9.

2   Related Work

Behaviors of real-world botnets have been analyzed to provide insights into how botnets operate in reality [4,12,13]. Complementary to these efforts, our work sheds light on the potential challenges regarding enumerating zombie machines in P2P botnets accurately. In spirit, our work is similar to that of Rajab *et al.* [18] as both explore the challenges of estimating botnet sizes, but ours focuses on P2P botnets rather than IRC botnets. Some previous work has shown that multiple factors contribute to inaccurate botnet size estimation, including DHCP and NAT effects [24]. Our results show that even if advanced techniques are deployed to sift out these effects [13,29], the botnet can still adopt sophisticated obfuscation techniques to make it a difficult task to estimate its size accurately.

A plethora of botnet detection techniques have been developed recently. Gu *et al.* have proposed a series of bot detection methods exploiting spatial-temporal correlation inherent in bot activities [11,10]. Other botnet detection techniques include DNS-based methods [19], ISP-level analysis [14], signature-based approaches [9], and flow-level aggregation and mining [31]. Our work is orthogonal to these efforts and focuses on anti-enumeration tactics.

Hypothetical botnets proposed previously include Super-Botnet [27], Overbot [20], AntBot [30], and hybrid P2P botnets [28]. Our work differs from these work on two aspects. First, our work focuses specifically on hypothetic P2P botnets that aim to inflate the adversary's estimation of botnet sizes. Second, we have used large-scale high-fidelity simulation to quantify the estimation errors under diverse settings rather than presenting the design from a conceptual level.

The design and implementation of RatBot presented in this work is based on the Storm botnet, which used the KAD protocol. Besides the Storm botnet, a few other botnets also applied the P2P protocol to organize their bots, such as Nugache [26], Waledac [23], and Conficker [17]. Although none of these botnets have applied anti-enumeration techniques to inflate the number of bots they have, some methods developed for RatBot can be borrowed to enhance their resilience against enumeration by the adversaries. However, as we shall discuss later, there is a tradeoff among operational flexibility, local detectability, and resilience against enumeration in the design space of P2P botnets.

3   Threat Model

In this work, we consider two families of P2P botnets: *immersive P2P botnets* and *exclusive P2P botnets*. For an immersive P2P botnet, the botmaster delivers C&C information through a P2P network that has normal P2P nodes in addition to bots. The original Storm botnet, for instance, was an immersive P2P botnet because the C&C information was delivered to the Storm bots through the Overnet network. An exclusive P2P botnet, by contrast, has bots exclusively as its peers and thus does not have any normal P2P user traffic in it. Since the Overnet network was shut down, the Storm botnet became an exclusive P2P botnet dubbed Stormnet because only bots can participate in the botnet.

The two primitive operations in a P2P network are *publish* and *search*. The *publish* primitive is used to publish a data item either on the machine used by the caller itself (*e.g.*, in an unstructured P2P network) or on a machine with an identifier that is close to that of the data object (e.g., in a structured P2P network). The *search* primitive is used by a peer node to search for data items that satisfy some specific conditions, such as containing certain keywords or producing a certain hash digest. In this work, we assume that in the P2P network *search* operations are *spoofable*, that is to say, a peer node can request a peer to find a data item using a spoofed source IP address. This holds for many P2P networks, which use UDP to implement the request/response mechanism in a search operation. For instance, the widely deployed KAD protocol uses UDP for signaling and TCP for data transfers [16].

It will be seen later that spoofable search operations play a key role in the design of RatBot for hiding authentic search operations. It is, however, noted that these constraints limit the design of RatBot only when it is implemented as an immersive P2P botnet. For an exclusive P2P botnet, as bots do not require an existing P2P network for their C&C communications, the botmaster has more freedom on the implementation of spoofable search operations.

In this work, we assume a reasonable adversarial model from the attacker's standpoint. First, we do not assume that the P2P botnet deploys a strong authentication scheme. As evidenced by previous efforts of successfully reverse-engineering the Storm bot executable, it is possible for white-hat security analysts to reveal secret keys used for bot communications through static or dynamic malware analysis, and create fake bots to infiltrate into the P2P botnet [12,13]. Second, we also assume that the white-hat security analyst, through thorough static code analysis, possesses full knowledge about the functionalities of an authentic bot, including its communication protocol and anti-enumeration techniques. Third, we assume that the behaviors of a fake bot and an authentic bot are indistinguishable to the bots. A fake bot can intercept any message that passes through it, thus obtaining the source IP address it has used. Fourth, a fake bot may stay in the P2P botnet for a long time so that for some P2P protocols (e.g., KAD) a large number of peer nodes would add it to their contact lists, or actively crawl the P2P network to obtain a list of observed P2P nodes.

In the paper, we use the *adversary* and the *white-hat security analyst* interchangeably. Next, we shall present the design of RatBot.

4   RatBot Design

The key idea of RatBot is the existence of an army of *obscure bots*, each of which creates a list of fake identities to hide itself. In this work, we assume that the identity of a bot is manifested as the IP address that it uses to communicate with other peers in the network. Although the P2P identifier (e.g., KAD ID) of a bot can also be used for enumeration purpose, these identifiers sometimes can be changed by bots, thus leading to inaccurate estimate of the botnet size. Moreover, a compromised machine can run multiple instances of bot executable and counting each instance as a bot overestimates the size of a botnet.

As opposed to obscure bots, we say the remaining bots are *explicit bots*. By their nature, explicit bots can be enumerated. In Figure 1, we present the architecture of RatBot in the form of an immersive P2P botnet. If RatBot is an exclusive P2P botnet, no normal peers would exist.

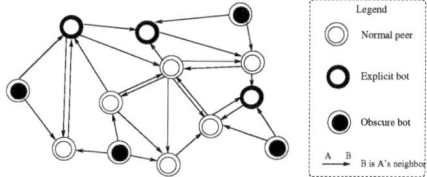

Fig. 1. RatBot Architecture

4.1   Obscure Bot Selection

When a machine is infected and becomes a bot, it decides whether it should be an obscure bot. As an obscure bot uses spoofed IP packets to hide its true identity, an obscure bot must be able to spoof IP packets. Not every end host in the Internet, however, possesses such a capability due to reasons such as NAT deployment and blocking of spoofed packets by firewalls or the host operating systems [5]. We thus let each bot contact a dedicated server during its bootstrapping phase. The server is hardcoded in the bot executable code[1]. When a bot contacts a server, it generates a UDP *query* packet with an arbitrary spoofed source; the payload of the packet carries the authentic IP address of the bot. If the packet arrives at the server, it means that the bot is capable of spoofing. The server decides whether the bot should become an obscure bot and if so, sends back a *response* packet to the bot using its authentic IP address carried in the query packet. If the bot receives the response packet within a certain period of time, it becomes an obscure bot; otherwise, it is an explicit bot.

How does the server decide whether a bot should be an obscure bot? Suppose that it knows the size of the current botnet; this can be done by simply letting each newly infected bot report to it using their authentic IP addresses. The server then makes its decision by aiming to have a fraction ξ of the entire botnet as obscure bots. ξ is *not* hardcoded in the bot executable and it is thus not known to the adversary. Hence, the adversary cannot estimate the botnet size as $m/(1 - \xi)$, where m is the number of explicit bots that he has observed.

---

[1] To improve the resilience of the botnet, multiple servers can be specified in the executable code. Also, fast flux techniques can be used to prevent easy disruption.

4.2 Identity Obfuscation

Once a bot decides that it is an obscure bot, it randomly generates a list of spoofing IP addresses that it will use to obfuscate its own IP address later in P2P communications. The spoofing IP addresses should be chosen to be difficult for the adversary to verify their validity, even if the adversary is able to reverse-engineer the bot code. For example, these spoofing IP addresses should avoid using those from the dark IP address subspace, and being too concentrated in a small IP address subspace. The detail of such algorithm is beyond the scope of this work. For a given obscure bot, how many spoofing IP addresses does it create? The answer provides a key role in the level of difficulty for the adversary to infer the correct botnet size. Consider a simple scheme in which each obscure bot generates a constant number k of spoofing IP addresses. As explained later, a distinguishing feature of an obscure bot is that it does not respond to any request by another peer. Suppose that the adversary can enumerate the entire list of IP addresses S that do not respond to any normal P2P requests. Then, the number of obscure bots can be estimated at $|S|/(k+1)$ if it is assumed that spoofing IP addresses do not overlap.

Two observations are worth noting here. First, as obscure bots generate spoofing IP addresses independently, these spoofing IP addresses may overlap in practice. But given that the large IP address space to spoof, the probability of such overlapping should be low. Second, due to the P2P structure of the botnet and independent generation of spoofing IP addresses by individual bots, compromising a small number of bots, although helping the adversary rule out the spoofing addresses used by these bots, does not prevent the overall size of the botnet from still being overestimated.

We now discuss how RatBot chooses the number of spoofing IP addresses per bot. Consider a botnet with n obscure bots. Let X_i denote the number of spoofing IP addresses obscure bot i generates. RatBot uses two levels of obfuscation. For the first level (**distribution-level obfuscation**), RatBot uses a distribution with high variation to generate X_i, such as the Pareto distribution with PDF:

$$f(x) = \begin{cases} \frac{\alpha x_m^\alpha}{x^{\alpha+1}} & \text{for } x \geq x_m, \\ 0 & \text{for } x < x_m, \end{cases}$$

where x_m and α are the *cutoff* and *scale* parameters, respectively. The mean of the Pareto distribution is $\alpha x_m/(\alpha-1)$ and its variance is $(x_m/(\alpha-1))^2 \cdot \alpha/(\alpha-2)$. It is noted that when $\alpha \leq 2$, the variance becomes infinite. If we set $\alpha \leq 2$, then we cannot apply the central limit theorem on $\sum_{i=1}^n X_i$ due to the infinite variance. It is noted that X_i drawn from the Pareto distribution is a float number. In practice, we generate $\lfloor X_i \rfloor$ spoofing IP addresses for sure, where $\lfloor x \rfloor$ denotes the largest integer no greater than x, and an extra one with probability $X_i - \lfloor X_i \rfloor$.

In Section 5, we shall present the rationale behind using the Pareto distribution for generating X_i and also its limitation. To make size estimation even more difficult, RatBot employs another level of obfuscation in generating X_i (**parameter-level obfuscation**). Instead of using a fixed mean for X_i, the mean of X_i on the i-th obscure bot actually depends on certain attributes of the bot itself. Measurements from the Storm botnet suggest that bot infection is not uniformly distributed either over different ASes or geographically [6]. Hence, we let the mean number of spoofing IP addresses generated by an obscure bot be a function of the time zone where the bot is located. In previous works, security analysts used the observed IP addresses to derive their geographic locations using IP geolocation tools [1] and thus their corresponding time zones. Now that spoofed IP addresses are used, it is difficult to accurately infer the time zone of each bot, which renders it hard to estimate the mean of each X_i.

An obscure bot may use a dynamic IP address to communicate with other peers. Whenever the obscure bot observes that the IP address of the hosting machine has changed, it regenerates its spoofing IP addresses as above.

4.3   Bot Behavior Description

In a typical P2P protocol, a packet between two peers can be classified into three categories: request, response, and data transfer. TCP makes spoofing difficult because it requires handshaking between peers. In many normal P2P networks, request and response signaling packets are delivered through UDP and data transfer uses TCP. We consider the two cases in the following. (1) If the P2P botnet is an exclusive P2P botnet, UDP can be chosen by design for delivering all request, response and data transfer packets. (2) If the P2P botnet is an immersive one, the botmaster does not have the freedom to choose the transport layer protocol. In this study, we assume that request and response signaling packets use UDP. If bot communications do not involve any data transfer packets, spoofing becomes much easier; however, if the P2P protocol uses TCP for data transfer *and* bots need data transfer for C&C messages, it leaves a door for more accurate bot size estimation by the adversary, as will be explained in Section 8.

For an explicit bot, its behavior conforms to the standard P2P protocol. For an obscure bot b, let $\mathcal{I}(b)$ denote the set of spoofing IP addresses associated with it. The behaviors of an obscure bot are given as follows.

Response packets. An obscure bot does not respond to any request by another peer. On the arrival of a request packet, it silently drops the packet. As the packet is delivered through UDP, which is connectionless, the origin of the request packet does not know whether the recipient receives the packet or not.

Request packets. We first consider a naive *packet-level* obfuscation scheme for request packets. When an obscure bot b needs to send out a request packet to peer A at time t, it replicates the packet for $|\mathcal{I}(b)|$ times and each of these packets uses a distinct source IP address from set $\mathcal{I}(b)$. Including the original request packet, there are in total $|\mathcal{I}(b)|+1$ packets to be sent to peer A. For each obscure bot, we define its obfuscation window as w time units. We *randomly* reorder the $|\mathcal{I}(b)|+1$ packets as $p_0, p_1, ...,$ and $p_{|\mathcal{I}(b)|}$. Packet p_0 is sent out at time t. The interval between the sending times of packet p_i and p_{i+1} where $i = 0, 1, ..., |\mathcal{I}(b)|$ is drawn from an exponential distribution with mean $w/|\mathcal{I}(b)|$.

As the order of the packets is random, the recipient peer, if a monitoring node by the adversary, cannot determine which packet carries the authentic source IP address. However, every time a request packet with an authentic source IP is sent, packets with all associated spoofing IP addresses are also sent to the recipient. Hence, if the recipient is a monitoring node deployed by the adversary, she can cluster IP addresses with the same (or approximately the same) number of appearances within w time units. It is highly unlikely that source IP addresses in normal request packets would show such strong correlation as in the naive obfuscation scheme. As such, even though the adversary does not know exactly which source IP address is authentic, he can still infer the actual size of the botnet by assuming that IP addresses frequently appearing in the same interval of w time units would come from the same obscure bot.

It is noted that request packets are usually used by a bot to search for C&C messages from the botmaster. Hence, to prevent correlation-based analysis, Rat-Bot uses a *session-level* obfuscation scheme for each search operation. Figure 2 illustrates the difference between packet-level and session-level obfuscation. Suppose that an obscure bot needs to find a data item with key $\mathcal{K}$. We call it an *authentic session*, which contains the whole sequence of the peer nodes this bot has contacted in order to accomplish this search operation.

For each of its spoofing IP addresses, the obscure bot will create a *spoofing session*, which contains a sequence of peer nodes that are randomly drawn from a local peer node repository. This repository, denoted $\mathcal{R}$, contains peers that were observed in the past authentic sessions and also the current neighbors that the obscure bot knows. It is noted that peers in an authentic session may appear with a certain order. For instance, when a bot searches a data item with key $\mathcal{K}$ in a DHT P2P network, peers in the authentic session are ordered (or partially ordered) in their distances from key ID $\mathcal{K}$. Hence, when constructing the sequence of peers in a spoofing session, such orders are also mimicked.

The intervals between the starting times of sessions, including both authentic and spoofing ones, are randomly drawn from an exponential distribution with mean γ time units. The order of the starting times of spoofing sessions is randomized. The authentic session is inserted among the top ϕ spoofing sessions, if there are so many, and its place is also randomly chosen. The decision on ϕ

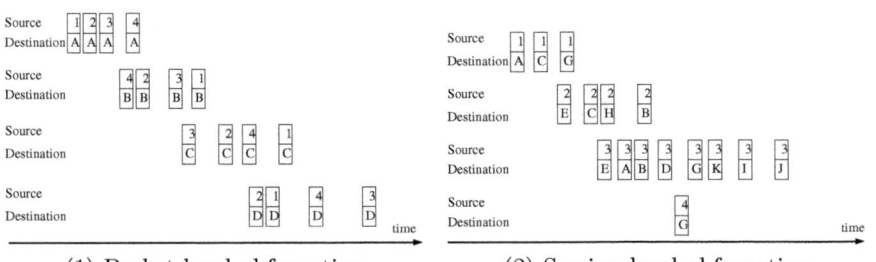

(1) Packet-level obfuscation   (2) Session-level obfuscation

Fig. 2. Obfuscation comparison (*In packet-level obfuscation, each authentic packet is mixed with a number of packets with spoofed sources but the same destination; in session-level obfuscation, each authentic session is mixed with a number of sessions with spoofed sources and previously observed peers as destinations*) should make it difficult to tell which session is authentic but meanwhile ensure that the start of the authentic session would not be postponed significantly due to obfuscation. In our implementation, we let ϕ be 5.

Let Ψ denote the empirical distribution of the number of request packets sent in an authentic session. For each spoofing session, we use Ψ to generate the number of request packets. Each of these request packet carries the spoofing IP address as its source IP and search key $\mathcal{K}$, and is sent to every peer node in the corresponding spoofing session. The interval between two request packets is randomly drawn from the empirical distribution of the intervals between request packets in the past authentic sessions. We use Γ to denote this distribution.

Data transfer packets. If botnet C&C information is stored as a file, each bot needs to fetch the file from the host machine. If RatBot is designed to be an exclusive P2P botnet, UDP can be chosen for data transfer. Otherwise, if it is an immersive P2P botnet, RatBot makes its decisions in the following order: (1) If the C&C information can be spread without involving data transfer, RatBot will not use data transfer. For instance, C&C information can be stored as metadata tags in a KAD-based P2P network. (2) If the P2P network allows UDP for data transfer, RatBot will use UDP instead of TCP for data transfer. (3) Only if the P2P network uses only TCP for data transfer, RatBot would use TCP. It is noted that the third option exposes the identity of obscure bots if the peer hosting the C&C information is actually a monitoring node deployed by the adversary. This is because TCP requires a three-way handshake between the obscure bot and thus the host machine and the connection cannot be spoofed.

5   Rationale

In this section, we explain why a high variance distribution such as the Pareto distribution is used to generate X_i in Section 4.2. As we assume an adversarial model in which the adversary knows the distribution used to generated X_i, we must ensure that the adversary's knowledge does not lead to a good estimation of the botnet size. The adversary also knows that an observed IP address cannot be from an explicit bot if it is used in response packets. Let M be the number of IP addresses observed by the adversary that never respond to any requests. The challenge is: can the adversary infer the number of obscure bots provided that he knows the distribution used to generate X_i?

If only the distribution-level obfuscation is used, all X_i are independent and identically-distributed random variables. According to the *law of large numbers*, $\sum_{i=1}^{n} X_i$ always approaches $n\mu$, where μ is the mean of X_i, when n is large. As the adversary knows the distribution and thus μ, he can estimate the botnet size as $M/(\mu+1)$. To defeat this type of inference, it is necessary to use a distribution that converges so slowly that $\sum_{i=1}^{n} X_i$ can still be far away from $n\mu$ at reasonable scales of botnet sizes.

The *Chebyshev's inequality* tells us that $\mathbb{P}\{|Y - \bar{Y}| \geq t\} \leq t^{-2}Var(Y)$, where $\bar{Y}$ and $Var(Y)$ are the mean and variance of random variable Y, respectively. Hence, the convergence speed of $\sum_{i=1}^{n} X_i$ is affected by the variation of X_i. That explains our choice of the Pareto distribution: for $\alpha < 2$, its variation is infinite and thus slows down the convergence of $\sum_{i=1}^{n} X_i$.

Suppose that there are $100,000$ obscure bots and the average number of spoofing IP addresses an obscure bot generates is 20. We consider four different settings for the scale parameter: $\alpha = 1.01, 1.1, 1.5$, and 1.8. We set the cutoff parameter accordingly to obtain the same mean for X_i. We simulate 1000 cases with different random number generation seeds. In each case, we assume that the adversary sees all the obscure and spoofed IP addresses. Let the observed total number be M. The adversary estimates the number of actual obscure IP addresses as $M/21$ as each obscure IP address has 20 spoofed ones. The following table shows the mean and the standard deviation of the adversary's estimation:

α	1.01	1.1	1.5	1.8
mean	23596.80	81758.83	99854.08	99962.19
standard deviation	83014.82	91258.15	4553.54	1262.98

From the table, it is clear that when α is close to 1, the variability of the estimated bot size becomes more significant. For instance, when $\alpha = 1.01$, even after 1000 sample runs, the derived mean is still far away from the actual one, which is 100000. In reality, the adversary witnesses the result of only one sample; hence, if α is small and thus the variability is very high, the adversary will get an estimate on the botnet size with high variation.

Using heavy tailed distributions such as the Pareto distribution to generate X_i does have its limitation, even though they can produce highly variable results. The high variation of these distributions actually results from their high skewness in their probability density functions. Figure 3 depicts the probability density function of the Pareto distribution when $\alpha = 1.01$ and the mean is 20. Clearly, it is highly skewed as $\mathbb{P}(X_i \leq 1) = 0.805$, which means that around 80% of the data points, if drawn from this distribution, would stay below 1.

To see how this would help the adversary's estimation, we simulate the observed number of spoofing IPs when there are 1000, 10000, 100000, and 1000000 obscure bots. Each obscure bot uses the Pareto distribution with mean 20 and scale parameter 1.01 to generate the number of spoofing IP addresses. For each scenario, we simulate 1000 times. The results are shown in Figure 4, where each data point represents the number of observed spoofing IPs. Note that for each scenario, the number of observed spoofing IPs is highly clustered among the 1000 sample runs. Suppose that the adversary has observed 3000 spoofing IP addresses. Then, he can infer that the real size of the botnet is likely to lie between 10000 and 100000. Hence, RatBot uses another level of obfuscation (i.e., parameter-level obfuscation) to defeat such kind of statistical inferences.

6   Kad-Based RatBot Implementation

In this section, we discuss how to implement RatBot based on KAD, which extends from the Kademlia protocol proposed by Maymounkov and Mazieres [15]. Our implementation of RatBot is based on a popular KAD client, *aMule*[2]. UDP

---

[2] The version we used in our study is aMule 2.1.3.

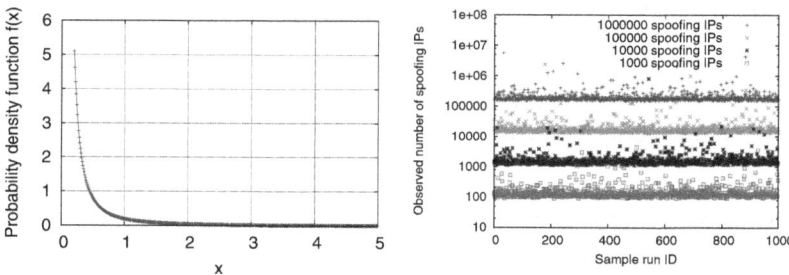

Fig. 3. PDF of Pareto distribution   **Fig. 4.** Observed spoofing IPs

is used in aMule for searching and publishing data objects. If it is an explicit bot, we keep the original implementation intact. Otherwise if it is an obscure bot, we make the following modifications. First, when the bot receives a request message, it drops the message immediately. A request message in KAD carries some special operation types, such as KADEMLIA_HELLO_REQ, KADEMLIA_SEARCH_REQ, KADEMLIA_REQ, KADEMLIA_PUBLISH_REQ, etc.

Second, in the KAD protocol peers regularly send KADEMLIA_HELLO_REQ messages to each other to exchange liveness information. It is noted that the adversary can use such messages to determine whether a peer is an obscure bot or just a spoofed IP address. There are two solutions to this. One option is that the obscure bot obfuscates these messages as well, using spoofing IP addresses. The flip side of this approach is that peers may inject those spoofed IP addresses into their routing tables, thus affecting normal routing operations. The other solution is that an obscure bot does not send out such messages at all. Even though obscure bots and their spoofed IP addresses may still be inserted into their neighbors' routing tables when their neighbors receive search requests from them, the lack of liveness messages makes them less likely to be chosen in a search process because KAD prefers long-lived nodes when forwarding search requests. Also, when a peer node finds that a neighbor has not been alive for a certain period of time, it removes that neighbor from its routing table. Given these considerations, we adopt the second approach in our implementation.

Third, as obscure bots do not send out KADEMLIA_HELLO_REQ messages to their peers, their peers do not send back response messages with type KADEMLIA_HELLO_RES. According to the standard KAD protocol, obscure bots' routing tables would shrink faster because neighbors without liveness messages are removed from the routing table after a certain period of time. To avoid this, we increase the longevity of each neighbor without liveness messages in an obscure bot's routing table from the original two minutes to two hours.

Fourth, a KAD node initiates some random search operations when a bucket does not have enough contacts in its routing table. For an obscure bot, it has to use its authentic IP address for such random lookups. It is necessary to obfuscate these searches also, because otherwise the adversary can infer whether an observed IP address is authentic or not by how many unique keys it uses for searching. In our implementation, we obfuscate these random searches as well.

Finally, we let RatBot use the metadata tags in KAD, such as filenames, to hide C&C information. Hence, no data transfer is needed for normal bot operations. Also, obscure bots never publish any information into the P2P network; they only passively search commands given from the botmaster. The botmaster uses only explicit bots to publish his C&C information.

7   Experimental Evaluation

We now evaluate the effectiveness of RatBot in preventing the adversary from obtaining an accurate estimate on the botnet size. Due to the destructive nature of RatBot, we do this in a simulated environment to avoid legal and ethical issues. Our KAD-based implementation of RatBot used the actual implementation code of aMule. We further intercepted all system calls in it, such as time-related and socket functions and replaced them with simulated function calls specific to our local distributed simulation platform. According to the literature, behaviors of both normal P2P users and bots exhibit strong time zone effects [21,8]. To incorporate these details into our simulation, we model the geographic distribution of normal KAD peers based on previous measurements on the KAD network [21] and that of bots according to the Storm botnet IP distribution [6].

Our model of normal P2P user behaviors is based on the observations on the online patterns of normal KAD users [22]. The starting time of a normal peer being online is modeled with a Gaussian distribution with mean at 7:00pm and standard deviation at 2 hours, and the duration of an online session is generated with a three-parameter Weibull distribution. The online activity model of a bot machine is simply defined as follows: the starting time of it being online is drawn from a Gaussian distribution with mean at 8:00am and the end time is drawn from a Gaussian distribution with mean at 6:00pm; for both distributions, the standard deviation is one hour. This model reflects people's normal work hours.

The number of spoofing IP addresses corresponding to an obscure bot is generated from a Pareto distribution whose parameters are set as follows. Let us number the 24 time zones from 1 to 24. The mean of the Pareto distribution is drawn from a Gaussian distribution with mean and standard deviation set as $2z$ and $4z$, respectively, where z is the time zone number of the obscure bot. The scale parameter of the Pareto distribution is 1.05 and its cutoff parameter can be calculated accordingly from its mean. In each experiment of this study, we use 100 processors from a cluster machine to simulate the behaviors of RatBot.

7.1   Exclusive RatBot

In the first set of experiments, we study the behavior dynamics of exclusive RatBots. We let the botmaster send out a command every day. To improve the reachability of the command to individual bots, the botmaster uses five bots to publish it with 32 keys (as in the Storm botnet) periodically every 100 seconds. Each individual bot, when online, periodically searches the command every 100 seconds with these 32 keys until it gets the command successfully. We simulate

10,000 bots and vary the number of obscure bots among $\{1000 \times i\}_{i=0,1,2,3,4,5}$. Among the 10,000 bots, 10% of them are P2P servers that always stay online. We assume an adversarial model in which the adversary controls 10 servers that can be used to monitor bot traffic. We simulate the botnet for two days: the first day is used as a ramp-up phase for each obscure bot to obtain some empirical distributions, and the second day is used for testing. For each scenario, we simulate it for 20 times with different random number seeds.

We first verify our implementation to ensure that behaviors of spoofing sessions are close to those of authentic sessions. In Figure 5, we depict the frequency histogram of the number of appearances of packets from spoofing and authentic sessions observed by the monitors, respectively, in five runs when there are 1000 obscure bots. There is no obvious systematic difference between authentic and spoofing sessions that can be exploited to differentiate them. From the simulation results, we also note that regardless of the number of obscure bots in the RatBot, almost every individual bot gets the command eventually. Hence, the existence of obscure bots does not affect the utility of the P2P botnet.

Figure 6 gives the median, smallest, and largest number of IP addresses observed by the adversary in 20 sample runs eventually and after one day, respectively, under different number of obscure bots. In the eventual results, we show the total number of spoofing IP addresses generated by obscure bots plus the number of actual bots. We notice that after one day, the adversary observes a large fraction of both actual and spoofing IP addresses. This is because we assume the adversary is able to deploy monitors among the core servers of the P2P botnet and the bots search the command frequently.

Unsurprisingly, if we increase the number of obscure bots, the number of observed IP addresses by the adversary also increases. When there are 4000 or 5000 obscure bots, there are cases where the total number of IP addresses observed by the adversary exceeds 100,000, suggesting that the obfuscation technique of RatBot can lead to an overestimation more than 10 times of its actual size. On the other hand, given the same number of obscure bots, the observed number of IP addresses also varies significantly among different runs. In some scenarios, the largest number of IP addresses observed is twice as much as the smallest number of IP address observed in the 20 sample runs. It is also noted that the median

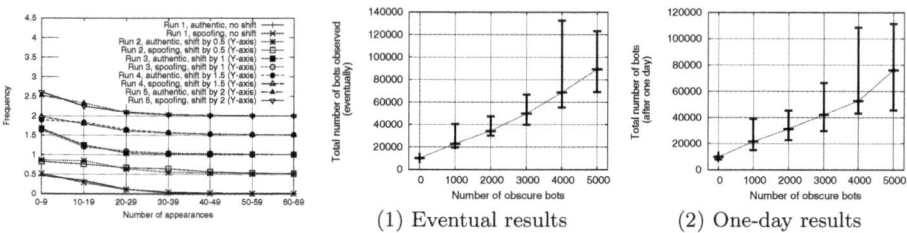

(1) Eventual results   (2) One-day results

Fig. 5. Frequency histogram of the number of appearances in five runs

Fig. 6. Total number of bots observed by the monitors, including explicit, obscure, and spoofing bots

tends to be close to the minimum due to the fact that the Pareto distribution is skewed towards its cutoff parameter at its lower end.

7.2   Immersive RatBot

We now evaluate how immersive RatBot affects the accuracy of botnet size estimation. We simulate a P2P network with 7,000 normal peers and 3,000 bots. The botmaster uses five bots to publish commands with 32 keys periodically every half hour. Each bot uses these 32 keys to search for the current command every half hour until it obtains the command successfully. Here, we let bots perform publish and search operations less frequently than those in exclusive RatBot because normal P2P peers may treat these bots performing frequent operations as abnormal and thus limit interations with them. Among 7,000 normal peers, 990 of them always stay online as servers. We assume the adversary deploys 10 monitors in the network and they appear as servers always online. Each monitor is also a captured bot and can be used to reveal the 32 keys used by the bots to search the current command. The monitor identifies a peer as a bot if it observes that the peer uses any of these keys to search or publish a data item in the P2P network. We vary the number of obscure bots among 0, 1000, 2000, and 3000. For each scenario, we simulate it for four days, the first of which is used as a ramp-up phase for each obscure bot to obtain some empirical distributions and the remaining days are used for testing. We simulate each scenario 20 times.

Figure 7 depicts the number of bots observed by the adversary under different numbers of obscure bots. For visual clarity, we shift the points horizontally slightly to prevent overlapping. For each scenario, we show the median, minimum, and maximum among the 20 sample runs. The results corresponding to "Eventually" show the sum of both the number of authentic bots (including obscure bots) and the total number of spoofing IP addresses generated by all obscure bots.

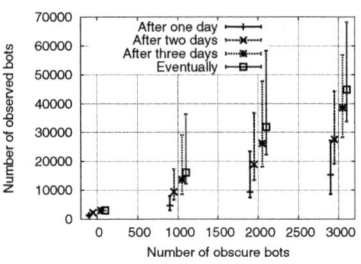

Fig. 7. Number of bots observed by monitors under different numbers of obscure bots (0, 1000, 2000, 3000)

According to the results, we make the following observations. First, the existence of obscure bots produces estimated botnet sizes with high variation. For instance, after three days, if there are no obscure bots, the ratio of the maximum and the minimum of observed bots is 1.016; when we introduce 1000, 2000, and 3000 obscure bots, the ratio becomes 3.405, 2.637, and 2.006, respectively. Such high variation renders it difficult for the adversary to infer the true size of the botnet. Second, it is obvious that increasing the number of obscure bots helps inflate the number of observed bots by the adversary. When there are 1000 obscure bots, the ratio of the median number of observed bots after three days to the true size of the botnet is only 4.5, but when there are 3000 obscure bots, this number becomes 12.8. Hence, the botmater can use the fraction of obscure bots to control the error in the adversary's estimation.

8   Countermeasures

Given the disruptive nature of RatBot, it is important for us to understand its weakness and potential methods to mitigate it. In this section, we present a few countermeasures that can defeat the obfuscation techniques deployed by RatBot. First, RatBot requires each bot to contact a central server initially to decide whether it should work as an obscure bot. The server can easily become a single point of failure. If the adversary manages to monitor traffic from and/or to this server, the identities of true bots can be revealed. With regard to this, it is noted that each bot only needs to contact this server during the bootstrapping phase. As there is little communication for this purpose, it is a difficult task to monitor such traffic. Moreover, existing botnets commonly apply distributed server farms and fast-flux techniques to improve resilience of their services. These techniques can also be applied here to prevent the single failure of the server.

In order for RatBot to operate, the search operation must be spoofable. Hence, if a P2P network deploys anti-spoofing techniques, RatBot cannot survive in it. For example, the P2P network can simply use TCP for all signaling and data transfers. Even if UDP is used for signaling, the P2P network can add a level of anti-spoofing mechanism in a query: when Peer A receives a query from Peer B, it sends back a confirmation request to Peer B and only answers Peer B's query after receiving a reply from Peer B on its request. It is noted that this countermeasure works only against immersive RatBot because the botnet has to be blended into an existing P2P network. Albeit effective in defeating the anti-enumeration scheme by RatBot, fully deploying anti-spoofing techniques in all enterprise networks and ISPs still has a long way to go [5]. For instance, the recent analysis of 5,000 DDoS attacks suggests that a significant fraction of them still used spoofing techniques to generate large volumes of attack traffic [3].

If the RatBot needs TCP data transfer to fetch the command, the adversary can deploy monitors in the P2P network and place those command data on them. By monitoring which machines fetch the command data, the adversary can obtain a list of authentic bots as the three-way handshaking mechanism in TCP cannot be spoofed with spurious IP addresses.

Another effective approach to defeat RatBot is deploying anti-spoofing techniques in the whole Internet. The degree to which the RatBot can obfuscate its size depends on how many obscure bots it has to perform spoofing operations. If the majority of Internet addresses cannot be spoofed, we can still obtain a good estimate on the size of RatBot by simply ignoring those obscure bots.

RatBot's relying on spoofing packets for obfuscation introduces another weakness: enterprise networks and ISPs can detect the existence of bots in their networks by looking for hosts that send out spoofed packets. RatBot prevents enumeration by the adversary at the global level at the price of increased vulnerability to detection at the local level due to its use of spoofing packets.

9   Conclusions

In recent years, botnets, which emerge as a major cyber threat, have been widely used to send spamming emails and launch DDoS attacks. In a botnet war, a botnet owner may want to bluff his botnet size in order to intimidate the adversary, gain media attention, or win a contract. In this work, we explore the tactics that a botnet may use to achieve this goal. We present the design of a type of P2P botnets called RatBot, which applies obfuscation techniques to defeat standard enumeration techniques, and use large-scale high-fidelity simulation to evaluate its performance. We hope our work will raise the awareness of white-hat cyber-security practitioners on the challenges of estimating the sizes of botnets accurately and adopt effective countermeasures in practice.

References

1. http://www.ip2location.com/
2. http://www.amule.org
3. http://asert.arbornetworks.com/2010/12/the-internet-goes-to-war/
4. Barford, P., Yegneswaran, V.: An Inside Look at Botnets. In: Malware Detection. Advances in Information Security, vol. 27. Springer, US (2007)
5. Beverly, R., Berger, A., Hyun, Y., Claffy, K.: Understanding the efficacy of deployed Internet source address validation filtering. In: Proceedings of ACM IMC 2009 (2009)
6. http://isisblogs.poly.edu/2008/05/19/storm-worm-ip-list-and-country-distribution-statistics
7. http://www.net-security.org/secworld.php?id=8858
8. Dagon, D., Zou, C.C., Lee, W.: Modeling botnet propagation using time zones. In: Proceedings of NDSS 2006 (2006)
9. Goebel, J., Holz, T.: Rishi: identify bot contaminated hosts by IRC nickname evaluation. In: Proceedings of HotBots 2007 (2007)
10. Gu, G., Perdisci, R., Zhang, J., Lee, W.: BotMiner: Clustering analysis of network traffic for protocol- and structure-independent botnet detection. In: Proceedings of USENIX Security 2008 (2008)
11. Gu, G., Porras, P., Yegneswaran, V., Fong, M., Lee, W.: BotHunter: Detecting malware infection through ids-driven dialog correlation. In: USENIX Security 2007 (2007)
12. Holz, T., Steiner, M., Dahl, F., Biersack, E., Freiling, F.: Measurements and mitigation of peer-to-peer-based botnets: a case study on storm worm. In: LEET 2008 (2008)
13. Kang, B.B., Chan-Tin, E., Lee, C.P., Tyra, J., Kang, H.J., Nunnery, C., Wadler, Z., Sinclair, G., Hopper, N., Dagon, D., Kim, Y.: Towards complete node enumeration in a peer-to-peer botnet. In: Proceedings of ACM ASIACCS 2009 (2009)
14. Karasaridis, A., Rexroad, B., Hoeflin, D.: Wide-scale botnet detection and characterization. In: Proceedings of HotBots 2007 (2007)
15. Maymounkov, P., Mazières, D.: Kademlia: A peer-to-peer information system based on the XOR metric. In: Proceedings of IPTPS 2001 (2001)
16. Pietrzyk, M., Urvoy-Keller, G., Costeux, J.-L.: Digging into kad users' shared folders. In: Posters of ACM SIGCOMM 2008 (2008)

17. Porras, P., Saidi, H., Yegneswaran, V.: Conficker C P2P protocol and implementation (September 2009), http://mtc.sri.com/Conficker/P2P/
18. Rajab, M.A., Zarfoss, J., Monrose, F., Terzis, A.: My botnet is bigger than yours (maybe, better than yours): why size estimates remain challenging. In: HotBots 2007 (2007)
19. Ramachandran, A., Feamster, N., Dagon, D.: Revealing botnet membership using dnsbl counter-intelligence. In: Proceedings of SRUTI 2006 (2006)
20. Starnberger, G., Kruegel, C., Kirda, E.: Overbot: a botnet protocol based on kademlia. In: Proceedings of SecureComm 2008 (2008)
21. Steiner, M., En-Najjary, T., Biersack, E.W.: A global view of kad. In: IMC 2007 (2007)
22. Steiner, M., En-Najjary, T., Biersack, E.W.: Analyzing peer behavior in kad. Technical Report EURECOM+2358, Institut Eurecom, France (October 2007)
23. Stock, B., Gobel, J., Engelberth, M., Freiling, F.C., Holz, T.: Walowdac - analysis of a peer-to-peer botnet. In: Proceedings of the 2009 European Conference on Computer Network Defense (2009)
24. Stone-Gross, B., Cova, M., Cavallaro, L., Gilbert, B., Szydlowski, M., Kemmerer, R., Kruegel, C., Vigna, G.: Your botnet is my botnet: Analysis of a botnet takeover. In: Proceedings of the ACM CCS 2009 (2009)
25. http://www.neoseeker.com/news/7103-worm-storm-gathers-strength/
26. Stover, S., Dittrich, D., Hernandez, J., Dietrich, S.: Analysis of the storm and nugache trojans: P2p is here. Login 32(6) (December 2007)
27. Vogt, R., Aycock, J., Jacobson, M.J.: Army of botnets. In: NDSS 2007 (2007)
28. Wang, P., Sparks, S., Zou, C.C.: An advanced hybrid peer-to-peer botnet. In: Proceedings of HotBots 2007 (2007)
29. Xie, Y., Yu, F., Achan, K., Gillum, E., Goldszmidt, M., Wobber, T.: How dynamic are ip addresses? In: Proceedings of ACM SIGCOMM 2007 (2007)
30. Yan, G., Ha, D.T., Eidenbenz, S.: AntBot: Anti-pollution peer-to-peer botnets. Computer Networks 55(8) (June 2011)
31. Yen, T.-F., Reiter, M.K.: Traffic aggregation for malware detection. In: Zamboni, D. (ed.) DIMVA 2008. LNCS, vol. 5137, pp. 207–227. Springer, Heidelberg (2008)

Detecting Near-Duplicate SPITs in Voice Mailboxes Using Hashes

Ge Zhang and Simone Fischer-Hübner

Karlstad University, Universitetsgatan 2, 65188, Karlstad, Sweden
{ge.zhang,simone.fischer-huebner}@kau.se

Abstract. Spam over Internet Telephony (SPIT) is a threat to the use of Voice of IP (VoIP) systems. One kind of SPIT can make unsolicited bulk calls to victims' voice mailboxes and then send them a prepared audio message. We detect this threat within a collaborative detection framework by comparing unknown VoIP flows with known SPIT samples since the same audio message generates VoIP flows with the same flow patterns (e.g., the sequence of packet sizes). In practice, however, these patterns are not exactly identical: (1) a VoIP flow may be unexpectedly altered by network impairments (e.g., delay jitter and packet loss); and (2) a sophisticated SPITer may dynamically generate each flow. For example, the SPITer employs a Text-To-Speech (TTS) synthesis engine to generate a speech audio instead of using a pre-recorded one. Thus, we measure the similarity among flows using local-sensitive hash algorithms. A close distance between the hash digest of flow x and a known SPIT suggests that flow x probably belongs the same bulk of the known SPIT. Finally, we also experimentally study the detection performance of the hash algorithms.

1 Introduction

Email spam has been a serious problem to annoy Internet providers and users for many years. It practically costs little for sending out massive junk emails by using an automatic tool without human interaction. The huge volume of spam introduces a significant overload on the network infrastructures and also consumes storage resources. The same phenomenon is foreseen to appear on Voice over IP (VoIP) as well, since the cost of making a VoIP call is low and the user-equipments of VoIP can be programmed. It is known as Spam over Internet Telephony (SPIT), which could automatically launch calls to a number of VoIP users and then play a pre-recorded audio in a conversation. Like Email spam, the SPIT has been predicted to be a serious problem when VoIP is widely accepted [1]. Considering the nature of telephony, there might be two kinds for a SPIT, namely *online SPIT* and *offline SPIT*. In online SPIT, the callee of a SPIT is available and thus the callee needs to decide whether to answer it or not. Therefore, the online SPITs annoy users by continuously drawing their attentions to answer the calls. In contrast, the offline SPIT means that the callee of a SPIT is not available and cannot make an answer personally. In this case,

X. Lai, J. Zhou, and H. Li (Eds.): ISC 2011, LNCS 7001, pp. 152–167, 2011.

the SPIT will be redirected to and answered by the callee's voice mailbox server. The voice mailbox server then stores the SPIT flow and later plays it to the callee. As a result, a user's voice mailbox might be filled up with junk voice messages and leaves no room for useful ones. Some previous anti-SPIT solutions designed for online SPIT prevention [2,3,4] by monitoring the callee's interaction. For example, a call is likely to be a SPIT if the callee quickly hangs it up, or there is no alternate-greeting at the beginning of the conversation. To avoid being detected by these methods, offline SPIT is an alternative since no callee's interaction is involved. Therefore, SPITs can be mounted when the callees are unavailable (e.g., during the midnight).

The profit model of spammers (SPITers) requires them to flood the same information to a number of users. Therefore, a collaborative detection architecture can tell whether a new email (VoIP flow) is a spam (SPIT) by comparing it with a list of known spams (SPITs), which are shared by users or other service providers. A new email (VoIP flow) is considered as a spam (SPIT) if it matches one of the known spams (SPITs). However, sophisticated spammers [5,6] might dynamically generate near-duplicate emails rather than sending the same copies in case of being detected. For instance, a script can help them to append a few random strings or replace some words with their synonyms in a spam message before sending it out. In this case, the generated spams are similar but not identical any more. One countermeasure is to use a local-sensitive hash algorithm, which takes a text message as input and produces a binary hash digest to identify this message. Different to traditional hash algorithms, local-sensitive hash algorithm generates digests with a close Hamming distance to similar messages. Therefore, near-duplicate email spams can be detected within the architecture by measuring the distance between digests of messages.

Our work is focused on detecting near-duplicate offline SPITs within a collaborative detection architecture. To the best of our knowledge, there is no previous research especially focused on this field. The detection is based on comparing the sequence of packet sizes of a flow. It is inspired by previous work which shows that the variation of VoIP packet sizes mostly depend on the utterance [7,8]. Thus, the same SPIT audio leads to the same variation of packet sizes in SPIT flows. Nevertheless, there are many methods to generate near-duplicate SPITs in practice: First, a SPIT might be accidently altered by network impairments like delay jitter and packet loss, or be truncated due to the maximum limitation of a voice mail message; In addition, a sophisticated SPITer might use a text-to-speech (TTS) synthesis engine to dynamically generate utterance audio, which leads to different flow patterns. The near-duplicate SPITs convey similar content but without identical flow patterns. To solve the problem, we evaluated two local-sensitive hash algorithms, which take flow patterns (mainly the sequence of packet sizes) as input and produce a hash digest to identify a flow. Two digests with close Hamming distance suggests that their identified flows have similar content.

In summary, the contributions of our paper include: (1) We discuss the methods for near-duplicating SPITs; and (2) We evaluate the detection performance of two local-sensitive hash algorithms on this problem. The remainder of this paper is organized as follows: Section 2 introduces background on VoIP. We introduce our research problem, collaborate detection architecture and near-duplicating methods in Section 3; The two local-sensitive hash algorithms are described in Section 4, with the experimental results for performance evaluation in Section 5. In Section 6, we summarize related work in SPIT and near-duplicating email spam. Finally, we conclude this paper in Section 7.

2   Background

VoIP relies on two kinds of protocols: a signaling protocol for call setup and termination (e.g., the Session Initiation Protocol (SIP) [9]) and a media delivery protocol for voice packets transmission (e.g., the Realtime Transport Protocol (RTP) [10]). A caller sends a call request (an INVITE message) to a callee and waits for the response. If the callee agrees to accept the request, he/she will reply a positive answer (a 200 OK message) to the caller. Then callee replies a ACK message and a session based on RTP protocol will be established between them. If there is no answer for a call request, the service provider can redirect the request to the callee's voice mailbox server which records the session and later plays it to the callee.

In a RTP session, the communication partners constantly send RTP packets with each other in a fixed time interval (e.g., 20 ms). The payloads of RTP packets are encoded and decoded from analog audio signals using a codec algorithm (e.g., G.711 [11] and Speex [12]). Utterance is sampled at 8-64k samples per second (Hz) by a user-agent. As a performance requirement, the packet inter-arrival time of voice flow is fixedly selected between 10 and 50 ms, with 20 ms being the common case. Thus, given a 8kHz voice source, we have 160 samples per packet with 20 ms packets interval. The size of each RTP packet payload depends on the encoding bit rate of the selected codec, which can be classified as: *Fixed Bit Rate (FBR)* and *Variable Bit Rate (VBR)*. A FBR codec (e.g., G.711) always adapts a constant bit rate and thus the user-agent produces RTP packets with a equal size. On the other hand, a VBR codec (e.g., Speex) dynamically selects the most appropriate bit rate for the input audio based on a scheme known as *code-excited linear prediction* (CELP) [13]. An empirical statistic result [7,8] shows that the selected bit rates

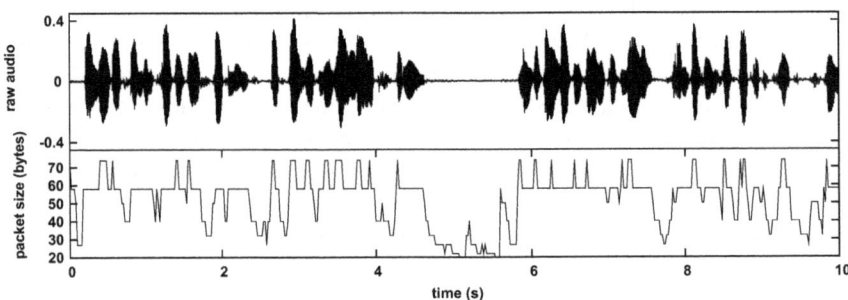

Fig. 1. The audio signal and corresponding RTP packet sizes

are significantly lower for fricative sounds (e.g., "f") than vowel sounds (e.g., "aa").
Currently many VoIP providers (including skype [14]) use VBR codecs to preserve
network bandwidth while maintaining call quality. For example,
Figure 1 shows an utterance audio input with the resulting RTP packet sizes. We
can observe their relationship: the less amplification, the less packet size.

Thus, the packet size variation can be used to identify a VoIP flow. In this paper, we define a *VoIP flow* containing a sequence of RTP packets in the received
order. A *SPIT* is a unsolicited VoIP flow and a *SPITer* is a tool which generates
SPITs. In contrast, a *ham VoIP flow* denotes a VoIP flow from a legitimate user.

3   The Problems

This section introduces a general collaborative detection architecture. We also
describe the challenges brought by near-duplicate SPITs.

3.1   Collaborative Detection Architecture

The general detecting architecture is illustrated in Figure 2. A SPITer aims to send
SPITs to users' voice mailboxes. He initializes calls at the time when the callees are
probably not available (e.g., at midnight). Since there is no answer from callees, a
SIP proxy will redirect SPITs to a voice mail server. A RTP session will be built
between the SPITer and the voice mail server. The voice mail server stores the received RTP packets (e.g., in .pcap format) and later replays them to authorized
users on demand. Note that the voice mail server here does not decode its captured
flow due to privacy protection. It only stores received VoIP flows. We refer a stored
flow as *a voice mail message record*. Later on, users may find SPIT in their mailboxes
and thus report to an anti-SPIT server. The anti-SPIT server maintains a database
containing the reported SPITs. It also scans each unread voice mail message record
in the voice mailboxes. A record is probably a SPIT if its pattern matches that of
the known SPITs. When a CELP VBR codec is applied, the generated RTP packet
sizes varies depending on the input audio. Thus, the same audio leads to the same
variation of packet sizes. We take it as the matching pattern to detect SPITs. Reactions can be performed based on the detection result (e.g., label it as a SPIT or
remove it from the voice mailbox).

Some requirements for this architecture are shown as follows:

1. **Req.1 Privacy:** The anti-SPIT server is not authorized to read the details
   of a voice mail message record which is going to be checked. At most, it
   is allowed to read the packet size and packet arrival time of a record. This
   requirement is set to protect the users' privacy.
2. **Req.2 Efficiency:** The detection should not take much system resources
   (e.g., CPU, memory storage). In addition, the time cost for each comparison
   should be acceptably small.

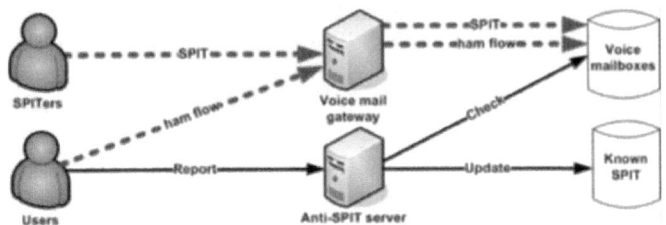

Fig. 2. The collaborative SPIT detection architecture

3. **Req.3 Accuracy:** The detection scheme should be robust enough against near-duplicate SPITs, which will be introduced in the next Section. The false detection rates including the false positive rate and false negative rate should be low.

3.2   Near-Duplicate SPIT

The collaborative detection architecture works based on the pattern match between unknown records with known SPITs. Thus, the design of the matching scheme is critical. Precisely matching is unrealistic since the patterns of a SPIT might be changed due to unintentional and intentional reasons. We name the SPIT whose patterns have been changed in the following ways as *near-duplicate SPIT*.

Unintentional Near-Duplicating. It indicates that patterns of a SPIT is accidentally modified due to the following reasons:

- **Delay jitter:** Delay jitter indicates the variations of inter-packet delays in a flow. A serious variation of inter-packet delay may distort the original sequence of packets. For example, a VoIP flow with initiated sequence packets (p_1, p_2, p_3) might be reordered to (p_1, p_3, p_2) in the recipient side.
- **Packet loss:** Packet loss indicates the amount of packets which are accidentally dropped in the transmission. For example, a VoIP flow with initiated sequence (p_1, p_2, p_3) might arrives as (p_1, p_3) at the recipient, with p_2 being lost. In this example, the packet loss rate is 33%.
- **Maximum duration control:**Users might set a policy to limit the maximum duration of a voice mail message record (e.g., 5 seconds). The exceeding parts will be truncated and discarded if the duration of a voice mail message record is larger than the limitation. For example, a VoIP flow with initiated sequence (p_1, p_2, p_3) turns to (p_1, p_2) in the voice mailbox with the maximum duration as 40 ms (2×20ms/packet). The p_3 has been discarded due to the policy.

Intentional Near-Duplicating. It indicates that a SPITer modifies the patterns of a SPIT on purpose. For example, a SPITer slightly changes the information to be conveyed and employs a new speaker to read them. The utterance is changed and thus the packet size sequence of the new SPIT is different to the previous ones. However, it defies the profit-model of SPITing since it is costly and not scalable. We define *the SPIT dilemma: A SPITer would like to change the pattern of a SPIT to avoid being detected. However, the SPITer has to avoid high cost and inconvenience. In addition, the SPITer does not want to change too much about the conveyed information.*

To tackle this problem, a SPITer can use a Text-To-Speech (TTS) synthetic engine, which generates audio from a normal language text. A TTS synthetic engine first finds pronunciations of the words in the text and assigns prosodic structure to them (e.g., phrasing, intonation). Then, it generates an audio waveform by mapping and concatenating the symbolic linguistic to pieces of recorded speech. A popular open source implementation of TTS engine is Festival [15] developed by University of Edinburgh and Carnegie Mellon University. It contains several recorded speech databases contributed by English speakers with different accents. Users can setup the speed and accent of the generated audio speech. By using a TTS engine, a SPITer can dynamically generate utterance audio from a text rather than preparing a recorded one. The following methods can automatically generate near-duplicate SPITs from a text message (*a master text message*).

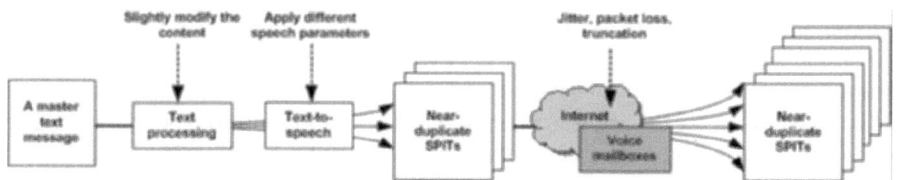

Fig. 3. An overview of methods to generate near-duplicate SPITs

- Modify the master text: A TTS engine pronounces the words in a given text. Thus different texts will lead to different utterance audio.
  - Append random strings: A SPITer can append random words to a master text. It will affect less the users' understanding, since an audio speech is played from the beginning. A user is assumed to get the meaningful information and then neglect the appended part.
  - Words replacement: Some words of the master text can be replaced by synonyms defined in a thesaurus without altering too much information of the text. A SPITer can randomly choose a synonym to substitute an arbitrary selected word in the master text.
- Tune up the TTS parameters: Some parameters of the TTS engine can be tuned up to generate a different audio even with the same master text input.
  - Select speech speed: A SPITer can slightly tune the speech speed for the generated audio.
  - Select speech accent: A SPITer can select the speech voice with different accents for synthesis. Festival provides some build-in accents.

These methods can effectively change the patterns of a generated SPIT, while have little impact on the information conveyed by the SPITs. Figure 3 illustrates an overview picture of producing near-duplicate SPITs.

4   Matching Algorithms

In this section, we focus on the matching approach of the collaborative detection architecture, especially, the matching approach should satisfy the requirements listed in Section 3. We consider to employ local-sensitive hash algorithms to fulfill the requirements. The local-sensitive hash algorithm takes the variation of packet sizes of a VoIP flow as input and produces a binary hash digest with a fixed length. As shown in Section 2, the variation of packet sizes of a VoIP flow is a unique feature to identify an speech audio. Moreover, different to traditional hash algorithms, a local-sensitive hash algorithm produces close-distanced hash digests for similar inputs. In this way, the anti-SPIT server generates a hash digest for an unknown voice mail record and compares it with each hash digest of known SPITs by calculating the distance between them.

There are advantages to use local-sensitive hash algorithms: (1) The anti-SPIT server does not need to read the actual content of a voice mail record. Only the variation of packet sizes is needed to generate hash digests for matching. (2) The efficiency of matching is enhanced by just measuring the distance between hash digests, which are with short and fixed length. (3) We measure the similarity between a voice mail record and a known SPIT. It is robust to counteract SPIT near-duplicating. Thus, it fulfills Req. 1 and Req. 2 proposed in Section 3. We do not know whether it meets Req. 3 yet. For unintentional near-duplicating SPITs, parts of a SPIT remain unchanged. If a SPIT employs a TTS synthetic engine and modifies the master text to produce SPITs, the modified text and the master text still have some overlap. Even if a SPIT applies different accents of a TTS synthetic engine, the generated SPITs might have the similar packet size variation since [7,8] suggest the generated packet sizes for some phonemes mostly depends on the phonemes themselves. However we do not know whether these similarities on packet size variation are enough for detection. How the method actual fulfills Req. 3 will be examined in experiment described in Section 5. The rest of this section introduces two local-sensitive hash algorithms, namely Coskun algorithm and Nilsimsa algorithm.

4.1   Coskun Hash Algorithm

Coskun et al. [16] proposed a local-sensitive hash algorithm to correlate VoIP flows for user tracking. The algorithm takes packet sizes and the packet arrival time of a VoIP flow as input. Given a *VoIP flow* containing P packets, let T_i indicate the arrival time of the ith packet and let B_i denote the payload size of the ith packet, where $i = 0, 1, \cdots, P - 1$. h is the hash digest with L bits and H is a projection array containing L integers. $R_1(), \cdots, R_L()$ are L smooth pseudorandom functions (e.g., we use $R_l(x) = \sin(x + l) \tan(x + l)$). Initially, all elements in H are initialized with 0. For each packet from 1 to $P - 1$, the algorithm calculates its size difference from the previous one (as $B_i^\Delta = B_i - B_{i-1}$) and the relative arrival time since the arrival time of the first packet (as $\hat{T}_i = T_i - T_0$). Then, the algorithm projects $\hat{T}_i$ on the smooth pseudorandom functions. The elements in H are updated using the B_i^Δ multiplied by the projecting result.

Finally, each bit of h is produced depending on the signs of the corresponding integers in H:

$$h_l = sign_1(H_l) \begin{cases} 1, & \text{if } H_l \geq 0 \\ 0, & \text{if } H_l < 0 \end{cases} \tag{1}$$

where $l = 1, 2, \cdots, L$. The detailed algorithm of Coskun hash algorithm is shown below in Algorithm 1.

$H \leftarrow [0, 0, 0, \cdots, 0]$ // initialize $H_1, H_2, \cdots, H_L$
for all captured packet i with $i = 0, 1, \cdots, P - 1$ **do**
   if $i = 0$ **then**
      $flowStart \leftarrow T_i$ // arrival time of the first packet
   else
      $\hat{T}_i \leftarrow T_i - flowStart$ // relative arrival time
      $B_i^\Delta = B_i - B_{i-1}$ // packet size different
      $H \leftarrow H + B_i^\Delta [R_1(\hat{T}_i), \cdots, R_L(\hat{T}_i)]$ // $R_l(\hat{T}_i) = \sin(\hat{T}_i + l) \tan(\hat{T}_i + l)$
   end if
end for
$h = sign_1(H)$

Algorithm 1. The Coskun flow hash algorithm [16]

4.2 Nilsimsa Hash Algorithm

Nilsimsa algorithm [17] computes a hash digest for a text by taking a trigram of characters within a sliding window moving over the text as the input. We reuse some notations from Section 4.1: Given a *VoIP flow* containing P packets, let B_i denote the payload size of the ith packet. h is the produced hash digest with L bits and H is a projection array containing L integers. The size of the slide window is w. For a given text "drugdeal", the slide window (with $w = 5$ as a default) is first located at "drugd". The algorithm then emulates all trigrams of characters from the window. The trigrams are not necessary to be adjacent in the text, but should be in the same sequence. For instance, "dru, drg, drd, dug, dud, dgd, rug, rud, ugd" are the trigrams for "drugd". Then all the trigrams are hashed using a traditional hash algorithm (e.g., MD5) and one hash digest is mapped from 1 to l to hit one element in H, whose value is increased by 1. The slide window thus moves from left to right and the operations are repeated. Finally, each bit of h is generated depending on the corresponding integers in H:

$$h_l = sign_2(H_l) = \begin{cases} 1, & \text{if } H_l \geq \phi \\ 0, & \text{if } H_l < \phi \end{cases} \tag{2}$$

where $l = 1, 2, \cdots, L$ and the ϕ is the median of the values in H elements. The Nilsimsa algorithm has been applied in Spamassassin [18], a widely used open-source email spam filter to find clusters of similar email spam messages. The evaluation results in [6,5] show that this algorithm performs well to counteract near-duplicate email spams. *Different to the traditional Nilsimsa algorithm, we takes the packet sizes as input instead of text characters.* The detailed algorithm of Nilsimsa hash algorithm for VoIP flows is shown below in Algorithm 2.

$H \leftarrow [0, 0, 0, \cdots, 0]$ // initialize $H_1, H_2, \cdots, H_L$
for all captured packet i with $i = 0, 1, \cdots, P - 1 - w$ **do**
  for all trigram from window $\{B_i, \cdots, B_{i+w-1}\}$ **do**
    $H[Hash(trigram) \mod l] + +$ // trigram $< B_i, B_{i+1}, B_{i+2} >, \cdots$
  end for
end for
$h = sign_2(H)$

Algorithm 2. The Nilsimsa flow hash algorithm [17]

5   Experimental Results

To evaluate the performance of the two algorithms, we did a series of experiments. In this section, we first introduce how we selected the samples for evaluation and then explain the process of the experiments. Finally, we show the test results.

5.1   Sample Collection

We assume that SPITs contain brief information, thus we choose Short Message Service (SMS) spam samples as master messages to generate our near-duplicate SPITs. We employ the SMS spam corpus [19] provided by Almeida and Hidalgo. The original corpus contains both legitimate messages and spams. We only extract spams for testing. However, we found that the corpus contains similar or even identical spams. Thus, we filtered the similar or identical ones by applying the traditional Nilsimsa algorithm. As suggested in [17], if the Hamming distance between 2 Nilsimsa digests is more than 24, the two messages are probably not independently generated. Thus, we only keep the spams whose largest Nilsimsa distance with any others are less than 24. In total 152 spams left after the filtering. Figure 4(a) shows the cumulative distribution function of the maximum Nilsimsa distance for the message with others. Then we generate SPITs from those SMS spam messages using Festival and wav2rtp[1] with default setup. Figure 4(b) shows that distribution of generated utterance audio durations for each SMS spam, scaled mostly from 10 to 25 seconds.

---

[1] The wav2rtp is an open source tool to convert a .wav file to a VoIP flow containing sequence RTP packets

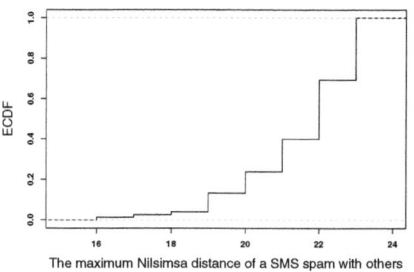

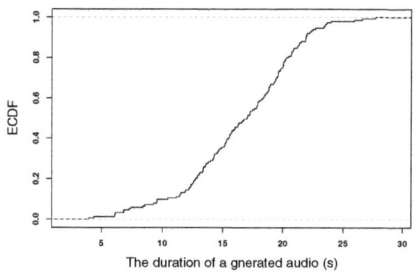

(a) distance distribution of SMS spam

(b) duration distribution of generated audio

Fig. 4. Collected samples for testing

5.2 Experiment and Evaluation Method

We first generate SPITs for each of the 152 SMS spams using Festival and wav2rtp with the default setup. These SPITs simulate known SPIT samples reported by users. We then employ the hash algorithms to generate 128-bit digests for the samples. To evaluate efficiency of producing a hash digest by using the two algorithms, we logged the time cost for each hashing and the CDF of them is shown in Figure 5(a). We can observe that Nilsimsa shows a better efficiency, but the time cost of both algorithms are mostly less than 0.5 second. After collecting the SPIT sample digests, we repeatedly generate near-duplicate SPITs by using the following near-duplicating options independently.

- We tune the packet loss rate (1% or 10%.) in wav2rtp.
- We tune the Gamma delay parameter in wav2rtp to simulate the delay jitter. The Gamma delay is parameterized in terms of a shape parameter k and scale parameter θ, with the probability density function (PDF):

$$f(x : k, \theta) = x^{k-1} \frac{e^{-x/\theta}}{\theta^k \Gamma(k)}$$

In the test, we fixedly set $k = 500$ (ms) and variably set θ from 10 to 30. The larger θ indicates larger delay jitter.
- We only keep the first $1 - x\%$ packets of a SPIT and drop the remaining ones to simulate the truncation. The x is selected to 10, 50, or 90.
- Given a SMS spam with length of l, we append random words with lengths of $x\%$ of l to the original SMS spam. The x is selected to 100, 300, or 500.
- We randomly substitute x words in the original SMS spam to simulate synonym replacement. The x is selected to be 1, 2, or 3.
- We tune the TTS speech speed parameter in Festival to apply different speech speed in SPIT, within ± 0.1 or ± 0.01 of the default one. We invited 4 subjects and played TTS converted audio to them by tuning speech speed within ± 0.1 of the default speed. There was no problem for subjects to understand these audio.

– We tune the TTS speech accent parameter in Festival, other than the default one, like "diphone_ked", or "diphone_don".

Please note that every time we applied only *one* of the above options for one near-duplicate SPIT. We did not apply the mixed options since we tried to find out the detecting performance for each option. We then produce the hash digests again and calculate the distance between them and the sample ones. If their distance is less than a threshold, it is detected as a near-duplicate SPIT of the sample one. Otherwise, it is not. We evaluate the performance by counting the true positive rate and the false positive rate, and plot its Receiver Operating Characteristic (ROC) curve. The result is provided in the next section.

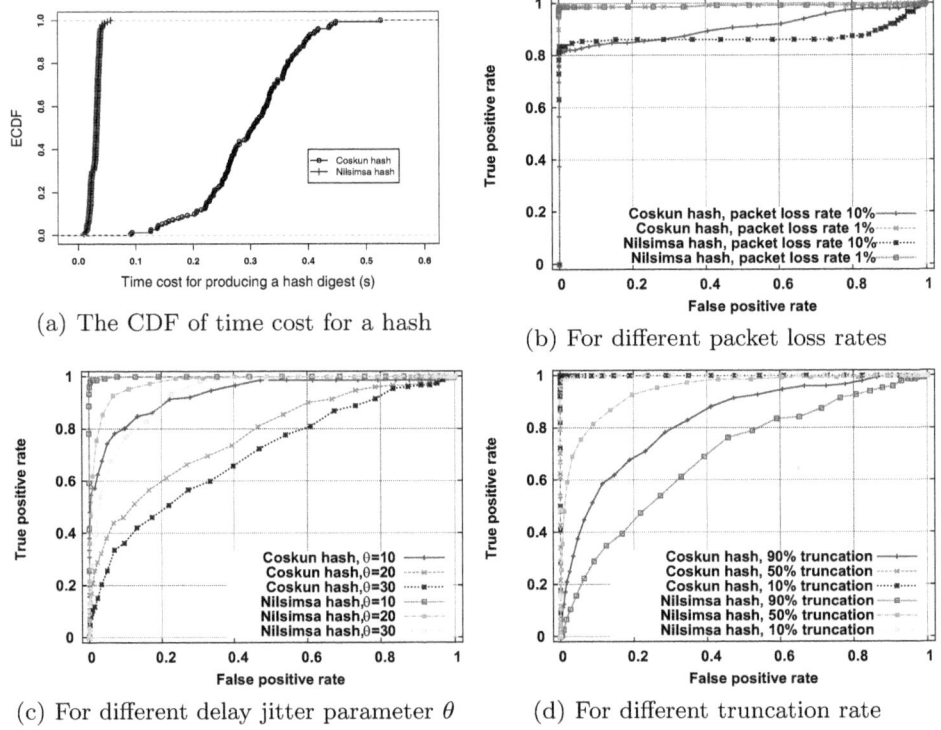

(a) The CDF of time cost for a hash

(b) For different packet loss rates

(c) For different delay jitter parameter θ

(d) For different truncation rate

Fig. 5. Performance of hash algorithms and the ROC curves for detecting unintentional near-duplicate SPITs

5.3 Result to Detect Unintentional Near-Duplicate SPITs

We present the resulting ROC curves for unintentional near-duplicate SPITs detection in Figure 5(b), 5(c) and 5(d). The two algorithms show similar performance to resistant packet loss. With 1% packet loss rate, both algorithms provide an Equal Error Rate (EER) around 0.5%. Even with 10% packet loss rate, still 18% EER is supported. Figure 5(c) shows that Nilsimsa gives better performance than Coskun with delay jitter. The result is as we expected, since

Nilsimsa only takes the packet sizes as the feature, while Coskun takes not only packet sizes, but also packet inter-arrival time. Thus, delay jitter should impact more on Coskun algorithm. Figure 5(d) shows Nilsimsa also performs better than Coskun to resistant truncation. The results also suggest that both of the two algorithms are suitable candidate to detect unintentional near-duplicate SPITs. VoIP does not work with a larger packet loss rate (e.g., usually only 1-2% packet loss rate can be accepted for VoIP conversation). Also, 90% truncation rate for a voice mail message record is unusual in reality. With less parameters, both algorithms are qualified.

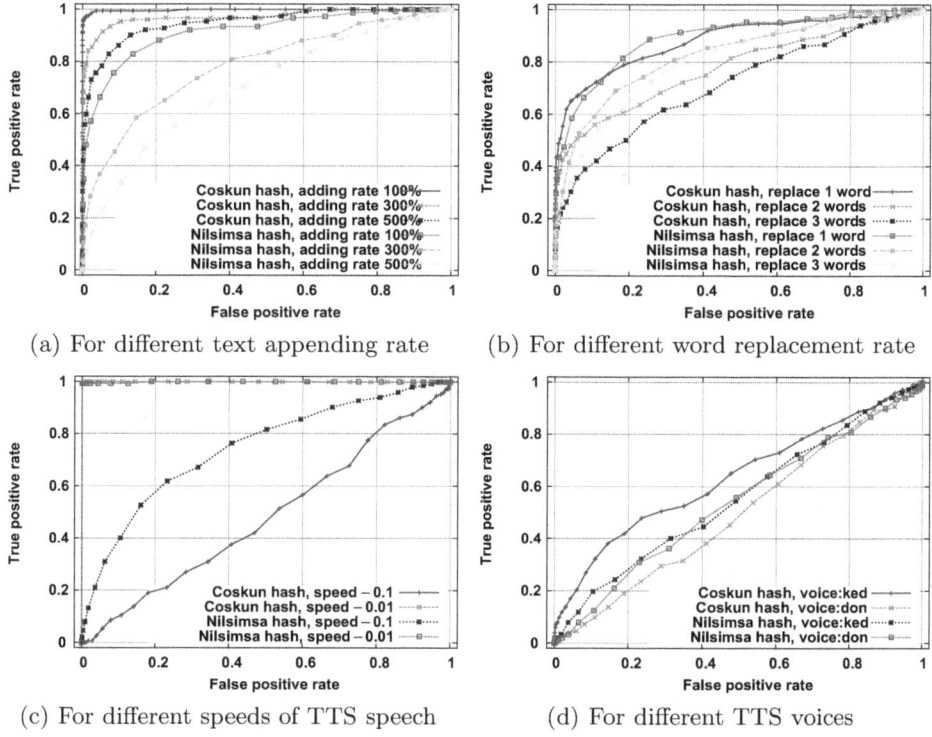

(a) For different text appending rate   (b) For different word replacement rate

(c) For different speeds of TTS speech   (d) For different TTS voices

Fig. 6. ROC curves for detecting intensional near-duplicate SPITs

5.4   Result to Detect Intentional Near-Duplicate SPITs

The ROC curves for intentional near-duplicate SPITs detection is shown in Figure 5. Coskun algorithm provides better performance than Nilsimsa to random text appended of the master messages. With the adding rate 100%, the EER of Coskun is around 3% and the one of Nilsimsa is around 18%. The two algorithms show similar performance on the test for synonyms replacement of master messages. With one word replacement on the master messages, the EERs of detection are both around 20%. The EERs increase to 38% if we replace three words as the worst case. We found that both algorithms perform terribly when the SPITs are near-duplicated by tuning up TTS parameters. Although the algorithms perform well when the speed was tuning around ±0.01, the performance turns worse when the speed tuning range is within ±0.1. Nilsimsa algorithm works better than Coskun algorithm in this case, but still with 35% EER. Both the algorithms perform terribly when the SPITs are near-duplicated by applying different accents, with around 55% EER, which means that the algorithms totally fail to detect near-duplicated SPITs if SPITers select different speech accent parameters in a TTS engine. In this case, a new generated SPIT will be considered as a different one despite they are converted from the same text document. However, the result is still acceptable for SPIT prevention if we consider the fact that the number of supported speech accents in a TTS engine is limited: A SPITer is unable to generate bulk SPITs, each of which is converted with a totally different speech accent parameter.

6   Related Work

This section summarizes proposed SPIT countermeasures and near-duplicate email spam detection. Firstly, there are several types of related work on SPIT countermeasures:

List: It labels callers with different trust level. For instance, all calls from the callers in a blacklist should be blocked and those from the callers in a whitelist can be accepted. The unclassified calls from the users in a gray-list can be temporarily rejected [20,21]. For instance, Skype users [14] can customize their configurations to allow being called by anyone or only by the users in their buddy lists.

Reputation: This approach prevents the calls from those callers who have low reputation values. There are different ways to calculate reputations: For example, Balasubramaniyan et al. [2] generate reputations for VoIP users based on the call durations of their previous calls. It is motivated by the observation that a legitimate user typically makes longer calls than a SPITer. Similarly, Zhang et al. [22] use cumulative online duration to calculate the reputation value. The less time a VoIP user is online, the less calls he/she can launch. This scheme prevents the SPITers who register new accounts for SPITing.

Turing test: A Turing test tells whether the caller is a human or an automatic SPIT generator. Markkola et al. [23] implemented a prototype of audio CAPTCHA [14]. It says 5 random digits and requires a caller to correctly input them for the call being processed. Soupionis et al. [24] evaluated existing audio CAPTCHA implementations and showed most of them are vulnerable to automatic analysis. To solve the problem, they proposed a new audio CAPTCHA implementation including different kinds of noises (e.g., random background noise) to prevent automatic analysis. Quittek et al. [4] proposed a hidden Turing test based on the factor that people usually greet each other at the beginning of a telephone conversation, which results in alternative short periods of silent and speech. SPITers typically do not react to greeting, and then can be detected.

Decoy: SPITers need to collect a number of callee addresses as targets. Salehin et al. [25] proposed a method to block SPITs using decoys. They first publish decoy VoIP addresses in the Internet in the way that a human can tell that they are decoy addresses, but an automated harvester would not know the difference and take them as the targets for SPITing. Ideally, a human will not hit the decoy addresses but only SPITers. In this way, a server can identify those who hit a decoy as SPITers.

As far as we know, there is no research on detecting near-duplicate SPIT yet. Due to the fact that email spam bears similarities with SPIT, we introduce several work on detecting near-duplicate email spam. Mehta et al. [26] studied a way to detect near-duplicate image spam by using the extracted visual features (e.g., color, texture and shape) from images. Based on the features, they build probabilistic models of images supported by a Gaussian Mixture Model (GMM). After all the GMMs have been learnt, they cluster similar images together by applying Jensen-Shannon divergence as the distance measure. In addition, HTML content is available in emails and provides sufficient information about an email layout structure. Tseng et al. [27] proposed a method to detect near-duplicate email spam based on the similarity of HTML tag sequences. Near-duplicate emails should have similar visual layout and thus result in similar HTML tag sequences.

Our work, focused on near-duplicate SPIT detection in voice mailboxes, is different with the related work above. Our approach takes the variance of packet sizes or packet arrival times in a VoIP flow as features to generate local-sensitive hash digest for each VoIP flow, and then compare them with known SPITs by calculating the distance between the hash digests.

7 Conclusion

A SPITer typically launches unsolicited calling requests to a number of users and plays a pre-recorded audio in conversations. Ideally, these SPIT flows have the same sequence of packet sizes. In a collaborative detection framework, a server compares a VoIP flow with a list of known SPITs: A match indicates that a particular SPIT is probably being played at the moment. Nevertheless, various network impairments, such as packet loss and delay jitter, accidently alter the patterns of a SPIT flow. Moreover, a SPIT flow might be partly truncated due to the maximum duration of a received voice mail message record. Furthermore, an advanced SPITer may employ a text-to-speech (TTS) synthetic engine to produce audio from a text document rather than using a pre-recorded audio. In this way, SPITer can slightly modify the content of the document or some configuration parameters of the TTS engine to generate near-duplicate SPITs. It brings challenges to the SPIT detection. We investigate two local-sensitive hash algorithms on solving the problems. The local-sensitive hash algorithms take the packet inter-arrival time or the variation of packet size in a VoIP flow as input and produce a binary hash digest. Different to traditional hash algorithms, they generate hash digests within a certain distance for similar inputs. Thus we can measure the similarity of two flows by calculating the distance between the hash digests. The smaller of the distance, the higher probability of the incoming VoIP flow is a particular SPIT. Our experiments show that the two algorithms are robust to detect the SPITs near-duplicated unintentionally, with EERs are less than 5% if the packet loss rate, delay jitter and truncation rate are reasonable. When a SPITer employ a TTS synthetic engine to intentionally create near-duplicate SPITs, the detection performance is reduced. Random appending and word replacement can increase the EER up to 18% and 40%. In addition, the detection totally fails if SPITs are generated by applying a different speech accent on the TTS. Nevertheless, the proposed detection method still can circumscribe near-duplicating SPITs since the accents applied in a TTS is still limited. It is difficult for a SPITer to apply different accent for each generated SPIT.

References

1. SPIT on VoIP. Communications News (January 2005),
   http://findarticles.com/p/articles/mi_mOCMN/is_1_42/ai_n27865818/?tag=mantle_skin;content (visited at May16, 2011)
2. Balasubramaniyan, V.A., Ahamad, M., Park, H.: Callrank: Using call duration, social networks and pagerank to combat SPIT. In: Proceedings of CEAS 2007. ACM, New York (2007)
3. Bai, Y., Su, X., Bhargava, B.: Adaptive voice spam control with user behavior analysis. In: Proceedings of HPCC 2009. IEEE Computer Society, Los Alamitos (2009)
4. Quittek, J., Niccolini, S., Tartarelli, S., Stiemerling, M., Brunner, M., Ewald, T.: Detecting SPIT calls by checking human communication patterns. In: Proceedings of ICC 2007. IEEE Communication Society, Los Alamitos (2007)
5. Sarafijanovic, S., Perez, S., Boudec, J.L.: Improving digest-based collaborative spam detection. In: Proceedings of MIT Spam Conference 2008 (2008)
6. Damiani, E., Vimercati, S.D.C., Paraboschi, S., Samarati, P.: An open digest-based technique for spam detection. In: Proceedings of ISCA PDCCS 2004. ISCA (2004)
7. Wright, C., Ballard, L., Coull, S., Monrose, F., Masson, G.: Spot me if you can: Uncovering spoken phrases in encrypted VoIP conversations. In: Proceedings of S&P 2008. IEEE Computer Society, Los Alamitos (2008)
8. White, A.M., Matthews, A.R., Snow, K.Z., Monrose, F.: Phonotactic reconstruction of encrypted VoIP conversations: Hookt on fon-iks. In: Proceedings of S&P 2011. IEEE Computer Society, Los Alamitos (2011)
9. Rosenberg, J., Schulzrinne, H., Camarillo, G., Johnston, A., Peterson, J., Sparks, R., Handley, M., Schooler, E.: SIP: Session Initiation Protocol, RFC 3261 (2002)
10. Schulzrinne, H., Casner, S., Frederick, R., Jacobson, V.: RTP: A transport protocol for real-time applications, RFC 3550 (2003)
11. G.711, http://www.itu.int/rec/T-REC-G.711/e (visited at May 15, 2011)
12. Speex, http://www.speex.org/ (visited at May 15, 2011)
13. Schroeder, M., Atal, B.: Code-excited linear prediction (CELP): High-quality speech at very low bit rates. In: Proceedings of ICASSP 1985. IEEE Signaling Proceesing Society (1985)
14. Skype, www.Skype.com (visited at May 15th, 2011)
15. Festival, http://www.cstr.ed.ac.uk/projects/festival/ (visited at May 16, 2011)

16. Coskun, B., Memon, N.: Tracking encrypted VoIP calls via robust hashing of network flows. In: Proceedings of ICASSP 2010. IEEE Signaling Proceesing Society (2010)
17. Nilsimsa, `http://ixazon.dynip.com/ cmeclax/nilsimsa.html` (visited at May 16th, 2011)
18. Spamassassin, `http://spamassassin.apache.org/` (visited at May 16, 2011)
19. SMS spam corpus, `http://www.dt.fee.unicamp.br/ tiago/smsspamcollection/` (visited at May 16, 2011)
20. Rosenberg, J., Jennings, C.: The Session Initiation Protocol (SIP) and Spam, RFC 5039 (2008)
21. Shin, D., Ahn, J., Shim, C.: Progressive multi gray-leveling: a voice spam protection algorithm. IEEE Networks 20(5), 18–24 (2006)
22. Zhang, R., Gurtov, A.: Collaborative reputation-based voice spam filtering. In: Proceedings of DEXA Workshop 2009. IEEE Computer Society, Los Alamitos (2009)
23. Markkola, A., Lindqvist, J.: Accessible voice CAPTCHAs for internet telephony. In: Proceedings of SOAPS 2008. ACM, New York (2008)
24. Soupionis, Y., Gritzalis, D.: Audio CAPTCHA: Existing solutions assessment and a new implementation for VoIP telephony. Computers & Security 29(5), 603–618 (2010)
25. Salehin, S.M.A., Ventura, N.: Blocking unsolicited voice calls using decoys for the IMS. In: Proceedings of ICC 2007. IEEE Communication Society (2007)
26. Mehta, B., Nangia, S., Gupta, M., Nejdl, W.: Detecting image spam using visual features and near duplicate detection. In: Proceeding of WWW 2008. ACM, New York (2008)
27. Yeh, C., Lin, C.: Near-duplicate mail detection based on URL information for spam filtering. In: Chong, I., Kawahara, K. (eds.) ICOIN 2006. LNCS, vol. 3961, pp. 842–851. Springer, Heidelberg (2006)

Multi-stage Binary Code Obfuscation Using Improved Virtual Machine

Hui Fang[1], Yongdong Wu[1], Shuhong Wang[2], and Yin Huang[2]

[1] Institute for Infocomm Research,
1 Fusionpolis Way, 21-01, Singapore 138632
{hfang,wydong}@i2r.a-star.edu.sg
[2] Sumavision Soft Tech Co., Ltd.,
15 Kaituo Road, Shangdi District, Beijing, 100085, China
{wangshuhong,huangyin}@sumavision.com

Abstract. A software obfuscator transforms a program into another executable one with the same functionality but unreadable code implementation. This paper presents an algorithm of multi-stage software obfuscation method using improved virtual machine techniques. The key idea is to iteratively obfuscate a program for many times in using different interpretations. An improved virtual machine (VM) core is appended to the protected program for byte-code interpretation. Adversaries will need to crack all intermediate results in order to figure out the structure of original code. Compared with existing obfuscators, our new obfuscator generates the protected code which performs more efficiently, and enjoys proven higher level security.

1 Introduction

Software obfuscation refers to transformations on the code which becomes hard to understand while preserving all functionalities. It plays an importance role in protecting confidential data and algorithms from reverse engineering or virus modification [12,11,22,8]. Ideally, an adversary possessing a well-obfuscated program should be only able to learn program input/output like a black-box access. Due to this, software obfuscation has received many research interests for the last ten years [3,33,28,39,21,24,2,4,10].

The challenge in software obfuscation lies in whether or not guaranteed security and fair performance can be provided for obfuscated binary code. Specifically, code security implies resistance to static analysis and even dynamic analysis, and code efficiency implies that the obfuscated code should not run much slower than the original code. Up to now, some practical metrics for software obfuscation have been proposed in the literature [25,21,22,27,2,9]. Meanwhile, obfuscation on Turing machine programs with formal definitions has been researched intensively as well [3,28,15,42,6,5,17,7]. Unfortunately most practical obfuscation techniques lack a well-founded theoretical base, and thus it is unclear how effectively they perform. We take consideration of both practical and

X. Lai, J. Zhou, and H. Li (Eds.): ISC 2011, LNCS 7001, pp. 168–181, 2011.
© Springer-Verlag Berlin Heidelberg 2011

theoretical obfuscation metrics, and design our obfuscation algorithm align to theoretical definitions in principle.

We address the challenge by presenting an algorithm of multi-stage software obfuscation using improved virtual machine. The key idea is to obfuscate a software for many times while each time applying different interpretations in order to improve security. To fulfil the purpose, an improved virtual machine core responsible for byte-code interpretation is appended to the protected software. Under this design, an adversary must crack all intermediate results in order to figure out the structure of original code. Compared with existing obfuscators, our new obfuscator creates obfuscated code which performances more efficiently, and enjoys a higher security level.

The paper is organized as follows. Section 2 introduces the related work on software obfuscation and virtual machine. Section 3 describes our approach in two steps: block-to-byte virtual machine and multi-stage code obfuscation. Section 4 analyzes the security of our new software obfuscation algorithm. Section 5 provides experimental results. Finally, Section 6 draws a conclusion.

2   Related Work

Most existing obfuscation techniques on binary code fall into three categories:

- data transformation, such as name renaming and string encryption.
- instruction transformation, which replaces binary instructions using a library of equivalent instructions.
- control flow transformation, which transforms the graph structure of program control flow.

Data transformation does not alter program controls. Even the encrypted data will have to be decrypted inside the program for use. The code for decryption again faces the attack from reverse engineering. Therefore data obfuscation is usually applied together with other complicated obfuscation techniques to increase security [26,16,35].

Control flow transformation is relatively complicated [41,18,14,30,1]. Typically a control flow flattening method puts all basic blocks into a single switch statement which maintains whole control flow. It obfuscates the order in which the computations are carried out, in order to stand against static analysis. However, constant propagation on the switch variable will expose the next block to be executed. Besides, one large switch statement will generate many jumps which decreases program performance. Opaque predicates are boolean expressions whose values are known to the obfuscator but difficult for adversary to deduce. Junk codes are usually inserted into the dead path of an opaque predicate. However, for the same reason as above, there still exists risk that an adversary may figure out the value of an opaque predicate by static analysis.

Instruction transformation refers to replacement of protected binary instruction with a block of instructions which is functionally equivalent [20,19,23,29,32]. The introduced blocks representing native instruction are written as byte-codes into the program. Those byte-codes are often maintained by a virtual machine integrated with the obfuscated program. In practice, instruction transformation works well against static analysis except for runtime disassembly. However, little theoretical work has been carried out to show guarantee on its security and performance on obfuscated software.

Virtual machine (VM) based obfuscation recently becomes popular for software obfuscation, and it is probably the most sophisticated in the literature [36,34,32]. It usually integrates several obfuscation techniques including data permutation, instruction institution, and control flow transformation. As a result, VM obfuscation is fairly good against dynamic analysis in practice [40,37,31]. We observe the common way how VM obfuscator works, and summarize a general code structure for the program before and after obfuscation as shown in Figure 1. Generally speaking, a VM section will be appended to the original program, and the protected binary code will be transformed to byte-code, which is interpreted by a VM core. Finally, the entry point of the program will be redirected into VM code. To fulfil the byte-code fetching, VM core still needs to save all registers and flags in its own context, and to restore upon exiting byte-code interpretation.

Classical VM obfuscators suffer two drawbacks. Firstly, they generate obfuscated software which runs much slower than the original one. It is largely because of byte-code interpretation working style [37,40]. Secondly, the security of VM obfuscated program relies merely on an uncustomized VM core integrated with program rather than each individual program. VM does not restore byte-codes to original instructions any more. Therefore success of attacking obfuscated program requires two steps: understanding VM code, and decoding mapping between binary instructions and byte-codes. One round VM obfuscation will output relatively intelligible mapping, which allows an adversary to perform instruction level analysis, and further to reconstruct the structure of original software [34,32].

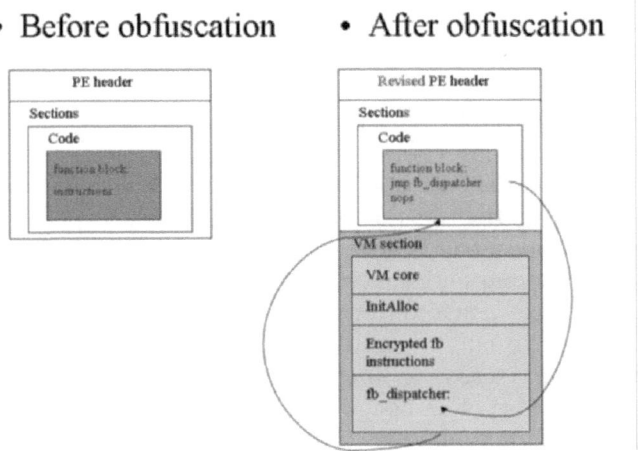

Fig. 1. Virtual machine based obfuscation

The existing works are promising under certain situations. However, the danger of software cracking is always changing and increasing [38,24]. Therefore we propose a new approach on software obfuscation in next section, introducing a more light-weighted obfuscator which generates harder understanding codes.

3   Our Approach

In this section we firstly introduce the concept of black box security, then present new design of block-to-byte virtual machine, and describe a framework of multi-stage code obfuscation based on improved virtual machine.

A program obfuscator is often regarded as a processor on computer programs, which outputs a new program of the same functionality but with unreadable code structure [28,10]. More precisely, a program obfuscator O is theoretically defined to be a probabilistic Turing machine or Boolean circuit, which satisfies three requirements [3]:

- (Functionality Equivalence) For every TM/circuit P and for every input $x : P(x) = O(P)(x)$.
- (Polynomial Slowdown) There exists a polynomial $q(.)$ such that for every TM/circuit P, $|O(P)| \leq q(|P|)$. TMs are additionally required that for every input x, if P halts after t steps on x then $O(P)$ halts within $q(t)$ steps on x.
- (Virtual Black Box) For any PPT A, there is a PPT oracle machine S and a negligible function $negl(.)$ such that for all TM/circuit P: $|Pr[A(O(P)) = 1] - Pr[S^P(1^{|P|}) = 1]| < negl(|P|)$.

Although Barak et al. [3] further proved that this kind of universal black box obfuscator does not exist, the theoretical concept is still useful in evaluating performance of code obfuscators. In other words, a good obfuscator shall as best as possible promise three properties: function equivalence, code efficiency, and black box security. In light of these requirements we present our customized VM obfuscator below.

3.1   Block-to-Byte Virtual Machine

The core of a virtual machine(VM) is a *dispatcher* which transforms byte-code to an implementation of binary instructions. To adapt to the purpose of program obfuscation, virtual machine must have *byte-codes* populated in and contain the *implementations* of all byte-codes for the program to protect. Specifically, a virtual machine will fetch byte-code one by one, position the target address in its *jump table*, and give control to the instruction in that address. So a complete virtual machine to be appended to the obfuscated program will be

$$V := \{Bytecodes, Impl, Jmptable, Dispatcher\}.$$

Classical VM obfuscator will map each binary instruction to a byte-code, together with its implementation (as described in Algorithm 1). We revise the design and present a block-to-byte VM obfuscation algorithm, as shown in Algorithm 2. The major difference lies in that a control flow graph (CFG) of the program is set up in prior, and then the obfuscator maps each basic block of the graph into a byte-code based on which the obfuscation is carried out.

Input: Original program P.
Output: Obfuscated program Q.
1 create a virtual machine V for P;
2 $V.Impl = \{\}$;
3 $V.Bytecodes = \{\}$;
4 **for** *binary instruction* $b \in P$ **do**
5   translate b into byte-code B with implementation $I(b)$;
6     $b = $ instruction "jump to V";
7     $I(b)$'s last instruction $= $ "jump to next to b";
8     $V.Jmptable[B] = I(b)$;
9     $V.Bytecodes+ = B$;
10    $V.Impl+ = I(b)$;
11 **end**
12 output $P + V$;

Algorithm 1. Classical VM based obfuscation

Input: Original program P.
Output: Obfuscated program Q.
1 construct control flow graph, $CFG(G)$;
2 create a virtual machine V for P;
3 $V.Impl = \{\}$;
4 $V.Bytecodes = \{\}$;
5 **for** *block* $BL \in CFG(P)$ **do**
6   translate BL into byte-code B with $I(BL) = \sum_{b \in BL} I(b)$;
7   BL's first instruction $= $ "jump to V";
8     $I(BL)$'s last instruction $= $ "jump to last of BL";
9     $V.Jmptable[B] = I(BL)$;
10    $V.Bytecodes+ = B$;
11    $V.Impl+ = I(BL)$;
12 **end**
13 output $P + V$;

Algorithm 2. Block-to-byte VM based obfuscation

Figure 2 shows the format for binary instructions and VM byte-codes respectively. It also gives an example how a binary instruction was transformed into byte-code together with an implementation.

VM dispatcher works on stack based style: it saves registers for native code and create own VM stack. The return value of last execution for each byte-code was saved in VM registers (*var_RegEip* and *var_RegDI* in Figure 3) for next byte-code execution. VM dispatcher then obtains the target address by searching a jump table using byte-code as index. Target address is the location that current instruction will transfer to. VM obfuscator retrieves all target addresses of the

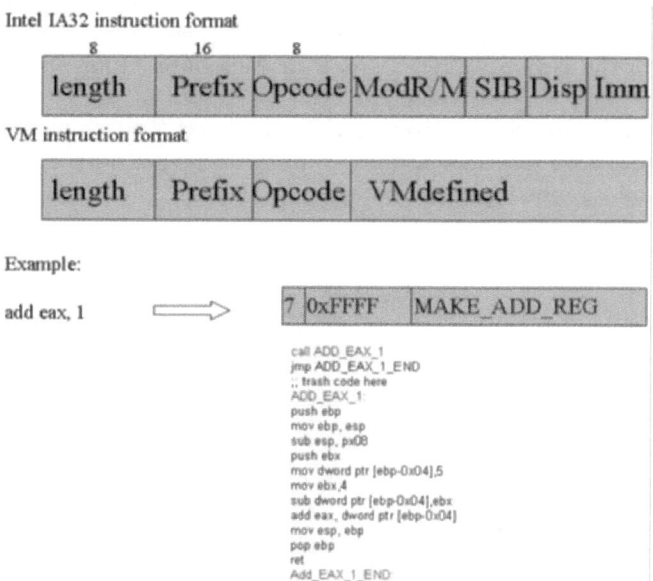

Intel IA32 instruction format

VM instruction format

Example:

add eax, 1

Fig. 2. Format of VM byte-code instruction and an example of implementation

original program in four different ways: for direct jump, target address is specified in the original instruction; for conditional jump, there are two target addresses with a predicate; for call instruction, one target address is set for called function, and another one for return address; and for return instruction, target address is stored on the stack.

3.2   Multi-staged Code Obfuscation

In this section we extend the technique of block-to-byte virtual machine to a multi-stage obfuscation. The idea of multi-stage obfuscation algorithm is described as follows. Given an original program P, we choose a random number n

```
00401060 VM_procedure  proc near
...
004010BC VM_Entry:    ; return here upon completion of each bytecode
004010BC      inc    [ebp+var_RegEip]
004010BF      mov    eax, [ebp+var_RegEip]
004010C2      mov    al, [eax]       ; fetch one byte from pseudo-code
004010C4      mov    [ebp+var_RegDI], al
004010C7      mov    eax, offset lpJumpAddrTable
004010CC      movzx  ebx, [ebp+var_RegDI]
004010D0      shl    ebx, 2   ; x4
004010D3      add    eax, ebx        ; look up jmp table
004010D5      jmp    dword ptr [eax] ; going to interpretation
004010D5 VM_procedure  endp
```

Fig. 3. VM byte-codes are executed by a dispatcher

to be the number of obfuscation stages, a one-way function f, and an obfuscation function Obf. Then we calculate multiple copies $\{P_0, P_1, ..., P_n\}$ of the program together with the keys $\{K_0, K_1, ..., K_n\}$ for each obfuscation stage, as shown in Figure 4.

We iteratively obfuscate program P for n times. The obfuscation key K_i is generated from each intermediate program P_i of the previous obfuscation stage, and K_i is again applied to P_i to compute P_{i+1}.

$$K_i = f(P_i),$$
$$P_{i+1} = Obf(P_i, K_i).$$

The function f maps any program into a key in binary string, satisfying that: f must have one-way hardness, and the output key can characterize the program. The examples of this type of function include: MD5 hash value of program where the program is feed as data, or the number of nodes in program's control flow graph.

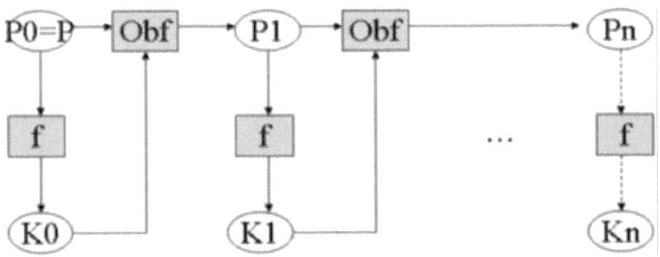

Fig. 4. The multi-stage obfuscation algorithm. P_n is output

The obfuscation of program requires to hide program's data and/or control flow while preserving all the functionalities. In other words, each copy P_i of the program must be executable and function normally. Our idea is to extract all $jmp/jcc/call$ points of P, and transform such information into a jump table. Then the jump table is obfuscated given a particular K and some dummy codes. Original program P is thus modified accordingly to jump table to preserve correct control. In other words, a separate hidden jump table will take control over program's running. Adversaries need to crack all intermediate obfuscated programs in order to recover original code's control flow.

For intra-block instructions or a single instruction, we use a revised tree structure to describe the whole process of multi-stage obfuscation. In this tree structure, each node represents a list of binary instructions (as shown in example of Figure 5). The root node x_1 refers to only one binary instruction, denoted by a circle. It links to its three children, V_1, V_2, V_3, which are different implementations of x_1. The children are called byte-codes, drawn in rectangles. Each byte-code, e.g. V_1, contains a list of binary instructions, e.g. $y_1 \rightarrow y_2 \rightarrow y_3$. In Stage-1 obfuscation, x_1 is assumed to be mapped into byte-code V_2; further in Stage-2, y_4 and y_5 of V_2 are mapped into V_5 and V_6 respectively. The path

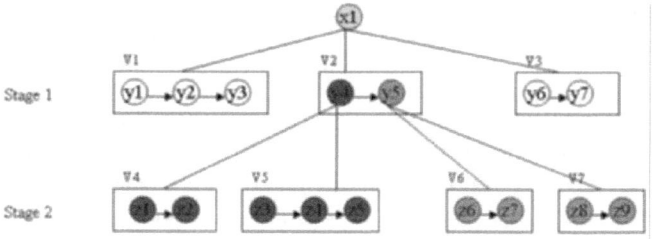

Fig. 5. Tree structure used in multi-stage obfuscation

selection from an earlier stage to next stage is determined by K_i. In the example case, a formal induction of resulted code would be

$$x_1 = V_2$$
$$= y_4 \rightarrow y_5$$
$$= V_5 \rightarrow V_6$$
$$= (z_3 \rightarrow z_4 \rightarrow z_5) \rightarrow (z_6 \rightarrow z_7)$$
$$= z_3 \rightarrow z_4 \rightarrow z_5 \rightarrow z_6 \rightarrow z_7.$$

4   Security Analysis

This section analyzes the security of multi-stage obfuscated program in two aspects: code efficiency and black box security. Specifically we strengthen the black box security by introducing code polymorphism during multi-stage obfuscation, and improve the code efficiency by removing unnecessary jump instructions during block-to-byte VM obfuscation.

4.1   Multi-stage Polymorphism

Polymorphism refers to that one binary instruction could have many byte-code interpretation with equivalent function. It is often used in code obfuscation to improve the difficulty in reversing program to original status.

When one instruction was obfuscated over twice, the mapping relationships from binary to byte codes become unrecognizable, due to many possible instruction combinations. Given an instruction sequence $z_3 \rightarrow z_4 \rightarrow z_5 \rightarrow z_6 \rightarrow z_7$, an adversary needs to separate them into byte-codes to understand the original program structure. In other words, one cannot easily split a sequence of instructions into correct $\{V_5, V_6\}$, and further obtain byte code V_2 which refers to x in first stage. Generally speaking, the fan-out width W of each binary node and the block size L of byte-code node for each stage determine the obfuscation complexity. In addition, the number n of stages is randomly chosen to control the complexity. The complexity of guessing increases exponentially with the number of stages. In this sense, multi-stage polymorphism makes the obfuscation of software more secure than the one obfuscated by single VM obfuscation. This claim is proved in Theorem 1.

Theorem 1. *An n-stage polymorphism tree provides $C(n)$ possible implementations for root node given constant W and L, where $C(n) = W^{L^{n-1}+...+L+1}$.*

Proof. Use mathematical induction. When $n = 1$, root node links to W children which are all available choices. So $C(1) = W$ satisfies the equation. Assume $C(k) = W^{L^{k-1}+...+L+1}$, and consider the case when $n = k + 1$. Firstly we notice that the number of choices owned by a binary component of each stage-1 node is $C(k)$. Since each node has L components, there will be $C(k)^L$ choices for solution passing through this node. Secondly we notice that the root node can choose path from its W children. So the total possible paths will be

$$\begin{aligned} C(k+1) &= W * C(k)^L \\ &= W * (W^{L^{k-1}+...+L+1})^L \\ &= W * (W^{L^k+...+L^2+L}) \\ &= W^{L^k+...+L^2+L+1}, \end{aligned}$$

which completes the proof.   □

4.2   Improved Execution Efficiency

The classical VM obfuscator transforms protected code into byte-codes. The resulted obfuscated program then interprets byte-codes sequentially, and runs the implementation of byte-codes accordingly. However, the program control will be unconditionally switched to VM dispatcher every time when one byte-code interpretation is completed. The number of *jmp*s inserted for byte-code interpretation is proportional to the number of binary instructions. It is well known that the jump operations block the instruction streamline for execution.

In contrast, our block-to-byte VM obfuscation chooses a "basic block" to execute before jumping back to VM dispatcher. There will be no new *jmp/jcc/call* instruction inserted inside one basic block. The obfuscated program only needs to interpret bytes representing basic blocks and follows the original control flow of the program. So the number of *jmp*s inserted for byte-code interpretation is only proportional to the number of nodes in program control flow graph. By interpreting a block of instructions into only one byte-code, our multi-stage VM obfuscator is able to reduce those unnecessary jumps during code obfuscation.

The number of *jmp* instructions in the program plays a heavy part in slowing down the program execution time. Given an average block size L of control flow graph of the program, our block-to-byte VM obfuscator will generate only $\frac{1}{L}$ the number of *jmp* instructions by the classical one.

5   Experiments

The testing experiment on our multi-stage VM obfuscation module was carried out on WinXP 2.4GHz CPU and 1G RAM platform. A demo of obfuscation out is given in Appendix A. Three parameters are take into consideration: structure of control flow graph, program size, and running time of obfuscated program. We adopt IDApro [13], a disassembly tool to facilitate view on IA-32 executables. VMprotect [40], a popular VM obfuscation software, was chosen for empirical comparison.

5.1 Control Flow Graph

The complexity of a program's control flow graph reflects program intelligibility to certain extent. We capture the number of nodes and edges in graph as an indicator of graph complexity. Accordingly, the *obfuscation level* is hereafter defined as the ratio of number of nodes or edges in CFG before and after obfuscation. Table 1 presents the obfuscation level for programs using multi-stage VM obfuscation. It implies that the control flow graph becomes interleaved which leads to high obfuscation level of program.

Table 1. The number of nodes and edges of control flow graph before and after obfuscation

Program	Original		Obfuscated		Obfuscation Level	
	#nodes,N	#edges,E	#nodes,N_2	#edges,E_2	N_2/N	E_2/E
md5	437	164	581	353	1.33	2.15
calc	458	175	746	308	1.63	1.76
draw	397	96	1439	258	3.62	2.69
crc32	151	47	354	125	2.34	2.66
aes	1908	517	3465	1392	1.82	2.70

5.2 Program Size

Program size is measured in two parameters: the number of instructions, and the size of program sections in bytes. Table 2 shows the program size of several programs before and after obfuscation. It tells that the number of instructions will normally increase at least four times after obfuscation, which implies the slowdown of obfuscated program.

5.3 Running Time

Table 3 provides the execution time of several x86 programs on average of 10000 times. It shows that our block obfuscator generates more efficient obfuscated

Table 2. Program size before and after obfuscation

Program	Original		Obfuscated		Increment Factor
	#instr, I	bytes	#instr, I_2	bytes	I_2/I
md5	675	1776	2837	9456	4.20
calc	485	825	2051	9559	4.23
draw	983	2109	8012	2935	8.15
crc32	231	583	1143	5665	4.95
aes	12302	32369	77748	314572	6.32

Table 3. Execution time (secs) of obfuscated programs

Program	Original T	VMprotect T_0	BlockVM T_1	MultiBlockVM($n = 2$) T_2	Slowdown T_2/T
md5	0.34	3.85	2.67	6.03	17.73
calc	0.12	3.40	2.34	8.73	72.75
draw	0.58	6.81	6.21	15.95	27.50
crc32	0.15	2.54	2.31	8.59	57.27
aes	0.23	4.59	5.43	11.15	48.48

code than classical VM obfuscator in one stage. However when given multi-stage obfuscation, the execution time of obfuscated program increases quickly due to more complicated obfuscation.

6 Conclusion

We have presented a new method to obfuscate code in multiple stages to protect software from reverse engineering. The key idea is to implement a block-to-byte virtual machine to interpret byte-codes, while modifying program structure iteratively. Block obfuscation hides the binary details into byte-codes while improving the program execution efficiency; multi-stage obfuscation hides the control flow of program in a more complicated level by using a polymorphism tree. Literally, an adversary will have to decode all n variants of program to obtain the structure of original program. Meanwhile compared with classical byte-code virtual machine obfuscation, block obfuscation makes the program run more efficiently by removing unnecessary jump instructions.

Acknowledgements. This paper is sponsored by the joint research project of MOST(2010DFA11110). We are grateful to Huang Xinyi for very helpful discussions and comments.

References

1. Abadi, M., Plotkin, G.: On protection by layout randomization. In: 23rd IEEE Computer Security Foundations Symposium, pp. 337–351 (2010)
2. Anckaert, B., Madou, M., De Sutter, B., De Bus, B., De Bosschere, K., Preneel, B.: Program obfuscation: a quantitative approach. In: ACM Workshop on Quality of Protection, pp. 15–20 (2007)
3. Barak, B., Goldreich, O., Impagliazzo, R., Rudich, S., Sahai, A., Vadhan, S., Yang, K.: On the (Im)possibility of obfuscating programs. In: Kilian, J. (ed.) CRYPTO 2001. LNCS, vol. 2139, pp. 1–18. Springer, Heidelberg (2001)
4. Beaucamps, P., Filiol, E.: On the possibility of practically obfuscating programs towards a unified perspective of code protection. Journal in Computer Virology 3, 3–21 (2007)
5. Bitansky, N., Canetti, R.: On Strong Simulation and Composable Point Obfuscation. In: Rabin, T. (ed.) CRYPTO 2010. LNCS, vol. 6223, pp. 520–537. Springer, Heidelberg (2010)

6. Canetti, R., Dakdouk, R.R.: Obfuscating Point Functions with Multibit Output. In: Smart, N.P. (ed.) EUROCRYPT 2008. LNCS, vol. 4965, pp. 489–508. Springer, Heidelberg (2008)
7. Canetti, R., Tauman Kalai, Y., Varia, M., Wichs, D.: On Symmetric Encryption and Point Obfuscation. In: Micciancio, D. (ed.) TCC 2010. LNCS, vol. 5978, pp. 52–71. Springer, Heidelberg (2010)
8. Cappaert, J., Preneel, B., Anckaert, B., Madou, M., De Bosschere, K.: Towards tamper resistant code encryption: Practice and experience. In: Chen, L., Mu, Y., Susilo, W. (eds.) ISPEC 2008. LNCS, vol. 4991, pp. 86–100. Springer, Heidelberg (2008)
9. Ceccato, M., Di Penta, M., Nagra, J., Falcarin, P., Ricca, F., Torchiano, M., Tonella, P.: The effectiveness of source code obfuscation -an experimental assessment. In: The 17th IEEE International Conference on Program Comprehension (ICPC), pp. 178–187. IEEE Computer Society, Los Alamitos (2009)
10. Collberg, C.: Tutorial: code transformation techniques for software protection. In: ACM SIGPLAN 2009 Conference on Programming Language Design and Implementation, PLDI 2009 (2009)
11. Collberg, C., Thomborson, C.: Watermarking, tamper-proofing, and obfuscation - tools for software protection. IEEE Transactions on Software Engineering 28, 735–746 (2002)
12. Collberg, C., Thomborson, C., Low, D.: A taxonomy of obfuscating transformations. Technical report (1997)
13. DataRescue. The ida pro disassembler and debugger (2005), http://www.hex-rays.com/idapro/
14. Ge, J.: Control flow based obfuscation. In: Proceedings of the 5th ACM Workshop on Digital Rights Management (DRM), pp. 83–92. ACM Press, New York (2005)
15. Goldweisser, S.: On the impossibility of obfuscation with auxiliary input, pp. 553–562. IEEE Computer Society, Los Alamitos (2005)
16. Hohenberger, S., Rothblum, G.N., Shelat, A., Vaikuntanathan, V.: Securely Obfuscating Re-encryption. In: Vadhan, S.P. (ed.) TCC 2007. LNCS, vol. 4392, pp. 233–252. Springer, Heidelberg (2007)
17. Hohenberger, S., Waters, B.: Constructing Verifiable Random Functions with Large Input Spaces. In: Gilbert, H. (ed.) EUROCRYPT 2010. LNCS, vol. 6110, pp. 656–672. Springer, Heidelberg (2010)
18. Jhala, R., Majumdar, R.: Path slicing. In: Proceedings of ACM SIGPLAN Conference on Programming Language Design and Implementation, PLDI 2005, pp. 38–47. ACM, New York (2005)
19. Kanzaki, Y., Monden, A., Nakamura, M.: A software protection method based on instruction camouflage. IEICE Transactions on Fundamentals of Electronics, Communications and Computer Sciences (Japanese Edition) J87-A(6):755-767, 47–59 (2004)
20. Linn, C., Debray, S.: Obfuscation of executable code to improve resistance to static disassembly. In: ACM Conference on Computer and Communications Security (CCS), pp. 290–299. ACM Press, New York (2003)
21. Lynn, B., Prabhakaran, M., Sahai, A.: Positive Results and Techniques for Obfuscation. In: Cachin, C., Camenisch, J.L. (eds.) EUROCRYPT 2004. LNCS, vol. 3027, pp. 20–39. Springer, Heidelberg (2004)
22. Madou, M., Anckaert, B., De Bus, B., De Bosschere, K.: On the effectiveness of source code transformations for binary obfuscation. In: Proc. of the Int'l Conf. on Software Engineering Research and Practice (SERP 2006), pp. 527–533 (2006)

23. Madou, M., Anckaert, B., Moseley, P., Debray, S.K., De Sutter, B., De Bosschere, K.: Software protection through dynamic code mutation. In: Song, J.-S., Kwon, T., Yung, M. (eds.) WISA 2005. LNCS, vol. 3786, pp. 194–206. Springer, Heidelberg (2006)
24. Madou, M., Van Put, L., De Bosschere, K.: Understanding obfuscated code. In: 14th IEEE Int'l Conf. on Program Comprehension (ICPC), pp. 268–274 (2006)
25. Mit, M.E., Ernst, M.D.: Static and dynamic analysis: synergy and duality. In: WODA 2003: ICSE Workshop on Dynamic Analysis, pp. 24–27 (2003)
26. Monden, A., Monsifrot, A., Thomborson, C.: Security improvements for encrypted interpretation. In: Proc. 3rd Workshop on Application Specific Processors (WASP) Digest, pp. 19–26 (2004)
27. Naeem, N.A., Batchelder, M., Hendren, L.: Metrics for measuring the effectiveness of decompilers and obfuscator. In: 15th IEEE Int'l. Conf. on Program Comprehension, pp. 253–258 (2007)
28. Ogiso, T., Sakabe, Y., Soshi, M., Miyaji, A.: Software obfuscation on a theoretical basis and its implementation. IEICE Transactions on Fundamentals of Electronics, Communications and Computer Sciences E86-A(1), 176–186 (2003)
29. Popov, I.V., Debray, S.K., Andrews, G.R.: Binary obfuscation using signals. In: USENIX Security Symposium (2007)
30. Dalla Preda, M., Madou, M., De Bosschere, K., Giacobazzi, R.: Opaque Predicates Detection by Abstract Interpretation. In: Johnson, M., Vene, V. (eds.) AMAST 2006. LNCS, vol. 4019, pp. 81–95. Springer, Heidelberg (2006)
31. Rolles, R.: X86 virtualizer (2008), http://rewolf.pl/
32. Rolles, R.: Unpacking virtualization obfuscators. In: Proceedings of the 3rd USENIX Conference on Offensive Technologies, WOOT 2009, p. 1. USENIX Association (2009)
33. Schwarz, B., Debray, S.K., Andrews, G.R.: Disassembly of executable code revisited. In: 10th Working Conference on Reverse Engineering, pp. 45–54 (2002)
34. Sharif, M., Lanzi, A., Giffin, J., Lee, W.: Automatic reverse engineering of malware emulators. In: Proceedings of the 30th IEEE Symposium on Security and Privacy, pp. 94–109. IEEE Computer Society, Los Alamitos (2009)
35. Sivadasan, P., Sojan Lal, P.: Jconsthide: a framework for java source code constant hiding. CoRR (2009)
36. Smith, J.E., Nair, R.: Virtual machines: versatile platforms for systems and processes. Morgan Kaufmann, San Francisco (2005)
37. Oreans Technologies. Code virtualizer, http://oreans.com/codevirtualizer.php
38. Udupa, S.K., Debray, S.K., Madou, M.: Deobfuscation: reverse engineering obfuscated code. In: 12th Working Conference on Reverse Engineering, pp. 45–54 (2005)
39. van Oorschot, P.C.: Revisiting Software Protection. In: Boyd, C., Mao, W. (eds.) ISC 2003. LNCS, vol. 2851, pp. 1–13. Springer, Heidelberg (2003)
40. VMPsoft. Vmprotect software, http://www.vmprotect.ru/
41. Wang, C., Hill, J., Knight, J.C., Davidson, J.W.: Protection of software-based survivability mechanism. In: Proceedings of the International Conference on Dependable Systems and Networks (formerly: FTCS), DSN 2001, pp. 193–202. IEEE Computer Society, Los Alamitos (2001)
42. Wee, H.: On obfuscating point functions. In: Proceedings of the 37th Annual ACM Symposium on Theory of Computing, STOC 2005, pp. 523–532. ACM, New York (2005)

A   Sample Output of Obfuscation

A function named *modexp* is to be obfuscated:

```
// modular exponentiation = base^exp % mod
int modexp (int base, int exp, int mod)
{
    int c = 1, expNum = 0;
    do
    {
        expNum++;
        c = (base * c) % mod;
    }
    while (expNum < exp);
    return c;
}
```

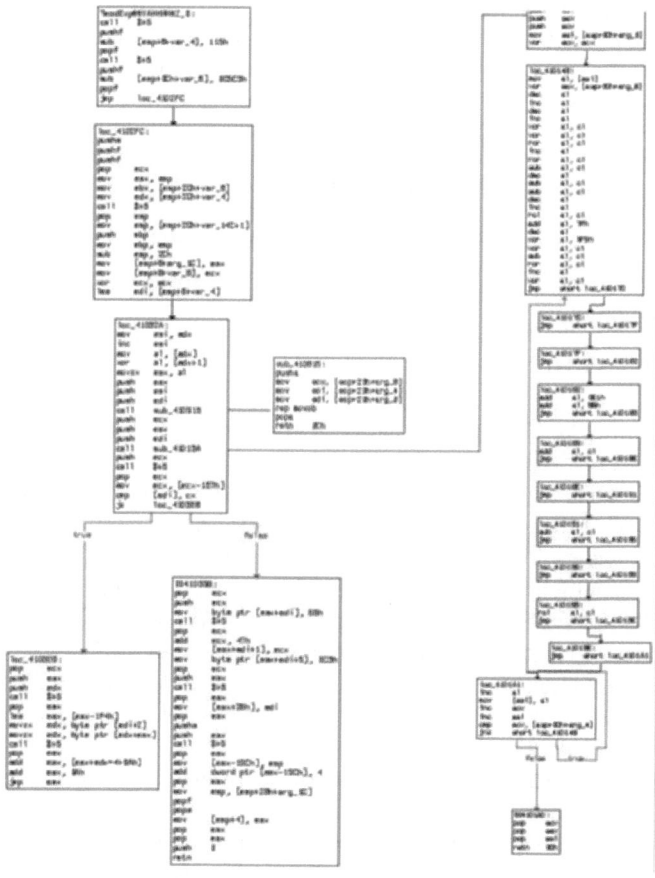

Fig. 6. CFG of obfuscated *modexp* function

Detection and Analysis of Cryptographic Data Inside Software*

Ruoxu Zhao, Dawu Gu, Juanru Li, and Ran Yu

Lab of Cryptology and Computer Security,
Dept. of Computer Science, Shanghai Jiao Tong University, Shanghai, China

Abstract. Cryptographic algorithms are widely used inside software for data security and integrity. The search of cryptographic data (include algorithms, input-output data and intermediated states of operation) is important to security analysis. However, various implementations of cryptographic algorithms lead the automatic detection and analysis to be very hard. This paper proposes a novel automatic cryptographic data detection and analysis approach. This approach is based on execution tracing and data pattern extraction techniques, searching the data pattern of cryptographic algorithms, and automatically extracting detected Cryptographic algorithms and input-output data. We implement and evaluate our approach, and the result shows our approach can detect and extract common symmetric ciphers and hash functions in most kinds of programs with accuracy, effectiveness and universality.

Keywords: Cryptographic data, Symmetric Cipher, Hash Function, Data Pattern, reverse engineering.

1   Introduction

The use of cryptographic algorithms to protect private information is common in software. Software dealing with huge amount of data such as Archive and compression tools, Disk encryption tools, Instant Messengers often use symmetric ciphers and hash functions to encrypt, decrypt and verify the data.

In practice, the complexity of binary program understanding makes analysts hard to identify which ciphers are inside the software, even only standard algorithms such as the AES, RC4 or SHA-1 are used. What's more, many programs achieve security through obfuscation. For instance, SkyPE uses RC4 algorithm while obfuscating it so that analysts spent years to understand[5]. It is important to develop automatic techniques for the analysts to detect specific cryptographic algorithms before security analysis.

Compared to the theoretical analysis of cryptographic algorithms, the analysis of the implementation is at most a craft rather than a science [12] [8] [9]. The main difficult is that the implementations of one algorithm might be various even if the mathematical abstraction is the same. For instance, the AES

---

* Supported by SafeNet Northeast Asia grant awards.

takes different implementations on 8-bit platform and 32-bit platform. Many cryptography libraries also use loop unwinding to optimize the algorithms and yet change the form of implementations. Malicious program even modifies or obfuscates the code, trying to fail the analysis. How to identify cryptographic algorithm inside program accurately and effectively is still an open problem that the existing tools cannot solve perfectly.

In this paper, we take the first leap toward cryptographic data detection and analysis based on the *data pattern*. To the best of our knowledge, all existing cryptographic algorithm identification techniques focus on program analysis or memory dump analysis [11] [6]. These techniques try to recover the abstract structure of algorithms inside programs or dumped memory and judge the existence of certain ciphers. Our approach, however, observes the data feature and dependency of specific ciphers during runtime information. We do program tracing first and conduct data analysis to extract the so called *data pattern*, which is the input-output of certain instruction collection. The pattern gives the analyst clues to quickly detect and understand the encryption process of the program. We implement an analysis system to achieve the goal of automatic identification, and the results show that our system can not only detect symmetric ciphers and hash functions in most kinds of programs with high accuracy, but could also extract cryptographic parameters such as expanded round key automatically.

The approach we proposed is able to reduce manual work significantly in debugging, forensic analysis and reverse engineering. Furthermore, the universal model we adopted is expected to be applied to different implementations of the same cipher regardless of the programming language.

2   Background and Related Work

Identification of cryptographic algorithm and data is an important yet seldom discussed topic in program analysis research. State of the art tools and techniques for cryptographic algorithm identification are divided into static based and dynamic based. In this section we summarize existing works and discuss their inadequacy.

2.1   Static Analysis Based Approaches

Static analysis based cipher identification is the most widely used technique. Many tools have been developed to help analyzing such as Krypto Analyzer (KANAL) and SnD Crypto Scanner. The key step of static analysis is to parse binary or source code of the program and try to find unique pattern of specific ciphers. Static analysis based approaches strongly rely on signatures, which are often the constant values related to certain ciphers or specific instruction sequence related to certain version of cryptographic libraries. The static based approaches have many defects. First, they often rely on pre-build signature libraries, and detection fails when the signature changes with software updating or code re-compile. Second, they cannot deal with packed programs because the normal code is compressed or encrypted before execution. Finally, static analysis based approaches only detect the existence of ciphers, but cannot analyze particular encryption and decryption data.

2.2 Dynamic Analysis Based Approaches

Dynamic analysis of software is the hot topic of security analysis in recent years especially using emulation technique or program instrumentation. Although many approaches and tools have been developed to do universal analysis[2] [4], Felix Gröbert's work[7] is the first significant dynamic analysis focusing on cipher identification. His work uses PIN tools[10] to dynamic trace the program, and then mixes signature based searching with simple memory reconstruction and searching. However, the model adopted by [7] takes advantage of many observation. For instance, the proposed signature-based and generic bitwise-arithmetic/loop based identification methods are all based on signatures or unique tuples, which are not so dynamic and universal. The only general identification method in [7] is generic memory-based identification method. The method is focused on memory data and uses verifiers to confirm an XOR encryption or a relation- ship between the input and output of a permutation box. A set of possible key, plaintext, and ciphertext candidates are passed to a reference implementation of the particular algorithm. If the output of the algorithm matches the output in memory, the verifier has successfully identified an instance of the algorithm including its parameters. Although this method exploits relationship between plaintext and ciphertext, many potential information is ignored. On the other hand, in digital forensic research, novel methods for cryptographic key identification in RAM are proposed[11] which relies on the property that the keys in memory is far more structured than previously believed.

In this paper we combine the dynamic analysis with structured key data. First we define the input-output of certain instruction collection as *data pattern*, then using the concept of data pattern we can easily analyze the uniqueness of cryptography algorithms by exploit details about the algorithms. For instance, according to the key concept of Symmetric ciphers - *pseudorandomness*, we can search and find the pseudorandom data pattern and test if the data pattern correspond to certain Symmetric ciphers.

3 Cryptographic Data Pattern Analysis

Our goal is to automatically detect cryptographic data, which includes the executed instructions of cryptographic primitives, encrypted data and secret keys. The main idea is to analyze a program's runtime data[1] rather than instructions and generate some data patterns matching to certain cryptographic primitives.

Although the mathematical definition of cryptographic primitives are determinate, the implementations of the same cryptographic algorithm can be quite different. For example, real-world programs may use optimizations like table lookup to speed up the cryptographic algorithms(e.g., AES fast implementation). Common cryptographic libraries such as OpenSSL and Crypto++ also take different approaches to implement the same algorithm. Programmers may even have their custom implementations. What's more, code obfuscation is often used to protect software, which makes the obfuscated code extremely difficult to analyze. However, we discovered that the input and output data must fulfill certain relations for deterministic algorithms. That is, if the input is given, there should be only one single possible output to a deterministic algorithm. By verifying if the input and output data match the pattern of a certain algorithm, we can say that a program execution contains the data characteristics of a certain algorithm with high credibility. Even if the program is obfuscated, the input and output data has to be present in the program execution data, which can be then analyzed regardless how the data is processed.

Because the size of traced data is usually very large, we have to determine the data sampling points. We found that modern computer programs are highly structured. The control flow of a traced program tells us how a program is executed. Although instructions are not important to data analysis, they help to build up high-level structures of traced programs. In our analysis, we have four levels of data representations during the analyzing process. The structure of these high-level representations are shown in Figure 1.

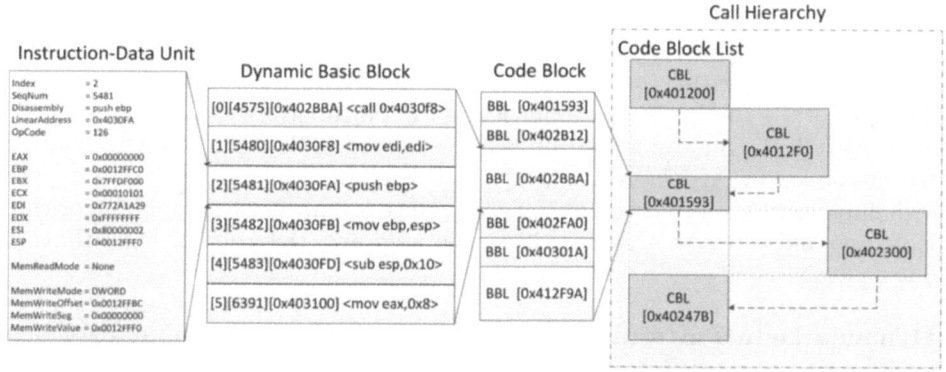

Fig. 1. Data Representations

Instruction-Data Unit. An Instruction-Data Unit is the basic unit of program tracing. It contains the instruction binary data, register values, linear address, memory access information, etc. A traced file usually contains millions of Instruction-Data Units.

Dynamic Basic Block. A Dynamic Basic Block(Dynamic BBL) contains a sequence of instructions to form a fixed group which has only one entering and one exiting instruction. The Dynamic Basic Block Generation algorithm is a little different than static ones, because we have to determine the control flow according to the actual traced result.

Code Block. A Code Block(CBL) consists of a sequence of continuous Dynamic BBLs that are executed without calling other functions. That is, a new CBL is generated when a call instruction is executed. Code Blocks are used to construct the Call Hierarchy of a function.

Call Hierarchy. A Call Hierarchy of a traced program is a recursive structure of Code Blocks, represented by a Code Block List(CBL List). A CBL List contains a single function call, which may call other functions during its execution, thus a CBL List may contain other CBL Lists to build up the recursive structure of a function call.

3.1   Data Patterns

Based on the concept of Instruction-Data Flow, the analyst could extract Intermediate Data State at any arbitrary time of execution. The Intermediate Data State contains the virtual memory state after certain instructions are executed. As mentioned earlier, we made the assumption that the parameters of cryptographic primitives must appear in memory during its execution. Our goal is to verify the existence of cryptographic algorithms by examining if the input and output parameters which match a certain pattern are contained in memory, and therefore we can extract the parameters.

A data pattern are defined as the mathematical relationship between the input and output data of a particular cryptographic algorithm. We know that for deterministic algorithms, output data is determined once the input data is given. This pattern is the key feature that is used in our analysis. It is unnecessary to know specifically how an algorithm is implemented in the target program; all we have to do is to verify the relationships between input and output data using our own implementation. Basically, the data pattern for any cryptographic algorithm is unique and concrete, therefore it can be used as a signature of algorithms in our analysis.

Dynamic Data Patterns. A dynamic data pattern is a group of data that matches one or more data templates at runtime. Dynamic data patterns must be verified at runtime, because the content of data cannot be pre-defined. Some examples of dynamic data patterns are:

- Feistel cipher
  Feistel cipher encryption takes plaintext and a group of keys as input, and ciphertext as output. Here the plaintext, ciphertext and keys are not pre-defined, but can be verified during runtime. We describe the Feistel cipher encryption calculation as $F(pt, k)$, which takes the plaintext(pt) and keys(k) as input parameters, and outputs the ciphertext(ct). By dynamically verifing if pt, k and ct satisfy $ct = F(pt, k)$, we can verify the existence of the data pattern of Feistel cipher encryption.
- Rijndael key expansion
  Rijndael key expansion is used to expand a short key into a group of separate round keys. Also, neither the input key nor the expanded keys can be predefined.

- RC4 key scheduling
  Similar to Rijndael key expansion, the input and output of RC4 key scheduling is unknown until runtime.

Static Data Patterns. Unlike dynamic data patterns, static data patterns can be pre-defined. They can be simply a block of data with known content. A good example of static data patterns is the 256-byte S-box and inverse S-box for AES. Their content is pre-determined and usually directly appear in memory. Another example is the constants in hash functions such as SHA-1. In our analysis, we do not directly use constant signatures as direct evidence of existence of cryptographic algorithms, but they can be used to locate the cryptographic routines.

Data Element Formats. The Intermediate Data State of a program trace at any arbitrary time consists of a group of memory chunks. A memory chunk is a block of memory that is continuous in its linear address, demonstrated in Figure 2. Data elements are extracted from these memory chunks. There are three kinds of data elements in our analysis:

- Fixed length. For example, a 128-bit memory chunk containing a block of AES plaintext.
- Variable length. For example, the expanded key in RC4 can be either 256 bits(8-bit each element) or 1024 bits (32-bit each element).
- Arbitrary length. For example, the input key in Blowfish can be from 1 bit to 448 bits.

```
0012FE78   09 57 6E 7F 0C FE 12 00 34 FF 12 00 20 33 40 00    .Wn..þ..4ÿ.. 3@.
0012FE88   00 1C 40 00 14 FF 12 00 31 57 6E 7F 50 FF 12 00    ..@..ÿ..1Wn.Pÿ..
0012FE98   C8 5F 39 00 14 FF 12 00 C8 5F 39 00 90 FE 12 00    È_9..ÿ..È_9..þ..
0012FEA8   34 FF 12 00 75 32 40 00 00 00 00 00 7C FF 12 00    4ÿ..u2@.....|ÿ..
0012FEB8   20 16 40 00 E8 5F 39 00 14 FF 12 00 00 14 40 00     .@.è_9..ÿ....@.
0012FEC8   00 14 40 00 FF FF FF FF 00 00 00 00 00 14 40 00    ..@.ÿÿÿÿ......@.
```

Fig. 2. Memory Chunk

Another thing we should pay attention to is that data elements are usually aligned. In x86 architecture, a block of memory usually has a 32-bit alignment to get maximum performance. So we treat alignment as another property for data elements.

4   Implementation

Our whole program analysis system consists of two parts: the front end is a tracing engine called Fochs, and the back end is a program analyzer called Lochs. A system architecture overview is shown in Figure 3.

To conduct an analysis, the testing programs are first executed in Fochs program tracer. The trace is done manually, and its result is saved to trace database for analysis. Then traced data is analyzed in Lochs program analyzer, where possible cryptographic algorithms are examined. After the analysis, a report is generated with the analysis results.

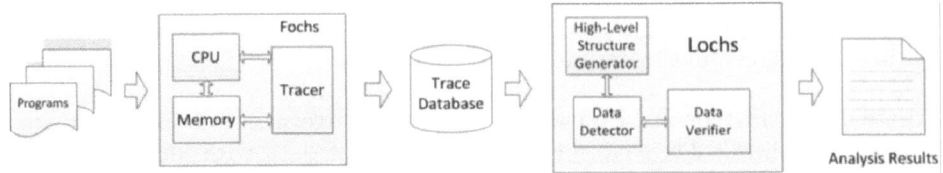

Fig. 3. System Architecture Overview

4.1 Fochs: Data Tracing System

The data tracing system we use is the digital forensic analyzer called Fochs. Fochs is based on the open-source Bochs x86 PC emulator[3]. The reason why we choose Bochs is that Bochs performs full-system emulation, and we can conveniently access the CPU status and memory status. We modifies Bochs so that it can trace program execution including its context, and save the trace result for further analysis. The structure of Fochs is shown in Figure 4.

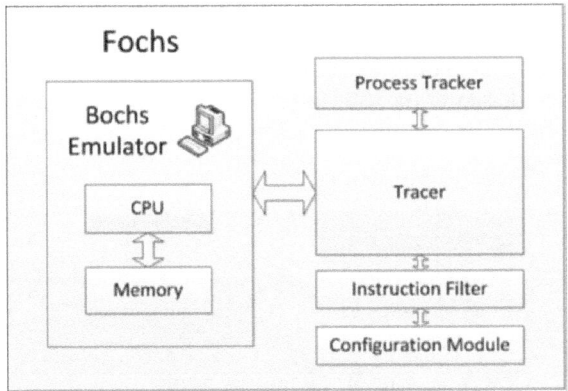

Fig. 4. Fochs program tracer

To get a valid and usable trace result, there are three major problems we should solve: what context data should be traced during program execution, which instructions should be traced, and how can we pick the process we want to trace in a multi-process environment. We analyzed our requirements and came up with solutions to these problems.

Execution Context. To get a valid program trace result, we have to log the register values and memory accesses for each instruction. We figured that only the register values before each instruction execution are necessary for our analysis, so only the values of general purpose registers before each instruction execution are traced. We also found that in common cases, each instruction has at most one memory reading and one memory writing in the current x86 architecture, thus one memory reading and writing is traced for each instruction.

Another important value is the linear address of each instruction, which is the Instruction Pointer register value. Also the instruction binary code is traced for disassembling.

Some repeat speedup instructions may have multiple memory accesses in a single instruction execution. In our implementation of Fochs, we trace the instruction whenever it has more than one memory access of the same type (reading/writing), acting like a single instruction executed multiple times. We also disabled the MMX and SSE instructions so that no more than 32-bit size of memory can be access during one cycle of instruction execution.

Instruction Filtering. If we trace every instruction that CPU executes, the result would be tremendous. We have to eliminate the number of instruction traced to focus on the instructions that contain our analyzing target. We do instruction filtering primarily based on the linear address. In Windows operating system, user space and kernel space are separated, where user space is in low address and kernel space is in high address. The address where an executable is loaded into memory can be easily found using any PE analysis tool or debugger. We limit the linear address that we trace to the bounds of the traced executables, and in this way we are able to ignore the unnecessary OS execution code such as process scheduling, and unnecessary user-space DLLs are also ignored. A configuration module is used to provide different configurations to trace different programs.

There are also times that only the instructions with memory accesses should be traced, because our analysis is based on memory data, and those instructions that have no memory access can be ignored. However, we still have to keep the branch instructions to build high-level representations such as Code Blocks and Call Hierarchy.

Process Tracking. Another critical feature that should be provided by the tracer is that only one single process is traced in a multi-process environment. In Windows operating systems, each process has a unique CR3 register value. CR3 register is used to locate page directory address for the current process. We track a process by filtering the CR3 register value: first, the entry address for each executable is manually obtained, and then whenever CPU runs to the entry point, the current CR3 register value is saved, which is the unique value for the desired process. In this way, we can successfully get rid of the interference of other unrelated processes.

By instruction filtering and process tracking, the trace can be focused on a single process. But still, the number of traced instructions can be quite large, usually 10^6 (100MB data) to 10^7 (1GB data). So the traced result has to be saved to disk for further analysis.

4.2   Lochs: Data Analysis System

The back end of our program analyzing system is called Lochs. Lochs analyzes cryptographic primitives of the traced results of Fochs automatically. There are three stages of data analysis. First, Lochs constructs high-level structures of the traces, including Dynamic Basic Blocks, Code Blocks and Call Hierarchies; and then, data reduction is performed to eliminate the unnecessary data to be analyzed; at last, Lochs does template verifications on the selected data to examine cryptographic algorithms and their parameters. The structure of Lochs is shown in Figure 5.

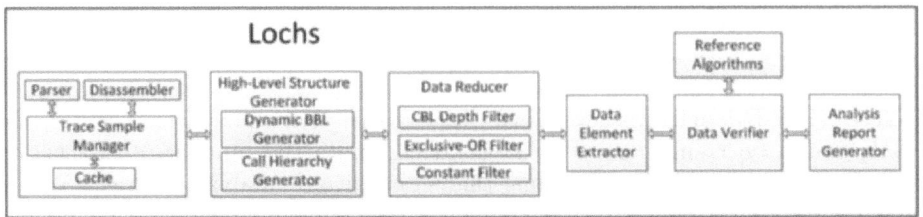

Fig. 5. Lochs program analyzer

High-Level Representations. Before data analysis, we have to extract data from the traces first. The points where Intermediate Data States are sampled are critical to our analysis, because we have to select the points where cryptographic data is most likely to appear. We may sample memory data at each instruction trace, but this is obviously impossible to analyze for common traces that contain 10^7 instructions. To solve this problem, we first construct high-level representations for the traces. Three levels of data representation are constructed: Dynamic Basic Blocks, Code Blocks and Call Hierarchies(CBL Lists). The algorithms to build these high-levels are mentioned earlier. These representations are constructed only once and then serialized to or deserialized from disk for future uses.

In the first stage of analysis, trace results are converted from binary data to CBL Lists. These CBL Lists are passed on to the second stage for further processing.

Heuristic Data Reduction. A complete trace of a program often contains huge amount of data unrelated to cryptography, even though these data is pre-filtered in the tracing process. These unrelated data may include program initialization, GUI operations, user input handling, error handling, etc. Therefore, a highly-optimized data reduction is performed in the second stage to reduce analysis time. After Call Hierarchies are constructed, heuristic data reductions are conducted on these Call Hierarchies which are represented by CBL Lists. Currently we have mainly three kinds of data reduction methods.

– CBL List Depth
  Functions that contain cryptographic primitives usually have a lower depth. That is, these functions usually have a single purpose, and they are less likely to call other functions because of performance issues. Therefore, cryptographic functions are most likely to appear in the inner CBL Lists, which have a low depth value.

During the data reduction procedure, we first filter the CBL Lists according to their depth value. A threshold of depth 6 is reasonable to most of the analysis.

– Exclusive-OR Instructions

Through observations to cryptographic algorithms, the exclusive-or operation is heavily used. The second data reduction method is based on the idea that cryptographic functions should contain a certain percentage of exclusive-or instructions. This heuristic method is applicable because we can safely judge that a function contains no cryptographic primitives if it has no exclusive-or operations.

– Characteristic Constants

Many cryptographic algorithms and their implementations contain characteristic constants. For example, in the fast implementation of AES, a 1k-size lookup table is commonly used; in hash functions like MD5 and SHA-1, several pre-defined constants are quite unique and must be used. These characteristics are used to filter CBL Lists for a specific algorithm, and those CBL Lists where characteristic constants appear are analyzed first.

Through data reduction, usually more than 90% of total data can be reduced. The reduced data is then passed to algorithm detectors to test if it contains a specific data pattern.

Data Verification. In the final stage of analysis, data is extracted from their high-level representations, and algorithm verifiers verify the extracted data to test if it satisfies a certain data pattern. Based on our earlier observation that program functionalities are implemented in the unit of functions, data analysis is conducted on Call Hierarchies. A Call Hierarchy is the representation of an entire function call, including its data. We compute the input and output for each function call, and verify if these data matches any data pattern of cryptographic algorithms.

First, for each CBL List l, we compute its input data $IN(l)$ and output data $OUT(l)$. Data exists in format of continuous memory chunks, which may contain cryptographic parameters.

Then, for a specific cryptographic algorithm, its data format is pre-defined. Possible data element combinations are extracted recursively, and passed to verifiers to test cryptographic algorithm existence and extract parameters.

At last, the verifier receives data elements, and test if they satisfy a pre-defined pattern. Each verifier implements a reference algorithm. This algorithm can be quite simple (testing constant existence), or rather complicated (AES block encryption). If a group of data elements matches a pre-defined pattern, the parameters are extracted from the data elements, and the detection procedure is successful. We can expand the usage of our system by implementing more reference algorithms, and add them to the Reference Algorithms module in Lochs, as shown in Figure 5. The extensibility of our analysis system is guaranteed by its modular architecture.

5   Evaluation

We do our experiments using real-world applications as well as custom programs. There are five kinds of testing programs:

- *Compression tools*, including RAR 3.93 (AES encryption, SHA-1 hashing) and FreeArc 0.666 (Blowfish encryption)
- *File encryption tools*, including AES Crypt 3.08 (AES file encryption) and TrueCrypt 7.0a (disk formatting using AES encryption)
- *Cryptography softwares*, including Putty 0.60 (login sessions with AES/ Blowfish encryption) and KeePass Password Safe 1.19b (password database saving)
- *Custom programs with different implementations*, including AES-OpenSSL (AES 128-bit and 256-bit block cipher), AES-OpenSSL-CBC (AES CBC mode cipher), MD5-OpenSSL (MD5 message digest), SHA1-OpenSSL (SHA-1 message digest), AES-Custom-Impl (a custom implementation of AES), RC4-OpenSSL (RC4 cipher), RC4-Custom-Impl1 (a custom implementation of RC4) and RC4-Custom-Impl2 (another different custom implementation of RC4).
- *Custom programs obfuscated by VMProtect and Themida*, including custom programs with AES, RC4, SHA-1 OpenSSL implementations that both the executable and OpenSSL libraries(libeay32.dll, ssleay32.dll) are obfuscated by VMProtect, and a custom program with AES OpenSSL implementation that only the executable is obfuscated by Themida and the OpenSSL libraries are original.

We implemented 8 reference algorithms, which are: AES 128-bit key expansion/block cipher, AES 256-bit key expansion/block cipher, Blowfish key scheduling, RC4 key scheduling, MD5 message digest and SHA-1 message digest. The block ciphers take a block of data and a group of expanded keys as input, and a block of data as output. The key expansions and key schedulings take a short key as input, and an expanded key as output. And the message digests take a block of data and an input message digest as input, and an updated message digest as output.

We run all of the reference algorithms on each of the traces of test programs. The testing results and performance analysis are shown in the following sections.

5.1   Accuracy

We successfully discovered the existing pre-known algorithms in all of the testing programs, and extracted the parameters including AES keys, plaintexts and ciphertexts, Blowfish keys, RC4 keys, MD5 input data and SHA-1 input data. There are some flaws that we failed to discover AES block cipher in Putty AES encrypted login session, and the AES block cipher in our custom implementation of AES. We also found a previously unknown SHA-1 algorithm in the first custom implementation of RC4. We also successfully discovered the underlying algorithms as well as the plaintexts, ciphertexts and secret keys in programs obfuscated by VMProtect[14] and Themida[13], and the analysis results are the same as the results without code obfuscation. The analysis results of testing programs are shown in Table 1.

Table 1. The Test Results

	algorithm(s)	key expansion/key scheduling
RAR 0	SHA-1	
RAR 1	AES(128-bit) encryption	AES-128 Key Expansion
FreeArc	-	Blowfish Key Scheduling
AES Crypt	AES(256-bit) encryption	AES-256 Key Expansion
TrueCrypt	-	AES-256 Key Expansion
Putty(AES)	-	AES-256 Key Expansion
Putty(Blowfish)	-	Blowfish Key Scheduling
KeePass AES	AES(256-bit) encryption	AES-256 Key Expansion
AES128-OpenSSL-ECB	AES(128-bit) encryption	AES-128 Key Expansion
AES256-OpenSSL-ECB	AES(256-bit) encryption	AES-256 Key Expansion
AES128-OpenSSL-CBC	AES(128-bit) encryption	AES-128 Key Expansion
AES-Custom-Impl	AES(128-bit) encryption	AES-128 Key Expansion
RC4-OpenSSL RC4	-	RC4 Key Scheduling
RC4-Custom-Impl1	-	RC4 Key Scheduling
RC4-Custom-Impl2	-	RC4 Key Scheduling
MD5-OpenSSL	MD5	-
SHA1-OpenSSL	SHA1	-
AES128-VMProtect	AES(128-bit) encryption	AES-128 Key Expansion
AES256-VMProtect	AES(256-bit) encryption	AES-256 Key Expansion
RC4-VMProtect	-	RC4 Key Scheduling
SHA1-VMProtect	SHA1	-
AES256-Themida	AES(256-bit) encryption	AES-256 Key Expansion

It's shown that our approach can successfully identify the same algorithm with different implementations. For example, one of the two custom implemented versions of RC4 uses a 32-bit memory to store an 8-bit value, while the other uses an 8-bit memory. Another example is that we use a regular implementation of AES as well as a fast implementation, which has optimizations such as table lookup. Our analysis doesn't rely on a specific implementation of a certain algorithm, so both implementations are successfully identified.

Real-world softwares may have countermeasures against this analysis method. For example, continuous memory chunks can be broken into smaller chunks to avoid matching a certain data pattern. However, these countermeasures require specific programming, and we did not find any software that uses such a countermeasure. To cope with these countermeasures, we can use non-perfect matching such as fuzzy matching, which gives a high possibility of the existence of certain algorithms.

False negatives may occur if the software contain countermeasures against this method. In our evaluation, 2 cases only identified AES key expansion process but not the AES encryption. One of the possible reasons is that our analysis optimization ignored deeper function calls which contain the encryption process. We can refine the optimization stage to resolve this issue. Because of the uniqueness of cryptographic data, false positives can be very rare. Real-world softwares hardly contain cryptographic data in both the input and output of a function call, which actually has no cryptographic primitive. We found no false positives during our analysis.

These test results demonstrated that our analysis is successful, in both real-world applications and custom implemented programs, and can be used to analyze obfuscated code.

5.2   Performance

The tracing process is usually manually operated, and the tracing time is trivial and can be ignored. The later analyzing process is fully automated, and the time of each analyzing stage is recorded and listed below, where Stage 1 represents the construction of high-level representations(Dynamic BBL, CBL, and CBL List), and the algorithm names represent time used to analyze each algorithm.

The performance evaluation is shown in Table 2, including the total number of instructions and file size of each trace, the analysis speed (in instructions per second), time used for each stage and algorithm, and the total time used. The results show that the average file size is about 500MB(5M instructions), the

Table 2. Performance

	Instructions	Size	Speed (instrs/sec)	Total Time
RAR 0	919k	77MB	19.4k	47s
RAR 1	1,359k	114MB	9.6k	2m22s
FreeArc	7,786k	653MB	34.6k	3m45s
AES Crypt	2,396k	201MB	14.1k	2m50s
TrueCrypt	12,800k	1,074MB	15.0k	14m11s
Putty (AES)	3,651k	306MB	14.5k	4m11s
Putty (Blowfish)	7,297k	612MB	18.2k	6m42s
KeePass	9,005k	755MB	22.5k	6m40s
AES-OpenSSL-128	5k	467KB	9.2k	< 1s
AES-OpenSSL-256	6k	500KB	7.9k	< 1s
AES-OpenSSL-CBC	6k	510KB	20.3k	< 1s
AES-Custom-Impl	11,294k	947MB	17.7k	10m36s
RC4-OpenSSL	10k	840KB	14.5k	< 1s
RC4-Custom-Impl1	298k	25MB	36.3k	8s
RC4-Custom-Impl2	10,078k	845MB	16.5k	10m13s
MD5-OpenSSL	4k	383KB	16.8k	< 1s
SHA1-OpenSSL	5k	465KB	23.7k	< 1s
AES128-VMProtect	6k	549KB	20.9k	< 1s
AES256-VMProtect	7k	581KB	6.6k	1s
RC4-VMProtect	10k	869KB	21.8k	< 1s
SHA1-VMProtect	7k	619KB	31.2k	< 1s
AES256-Themida	20k	1MB	14.4k	1s

average analysis speed is about 15k instructions per second, and the analysis time is usually within or around 10 minutes.

The performance results also show that the most time consuming part is the constructions of high-level representations, and that analysis for RC4 is most time consuming among all these algorithms, because there is no constant heuristic data reduction for RC4. The time spent for other algorithms is almost the same.

6 Conclusion

In this paper we have presented a novel approach of analysis of cryptographic data. We use a two-stage method to trace and analyze program data. First, dynamic data tracing is conducted based on full system emulation. The trace results are saved for further analysis. Then, we use an automatic analyzer to perform cryptographic data analysis on the trace results. The target of our analysis is to identify cryptographic algorithms and to extract their parameters. We studied the data patterns for symmetric ciphers AES, Blowfish, stream cipher RC4, and cryptographic hash functions MD5 and SHA-1, and implemented their reference algorithms. In the analysis phase, the high-level representations of traces are first constructed, including Dynamic Basic Blocks, Code Blocks and Call Hierarchies. Then, heuristic data reductions are conducted to reduce the size of data to be analyzed. And at last, we use the reference algorithms to verify the existence of certain algorithms and extract their parameters.

It is possible to extend our analysis method to asymmetric cryptographic algorithms. For example, it's possible to identify RSA encryptions/decryptions which comply with PKCS formats. However the analysis process can be much slower, because the asymmetric algorithms usually take much longer time than symmetric algorithms. It's quite difficult to identify custom asymmetric algorithms because of the irregularity of asymmetric cryptographic data.

We did our experiments on 22 Windows programs, including both real world applications and custom implemented programs, and some of them are obfuscated using VMProtect or Themida. We successfully identified the existing cryptographic algorithms in these programs, and extracted the keys or input data of these programs. For most of the programs that contain symmetric cipher we also extracted the plaintext and ciphertext. The programs with code obfuscation are also successfully analyzed. The analysis result showed the universality and effectivity of our analysis method.

References

1. Bhansali, S., Chen, W., De Jong, S., Edwards, A., Murray, R., Drinić, M., Mihočka, D., Chau, J.: Framework for instruction-level tracing and analysis of program executions. In: Proceedings of the 2nd International Conference on Virtual Execution Environments, p. 163. ACM, New York (2006)
2. Caballero, J., Yin, H., Liang, Z., Song, D.X.: Polyglot: automatic extraction of protocol message format using dynamic binary analysis. In: Proceedings of the 2007 ACM Conference on Computer and Communications Security, CCS 2007, Alexandria, Virginia, USA, October 28-31, pp. 317–329 (2007)

3. Chow, J., Pfaff, B., Garfinkel, T., Christopher, K., Rosenblum, M.: Understanding Data Lifetime via Whole System Simulation. In: USENIX Security Symposium, pp. 321–336. USENIX (2004)
4. Cui, W., Peinado, M., Chen, K., Wang, H., Irun-Briz, L.: Tupni: Automatic Reverse Engineering of Input Formats. In: Proceedings of the 15th ACM Conference on Computer and Communications Security, pp. 391–402. ACM, New York (2008)
5. De-obfuscating the RC4 layer of Skype, http://lukenotricks.blogspot.com/2010/08/de-obfuscating-rc4-layer-of-skype.html
6. Findcrypt plugin, http://www.hexblog.com/ida_pro/files/findcrypt2.zip
7. Gröbert, F.: Automatic Identification of Cryptographic Primitives in Software. Diploma Thesis, Ruhr-University Bochum (2010)
8. Janssens, D.: Heuristic methods for Locating Cryptographic Keys Inside Computer Systems. PhD thesis, Katholieke Universiteit Leuven (1999)
9. Janssens, D., Bjones, R., Claessens, J.: KeyGrab TOO - The search for keys continues..., Whitepaper, Utimaco Safeware AG, KU Leuven (2000)
10. Luk, C., Cohn, R., Muth, R., Patil, H., Klauser, A., Lowney, G., Wallace, S., Reddi, V., Hazelwood, K.: Pin: Building Customized Program Analysis Tools with Dynamic Instrumentation. In: Proceedings of the 2005 ACM SIGPLAN Conference on Programming Language Design and Implementation, pp. 190–200. ACM, New York (2005)
11. Maartmann-Moe, C., Thorkildsen, S., Årnes, A.: The persistence of memory: Forensic identification and extraction of cryptographic keys. Digital Investigation 6, 132–140 (2009)
12. Shamir, A., van Someren, N.: Playing Hide and Seek with Stored Keys. In: Franklin, M. (ed.) FC 1999. LNCS, vol. 1648, pp. 118–124. Springer, Heidelberg (1999)
13. Themida - Oreans Technology: Software Security Defined, http://www.oreans.com/themida.php
14. VMProtect Software Protection, http://vmpsoft.com

SudoWeb: Minimizing Information Disclosure to Third Parties in Single Sign-on Platforms

Georgios Kontaxis[1], Michalis Polychronakis[1], and Evangelos P. Markatos[2]

[1] Computer Science Department, Columbia University, USA
{kontaxis,mikepo}@cs.columbia.edu
[2] Institute of Computer Science,
Foundation for Research and Technology – Hellas, Greece
markatos@ics.forth.gr

Abstract. Over the past few months we are seeing a large and ever increasing number of Web sites encouraging users to log in with their Facebook, Twitter, or Gmail identity, or personalize their browsing experience through a set of plug-ins that interact with the users' social profile. Research results suggest that more than two million Web sites have already adopted Facebook's social plug-ins, and the number is increasing sharply. Although one might theoretically refrain from such single sign-on platforms and cross-site interactions, usage statistics show that more than 250 million people might not fully realize the privacy implications of opting-in. To make matters worse, certain Web sites do not offer even the minimum of their functionality unless the users meet their demands for information and social interaction. At the same time, in a large number of cases, it is unclear why these sites require all that personal information for their purposes.

In this paper we mitigate this problem by designing and developing a framework for minimum information disclosure across third-party sites with single sign-on interactions. Our example case is Facebook, which combines a very popular single sign-on platform with information-rich social networking profiles. When a user wants to browse a Web site that requires authentication or social interaction with his Facebook identity, our system employs, by default, a Facebook session that reveals the minimum amount of information necessary. The user has the option to explicitly elevate that Facebook session in a manner that reveals more or all of the information tied to his social identity. This enables users to disclose the minimum possible amount of personal information during their browsing experience on third-party Web sites.

1 Introduction

An emerging trend on the Web is "single sign-on" initiatives where users register and log on in multiple Web sites using a single account and an OAuth-like protocol [4]. Social networking sites, such as Facebook and Twitter, have been in the front lines of this initiative, allowing their users to utilize their social profiles in a plethora of third-party Web sites. This type of cross-site interaction enables,

X. Lai, J. Zhou, and H. Li (Eds.): ISC 2011, LNCS 7001, pp. 197–212, 2011.
© Springer-Verlag Berlin Heidelberg 2011

for instance, third-party Web sites to authenticate users based on their Facebook (or Twitter) identity. In addition, such sites may add a social dimension to a user's browsing experience by encouraging him to "like," share, or comment on their content using his social network capacity, i.e., automatically post respective favorable messages to his social profile and let his friends know about the site. To enable this social dimension, third-party sites request access and control over the user's information and account.

In other words, these sites request users to authorize Web applications specific to the third-party site, or API calls originating from the third-party site, to access and control part or whole of their social profile. Unfortunately, this process may have several disadvantages, including:

- **Loss of anonymity.** Even the simple act of signing on to a third-party Web site using the Facebook identity sacrifices the anonymous browsing of the user; his social identity usually contains his real name. In most cases it is unclear how this loss of anonymity is necessary for the site's purposes.
- **User's social circle revealed.** Several of these third-party Web sites install Web applications in the user's social profile or issue API calls which request access to a user's "friends." Although having access to a user's friends may improve the user's browsing experience, e.g., for distributed multi-player games, in most cases it is not clear why third-party Web sites request this information, and how based on it they are going to improve the user's browsing experience.
- **Loss of track.** Once users start to enable a torrent of third-party applications to have access to their personal contacts, they will soon lose control of which applications and sites have access to their personal data, and thus they will not be able to find out which of them may have leaked the data in a case of a data breach.
- **Propagation of advertising information in user's social network.** Several of these third-party sites request permission to access and act upon a user's social profile (e.g., upload content to it) even when the user is not accessing the third-party site. Such actions may frequently take the form of explicit or implicit advertisements, not necessarily approved by the user.
- **Disclosure of users' credentials to unauthorized applications.** Once a large number of applications starts receiving credentials to access a user's profile, such credentials may be subject to loss or theft, or accidental leakage. Indeed, recent reports by Symantec suggest that such Facebook applications accidentally leaked access to third parties [7].
- **Reverse Sign-on Semantics.** When a service prompts a user to sign on, he provides his credentials and gains access to data offered by that service. However, in the cases described above, the service is the one being given access to the data of the user, and from that data selects information that may be used to identify or authenticate the user.

Although a user could theoretically deny this single sign-on approach, and the installation of the third-party application, many Web sites respond to this disapproval usually by diminishing the user's browsing experience significantly, cutting the user off from the largest part of their sophisticated functionality. Although this might be less of a problem if only a handful of third-party Web sites used this single sign on mechanism, recent results suggest that more than two million Web sites have added Facebook social plug-ins [9]. To make matters worse, popular Web sites seem to adopt Facebook social plug-ins even more aggressively. Indeed, as of May 2011, as many as 15% of the top 10,000 most popular Web sites have adopted Facebook social plug-ins, a whooping 300% increase compared to May 2010 [1]. If this trend continues, as it appears to be, then it will be very difficult for users to browse a significant percentage of the Web sites without revealing their personal information.

In this paper, we propose a new way for users to interact with single sign-on platforms so as to protect their privacy; we propose that users surf the Web using downgraded sessions with the single sign-on platform, i.e., stripped from excessive or personal information, and with a limited set of privileged actions. Thereby, by default, all interactions with third-party Web sites take place under that privacy umbrella. On occasion users may explicitly elevate that session on-the-fly to a more privileged or information-rich state to facilitate their needs when appropriate. Our proposed concept is inspired by privilege separation among user accounts in operating systems, and the UNIX `su` command which upgrades the permissions of a user to those of the `super-user`, if and when the ordinary user needs to perform a privileged instruction. In UNIX even system administrators initially log in with their ordinary (i.e., non super-user) accounts and upgrade to super-user status if and when they need to execute privileged operations.

For instance, we propose that users may use two parallel and distinct sessions with the Facebook Connect platform, tied to respective social profiles; their primary profile, and a second "disposable" profile. Their primary session will be associated to a social profile where they will maintain all of their social contacts, photographs, and personal information. The primary profile will be their current profile, if they already have one. The "disposable" session will be associated to a profile that will be a stripped-down version of the primary one. It may contain no personal information, social contacts, or other sensitive information that the user is not comfortable sharing with a plethora of random third-party Web sites. By default, the user's browser will keep the appropriate state, i.e., active sessions and cookies, to maintain the "disposable" session alive. As a result, when the user employs the single sign-on mechanism just to bypass the registration step in various Web sites, he will surrender only a small portion of his information or no actual information at all, as the "disposable" session with Facebook will be used. If at any point the user wishes to activate his actual profile, because he actually wants to associate his identity with a third-party Web site or online application, he is able to elevate his browser session with Facebook, i.e., by switching to the primary session from above.

In summary, the contributions of this paper are the following:

- We identify and describe an increasing threat to the users' privacy: a threat which masquerades under the convenience of a single sign-on mechanism and gives third-party Web sites access to a user's personal information stored in social networks.
- We propose a new privacy-preserving framework for users to interact with single sign-on and OAuth-like platforms provided by social networks in their daily activities on the web.
- We implement a prototype of our framework as a browser extension for the Google Chrome browser. Our prototype supports the popular single sign-on mechanism "Facebook Connect" [2] and can be easily extended to support others, such as "Sign in with Twitter" [6].
- We evaluate our implementation and show that (i) it allows users to preserve their privacy when signing on with third-party Web sites and (ii) it does not affect any open sessions they might have with other third-party Web sites that use the same single sign-on mechanisms.

2   Background

In this section we provide some background on the OAuth protocol [4], which is the primary method for implementing single sign-on functionality across multiple Web sites. We also detail Facebook's single sign-on platform [2], which at the moment is the most popular single sing-on platform with more than 2.5 million Web sites using it [3].

2.1   OAuth Protocol

The OAuth or Open Authentication protocol [4] provides a method for clients to access server resources on behalf of a resource owner. In practice, it is a secure way for end users to authorize third-party access to their server resources without sharing their credentials.

As an example, one could consider the usual case in which third-party sites require access to a user's e-mail account so that they can retrieve his contacts in order to enhance the user's experience in their own service. Traditionally, the user has to surrender his username and password to the third-party site so that it can log into his account and retrieve that information. Clearly, this entails the risk of the password being compromised. Using the OAuth protocol, the third party registers with the user's e-mail provider using a unique application identifier. For each user that the third-party requires access to his e-mail account, it redirects the user's browser to an authorization request page located under the e-mail provider's own Web domain, and appends the site's application identifier so that the provider is able to find out which site is asking for the authorization. That authorization request page, located in the e-mail provider's domain, validates the user's identity (e.g., using his account cookies or by prompting him to log in), and subsequently asks the user to allow or deny information access to the third-party site. If the user allows such access, the third-party site is able to use the e-mail provider's API to query for the specific user's e-mail contacts. At no point in this process does the user have to provide his password to the third-party Web site.

2.2   Facebook Authentication

The Facebook authentication or Facebook Connect [2] is an extension to the OAuth protocol that allows third-party sites to identify users by gaining access to their Facebook identity. This is convenient for both the sites and the users; sites do not have to maintain their own accounting system, and users are able to skip yet another account registration and thereby avoid the associated overhead. A "login with Facebook" button is embedded in these third-party sites that, once clicked, directs the user's browser to `http://www.facebook.com/dialog/oauth ?client_id=THIRD_PARTY_SITE_ID` where the user's cookies or credentials are validated by Facebook. On successful identity validation, Facebook presents a "request for permission" dialog where the user is prompted to allow or deny the actions of the third-party Web site, i.e., social plug-in actions or access to account information. However, the user is not able to modify or regulate the third-party's request, for instance to allow access to only a part of the information the site is requesting. If the user grants permissions to the site's request, Facebook will indefinitely honor API requests, originating from that third-party site ID, that conform to what the user has just agreed upon.

2.3   Facebook Social Plug-ins

Facebook has implemented a platform of social plug-ins on top of the OAuth protocol to allow third-party sites to integrate the functionality of Facebook's social experience to their own pages [2]. In addition to authenticating a user, third-party developers are able to add "like" or "comment" Facebook buttons and forms in their site which, once clicked, update the user's Facebook profile with content from that third-party Web site, or allow the user to upload content to the site using his Facebook identity.

3   Related Work

Ardagna et al. [10] highlight the practice of Internet services requiring user information for accessing their digital resources. They coin the concept of a user portfolio containing personal data and propose the use of sensitivity labels that express how much the user values different pieces of information. Furthermore, they assume scenarios where an atypical negotiation takes place between the user and the server, in which the server prompts the user to choose among disclosing alternative pieces of information. The user decides on the type and amount of information disclosed in relation to the type and amount of digital resources being offered.

Facecloak [14] shields a user's personal information from a social networking site and any third-party interaction, by providing fake information to the social networking site and storing actual, sensitive information in an encrypted form on a separate server. At the same time, social functions are maintained.

Felt et al. [13] studied the 150 most popular Facebook applications and found that almost all of them required too much user information for their purposes. They propose the use of a proxy to improve social networking APIs such that third-party applications are prevented from accessing real user data while social functions are not affected.

The xBook [16] framework addresses threats against the privacy of social network users due to information leaked via the interaction with third-party applications. It provides a trusted hosting environment where untrusted applications are split into components with a manifest to detail security permissions in terms of user data access and communication between components or remote locations. xBook takes up the role of enforcing that manifest at run-time.

OpenID [5] is a platform supporting a federation of single sign-on providers. Its nature of operation has been described in section 2. An interesting feature of OpenID is the support for multiple identities per user; upon receiving a user-identification request from a third-party site and after authenticating with the user, it may decide to return a different identity for the same user to different third-party sites. PseudoID [12] is a privacy enhancement for single-sign-on systems like OpenID or Facebook Connect. As third-party sites interact with the single sign-on provider to acquire access to a user's identity, that provider is able to correlate a user's identity with the sites she logs into. In PseudoID users set up the profiles or identities they wish to use with third-party sites and employ the PseudoID's blind signature service to cryptographically blindly sign such tokens of information. When they need to identify themselves to a third-party site, just as before, that site interacts with PseudoID to retrieve the user's identity. Contrary to the traditional model, the user does not log in to PseudoID, thereby allowing the service to associate her with that particular third-party site request. The user presents to PseudoID a blindly signed identity and PseudoID, after checking the validity of the cryptographic signature, forwards that identity to the third-party site.

Concurrently and independently to our work, a user-friendly mechanism for users to switch between Google Chrome profiles is being developed [8]. At the moment, one is able to use multiple browser profiles by adding a data-directory flag when invoking the browser. Browser profiles contain their own cookie store, browser settings and installed extensions. By using different profiles, among other things, one is able to switch between cookie stores and therefore between Web site identities.

Our approach is similar to that feature of Chrome but at the same time bears significant differences. While Google is building a profile manager for browsers, we design a more generic privacy-preserving framework that describes information and privilege separation in Web sessions involving cross-site interaction. While Chrome's profiles bundle sessions with different sites in a single browser profile, we operate on a more flexible basis where we populate an isolated and distinct browser instance with the state of only those sessions that the user has explicitly activated. Moreover, while in the Chrome feature the user is responsible for switching between the different identities and profiles, we employ heuristics that automatically detect the need to switch to a downgraded Web session. Therefore, we do not have to rely on the user's alertness to protect his information.

4   Design

The modus operandi we assume in our approach is the following:

1. The user browses the Web having opened several tabs in her browser.
2. Then, the user logs in her ordinary Facebook account so as to interact with friends and colleagues.
3. While browsing the Web at some other tab of the same browser, the user encounters a third-party Web site asking her to log in with her Facebook credentials. At this point in time, our system kicks in and establishes a new and separate downgraded session with Facebook for that cross-site interaction. That session is tied to a stripped-down version of her account which reveals little, if any at all, personal information. Now:
   (a) The user may choose to follow our "advice" and log in with this downgraded Facebook session. Let there be noted that this stripped-down mode does not affect the browsing experience of the user in the tabs opened at step 2 above: the user remains logged in with her normal Facebook account in the tabs of step 2, while in the tabs of this step she logs in with the stripped-down version of her account. Effectively, the user maintains two sessions with Facebook:
      i. One session logged in with her normal Facebook account, and
      ii. One session logged in with the stripped-down version of her account.
   (b) Alternatively, the user may want to override our system's logic and log in with her normal Facebook account revealing her personal information; in that cases she performs a "sudo" on that particular cross-site interaction with Facebook and elevate the by-default downgraded Web session.

In the description of our system we assume the use of a single sign-on mechanism such as Facebook Connect [2,17]. However, our mechanisms can be extended to cover other single sign-on mechanisms as well.

Figure 1(b) shows the architecture of our system. To understand our approach we will first describe in Figure 1(a) how an ordinary Web browser manages session state. We see that the browser uses a default session store (**Session Store [0] (default)**) which stores all relevant state information, including cookies. Thus, when the user logs into Facebook (or any other site for that matter) using her ordinary Facebook account, the browser stores the relevant cookie in this default session store. When the browser tries to access Facebook from another tab (**Tab 3**

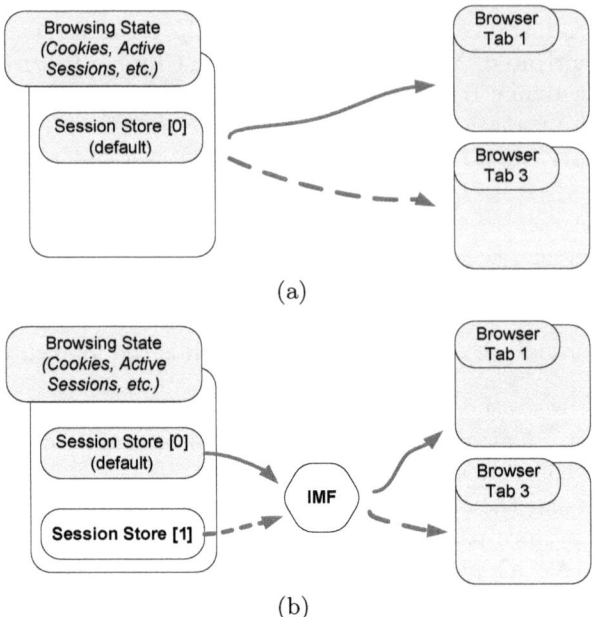

Fig. 1. Typical communication of session state to loaded pages (a), and how *SudoWeb* handles the same communication using multiple session stores (b)

in the figure), the cookie is retrieved from the default session store and the page is accessed using the same state as before.

In our design, we extend this architecture by including more than one session stores. Indeed, in Figure 1(b) (bottom left) we have added "**Session Store [1]**" which stores all relevant information, including cookies, for the stripped-down Facebook session. This gives us the opportunity to enable users to surf the web using two distinct and isolated sessions with Facebook at the same time: a session tied to the "normal" account is enabled in **Tab 1** while a stripped-down session is in effect in **Tab 3**. To select the appropriate account, our system (**IMF**) intercepts all URL accesses and checks their HTTP referrer field. If the URL points to a single sign-on platform (such as Facebook Connect) but the HTTP referrer field belongs to a different domain name, then our system suspects that this is probably an attempt from a third-party Web site to authenticate the user with her Facebook credentials.

Therefore, as it stands inline between the loading page and the browser's state store(s), it supplies the appropriate state (from **Session Store [1]** for the stripped-down Facebook session to be employed. This is an implicit privacy suggestion towards the user. If the user disagrees, she may choose to authenticate with her ordinary Facebook account, in which case, **Tab 3** will receive all cookies from **Session Store [0]**.

We consider the proposed concept as analogous to privilege separation in operating systems, i.e., different accounts with different privileges, such as root

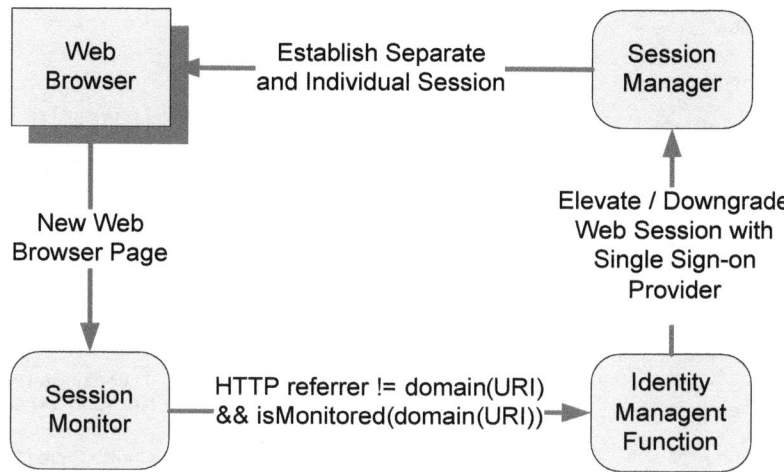

Fig. 2. *SudoWeb* extension modules

and user accounts. Our design can scale and evolve so that it accommodates different privacy-preserving scenarios in interaction with third-party Web sites.

Figure 2 shows the modules of our system. Initially, in the upper left corner, the user browses ordinary web pages (**Web Browser**). When a new browser page (i.e., tab or window) is created (**New Web Browser Page**), the **Session Monitor** kicks in to find whether this is an attempt to log in a single sign-on mechanism[1]. If (i) it is such an attempt (i.e., **isMonitorred(domain(URI))** is TRUE) and (ii) the attempt is from a third-party Web site (i.e., **HTTP referrer != domain(URI)**) then our system calls the **Identity Management Function** (IMF) which employs a downgraded, stripped-down from all personal information, session for the user. From that point onwards, the **Session Manager** manages all the active sessions of the user, in some cases different sessions with different credentials for the same single sign-on domain. Figure 3 shows the workflow of our system in more detail.

5   Implementation

We have implemented our proposed architecture as a browser extension for the latest version of Google Chrome[2] with support for the "Facebook Connect" single sign-on mechanism. We find that, due to its popularity, our proof of concept application covers a great part of single sign-on interactions on the Web. Moreover, we implicitly support Facebook's social plug-ins, such as *share site content* or *comment on site content*, described in section 2, that require the "connect" mechanism as a first step. Our browser extension can be seamlessly configured to support a greater variety of such cross-site single sign-on interactions.

---

[1] *SudoWeb* keeps a list with all single sign-on domain mechanisms monitored. If such a domain is monitored the **isMonitorred(domain(URI))** function returns TRUE.

[2] As we take advantage of generic functionality in the extension-browser communication API, we find it feasible to also port the extension to Mozilla Firefox.

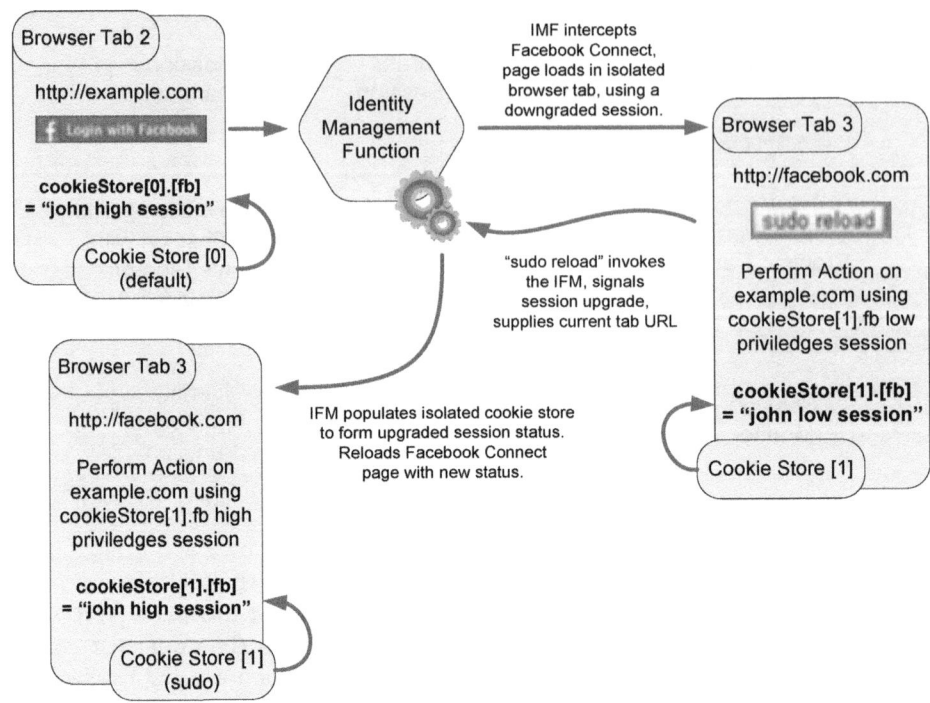

Fig. 3. Example Workflow of *SudoWeb*

5.1   *SudoWeb* Modules

Here we describe the modules that comprise our extension to the Google Chrome browser, in support of our proposed architecture.

Identity Management Function (IMF). In the heart of the extension lies the logic module offering the identity management function or IMF. This function is responsible for detecting the possible need for elevating or downgrading a current session with a single sign-on provider (here: Facebook). Such need is detected by identifying differences in the HTTP referrer domain and the URL domain of pages to be loaded. That is, when the user navigates away from a third-party Web site (identified by the HTTP referrer field) towards a single sign-on Web site (we keep a configuration file with all single sign-on sites supported), IMF steps in, instantiates a new, isolated and independent session store in the browser and instructs the *session manager module* to initialize it so that the browser receives such state that establishes a downgraded or stripped-down session with the single sign-on provider. Furthermore, it places a "sudo reload" HTML button on that page giving the user the opportunity to reload that page using an elevated session instead.

Session Monitor. Supporting role to the IMF plays the session monitoring module. If one considers our extension as a black box, the session monitor stands at its input. It inspects new pages opening in the Web browser and looks for cases where the page URL belongs to a monitored single-sign-on provider domain (here: Facebook) but the page has been invoked through a different, third-party domain. It does so by comparing that URL with the HTTP referrer. The referrer is an HTTP parameter supplied by the browser itself based on the URL of the parent tab or window that resulted in a child tab or window being spawned. The session monitor notifies the IMF of such incidents and supplies the respective page URL. We should note that recent research has revealed that the HTTP referrer field in several cases can be empty or even spoofed [11,15] undermining all mechanisms based on it. Although it is true at the network elements may remove or spoof the HTTP referrer field so that it will be invalid when it reaches the destination web server, our work with the HTTP referrer field is at the web *client* side, not at the web server side. That is, the HTTP referrer field is provided to *SudoWeb* by the web browser *before* it reaches any network elements which may remove it or spoof it.

Session Manager. This module also plays a supporting role to the IMF. If one considers our extension as a black box, the session manager stands at its output. Upon the installation of our extension, the session manager prompts the user of the Web browser to fill in his ordinary single sign-on (here: Facebook) account username and password, as well as his stripped-down one that is to be used for the downgraded integration with third-party Web sites. The session manager maintains in store the necessary state, e.g., cookies, required to establish the two distinct sessions with the single sign-on provider and is responsible for populating the browser's cookie store once instructed by the IMF. As a result, it stands at the output of our extension and between the browser's session store and the rendered pages that reside in tabs or windows. It affects the state upon which a resulting page rely on.

Our extension takes advantage of the incognito mode in Google Chrome to launch a separate browser process with isolated cookie store and session state so that when the session manager pushes the new state in the cookie store, the user is not logged off of the existing elevated session (here: with Facebook) that may be actively used in a different browser window.

5.2   Operation and Interaction of *SudoWeb* Modules

Following the use-case presented at the beginning of section 4, a user browsing the Web will eventually come across a third-party that wishes to interact with his Facebook identity via the cross-site single sign-on mechanism. As soon as the user clicks on the "login with Facebook" button, our system kicks in;

1. The *session monitor* detects the launch of a new Facebook page from a page under the domain of the third-party Web site. The *session monitor* notifies the *IMF* module of our extension and so the page launch is intercepted

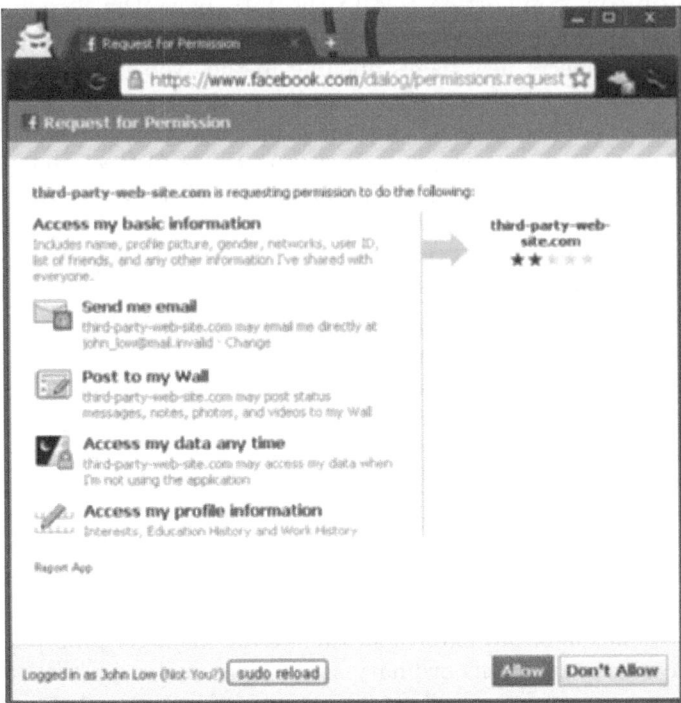

Fig. 4. Example screenshot of a Facebook "Request for Permission" page that has been invoked by fictional site `third-party-web-site.com` so that the Web user may authorize that site to access his Facebook account information. By default, a downgraded-status session is maintained with Facebook, using the appropriate state to be logged in with the disposable account "John Low." There is the option to switch to an elevated-status session via the "sudo reload" button at the bottom of the window.

    and loaded in an incognito window, i.e., an isolated browser process with a separate and individual session store.

2. The *IMF* coordinates with the *session manager module* so that this isolated environment is populated with the necessary state for a downgraded Facebook session to exist.

The entire process happens in an instant and the user is presented with a browser window similar to figure 4. In this figure, we have used `third-party-web-site.com` as the name of the third part Web site which wants to authenticate the user using her Facebook account. We see that in addition to authenticate the user, the third-party Web site asks for permission to (i) send the user email, (ii) post on the user's wall, (iii) access the user's data any time, and (iv) access the user's profile information. Although Facebook enables users to "Allow" or "Don't Allow" access to this information (bottom right corner), if the user chooses not not allow this access, the entire authentication session is over and the user will not gain access to the content of `third-party-web-site.com`.

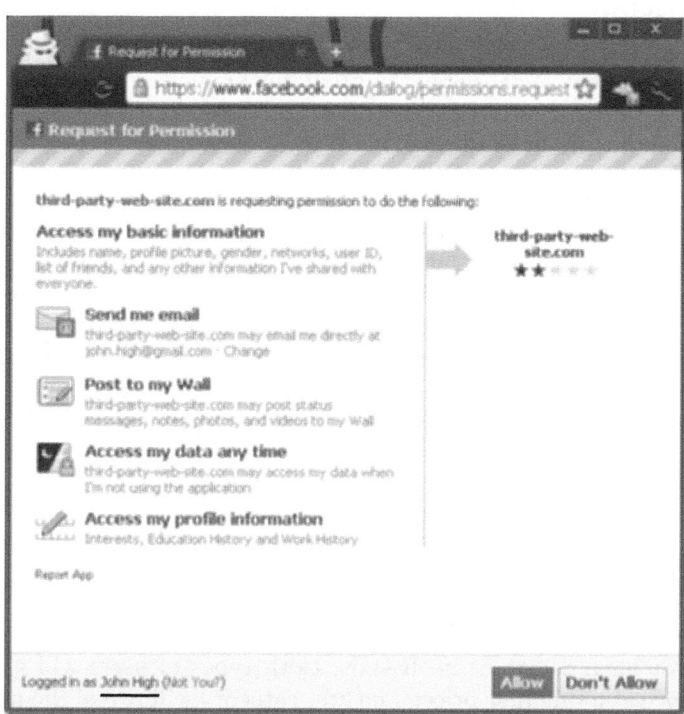

Fig. 5. Example screenshot of the previous Facebook page, after the "sudo reload" option has been selected by the Web user; the page has been reloaded using on-the-fly the necessary session state to maintain an elevated-status Facebook session using the account "John High".

Having intercepted this third-party authentication operation, *SudoWeb* brings the stripped-down account (i.e., John Low) forward, on behalf of the user. Therefore if the user chooses at this point to allow access to his information by the third-party site, only a small subset of his actual information will be surrendered. Note that a "sudo reload" button has been placed at the bottom of the page, allowing the user to elevate this session to the one tied to his actual, or a more privileged, Facebook identity.

Figure 5 presents an example screenshot of the browser window the user will see if he chooses to elevate his session. One may notice at the bottom of the browser's page that the user is no longer considered to be logged in as "John Low" but as "John High."

The Facebook session with which the user was surfing prior to engaging in this cross-site Facebook interaction remains intact in the other open browser windows since, as mentioned earlier, we take advantage of the browser's incognito mode to initiate an isolated session store in which we manage the escalation and de-escalation of user sessions. All the user has to do is close this new window to return to his previous surfing activity.

6   Discussion

Here we discuss how social networks providing single sign-on interaction could evolve to facilitate user needs and better protect their privacy. We also propose a series of requirements from third-party Web applications in terms of "fair play."

Fine-Grained Privacy Settings. Inspired by the privilege separation principles of UNIX, *SudoWeb* presents a step towards surfing the Web using several distinct sessions: each session with different privileges. We have implemented the philosophy of our system using parallel Web sessions tied to distinct Facebook accounts; each account revealing a different amount of information. We believe that the increasing privacy concerns of users will motivate single sign-on providers to offer more fine grained disclosure of user information, and more control over the user's privacy in a single account. If that happens, the concept of our system will still be valid, but implemented closer to the mechanics of single sign-on providers.

Fairness. Current single sign-on mechanisms in social networks are especially unfair to people with rich social circles. For example, if a third-party Web site wants to install an application that has access to all of a user's friends in return for a service, this is unfair to people who have lots of friends, compared to people who have (or have declared) no friends. Both types of users will get the same kind of service at a different price: The first category will reveal the names of lots of friends, while the second will reveal none. To make matters worse, this cost (and unfairness) seems to increase with time: as the user accumulates friends, the installed third-party application will continue to have access to all of them.[3] We believe that single sign-on mechanisms should: (i) Restrict themselves only to authentication and refrain from asking access to more personal information, such as friends and photos. (ii) If they do ask for more personal information, they should make clear how they are going to use it and how this will benefit the user. (iii) If the user denies the provision of more personal information, the single sign-on mechanisms should continue to function and provide their services to the users.

Terms of Use. It may seem that our approach may conflict with the terms of use for some sites. For example, maintaining multiple accounts is a violation of the terms of use of Facebook, while it appears not to be a violation of the terms of use of Google. We believe that this conflict stems from the fact that some sign-on mechanisms have not yet caught up with the changing needs of the users. For example, several users maintain two Facebook profiles: one personal profile with all their personal contacts, friends, and relatives, and one professional profile where their "friends" are their colleagues and business contacts. The postings that run on their personal profiles are quite different from the postings than run on their professional profiles. Even the language of these postings may be different. Forcing those users to have a single Facebook account will make their

---

[3] Unless the user explicitly uninstalls the application.

social interactions more difficult or will force them to move one of their profiles, e.g., the professional one, to another social network, such as LinkedIn. We believe that sooner or later most successful single sign-on sites will catch up with the changing user needs and will adapt their terms of use to suit the users. Otherwise, the users might adopt single sign-on sites which are closer to their needs.

7 Conclusion

Recent results suggest that hundreds of thousands of Web sites have already employed single sign-on mechanisms provided by social networks such as Facebook and Twitter. Unfortunately, this convenient authentication usually comes bundled (i) with a request to the user's personal information, as well as (ii) the request to act upon a user's social network on behalf of the user, e.g., for advertisement. Unfortunately, the user can not deny these requests, if she wants to proceed with the authentication.

In this paper, we explore this problem and propose a framework to enable users to authenticate on third-party Web sites using single sign-on mechanisms provided by popular social networks while protecting their privacy; we propose that users surf the Web using downgraded sessions with the single sign-on platform, i.e., stripped from excessive or personal information and with a limited set of privileged actions. Thereby, by default, all interactions with third-party Web sites take place under that privacy umbrella. On occasion, users may explicitly elevate that session on-the-fly to a more privileged or information-rich state to facilitate their needs when appropriate. We have implemented our framework in the Chrome browser with current support for the popular single sign-on mechanism Facebook Connect. Our results suggest that our framework is able to intercept attempts for third-party Web site authentication and handle them in a way to protect the user's privacy, while not affecting other ongoing Web sessions that the user may concurrently have.

Acknowledgments. This work was supported in part by the FP7-PEOPLE-2009-IOF project MALCODE and the FP7 project SysSec, funded by the European Commission under Grant Agreements No. 254116 and No. 257007. Evangelos Markatos is also with the University of Crete. Most of the work of Georgios Kontaxis was done while at FORTH-ICS.

Availability. The source code of *SudoWeb* is available at https://code.google.com/p/sudoweb/

References

1. BuiltWith - Facebook for Websites Usage Trends,
   http://trends.builtwith.com/javascript/Facebook-for-Websites
2. Facebook for Websites, https://developers.facebook.com/docs/guides/web/

3. Facebook Statistics, `https://www.facebook.com/press/info.php?statistics`
4. OAuth, `http://oauth.net/`
5. OpenID Foundation - OpenID Authentication 2.0 Specifications,
   `http://openid.net/specs/openid-authentication-2_0.html`
6. Sign in with Twitter, `http://dev.twitter.com/pages/sign_in_with_twitter`
7. Symantec Official Blog - Facebook Applications Accidentally Leaking Access to Third Parties,
   `http://www.symantec.com/connect/blogs/facebook-applications-accidentally-leaking-access-third-parties`
8. The Chromium Projects - Multiple Profiles,
   `http://www.chromium.org/user-experience/multi-profiles`
9. WebProNews - Million Sites Have Added Facebook's Social Plugins Since f8,
   `http://www.webpronews.com/2-million-sites-have-added-facebooks-social-plugins-since-f8-2010-09`
10. Ardagna, C.A., De Capitani di Vimercati, S., Foresti, S., Paraboschi, S., Samarati, P.: Supporting privacy preferences in credential-based interactions. In: Proceedings of the 9th Annual ACM Workshop on Privacy in the Electronic Society (2010)
11. Barth, A., Jackson, C., Mitchell, J.C.: Robust defenses for cross-site request forgery. In: Proceedings of the 15th ACM Conference on Computer and Communications Security (2008)
12. Dey, A., Weis, S.: PseudoID: Enhancing privacy in federated login. In: Hot Topics in Privacy Enhancing Technologies (2010)
13. Felt, A., Evans, D.: Privacy protection for social networking platforms. In: Proceedings of the Workshop on Web 2.0 Security and Privacy (2008)
14. Luo, W., Xie, Q., Hengartner, U.: Facecloak: An architecture for user privacy on social networking sites. In: Proceedings of the International Conference on Computational Science and Engineering (2009)
15. Meiss, M., Duncan, J., Gonçalves, B., Ramasco, J.J., Menczer, F.: What's in a session: tracking individual behavior on the web. In: Proceedings of the 20th ACM Conference on Hypertext and Hypermedia (2009)
16. Singh, K., Bhola, S., Lee, W.: xbook: redesigning privacy control in social networking platforms. In: Proceedings of the 18th Conference on USENIX Security Symposium (2009)
17. Stone, B.: Facebook aims to extend its reach across the web. New York Times (2008)

Hello rootKitty:
A Lightweight Invariance-Enforcing Framework

Francesco Gadaleta, Nick Nikiforakis, Yves Younan, and Wouter Joosen

IBBT-DistriNet
Katholieke Universiteit Leuven
Celestijnenlaan 200A B3001
Leuven, Belgium
francesco.gadaleta@cs.kuleuven.be

Abstract. In monolithic operating systems, the kernel is the piece of code that executes with the highest privileges and has control over all the software running on a host. A successful attack against an operating system's kernel means a total and complete compromise of the running system. These attacks usually end with the installation of a rootkit, a stealthy piece of software running with kernel privileges. When a rootkit is present, no guarantees can be made about the correctness, privacy or isolation of the operating system.

In this paper we present *Hello rootKitty*, an invariance-enforcing framework which takes advantage of current virtualization technology to protect a guest operating system against rootkits. *Hello rootKitty* uses the idea of invariance to detect maliciously modified kernel data structures and restore them to their original legitimate values. Our prototype has negligible performance and memory overhead while effectively protecting commodity operating systems from modern rootkits.

Keywords: rootkits, virtualization, detection, invariance.

1 Introduction

Operating systems consist of trusted software that executes directly on top of a host's hardware providing abstraction, arbitration, isolation and security to the rest of the software. Due to their prominent position, operating systems have been a common target of attackers who try to circumvent their protection mechanisms and modify them to their advantage. In the past, a program that allowed a user to elevate his access and become a system administrator ("root") was called a `rootkit`. Today the meaning of rootkits has changed and is used to describe software that hides the attacker's presence from the legitimate system's administrator. Kernel-mode rootkits[1] target the core of an operating system and thus they are the hardest to detect and remove. In extreme cases, a kernel-mode rootkit may be introduced by a software bug in the kernel, triggered by a

---

[1] Such rootkits appear very often in the form of device drivers (Microsoft Windows) or Loadable Kernel Modules (Linux kernel).

X. Lai, J. Zhou, and H. Li (Eds.): ISC 2011, LNCS 7001, pp. 213–228, 2011.

malicious or a benign but-exploitable process. Regardless of the way the rootkit is introduced, the result is malicious code running with operating system privileges which can add and execute additional code or modify existent kernel code. The activities resulting from a successful attack can range from spamming and key-logging to stealing private user-data and disabling security software running on the host. In the past, rootkits have also been used to turn their targets into nodes of a botnet as with Storm Worm [9] or to perform massive bank frauding [13].

Even rootkits that do not introduce new code, but rather make use of existing fragments of code to fabricate their malicious functions, need to somehow have these fragments executed in the order of their choosing [10]. Changing the control flow of the kernel involves either changing specific kernel objects such as function pointers or overwriting existing fragments of code with new code. Using the idea of modified kernel-data structures as a sign of rootkits, security researchers have developed several approaches to mitigate rootkits. Unfortunately, many of these are affected by considerable overhead [8,16] or miss a fundamental security requirement such as isolation[32].

Isolation is needed to prevent a countermeasure in the target system from being disabled/crippled by a potential attack.

Other countermeasures have been presented in which operating system kernels are protected against rootkits by executing only authenticated (or validated) kernel code [24,22,16]. The aforementioned rootkit [10] that doesn't introduce new kernel code and re-uses fragments of authenticated code bypasses such countermeasures. In [19] a countermeasure to detect changes of the kernel's control flow graph is presented; Anh et al. [14] uses virtualization technology and emulation to perform malware analysis and [29] protects kernel function pointers. Another interesting work is [23] which gives more attention to kernel rootkit profiling and reveals key aspects of the rootkit behavior by the analysis of compromised kernel objects. Determining which kernel objects are modied by a rootkit not only provides an overview of the damage inflicted on the target but is also an important step to design and implement systems to detect and prevent rootkits.

A rising trend in security research is the use of virtualization technology for non-virtualization specific purposes [7,21,5,4]. The property that makes virtualization particularly attractive, from a security perspective, is that isolation is guaranteed by the current virtualization-enabled hardware. Using the appropriate instruction primitives of such hardware, makes it straightforward to fully separate and isolate the target from the monitor system. In this paper we present *Hello rootKitty*, a lightweight invariance-enforcing framework to mitigate kernel-mode rootkits in common operating system kernels. We start from the observation that many critical kernel data structures are invariant. Many data structures used by rootkits to change the control-flow of the kernel contain values that would normally stay unchanged for the lifetime of a running kernel. Our protection system consists of a monitor that checks the contents of data structures that need to be protected at regular times and detects whether their contents have changed. If a change is detected, our system warns the administrator of the exploited kernel and corrects the problem by restoring the modified data structures to their original contents. Our monitor runs inside a hypervisor and protects operating systems that are being virtualized. Due to the hardware-guarantees of isolation that virtualization provides, an attacker has no way of disabling our monitor or tamper with the memory areas that our system uses. *Hello rootKitty* imposes negligible performance overhead on the virtualized system and it doesn't require kernel-wide changes other than a trusted module to communicate the invariant data structures from the guest operating system to the hypervisor. *Hello rootKitty* can be integrated with existing invariance-inferencing engines and protect commodity operating systems running on virtualized environments. Alternatively, our system can be used directly by kernel developers to protect their invariant structures from rootkit attacks.

The rest of the paper is structured as follows. Section 2 describes the problem of rootkits and presents our attacker model. Section 3 presents our solution. In Section 3.1 we present the architectural details of *Hello rootKitty* followed by its implementation in Section 3.2. We evaluate our prototype implementation in Section 4 and present its limitations in Section 5. We discuss related work on rootkit detection in Section 6 and conclude in Section 7.

2 Problem Description

In this section we describe common rootkit technology and we also present the model of the attacker that our system can detect and neutralize.

2.1 Rootkits

Rootkits are pieces of software that attackers deploy in order to hide their presence from a system. Rootkits can be classified according to the privilege-level which they require to operate. The two most common rootkit classes are: a) user-mode and b) kernel-mode.

User-mode rootkits run in the user-space of a system without the need of tampering with the kernel. In Windows, user-mode rootkits commonly modify the loaded copies of the Dynamic Link Libraries (DLL) that each application loads in its address space [28]. More specifically, an attacker can modify function pointers of specific Windows APIs and execute their own code before and/or after the execution of the legitimate API call. In Linux, user-mode rootkits hide themselves mainly by changing standard linux utilities, such as `ps` and `ls`. Depending on the privileges of the executing user, the rootkit can either modify the default executables or modify the user's profile in a way that their executables will be called instead of the system ones (e.g.,by changing the `PATH` variable in the Bash shell).

Kernel-mode rootkits run in the kernel-space of an operating system and are thus much stronger and much more capable. The downside, from an attacker's perspective, is that the user must have enough privileges to introduce new code in the kernel-space of each operating system. In Windows, kernel-mode rootkits are loaded as device-drivers and target locations such as the `call gate` for interrupt handling or the System Service Descriptor Table (SSDT). The rootkits change these addresses (hooking) so that their code can execute before specific system calls. In Linux, rootkits can be loaded either as a Loadable Kernel Module (LKM) or written directly in the memory of the kernel through the /dev/mem and /dev/kmem file [20]. These rootkits target kernel-data structures in the same way that their Windows-counterparts do. Although this paper focuses on Linux kernel-mode rootkits, the concepts introduced apply equally well to Windows kernel-mode rootkits. An empirical observation is that kernel-mode rootkits need to corrupt specific kernel objects, in order to execute their own code and add malicious functionality to the victim kernel. Studies of common rootkits [17,19] show that most dangerous and insidious rootkits change function pointers in the system call table, interrupt descriptor table or in the file system, to point to malicious code. The attack is triggered by calling the relative system call from user space or by handling an exception or, in general, by calling the function whose function pointer has been compromised. We report a list of rootkits which compromise the target kernel in Table 1.

Table 1. Hooking methods of common linux rootkits

Rootkit	Description
Adore, afhrm, Rkit, Rial, kbd, All-root, THC, heroin, Synapsis, itf, kis	Modify system call table
SuckIT	Modify interrupt handler
Adore-ng	Hijack function pointers of fork(), write(), open(), close(), stat64(), lstat64() and getdents64()
Knark	Add hooks to /proc file system

2.2 Attacker Model

In this work we first assume that the operating system which is being attacked is virtualized, i.e., it runs on top of a hypervisor which has more privileges than the operating system itself. Virtualization guarantees isolation thus we assume that the guest operating system cannot access the memory or code of the hypervisor. Our system detects the rootkit after it has been deployed, a fact which allows our model to include all possible ways of introducing a rootkit in a system. Thus, a rootkit can be introduced either by:

- A privileged user loading the rootkit as a Loadable Kernel Module
- A privileged user loading the rootkit by directly overwriting memory parts through the /dev/ memory interfaces
- An unprivileged user exploiting a vulnerability in the kernel of the running operating system which will allow him to execute arbitrary code

Finally, our system doesn't rely on secrecy so our model of the attacker includes him being aware of the protection system.

3   Hello rootKitty: Protecting Kernel Data against Rootkits

In this section we describe our approach to detect rootkits that compromise function pointers or data structures residing in the kernel.

3.1   Approach

By studying the most common rootkits and their hooking techniques one can realize that they share at least one common characteristic. In order to achieve execution of their malicious code, rootkits overwrite locations in kernel memory which are used to dictate, at some point, the control-flow inside the kernel. Most of these locations are very specific (see Table 1) and their values are normally invariant, i.e. they don't change over the normal execution of the kernel. Since these objects are normally invariant, any sign of variance can be used to detect the presence of rootkits. We use the terminology of "critical kernel objects" to name objects that can be used by an attacker to change the control-flow of the kernel. The approach of *Hello rootKitty* is, given a list of invariant critical kernel objects, to periodically check them for signs of variance. When our countermeasure detects that the contents of an invariant critical kernel object have been modified, it will report an ongoing attack. Invariant critical kernel objects have been identified in several contributions, such as [29,2,3,6]. The methods to detect invariance differ depending on the type of critical kernel object and are the following:

1. Static kernel objects at addresses hardcoded and not dependent on kernel compilation
2. Static kernel objects dependent on kernel compilation (e.g., provided by /boot/System.map in a regular Linux kernel)
3. Dynamic kernel objects allocated on the heap by kmalloc, vmalloc and the rest of the kernel-specific memory allocation functions

Identifying and protecting static kernel objects (type 1 and type 2) is straightforward. During the installation of the operating system to be monitored, a virtual machine installer would know in advance whether the guest is of Windows or Unix type. This is the minimal information required to detect kernel objects whose addresses have been hardcoded (type 1). Moreover, the Linux operating system provides System.map, where compilation-dependent addresses of critical kernel objects are stored (type 2). In contrast, identifying dynamic kernel objects (type 3) needs much more effort and depends on the invariance detection algorithm in place. Part of our countermeasure is a trusted module which operates in the guest operating system at boot time. Boot time is considered our root of trust. We are confident this to be a realistic assumption.[2] From this point on,

---

[2] Boot time ends right before calling *kernel_thread* which starts *init*, the first userspace application of the Linux kernel. At this stage the kernel is booted, initialized and all the required device drivers have been loaded.

the system is considered to operate in an untrusted environment and a regular integrity checking of the protected objects is necessary to preserve the system's safety. Given a list of invariant kernel objects, the trusted module communicates this data (virtual address and size) of the kernel objects to observe after boot, and stores them in the guest's address space. Then it will raise a hypercall in order to send the collected entries to the hypervisor. The hypervisor will checksum the contents mapped at the addresses provided by the trusted module and will store their hashes in its address space, which is not accessible to the guest. The trusted module is then forced to unload via a *end-of-operation* message sent by the hypervisor. *Hello rootKitty* doesn't accept objects after the kernel has booted, in order to prevent a possible Denial-Of-Service attack launched by an attacker who is aware of the presence of our system. It is important to point out that *Hello rootKitty* is not a invariance-detection system for critical kernel objects and thus it must be provided with a list of kernel objects on which it will enforce invariance. This list can be either generated by invariance detection systems [29,2,3,6] or manually compiled by kernel and kernel-module developers.

Although implementing countermeasures in a separated virtual machine or within the hypervisor increases the degree of security via isolation, it often leads to higher performance overhead than the equivalent implementation in the target system. A challenging task is that of checking integrity outside of the target operating system while limiting the performance overhead. We achieve this by exploiting the regular interaction of a Virtual Machine Monitor and the guest operating system. In a virtualized environment the guest's software stack runs on a logical processor in VMX non-root operation [11]. This mode differs from the ordinary operation mode because certain instructions executed by the guest kernel may cause a VMExit. A VMExit, is a transition from VMX non-root mode to VMX root mode. After a VMExit the hypervisor will gain control of the CPU and will handle the exception. When the handler terminates the hypervisor performs a VMEntry and returns the control to the guest which will load the latest state of the logical processor and resume execution in VMX non-root mode. Our countermeasure performs integrity checks every time the guest kernel writes to a control register (MOV_CR* event) which in-turn causes a VMExit. Trapping this event is strategic because when virtual addressing is enabled, the upper 20 bits of control register 3 (CR3) become the page directory base register (PDBR). This register is fundamental to locate the page directory and the page tables for the current task. Whenever the guest kernel schedules a new process (process switch) the guest CR3 is modified. Performing integrity checks on the MOV_CR* events is a convenient way to keep detection time and performance overhead to a minimum while guaranteeing a high level of security on protected objects. Moreover, this choice allows a constraint relaxation to improve performance even more by paying a small cost in terms of detection time. We provide more details for our constraint relaxation in Section 3.2. Alternatively the hypervisor could check integrity randomly during the execution of the guest operating system. But this would not scale according to the guest system load as our current approach does.

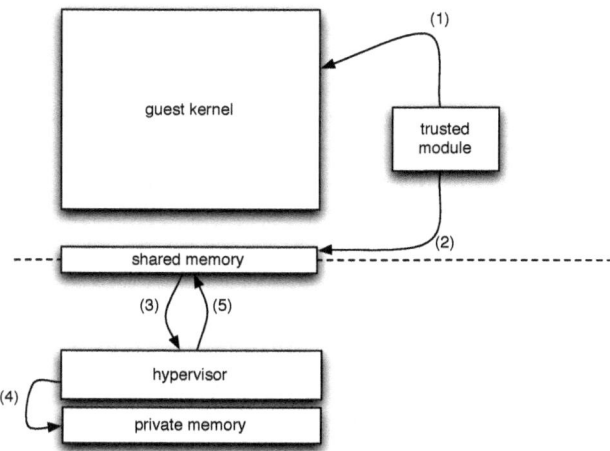

Fig. 1. High level view of trusted module-hypervisor interaction

3.2   Implementation

In this section we discuss the implementation details of *Hello rootKitty*. We consider the choice of the hypervisor of critical importance in order to limit the overhead of the entire system. In fact, countermeasures implemented in virtualized environments are usually affected by considerable overhead which often prevents their deployment in actual production systems. We developed a prototype of our countermeasure in BitVisor, a tiny Type-I hypervisor [26] which exploits Intel VT and AMD-V instruction sets. Our target system runs a Linux kernel with version 2.6.35 and the trusted module has been implemented as a loadable kernel module for the Linux kernel. Our choice of BitVisor is mainly due to its memory address translation features. In BitVisor, the guest operating system and the Virtual Machine Monitor share the same physical address space. Thus, the VMM does not need any complex mechanism to provide translations from guest to host virtual addresses. The guest operating system will rely on the guest page table to perform translations from virtual to physical addresses. This considerably reduces the size of the hypervisor's code and has a very low impact on the overall performance. Unfortunately, in this specific architecture, the VMM can not directly use the guest page table. Translations of guest virtual addresses to host virtual addresses are thus performed by the cooperation of the trusted module and the hypervisor, as explained later in this section.

Figure 1 presents a high-level view of *Hello rootKitty*. Our system detects illegal modifications to invariant critical kernel objects in three phases which are described below.

Communicating phase. The trusted module executes in the guest's address space and communicates the addresses and sizes of critical kernel objects to be protected. In order to test and benchmark our system in a realistic way, we created an artificial list of critical kernel objects by allocating synthetic kernel data. For each critical object the trusted module will retrieve its physical address by calling $\_pa(virtual\_address)$, a macro of the Linux kernel. If the kernel object is stored in one physical frame the trusted module will immediately collect the start address and the size. If the kernel object is stored on more than one physical frames the trusted module will store the relative list of physical addresses. When the virtual addresses of all objects have been translated, a hypercall is raised which signals the hypervisor to start the integrity checking.

Detection phase. In order to detect changes the hypervisor needs to access the contents at the physical addresses collected by the trusted module. This is achieved by mapping the physical address and size of each object in its private memory and computing the signature of its actual contents. When all objects have been checksummed an end-of-operation flag is set in a memory area shared with the trusted module, which in turn will be unloaded. The checksum is performed by a procedure which implements MD5. This cryptographic hash function provides the integrity guarantees needed for our purposes. While stronger hash functions exist, we believe that the security and collision rate provided by MD5 are strong enough to adequately protect our approach from mimicry attacks.

Repairing phase. When the hypervisor detects that the signature of a protected object is different from the one computed the first time, two different behaviors are allowed: a) the system will report an ongoing attack or b) the system attempts to restore the contents of the compromised object if a copy has been provided by the trusted module. Since the hypervisor and the guest share the same physical address space, the hypervisor can restore the original content by mapping the physical address of the compromised object in its virtual space. The untampered value is then copied and control returns to the guest. The restoration of modified critical data structures means that, while the rootkit's code is still present in the address space of the kernel, it is no longer reachable by the kernel control-flow and thus it is neutralized. Since switching from VMX-root to VMX-non-root causes a flush of the TLB, any code in the guest that was using the compromised object will perform the address translation and memory load again and will thus load the restored value. As previously mentioned, whenever task switching, the CR3 register's contents are changed. *Hello rootKitty* traps the MOV_CR* event and checks the integrity of the critical kernel objects. This checking occurs outside of the guest operating system and thus can't be influenced by it. Since the number of kernel objects might be high, the hypervisor will perform the integrity checking of only a subset of objects. Control is then returned to the guest kernel and another subset of critical kernel objects will be checked at the next MOV_CR* event. While considerably improving the performance overhead, this relaxation obviously comes at a cost in terms of security and detection time. We do believe however, that the resulting detection ability of *Hello rootKitty* remains strong, a belief which we explore further in Section 5.

4   Evaluation

We implemented *Hello rootKitty* in BitVisor (Ver 1.1) and the trusted module as a loadable kernel module of the Linux kernel. All experiments were performed on Intel Core 2 Duo 2 Ghz processor with 4GB of RAM.

4.1   Security Evaluation

In order to evaluate whether *Hello rootKitty* would detect a real rootkit we downloaded and installed a minimal rootkit [15] which hijacks a system-call entry, specifically the `setuid` systemcall, from the system-call table. When the `setuid` system call is invoked with the number 31337 as an argument, the rootkit locates the kernel structure for the calling process and elevates its permissions to "root". The way of hijacking entries in the system-call table is very common among rootkits (see Table 1) since it provides the rootkit a convenient and reliable control of sensitive system calls. The critical kernel object that the rootkit modifies is the system-call table which normally remains invariant throughout the lifetime of a specific kernel version. The Linux kernel developers have actually placed this table in read-only memory, however the rootkit circumvents this by remapping the underlining physical memory to new virtual memory pages with write permissions.

Before installing the rootkit, we gave as input to our trusted module, the address of the invariant system-call table and its size. Since *Hello rootKitty* is an invariance-enforcing framework and not an invariance-discovering system, the invariant critical kernel-objects and their size must be provided to it from an external source. This source can either be automatic invariance-discovering systems or kernel programmers who wish to protect their data structures from malicious modifications. Once our system was booted we loaded the rootkit in the running kernel. When the next `MOV` to control-registered occured, the system trapped into the hypervisor and *Hello rootKitty* detected the change on the invariant system-call table. After reporting the attack, the system repaired the system-call table by restoring the system-call entry with the original memory address. This means, that while the rootkit's code is still loaded in kernel-memory it is no longer reachable by any statement and thus inactive.

This shows, that *Hello rootKitty* can detect rootkits and repair the kernel provided that a) the kernel objects used by the rootkit to achieve control are invariant and b) the utilized kernel objects are included in the list of invariance that is given to our system.

4.2   Performance Benchmarks

According to *slabtop*, a Linux utility which displays kernel slab cache information, approximately 15,000 kernel objects are allocated during system's lifetime, 75% of which smaller than 128 bytes. These numbers are never exceeded in other detection systems. Thus in order to measure the overhead introduced by our countermeasure we instrumented the trusted module to create 15,000 kernel objects each of 128 bytes and then performed different types of benchmarks. In order to avoid checking all 15,000 objects at each VMExit we check each time a different subset of the object set. This parameter is configurable and its value depends on the priorities of each installation (performance versus detection time). We measure real (wall-clock) timings in a virtualized environment to compensate for the inaccuracy of time measurements within the virtual machine (i.e. the guest's timers are paused when the hypervisor is performing any other operation). We collected results from ApacheBench [1] sending requests on a local webserver running lighttpd (Table 3) and from SPECINT 2000 as macrobenchmarks to estimate the delay perceived by the user (Table 4). Lastly, we collected accurate timings of microbenchmarks from lmbench (Table 2). The macrobenchmarks show that our system imposes neglibible overhead on the SPEC applications (0.005%) allowing its widespread adoption as a security mechanism in virtualized systems.

Microbenchmarks show a consistent overhead on process forking, as expected. In Table 2 we do not report measurements of context switching latencies because the numbers produced by this benchmark are inaccurate [27,12]. An improvement of local communication bandwidth is due to the slower context switching which has the side effect of slightly increasing the troughtput of file or mmap re-reading operations.

4.3   Memory Overhead

Memory overhead is proportional to the number of protected objects. The data structure needed to store information for integrity checking is 20 bytes long (64-bit kernel object physical address, 32-bit kernel object size, 32-bit checksum, 32-bit support flags used by the hypervisor)[3]. Protecting 15,000 objects costs 193KB when the original content is not provided and 2168 KB otherwise. Moreover, every time a subset of the list of objects is checked the hypervisor needs to map each object from the guest physical space to its virtual space. In our proof of concept the hypervisor will map 100 objects of 128 bytes each every time a MOV_CR event is trapped. This has an additional cost of 13KB. Thus the overall cost in terms of memory is approximately 206KB (2181KB if a copy of the original content is provided). Since the regular hypervisor allocates 128MB at system startup, the memory overhead is 1.5%. The trusted module needs the same amount of memory. But since it will free previously allocated memory after raising the hypercall, that memory would be regularly used by the kernel.

4.4   Detection Time

Due to the relaxation of integrity checking introduced in the earlier sections, it is possible that the modified critical kernel object will not be in the current subset that *Hello rootKitty* checks. The current prototype of our system checks

---

[3] In order to repair the compromised object, the hypervisor needs to store the object's original content too. This may increase the memory overhead.

Table 2. Performance overhead of Hello rootKitty in action on lmbench benchmarks

Processes - times in microseconds - smaller is better							
	open clos	slct TCP	sig inst	sig hndl	fork proc	exec proc	sh proc
no counterm.	16.6	3.08	0.48	2.41	1222	4082	16.K
Hello rootKitty	16.5	3.09	0.48	2.47	1724	5547	18.K
overhead (%)	0.6%	0.3%	0%	2.5%	41.0%	35.8%	12.5%

File and VM system latencies in microseconds - smaller is better							
	0K File create	0K File delete	10K File create	10K File delete	Mmap latency	Prot fault	Page fault
no counterm.	26.0	21.5	99.9	28.2	62.2K	4.355	9.32010
Hello rootkitty	26.4	21.3	99.8	27.8	66.5K	4.444	9.84780
overhead (%)	1.53%	-0.93%	-0.1%	-1.43%	6.9%	2.0%	5.5%

Local Communication bandwidth in MB/s - bigger is better							
	TCP	File reread	Mmap reread	Bcopy(libc)	Bcopy (hand)	Mem read	Mem write
no counterm.	2401	313.0	4838.1	617.5	616.1	4836	698.7
Hello rootkitty	2348	313.2	4885.0	619.7	618.8	4842	697.8
overhead (%)	2.2%	-0.06%	-0.93%	-0.32%	-0.43%	-0.12%	0.12%

Table 3. Results of ApacheBench sending 100000 requests, 50 concurrently on local lighttpd webserver

Benchmark	no counterm.	Hello rootKitty	Perf.overh.
Time	7.153 (sec)	7.261 (sec)	1.50%
Requests per second	13981.10 (num/sec)	13771.43 (num/sec)	1.52%
Time per request	3.576 (ms)	3.631 (ms)	1.54%
Time per concurrent request	0.072 (ms)	0.073 (ms)	1.4%
Transfer rate	52534.36 (Kbytes/sec)	51746.51 (Kbytes/sec)	1.52%

Table 4. SPEC2000 benchmarks of Hello rootKitty in action

Benchmark	no counterm.(sec)	Hello rootKitty(sec)	Perf. overh.
164.gzip	204	204	0%
175.vpr	138	142	2.8%
176.gcc	88.7	89.0	0.3%
181.mcf	86.4	86.7	0.34%
197.parser	206	207	0.5%
256.bzip2	179	179	0%
300.twolf	229	229	0%
Average	161.6	162.4	0.005%

a subset of 100 objects at every change of a Control Register. For the total of 15,000 objects, this means that in the worst case scenario *Hello rootKitty* will detect the malicious modification 149 process switches later than the moment it happened. We found out that in a normally loaded system, this corresponds to approximately 6 seconds of wall-clock time. We believe that this is an acceptable security trade-off for the performance benefits that relaxation offers.

5   Limitations

In this section we describe the limitations and possible weak points of our countermeasure. Since *Hello rootKitty* checks the critical kernel objects for invariance at every change of a Control Register (CR*), an attacker can possibly compromise the scheduler of the operating system and avoid task switching, thus avoiding changes of the CR3. The problem with this attack is that it effectively freezes the system, since the control can't be returned back from the kernel to the running applications. A rootkit's main goal is to hide itself from administrators thus any rootkit behaving this way will a) reveal that there is something wrong in the kernel of the running operating system and b) will never be able to intercept system calls of running processes. These facts suggest that while the attack is possible, it is not probable.

Another attack might occur because of the relaxation explained in Section 3.2 which improves the performance overhead but comes with a cost in terms of detection time. Since at any MOV_CR event the hypervisor will check the integrity of a subset of objects, several malicious processes in the guest might compromise the kernel and restore the original contents before the hypervisor performs the checking. We consider such an attack hard to accomplish because, although the hypervisor performs integrity checking in a deterministic fashion, the attacker has no information about the position of the compromised object in the hypervisor's memory space. A possible mitigation for this kind of attack is the randomization of the sequence in which blocks are checked.

A third possible way to compromise the guest kernel would be by corrupting critical kernel objects whose values legitimately change during the kernel's lifetime. Such objects are not invariant and thus can't be included in the list of objects that *Hello rootKitty* checks since our system is unable to differentiate legitimate from non-legitimate changes. The majority though of existing kernel-mode rootkits modify invariant data structures thus our system reduces considerably the rootkit attack surface and prevents most rootkits from performing a successful attack.

Lastly, *Hello rootKitty* depends on invariance inference engines to provide an accurate list of invariant critical kernel objects. Thus, if the inference engine used doesn't provide all the invariant critical kernel objects (false negatives), *Hello rootKitty* will be unable to detect attacks that occur in the non-reported kernel objects.

6   Related Work

A number of efforts exist on detecting and preventing kernel malware. In this section we explore related work that attempts to protect a kernel using specialized hardware, virtualization, code integrity and profiling.

Hardware-based countermeasures. Copilot [18] is a kernel integrity monitor which detects illegal modications to a hosts kernel by fetching physical memory pages where kernel data and code is stored. Although the monitor has negligible overhead, it needs a separate PCI-card to fetch pages of the running kernel. Gibraltar [2] is a system to infer kernel data structure invariants by fetching snapshots of kernel memory in a way similar to Copilot, by using a PCI-card. The violation of the inferred invariants is reported as a potential attack. While detecting malicious behavior, those countermeasures are limited by the usage of special hardware. A wide deployment of such systems is harder to achieve.

Kernel code integrity. A countermeasure specifically designed to prevent the execution of unauthorized code is described in [24]. The system comes in the form of a tiny hypervisor which protects legacy OSes and ensures that only validated code of the guest can execute in kernel mode. Another rootkit prevention system is NICKLE [22], which prevents unauthorized kernel code execution via memory shadowing. The hypervisor maintains a shadow physical memory to store authenticated guest kernel code. At runtime it determines if the instruction fetch is for kernel or user mode. After verifying the code it will route the instruction fetch accordingly to shadow or standard memory. Kernel rootkit attacks would be detected and prevented since invalidated code would attempt to run in kernel mode.

A recent attack to bypass countermeasures against code injection attacks, such as the Non-Executable stack countermeasure, is Return Oriented Programming (ROP) [25]. In ROP, the attacker, instead of injecting malicious code in the address space of a vulnerable process, crafts his malicious payload by combining fragments of existing code. This method of attacking has been used to create return-oriented rootkits which re-use fragments of authorized kernel code for malicious purposes [10]. Such rootkits can bypass countermeasures like the ones proposed by [22,24].

Yin et al. [32] protect kernel function pointers from being compromised by rootkits. The approach consists of an analysis and a detection subsystems. The analysis subsystem keeps track of function pointer propagation in kernel memory via a whole-system emulator and generates the policy for hook detection. The detection subsystem resides on the target machine and detects violations of the inferred policy. Although the described countermeasure is binary-centric and can generate a hook detection policy without modifying any guest source code, it can be disabled by a rootkit attack since the detection system resides in the target machine. A countermeasure which protects kernel hooks dynamically allocated from the heap is described in [29]. Since protection of kernel objects needs byte-level granularity, obviously finer than the page-level granularity provided by commodity hardware, this countermeasure relocates kernel hooks to be protected to a dedicated page and then exploits the regular page-level protection of the MMU. Although the overhead is negligible, rootkit attacks that compromise non-control data would not be prevented.

Analysis and profiling systems. Malware analysis can reveal important information about the way a rootkit compromises data structures or how private data are stolen, allowing researchers to understand rootkit's behavior and design effective countermeasures. A virtual machine monitor designed for malware analysis as described in [14] extracts features like memory pages, system calls, disk and network accesses from the analyzed program running in the guest. Wang et al. [30] is an analysis tool against persistent rootkits, which compromise kernel hooks for hiding purposes. Those kernel hooks are first identified by monitoring the kernel-side execution path of system utility programs (e.g. ps, ls) and then reported as potential targets. Identification of kernel hooks and extraction of the hook implanting mechanisms via dynamic analysis is proposed in [31]. A whole-system emulator is used rather than a virtual machine monitor. A rootkit proler is described in [23]. A VMM is used to log the rootkit hooking behavior, to monitor targeted kernel objects, to extract kernel rootkit code and infer the potential impact on user-level programs.

Hello rootKitty can be easily integrated with the systems described above in order to perform integrity checking and detect illegal changes to those kernel objects collected by the analysis tools.

7 Conclusion

In this paper we demonstrated how the guaranteed isolation between a hypervisor and a guest operating system can be used to build a non-bypassable invariance-enforcing framework. We realized our idea by designing and implementing *Hello rootKitty*, a lightweight countermeasure to mitigate kernel-mode rootkits in common operating system kernels. Upon detection of a change of an invariant kernel object, *Hello rootKitty* alerts the administrator of the guest operating system and proceeds to repair the kernel by restoring the data structure to its original values. Due to this change, the rootkit's injected code is no longer reachable by any statements in the kernel and thus can no longer affect the running kernel's operations. The evaluation of our prototype showed that *Hello rootKitty* can detect control-flow changes used by modern rootkits with negligible performance and memory overhead, making it a viable countermeasure for protecting operating systems in virtualized environments.

Acknowledgements. This research is partially funded by the Interuniversity Attraction Poles Programme Belgian State, Belgian Science Policy, IBBT and the Research Fund K.U.Leuven.

References

1. Apachebench: A complete benchmarking and regression testing suite
2. Baliga, A., Ganapathy, V., Iftode, L.: Detecting kernel-level rootkits using data structure invariants (2010)
3. Carbone, M., Lee, W., Cui, W., Peinado, M., Lu, L., Jiang, X.: Mapping kernel objects to enable systematic integrity checking. In: ACM Conf. on Computer and Communications Security (2009)
4. Criswell, J., Lenharth, A., Dhurjati, D., Adve, V.: Secure Virtual Architecture: A Safe Execution Environment for Commodity Operating Systems. In: Proceedings of SOSP 2007 (2007)
5. Dewan, P., Durham, D., Khosravi, H., Long, M., Nagabhushan, G.: A hypervisor-based system for protecting software runtime memory and persistent storage. In: Proceedings of SpringSim 2008 (2008)
6. Dolan-Gavitt, B., Srivastava, A., Traynor, P., Giffin, J.: Robust signatures for kernel data structures. In: Proceedings of CCS 2009 (2009)
7. Gadaleta, F., Younan, Y., Jacobs, B., Joosen, W., De Neve, E., Beosier, N.: Instruction-level countermeasures against stack-based buffer overflow attacks. In: Eurosys (2009)
8. Garfinkel, T., Rosenblum, M.: A virtual machine introspection based architecture for intrusion detection (2003)
9. Holz, T., Steiner, M., Dahl, F., Biersack, E., Freiling, F.: Measurements and mitigation of peer-to-peer-based botnets: a case study on storm worm. In: Proceedings of LEET 2008 (2008)
10. Hund, R., Holz, T., Freiling, F.C.: Return-oriented rootkits: Bypassing kernel code integrity protection mechanisms. In: SSYM 2009: Proceedings of the 18th Conference on USENIX Security Symposium (2009)
11. Intel Corporation. Intel 64 and IA-32 Architectures Software Developer's Manual, vol. 3B (2007)
12. Open Kernel labs. Why lmbench is evil?
    http://www.ok-labs.com/blog/entry/why-lmbench-is-evil/
13. Mcaffee: 2010 threat predictions (2010),
    http://mcafee.com/us/local_content/white_papers/7985rpt_labs_threat_predict_1209_v2.pdf
14. Nguyen, A.M., Schear, N., Jung, H., Godiyal, A., King, S.T., Nguyen, H.D.: Mavmm: Lightweight and purpose built vmm for malware analysis. In: Annual Computer Security Applications Conference (2009)
15. oblique. setuid rootkit,
    http://codenull.net/articles/kernel_mode_hooking.tar.gz

16. de Oliveira, D.A.S., Felix Wu, S.: Protecting kernel code and data with a virtualization-aware collaborative operating system. In: Proceedings of ACSAC 2009 (2009)
17. PacketStorm, http://packetstormsecurity.org/UNIX/penetration/rootkits/
18. Petroni Jr., N.L., Timothy, Jesus, F., William, M., Arbaugh, A.: Copilot - a coprocessor-based kernel runtime integrity monitor. In: Proceedings of the 13th USENIX Security Symposium (2004)
19. Petroni Jr., N.L., Hicks, M.: Automated detection of persistent kernel control-flow attacks. In: Proceedings of CCS 2007 (2007)
20. Linux on-the-fly kernel patching without LKM, by sd and devik. Phrack Issue 58
21. QubesOS: Architecture Specification, http://qubes-os.org/files/doc/arch-spec-0.3.pdf
22. Riley, R., Jiang, X., Xu, D.: Guest-Transparent Prevention of Kernel Rootkits with VMM-Based Memory Shadowing. In: Lippmann, R., Kirda, E., Trachtenberg, A. (eds.) RAID 2008. LNCS, vol. 5230, pp. 1–20. Springer, Heidelberg (2008)
23. Riley, R., Jiang, X., Xu, D.: Multi-aspect profiling of kernel rootkit behavior. In: Proceedings of Eurosys 2009 (2009)
24. Seshadri, A., Luk, M., Qu, N., Perrig, A.: Secvisor: a tiny hypervisor to provide lifetime kernel code integrity for commodity oses. In: Proceedings of Twenty-First ACM SIGOPS Symposium on Operating Systems Principles (2007)
25. Shacham, H.: The geometry of innocent flesh on the bone: return-into-libc without function calls (on the x86). In: Proceedings of CCS 2007 (2007)
26. Shinagawa, T., Eiraku, H., Tanimoto, K., Omote, K., Hasegawa, S., Horie, T., Hirano, M., Kourai, K., Oyama, Y., Kawai, E., Kono, K., Chiba, S., Shinjo, Y., Kato, K.: Bitvisor: a thin hypervisor for enforcing i/o device security. In: Proceedings of VEE 2009 (2009)
27. Staelin, C., McVoy, L.: lmbench manual page
28. Symantec. Windows rootkit overview, http://www.symantec.com/avcenter/reference/windows.rootkit.overview.pdf
29. Wang, Z., Jiang, X., Cui, W., Ning, P.: Countering kernel rootkits with lightweight hook protection. In: Proceedings of CCS 2009 (2009)
30. Wang, Z., Jiang, X., Cui, W., Wang, X.: Countering Persistent Kernel Rootkits through Systematic Hook Discovery. In: Lippmann, R., Kirda, E., Trachtenberg, A. (eds.) RAID 2008. LNCS, vol. 5230, pp. 21–38. Springer, Heidelberg (2008)
31. Yin, H., Liang, Z., Song, D.: Hookfinder: Identifying and understanding malware hooking behaviors. In: NDSS 2008 (2008)
32. Yin, H., Poosankam, P., Hanna, S., Song, D.: HookScout: Proactive binary-centric hook detection. In: Kreibich, C., Jahnke, M. (eds.) DIMVA 2010. LNCS, vol. 6201, pp. 1–20. Springer, Heidelberg (2010)

Opacity Analysis in Trust Management Systems

Moritz Y. Becker[1] and Masoud Koleini[2]

[1] Microsoft Research, Cambridge, UK
[2] University of Birmingham, UK

Abstract. Trust management systems are vulnerable to so-called probing attacks, which enable an adversary to gain knowledge about confidential facts in the system. We present the first method for deciding if an adversary can gain knowledge about confidential information in a Datalog-based policy.

1 Introduction

In the trust management paradigm [7], authorization rules are specified in a high-level *policy language* (e.g. [18,4,11,17]). Access is granted only if the user's request complies with the policy in conjunction with the user-submitted *credentials*. There is a class of attacks on such systems, called *probing attacks*, that enables an adversary to gain knowledge about confidential information in a service's policy, by submitting a series of *probes*, i.e., access requests together with conditional credentials.

Here is an example of a simple probing attack on a policy written in Binder [11]. The service `Hospital` publishes the policy rule "`Hospital` says `canCreateAcc(x)` if `AgeCert` says `over21(x)`". This rule stipulates that any principal x can create a patient account if `AgeCert` says that x is an adult. `Hospital`'s policy also contains confidential facts that are not visible to the adversary `Eve`, for instance, whether `Bob` is a patient or not. (Our use of the term "policy" includes not only rules, but also all authorization-relevant facts.) Suppose `Eve` collaborates with `AgeCert`, and hence can get hold of the `AgeCert`-issued credential "`AgeCert` says `over21(Eve)` if `Hospital` says `patient(Bob)`". She starts her probing attack by submitting this credential together with a request to create a patient account. The service evaluates the corresponding query "`Hospital` says `canCreateAcc(Eve)`?" against its policy *in union* with the credential. Suppose it responds by granting access. Next, `Eve` submits a second probe, with the same access request, but with no supporting credentials. This time, access is denied.

From these two probes, `Eve` deduces that the submitted credential must have been crucial in making the access query succeed. But this is only possible if credential's condition "`Hospital` says `patient(Bob)`" is true in `Hospital`'s policy. She has therefore *detected* a confidential fact through probing.

Probing attacks present a serious problem to trust management systems, as policies commonly contain confidential information, and it takes little effort to conduct a probing attack with legitimately collected and self-issued credentials.

X. Lai, J. Zhou, and H. Li (Eds.): ISC 2011, LNCS 7001, pp. 229–245, 2011.
© Springer-Verlag Berlin Heidelberg 2011

At the same time, it is non-trivial to see, in more involved examples, if a piece of information can be detected through probing or not. It is therefore critical to have an automated method for verifying *non*-detectability – or, *opacity* – of information in credential systems.

This paper presents several significant contributions to the problem space. Firstly, on a foundational level, it provides the first formal framework for probing attacks that makes no assumptions on the structure of policies and of the credential evaluation mechanism, and that is thus general enough to encompass widely different languages such as XACML [18] and DKAL2 [14]. Based on an abstract notion of observational equivalence, this framework specifies precisely what it means for a piece of information to be detectable or opaque (Section 3).

Secondly, we present the first algorithm for checking opacity in policies written in Datalog, which is the basis of many existing policy languages (Section 5). The algorithm is not only sound, but also complete and terminating: if it fails to prove opacity (in which case failure is reported in finite time), then the given fact is provably detectable. This is a strong result, as the mere existence of a complete decision procedure for opacity in this context is far from obvious.

A particularly attractive feature of the algorithm is its constructiveness. Intuitively, a property is opaque in a policy if there exists some policy that behaves in the same way as the first policy with regards to the adversary's probes, but in which the property does not hold. If the property is opaque, the algorithm actually constructs such a "witness" policy. In fact, it can iteratively construct a finite sequence of such witnesses that subsumes the generally infinite set of *all* witnesses. What does this feature buy us? Without it, the algorithm would be merely *possibilistic*: the mere existence of a witness policy, no matter how pathological it may be, would be sufficient for opacity. But since the algorithm constructs the witnesses, they can be assigned probabilities, or the security analyst could interactively discard unlikely witnesses. The final result could therefore be interpreted as a degree of likelihood of the property being opaque.

Thirdly, we identify several optimization methods for cutting the high computational cost by pruning the search space, and show empirically that these render the opacity verification problem feasible in medium-sized cases, whereas the straightforward implementation of the algorithm is unusable for other than very small test cases (Section 6). Full proofs and extended examples are included in a technical report [5].

2   Related Work

Probing attacks on credential systems were first mentioned in [13]. One of the primary design goals of their policy language, DKAL, was to provide protection against probing attacks. However, they do not precisely define what they are protecting against, and indeed, it has been shown that DKAL2 [14] is susceptible to probing attacks [3].

Probing attacks were first defined in terms of opacity in [3]. However, in contrast to our general framework (Section 3), their definitions only apply to simple logic-based policy languages, but not to more complex languages such as XACML [18] or Ponder [10], in which policies have some (e.g. hierarchical) structure, or languages such as DKAL, where incoming credentials are filtered and transformed before being added to the policy. That paper also presents an inference system for analyzing detectability in Datalog policies, but it is incomplete and non-constructive in the sense that it does not map to a terminating algorithm, and thus cannot be used to check opacity. We present the first decision procedure for proving opacity (Section 5).

Research on information flow has mainly focused on stateful, temporal computations (see [19] for an overview). The current setting is very different as there is no notion of state, run or trace, and, most importantly, probes may contain credentials that are temporarily combined with the local policy during query evaluation – this is precisely what makes the analysis so hard. In contrast, the adversaries considered in computational information flow analysis typically cannot inject code into the program.

A policy could be seen as a database, and the probes as queries against the database. Detectability through probing could therefore be seen as related to the database inference problem, which is concerned with covert channels through which confidential information from a database can leak to a database user. A wide variety of such channels have been studied [15,12], mainly for relational databases. Bonatti et al. have studied the database inference problem in deductive databases [8], which are similar to the Datalog-based policies considered in the current paper. However, the problem considered in the current paper is harder, as it corresponds to users who can temporarily inject new rules and relations into the database (which is not natural in the typical database context).

There has been some work on formalizing and enforcing safety in Automated Trust Negotiaton (ATN) protocols [22,21], i.e., the property that no information about the presence or absence of credentials is prematurely leaked during a credential exchange [20]. This problem is quite different from the one we are considering here; e.g., we are interested in the confidentiality of internal properties of the policy rather than that of submitted credentials, and our credentials are not mere attributes, but may be conditional and may affect policy evaluation results.

3   A Framework for Probing Attacks

3.1   Abstract Framework

This section establishes the fundamental concepts for reasoning about probing attacks in credential systems.

Definition 1 (Policy language, probe). *A* policy language *is a triple* (**Pol**, **Prb**, ⊢), *where* **Pol** *and* **Prb** *are sets called* policies *and* probes, *respectively, and* ⊢ *is a binary infix relation from* **Pol** × **Prb**, *called* decision relation.

*Let $A \in$ **Pol**, $\pi \in$ **Prb**. If $A \vdash \pi$ we say that π is* positive in A; *otherwise (i.e., $A \nvdash \pi$), π is* negative in A.

Although Definition 1 does not prescribe the structure of probes, it helps to think of a probe as a pair containing a set of credentials that the adversary submits to the service under attack, and a query corresponding to some access request. A positive probe is one that leads to an access grant, whereas a negative leads to an access denial.

To illustrate Definition 1, we briefly sketch how it would be instantiated to the concrete policy languages SecPAL [4], DKAL2 [14], and XACML [18].

In SecPAL, a policy is a set of SecPAL *assertions* such as "Alice says x canRead if x canWrite" or "Bob says Eve cansay x canWrite". Access requests are mapped to SecPAL *queries*, which are first-order formulas over atoms of the form "$\langle Principal \rangle$ says $\langle Fact \rangle$". An inference system defines which queries are deducible from a policy. A user's access request is mapped to a query, and is granted only if the query is deducible from the *union* of the local policy *and* the set of credentials (which are also just assertions) submitted by the user together with the request. Therefore, a probe π is naturally defined as a pair $\langle A, \varphi \rangle$ containing a set A of credentials and a query φ. Then $A_0 \vdash \pi$ iff φ is deducible from $A_0 \cup A$. The definitions can be instantiated in a very similar fashion for other related languages such as RT [17], Cassandra [6], SD3 [16], and Binder [11].

In DKAL2, a policy is a set of so-called *infon terms* such as "Eve said Alice canRead" or "Bob implied Alice said Eve canWrite". As in SecPAL, a set of inference rules defines which other infon terms can be deduced from a policy. However, infon terms sent by the adversary (corresponding to submitted credentials) are not simply added verbatim to the local policy, but are converted depending on the term's shape. For example, the infon term "A canRead ⟵ B canWrite" submitted by Eve would be imported as the infon term "B canWrite → Eve implied A canRead". Why this is done and what this means is beyond the scope of this paper; the important point here is that there are credential systems where the access query is not simply evaluated against the union of the local policy and the submitted credentials, but where the latter are first modified according to some rules.

In XACML, a policy (as in Definition 1) would correspond to a *PolicySet*, which is a hierarchical structure containing other PolicySets or items called (XACML) *Policy*. The latter is a collection of *Rules*. XACML is thus an example of a system where a policy is not just a flat collection of assertions. Just as in DKAL2, incoming credentials (SOAP messages conveying *Attributes* [1]) may be transformed before evaluation. For example, the client-supplied Attribute may be written in SAML and would first be converted by an XACML Context Handler [2]. As in the case of DKAL2, the instantiation of $\vdash$ would have to take such transformations into account.

We now consider the adversary, i.e., the principal who mounts the probing attack against a service's policy A_0. Informally, the adversary has a *passive* and an *active* capability. A passively acting adversary only reads the visible part of A_0 that is presented to her by the service.

An active adversary can additionally evaluate probes against A_0 and observe whether they are positive or negative in A_0. Typically, the adversary does not have the power to evaluate arbitrary probes, but only the probes *available* to her. In the standard case where probes are pairs containing a set of credentials and a query, the availability of a probe is typically determined, firstly, by which credentials the adversary possesses or can create, and, secondly, by which queries the service allows her to run. For instance, SecPAL services define an Authorization Query Table that map access requests to SecPAL queries, so only these queries can be evaluated by clients. In DKAL2, incoming infon terms (corresponding to credentials) are filtered by a filtering policy, so not all credentials possessed by clients are available in probes. Our definition of available probes abstracts away such language-dependent details.

Definition 2 (Alikeness and available probes). *An* adversary *is defined by an equivalence relation* $\simeq \; \subseteq \mathbf{Pol} \times \mathbf{Pol}$, *and a set* $\mathbf{Avail} \subseteq \mathbf{Prb}$ *of* available probes. *If* $A_1 \simeq A_2$ *for two policies* A_1 *and* A_2, *we say that* A_1 *and* A_2 *are* alike.

The alikeness relation, which specifies the adversary's passive capability, is also kept abstract in Definition 2. Typically, a policy can be split into a publicly visible and a private part (relative to a particular adversary). A useful instantiation in this case would be that two policies are alike iff their visible parts are syntactically equal. Alternatively, one could adopt a more semantic instantiation, such that two policies are alike iff their visible parts are semantically equivalent.

We can now define the adversary's *active* capability.

Definition 3 (Observational equivalence). *Two policies* A_1 *and* A_2 *are observationally equivalent* $(A \equiv A')$ *iff*

1. $A_1 \simeq A_2$, *and*
2. $\forall \pi \in \mathbf{Avail}, A_1 \vdash \pi \iff A_2 \vdash \pi$.

Alikeness and observational equivalence induce two different notions of indistinguishability of policies. A passive adversary cannot distinguish policies that are alike. An active adversary can see the visible parts of a policy *and* run probes against it. These two capabilities are represented by conditions 1. and 2. in Definition 3. Hence an active adversary cannot distinguish policies that are observationally equivalent.

We are interested in whether the adversary can infer that some property Φ holds about policy A, just by looking at the policy's public parts and by running the probes available to her. If she can, then we say that Φ is detectable in A, otherwise Φ is opaque in A. This is formalized in the following definition, which again is implicitly relative to a given adversary.

Definition 4 (Detectability, opacity). *A predicate* $\Phi \subseteq \mathbf{Pol}$ *is* detectable in $A \in \mathbf{Pol}$ *iff* $\forall A' \in \mathbf{Pol} : \; A \equiv A' \Rightarrow \Phi(A')$.

A predicate $\Phi \subseteq \mathbf{Pol}$ *is* opaque in $A \in \mathbf{Pol}$ *iff it is not detectable in* A, *or equivalently, iff* $\exists A' \in \mathbf{Pol} : \; A \equiv A' \wedge \neg\Phi(A')$.

The definitions established in this section so far provide a general framework for reasoning about probing attacks in credential systems. For specific trust management frameworks, the definitions of policy language and alikeness need to be instantiated accordingly. We do this in the remainder of this section for Datalog. In Section 4, we discuss an example in Datalog, which will also help illustrate the definitions above.

3.2   Datalog-Based Policies

Datalog is not used as a policy language per se, but is the semantic basis for many existing policy languages, and many others can be translated into it (e.g. [4,17,6,16,11]). Reasoning techniques and analysis tools for Datalog therefore apply to a wide range of policy languages. We only give a very brief overview of Datalog. (For a more careful introduction, see e.g. [9].)

The central construct in Datalog is a *clause*. A clause a is of the form

$$P_0 \leftarrow P_1, ..., P_n,$$

where $n \geq 0$, and the P_i are *atoms* of the form $p(\vec{e})$ (where p is a predicate symbol, and $\vec{e}$ a sequence of variables and constants). (We usually omit the arrow if $n = 0$.) We write $\mathbf{hd}(a)$ to denote a's *head* P_0 and $\mathbf{bd}(a)$ to denote its *body* $\vec{P} = \langle P_1, ..., P_n \rangle$. Given a set of clauses A, we write $\mathbf{hds}(A)$ to denote the atom set $\{\mathbf{hd}(a) \mid a \in A\}$.

A *query* φ is either **true, false** or a ground (i.e., variable-free) boolean formula (i.e., involving connectives $\neg$, $\wedge$ and $\vee$) over atoms P. We write $\mathbf{Qry}$ to denote the set of all queries. A query φ is evaluated with respect to a set A of assertions. For atomic $\varphi = P$, we define that $A \vdash P$ holds iff there exists a ground (i.e., variable-free) instance $P \leftarrow \vec{P}$ of some clause in A and $A \vdash P_i$ for all $P_i \in \vec{P}$. The non-atomic cases are defined in the standard way, e.g. $A \vdash \neg\varphi$ iff $A \nvdash \varphi$.

Now we can instantiate the abstract Definitions 1 and 2. For evaluating probes, we adopt the simple model where the query of a probe is evaluated against the union of the service's policy and the credentials (i.e., clauses) of the probe.

Definition 5 (Datalog instantiation). *We instantiate* **Pol** *to the powerset of clauses,* $\wp(\mathbf{Cls})$. *A (Datalog) policy is hence a set* $A_0 \subseteq \mathbf{Cls}$.

A (Datalog) probe π *is a pair* $\langle A, \varphi \rangle$, *where* $A \subseteq \mathbf{Cls}$ *and* $\varphi \in \mathbf{Qry}$. *Hence* **Prb** *is instantiated to the set of all such probes. A probe is* ground *iff it does not contain any variables. We write* $\neg\langle A, \varphi \rangle$ *to denote the probe* $\langle A, \neg\varphi \rangle$.

The decision relation $\vdash \subseteq \mathbf{Pol} \times \mathbf{Prb}$ *is defined by* $A_0 \vdash \langle A, \varphi \rangle \iff A_0 \cup A \vdash \varphi$.

Definition 6 (Adversary, Datalog alikeness). *An* adversary *is defined by a set* **Avail** $\subseteq$ **Prb** *and a unary predicate* **Visible** $\subseteq$ **Cls**. *If* **Visible**(a) *for some* $a \in \mathbf{Cls}$, *we say that* a *is* visible. *We extend* **Visible** *to policies by defining the* visible part *of* A, **Visible**(A), *as* $\{a \in A \mid \mathbf{Visible}(a)\}$, *for all* $A \subseteq \mathbf{Cls}$.

Two policies $A_1, A_2 \subseteq \mathbf{Cls}$ *are* alike *(*$A_1 \simeq A_2$*) iff* **Visible**(A_1) = **Visible**(A_2).

Definitions 5 and 6 induce instantiations for the Datalog definitions of observational equivalence between policies, and of opacity and detectability. Recall that the latter two were defined for arbitrary properties of policies. Here, we are interested in a particular class of policy properties, namely whether a given probe (usually one that is not in **Avail**) is positive or negative.

Definition 7 (Probe detectability & opacity). *A probe* $\pi \in$ **Prb** *is detectable in* $A \in$ **Pol** *iff* $\forall A' \in$ **Pol** : $A \equiv A' \Rightarrow A' \vdash \pi$.

A probe $\pi \in$ **Prb** *is opaque in* $A \in$ **Pol** *iff it is not detectable in* A, *or equivalently, iff* $\exists A' \in$ **Pol** : $A \equiv A' \wedge A' \nvdash \pi$.

Note that this definition is just a specialization of Definition 4, with the predicate Φ instantiated to $\{A \subseteq$ **Cls** $\mid A \vdash \pi\}$.

4 Example

We illustrate the definitions from the previous section using an example of an authorization policy written in Datalog. The example also serves as the basis for the test cases in Section 6. Our example is taken from a grid computing scenario. A compute cluster allows users to run compute jobs. The execution of a job may require read access to data that is stored in an external data center. The cluster has a policy that governs who can run compute jobs, and the data center has a policy that governs who can access data. Both policies delegate authority over certain attributes to trusted third parties. The policies consist of the following seven clauses:

$$\mathsf{canExe}(\mathtt{Clstr}, x, j) \leftarrow \mathsf{mem}(\mathtt{Clstr}, x), \mathsf{owns}(\mathtt{Clstr}, x, j), \mathsf{canRd}(\mathtt{Data}, \mathtt{Clstr}, j). \quad (1)$$

$$\mathsf{owns}(\mathtt{Clstr}, x, j) \leftarrow \mathsf{owns}(y, x, j), \mathsf{isTTP}(\mathtt{Clstr}, y). \quad (2)$$

$$\mathsf{mem}(\mathtt{Clstr}, x, j) \leftarrow \mathsf{mem}(y, x, j), \mathsf{isTTP}(\mathtt{Clstr}, y). \quad (3)$$

$$\mathsf{canRd}(\mathtt{Data}, x, j) \leftarrow \mathsf{canRd}(y, x, j), \mathsf{owns}(\mathtt{Data}, y, j). \quad (4)$$

$$\mathsf{owns}(\mathtt{Data}, x, j) \leftarrow \mathsf{owns}(y, x, j), \mathsf{isTTP}(\mathtt{Data}, y). \quad (5)$$

$$\mathsf{isTTP}(\mathtt{Clstr}, \mathtt{CA}). \quad (6)$$

$$\mathsf{isTTP}(\mathtt{Data}, \mathtt{CA}). \quad (7)$$

Here, we adopt the convention that the first parameter of a predicate denotes the principal "saying" (i.e., vouching for) the predicate, and the second parameter denotes the subject of the predicate. For instance, $\mathsf{canExe}(\mathtt{Clstr}, x, j)$ intuitively means that $\mathtt{Clstr}$ says that x can execute job j.

According to Clause (1), anyone who is a member and owns a job (according to $\mathtt{Clstr}$) can execute that job (according to $\mathtt{Clstr}$), if the data center $\mathtt{Data}$ allows $\mathtt{Clstr}$ to read the data associated with that job. $\mathtt{Clstr}$ delegates authority over job ownership and membership to trusted third parties (2)–(3). The next clause implements a variant of discretionary access control: data center $\mathtt{Data}$ stipulates that owners y of data associated with job j can delegate read access to this data to other principals x (4). Just like $\mathtt{Clstr}$, $\mathtt{Data}$ delegates authority over ownership to third parties it trusts (5). Finally, both $\mathtt{Clstr}$ and $\mathtt{Data}$ specify certificate authority $\mathtt{CA}$ as a trusted third party (6)–(7).

Clstr has an interface that allows users to submit a job execution request. When some user Eve requests to execute a job Job, the corresponding query

$$\varphi_{\text{Eve}} = \text{canExe}(\text{Clstr}, \text{Eve}, \text{Job}) \tag{8}$$

is evaluated against the policy consisting of the clauses (1)–(7), in union with the (possibly empty) set of credentials submitted by Eve together with the request.

Eve, who plays the role of the adversary in our scenario, possesses four credentials:

$$\text{owns}(\text{CA}, \text{Eve}, \text{Job}). \tag{9}$$
$$\text{mem}(\text{CA}, \text{Eve}). \tag{10}$$
$$\text{canRd}(\text{Eve}, \text{Clstr}, \text{Job}). \tag{11}$$
$$\text{canRd}(\text{Eve}, \text{Clstr}, \text{Job}) \leftarrow \text{mem}(\text{Clstr}, \text{Bob}). \tag{12}$$

Credentials (9)–(10) are issued by CA, and the other two are self-issued. Eve is interested in finding out if Bob is a member, according to Clstr's policy. Of course, she does not have the authority to query this fact directly, so instead she hopes to be able to detect this fact using (12) in particular, stating that she is willing to give Clstr read access, provided that Bob is a member of Clstr.

Let A_0 be the policy consisting of clauses (1)–(7), and A_{Eve} be the set of clauses (9)–(12). A_{Eve} and φ_{Eve} together give rise to a set of $2^4 = 16$ available probes that Eve is able to run against A_0: $\textbf{Avail} = \{\langle A, \varphi_{\text{Eve}} \rangle \mid A \subseteq A_{\text{Eve}}\}$. For simplicity, we assume that $\textbf{Visible} = \emptyset$, i.e., Eve is not able to passively read any of the clauses in A_0. Based on this scenario, we make the following observations.

We have $A_0 \vdash \langle A_{\text{Eve}}, \varphi_{\text{Eve}} \rangle$, in other words, $A_0 \cup A_{\text{Eve}} \vdash \varphi_{\text{Eve}}$. The derivation goes roughly as follows: Credential (9) proves Eve's ownership over Job to both Data and Clstr. Hence Data allows Eve to delegate read access to Clstr using (11). Furthermore, (10) is sufficient for proving Eve's membership to Clstr, hence all body atoms of (1) are satisfied, which implies φ_{Eve}.

We also have $A_0 \vdash \langle \{(9)–(11)\}, \varphi_{\text{Eve}} \rangle$, since the derivation above only makes use of the clauses (9)–(11). But $A_0 \nvdash \langle A, \varphi_{\text{Eve}} \rangle$, for all $A \subseteq \{(9),(10),(12)\}$. In particular, replacing clause (11) in the probe in item 2) above by clause (12) produces a negative probe. Note that the two clauses only differ in the body.

All policies A_0' that are observationally equivalent to A_0, i.e., that exhibit the same behaviour as observed above, satisfy the property that $A_0 \nvdash \text{mem}(\text{Clstr}, \text{Bob})$. For suppose the contrary were the case. We observed that φ_{Eve} holds in $A_0 \cup \{(9)–(11)\}$. By assumption, the body of clause (12) is true in A_0, which means that replacing clause (11) by (12) in the probe cannot make a difference. But this contradicts the observation that φ_{Eve} does *not* hold in $\{(9),(10),(12)\}$.

It follows that the probe $\langle \emptyset, \neg\text{mem}(\text{Clstr}, \text{Bob}) \rangle$, which is not in $\textbf{Avail}$, is *detectable* in A_0. In other words, Eve can be sure that Bob is not a member of Clstr.

5 Verifying Opacity

This section presents an algorithm for verifying opacity. Given a set of available probes, the algorithm decides if a given probe is opaque (or detectable) in a given Datalog policy. The algorithm works with arbitrary input policies, but we restrict the input probes to ground ones, in order to simplify the problem. This restriction is reasonable, as attribute and delegation credentials are usually issued for one specific principal and purpose, and are thus ground anyway.

In the following, we assume as given a policy $A_0 \subseteq \mathbf{Cls}$, a ground probe $\pi_0 \in \mathbf{Prb}$, and an adversary defined by a set $\mathbf{Avail} \subseteq \mathbf{Prb}$ of ground probes and the visibility function $\mathbf{Visible}$. The algorithm should decide if π_0 is opaque in A_0, relative to the adversary specified by $\mathbf{Avail}$ and $\mathbf{Visible}$.

Overview. The algorithm is succinctly specified as the transition system in Fig. 1, but it is actually rather involved. We first give a high-level roadmap of the algorithm before proceeding to the details.

Recall that π_0 is opaque in A_0 iff there exists a policy A_0' that is observationally equivalent to A_0 (with respect to the probes in $\mathbf{Avail}$), but such that π_0 is negative in A_0'. To prove opacity, the algorithm attempts to construct such an *opacity witness* A_0'. Conversely, to prove detectability, it proves that no such A_0' exists.

A *state* in the transition system is a triple of the form $\langle \Pi^+, \Pi^-, A_1 \rangle$, and the (INIT) rule in Fig. 1 defines the set $\mathbf{Init}$ of *initial states*. Intuitively, an initial state is populated with sets Π^+, Π^- of probes that are required to be positive (negative, respectively) in the opacity witness A_0' to be constructed. At each (PROBE) transition, the system considers and discards one positive probe in Π^+, and adds a set of clauses to the *witness candidate* $A_1 \subseteq \mathbf{Cls}$. Theorem 1 states that π_0 is opaque iff the transition system reaches a state of the form $\langle \emptyset, \Pi^-, A_0' \rangle$, starting from some initial state. Furthermore, A_0' will be an opacity witness.

Our results also show that opacity checking is decidable. This is nontrivial, as the definition of opacity is quantified over the infinite set of all policies; and many other simple-looking quantified properties such as containment are undecidable. (Note that the set of predicate symbols and constants may be infinite.) The decidability of opacity checking essentially stems from a sort of topological compactness property of the set of policies A_0' that are observationally equivalent to A_0. More precisely, even though there may be infinitely many candidates for A_0', we only ever need to consider a finite number of them.

5.1 Initial States

The initial states $\mathbf{Init}$, defined declaratively by (INIT) in Fig. 1, are produced by transforming all available probes into equivalent disjunction- and negation-free ones. Consider a disjunctive probe $\pi = \langle A, \varphi_1 \vee \varphi_2 \rangle \in \mathbf{Avail}$ that is positive in A_0. The algorithm attempts to find a policy A_0' such that $A_0' \vdash \pi$ holds, in other words, $A_0' \cup A \vdash \varphi_1 \vee \varphi_2$. This is equivalent to either finding an A_0' such that $A_0' \vdash \langle A, \varphi_1 \rangle$,

$$\text{(INIT)} \frac{(\Pi^+, \Pi^-) \in \mathbf{flatten}_{A_0}(\mathbf{Avail}) \quad \pi_0 = \langle A, \varphi \rangle \quad (S^+, S^-) \in \mathbf{dnf}(\neg\varphi)}{\forall \pi \in \Pi^- \cup \{\langle A, \bigvee S^-\rangle\} : \mathbf{Visible}(A_0) \nvdash \pi}$$
$$\langle \Pi^+ \cup \{\langle A, \bigwedge S^+\rangle\}, \ \Pi^- \cup \{\langle A, \bigvee S^-\rangle\}, \ \mathbf{Visible}(A_0)\rangle \in \mathbf{Init}$$

$$\text{(PROBE)} \frac{\tilde{A} \subseteq A \quad \langle a_1, ..., a_n \rangle \in \mathbf{perms}(\tilde{A}) \quad \forall i \in \{1, ..., n\} : \vec{P_i} = \mathbf{bd}(a_i) \quad \vec{P}_{n+1} = \vec{P}}{A'' = \bigcup_{k=1}^{n+1} \bigcup_{P_k \in \vec{P}_k} \{P_k \leftarrow \mathbf{hds}(\{a_1, ..., a_{k-1}\})\} \quad \forall \pi \in \Pi^- : A' \cup A'' \nvdash \pi}$$
$$\langle \Pi^+ \cup \{\langle A, \bigwedge \vec{P}\rangle\}, \ \Pi^-, \ A'\rangle \xrightarrow{\langle A, \bigwedge \vec{P}\rangle} \langle \Pi^+, \ \Pi^-, \ A' \cup A''\rangle$$

Fig. 1. Transition system for verifying opacity

or finding one such that $A'_0 \vdash \langle A, \varphi_2 \rangle$. A disjunction in the query of a positive probe in **Avail** therefore corresponds to a branch in the search for A'_0.

What about probes in **Avail** that are negative in A_0? Since $A_0 \nvdash \pi$ is equivalent to $A_0 \vdash \neg\pi$, we can convert all negative probes in **Avail** into equivalent positive ones before dealing with the disjunctions in positive probes.

The function $\mathbf{flatten}_{A_0}$ (defined below) applied to **Avail** first performs the mentioned conversion of negative probes into equivalent positive ones. It then splits each probe into disjuncts of atomic and negated atomic queries. Finally, it produces a cartesian product of all these disjuncts that keeps the atomic and negated queries apart. The result is a set of pairs of disjunction-free probe sets; each such pair (Π^+, Π^-) corresponds to a disjunctive search branch. The problem of finding a policy A'_0 that is observationally equivalent to A_0 (*cf.* Def. 3) can then be reduced to finding an A'_0 and picking a pair $(\Pi^+, \Pi^-) \in \mathbf{flatten}_{A_0}(\mathbf{Avail})$ such that all probes in Π^+ are positive, and all probes in Π^- are negative in A'_0.

In the following, we write $\mathbf{dnf}(\varphi)$ to denote the disjunctive normal form of a query φ, represented as a set of pairs (S^+, S^-) of sets of atoms. For instance, if $\varphi = (p \wedge q \wedge \neg s) \vee (\neg p \wedge \neg q \wedge s)$, then $\mathbf{dnf}(\varphi) = \{(\{p, q\}, \{s\}), (\{s\}, \{p, q\})\}$.

Definition 8 (Flatten). *Let $\Pi \subseteq \mathbf{Prb}$. Then $\mathbf{flatten}_{A_0}(\Pi)$ is a set of pairs (Π^+, Π^-) of sets of probes defined inductively as follows:*

$$\mathbf{flatten}_{A_0}(\emptyset) = \{(\emptyset, \emptyset)\}.$$

$$\mathbf{flatten}_{A_0}(\Pi \cup \{\langle A, \varphi\rangle\}) = \{(\Pi^+, \Pi^-) \mid$$
$$\exists (S^+, S^-) \in \mathbf{dnf}(\tilde{\varphi}), (\Pi_0^+, \Pi_0^-) \in \mathbf{flatten}_{A_0}(\Pi) :$$
$$\Pi^+ = \Pi_0^+ \cup \{\langle A, \bigwedge S^+\rangle\} \ and \ \Pi^- = \Pi_0^- \cup \{\langle A, \bigvee S^-\rangle\},$$
$$where \ \tilde{\varphi} = \varphi \ if \ A_0 \vdash \langle A, \varphi\rangle, \ and \ \tilde{\varphi} = \neg\varphi \ otherwise.$$

Apart from the observational equivalence $A'_0 \equiv A_0$, opacity of π_0 in A_0 additionally requires that π_0 be negative in A'_0. Let $\pi_0 = \langle A, \varphi\rangle$. This is equivalent to finding a pair $(S^+, S^-) \in \mathbf{dnf}(\neg\varphi)$ such that $A'_0 \vdash \langle A, \bigwedge S^+\rangle$ and $A'_0 \nvdash \langle A, \bigvee S^-\rangle$.

We can then reduce the problem of proving opacity of π_0 in A_0 to constructing an A'_0 for some $(\Pi_0^+, \Pi_0^-) \in \mathbf{flatten}_{A_0}(\mathbf{Avail})$ and $(S^+, S^-) \in \mathbf{dnf}(\neg\varphi)$ such that all probes in $\Pi^+ = \Pi_0^+ \cup \{\langle A, \bigwedge S^+\rangle\}$ are positive, and all probes in $\Pi^- = \Pi_0^- \cup \{\langle A, \bigvee S^-\rangle\}$ are negative in A'_0. If such an A'_0 exists, we say that A'_0

is a *witness* (for the opacity of π_0 in A_0). We call Π^+ a set of *positive probe requirements*, and Π^- a set of *negative probe requirements*.

Furthermore, from the alikeness condition $A_0' \simeq A_0$, any witness must contain **Visible**(A_0). Hence (INIT) picks **Visible**(A_0) as the initial witness candidate for each initial state. The last premise in (INIT) filters out those witnesses candidates that fail to make all probes in Π^- negative.

These observations are formalized in Lemma 1, stating the correctness of (INIT).

Lemma 1. π_0 *is opaque in* A_0 *iff there exist* $\langle \Pi^+, \Pi^-, \mathbf{Visible}(A_0) \rangle \in \mathbf{Init}$ *and* $A_0' \subseteq \mathbf{Cls}$ *such that* $A_0' \supseteq \mathbf{Visible}(A_0)$ *and* $\forall \pi \in \Pi^+ : A_0' \vdash \pi$ *and* $\forall \pi \in \Pi^- :$ $A_0' \nvdash \pi$.

5.2   Finding Minimal Witnesses

Given **Init**, we now have to find a witness A_0' that satisfies the requirements from Lemma 1. Consider an initial or intermediate state $\langle \Pi^+ \cup \{\pi\}, \Pi^-, A' \rangle$. The transition rule (PROBE) from Fig. 1 picks the positive probe requirement π and adds to the current witness candidate A' a set A'' of clauses such that $A' \cup A'' \vdash \pi$. The monotonicity of $\vdash$ guarantees that adding A'' does not make any previously considered positive probe requirement negative. However, adding clauses may make negative probe requirements in Π^- positive, so we need to check $\forall \pi' \in \Pi^- : A' \cup A'' \nvdash \pi'$. If this fails, the algorithm backtracks to try out a different A''. Otherwise, the transition succeeds and produces the new state $\langle \Pi^+, \Pi^-, A' \cup A'' \rangle$. If we reach a finite state, i.e. one where Π^+ is empty, then π_0 is opaque, by Lemma 1, and moreover, the witness candidate of the final state is a genuine witness. This informally shows that the algorithm is sound.

Minimality. To ensure completeness, we have to consider *all* candidate extensions A'' such that $A' \cup A'' \vdash \pi$. It turns out that we can ignore A' and simply consider all A'' such that $A'' \vdash \pi$ (which then implies $A \cup A'' \vdash \pi$). However, there may be infinitely many such A''. At the same time, we want to ensure that the algorithm is a *decision procedure*, in other words, that it is both complete and terminating, which is necessary for proving that the goal π_0 is *not* opaque (i.e., detectable). We certainly do not want infinite branching. Fortunately, it turns out that we do not need to compute all candidate extensions. Instead, we only compute the *minimal* ones. This notion of minimality is based on Datalog containment.

Definition 9 (Containment). *A policy A is* contained *in a policy A' (we write: $A \preceq A'$) iff for all ground atoms P and all sets S of ground atoms:* $A \vdash \langle S, P \rangle \Rightarrow A' \vdash \langle S, P \rangle$.

So, to be more precise, the candidate extensions actually considered by the algorithm form a *finite* set S such that

  $- \forall A'' \in S : A'' \vdash \pi$, and

$- \forall \tilde{A}'' \subseteq \mathbf{Cls}: \quad \tilde{A}'' \vdash \pi \Rightarrow \exists A'' \in S: A'' \preceq \tilde{A}''.$

This property has two significant ramifications. Firstly, it ensures termination, since S is finite for each considered positive probe requirement π; furthermore, each initial state only has finitely many positive probe requirements, and there are only finitely many initial states in $\mathbf{Init}$.

Secondly, it ensures that the algorithm is complete: consider any $\tilde{A}'_0$ that makes all positive probe requirements Π^+ of some initial state positive. Then there exists A'_0 constructed by iteratively adding a minimal extension for each $\pi \in \Pi^+$, such that A'_0 also makes all probes in Π^+ positive and $A'_0 \preceq \tilde{A}'_0$. Therefore, if $\tilde{A}'_0$ is a genuine witness (i.e., it also makes all negative probe requirements in Π^- negative) then A'_0 is also a genuine witness, by anti-monotonicity of $\nvdash$. Hence if there is a genuine witness, the algorithm will find one that is at least as small, in finite time.

It remains to explain how (PROBE) computes the minimal extensions A''.

Relevant subprobes. To gain an intuition for this process, it helps to ask the question "how could $\pi = \langle A, \varphi \rangle$ possibly be positive in some policy A''?" If A is nonempty, there are multiple explanations. Perhaps φ is true in A'' anyway, so none of the clauses in A are necessary, or *relevant*, for making π positive in union with A''. Or perhaps all clauses in A are relevant, in that removing just one clause from A would result in a negative probe. The most general explanation would be that there exists some subset $\tilde{A} \subseteq A$ that is relevant, i.e., $A'' \vdash \langle \tilde{A}, \varphi \rangle$ but $A'' \nvdash \langle \tilde{A}', \varphi \rangle$ for all $\tilde{A}' \subsetneq \tilde{A}$.

We need to consider all of these $2^{|A|}$ possible cases, since, as we shall see, each different choice of $\tilde{A}$ results in a different set of minimal witness extensions A''. This source of branching is reflected in the condition $\tilde{A} \subseteq A$ in the transition rule (PROBE) in Fig. 1.

Derivation order. Having chosen $\tilde{A} \subseteq A$ to be relevant, there may still be multiple minimal solutions for A'' that makes $\pi = \langle A, \bigwedge \vec{P} \rangle$ positive. Since $\tilde{A}$ is relevant, every clause $P_0 \leftarrow P_1, .., P_n \in \tilde{A}$ is actively used at least once in the derivation $A'' \cup \tilde{A} \vdash \bigwedge \vec{P}$. But this is only possible if (i) the body atoms are also derivable, and (ii) the derivation of $\bigwedge \vec{P}$ depends on all the heads of clauses in $\tilde{A}$, i.e., $\mathbf{hds}(\tilde{A})$. We now attempt to solve this set of constraints for the unknown A''.

At first sight, a plausible requirement on A'' seems to be that (i) $\{P_1, ..., P_n\} \subseteq A''$, and (ii) $\{P \leftarrow \mathbf{hds}(\tilde{A}) \mid P \in \vec{P}\} \subseteq A''$. However, while this is a correct solution for A'', it is not the only one, and not even a minimal one. In general, A'' may contain the body atoms of just a subset of $\tilde{A}$'s clauses, and the heads belonging to these clauses combine with clauses in A'' to make the body atoms of other clauses in $\tilde{A}$ true; this oscillatory back and forth between A'' and $\tilde{A}$ continues until the query $\bigwedge \vec{P}$ is true. The simple solution above corresponds to the special case where the "oscillation" only has one stage. This process is best illustrated by an example.

Example 1. Suppose $\varphi = \vec{P} = z$ and $\tilde{A} = \{p \leftarrow q., r \leftarrow s., u \leftarrow v.\}$. We have to find all minimal A'' such that $A'' \cup \tilde{A} \vdash z$. In the case where the number of stages n is just 1, there is only one minimal solution for A'', containing four clauses:

$$A_1'' = \{q., s., v., z \leftarrow p, r, u.\}$$

In the case $n = 2$, there are six solutions, each containing four clauses; three in which A'' contains one of $\tilde{A}$'s body atoms, and three in which it contains two. Here are two of the six solutions:

$$A_2'' = \{q., s \leftarrow p., v \leftarrow p., z \leftarrow p, r, u.\}$$
$$A_5'' = \{q., s., v \leftarrow p, r., z \leftarrow p, r, u.\}$$

For $n = 3$, there are again six solutions, one for each permutation of $\tilde{A}$, resulting in $1 + 6 + 6 = 13$ solutions. Again, here are two of the 13 solutions:

$$A_8'' = \{q., s \leftarrow p., v \leftarrow p, r., z \leftarrow p, r, u.\}$$
$$A_{13}'' = \{v., s \leftarrow u., q \leftarrow r, u., z \leftarrow p, r, u.\}$$

But note that A_8'' is contained in $(\preceq)$ A_1'', A_2'', and A_5''. Indeed, for each solution from $\{A_1'',...,A_7''\}$, there exists a solution from $\{A_8'', ..., A_{13}''\}$ such that the latter is contained in the former. Hence only the solutions for the case $n = 3$ are minimal.
□

It turns out that this observation holds in the general case. Given a particular $\tilde{A} \subseteq A$, we can prove that we only need to consider the case $n = |\tilde{A}|$, which has $n!$ solutions. Let **perms** denote the function that maps a set S to the set of all permutations of S. Each permutation of clauses $\langle a_1, ..., a_n \rangle \in \mathbf{perms}(\tilde{A})$ gives rise to a unique witness candidate A'', constructed as in the example above. Let $\vec{P}_i = \mathbf{bd}(a_i)$, for $i \in \{1, ..., n\}$. Then for each $P_1 \in \vec{P}_1$, A'' contains P_1. For each $P_2 \in \vec{P}_2$, it contains the clause $P_2 \leftarrow \mathbf{hd}(a_1)$. For each $P_3 \in \vec{P}_3$, it contains the clause $P_3 \leftarrow \mathbf{hd}(a_1), \mathbf{hd}(a_2)$. In general, for each $P_k \in \vec{P}_k$, A'' contains the clause $P_k \leftarrow \mathbf{hds}(\{a_1, ..., a_{k-1}\})$. Finally, letting $\vec{P}_{n+1} = \vec{P}$, we have that for each $P_{n+1} \in \vec{P}_{n+1}$, A'' contains $P_{n+1} \leftarrow \mathbf{hds}(\{a_1, ..., a_n\})$.

The transition rule (PROBE) in Fig. 1 shows that nondeterministic branching is not only due to picking $\tilde{A} \subseteq A$, but also to picking a permutation from $\mathbf{perms}(\tilde{A})$. The rule constructs A'' as described, and then tests if the new candidate makes all probes in Π^- negative. If it does, then the state transition $\langle \Pi^+ \cup \{\pi\}, \Pi^-, A' \rangle \xrightarrow{\pi} \langle \Pi^+, \Pi^-, A' \cup A'' \rangle$ is valid.

Theorem 1 (Soundness and completeness). π_0 *is opaque in* A_0 *iff there exist* $\sigma_0 \in \mathbf{Init}$, $\Pi^- \subseteq \mathbf{Prb}$ *and* $A_0' \subseteq \mathbf{Cls}$ *such that* $\sigma_0 \rightarrow^* \langle \emptyset, \Pi^-, A_0' \rangle$.

Theorem 2. *The number of* $(\xrightarrow{\langle A, \bigwedge \vec{P} \rangle})$ *transitions from any state is bounded by* $\sum_{m=0}^{n} \frac{n!}{(n-m)!}$, *where* $n = |A|$.

Theorem 2 also implies that the transition system is finite, and hence the algorithm terminates, since **Init** is finite, and Π^+ in every initial state is finite.

6   Implementation with Optimizations

We implemented a prototype of the state transition system in Fig. 1 in F#. It first computes **Init** as a lazy enumeration, and then performs a backtracking depth first search based on the transition rule (PROBE). The back end is an implementation of Datalog's evaluation relation $\vdash$. It is not highly optimized, and even though it is the main bottleneck, we did not spent much effort making it more efficient, as we are more interested in algorithmic improvements of the search procedure.

The front end includes a parser for problem specifications (A_0, **Visible**(A_0), **Avail**, and π_0) and a GUI that displays the witness, if a final state has been found, or reports that no final state exists. In the former case, the user can choose to discard the found witness and continue the search for the next witness. We have found this to be an extremely useful feature, which helps to overcome some of the limitations of the strict possibilistic (as opposed to probabilistic) concept of opacity.

For example, we added the atomic clause mem(Clstr, Bob) to the policy in Section 4, and expected the fact that Bob now *is* a member to be detectable by Eve. After all, the probe containing clauses (9), (10) and (12) is positive, whereas the one containing only (9) and (10) is negative. This suggests that (12) is relevant, which is only possible if its body atom mem(Clstr, Bob) is derivable.

However, the prototype (correctly) reports that $\langle\emptyset, \text{mem}(\text{Clstr}, \text{Bob})\rangle$ is opaque. A closer look at the produced witnesses reveals that they all contain the rather "improbable" clause "mem(Clstr, Bob) $\leftarrow$ mem(CA, Eve), owns(CA, Eve, Job)".

Indeed, we can prove this hypothesis with our prototype: the weakened input probe $\langle\{(9), (10)\}, \text{mem}(\text{Clstr}, \text{Bob})\rangle$ *is* detectable. Thus, the constructiveness of the algorithm enabled us to form the informal judgement that the original input probe was detectable *with a high likelihood*.

Example 2. Here is another example that shows how the tool can be used interactively. Let $\pi_1 = \langle\{b \leftarrow a., \; d \leftarrow c.\}, e\rangle$, $\pi_2 = \langle\emptyset, \neg a\rangle$, and $\pi_1 = \langle\{d\}, \neg e\rangle$. Suppose that all three probes are positive in A_0, and **Visible**(A_0) = $\emptyset$. What does this tell us about c and a in A_0? This example is interesting because the answer is not obvious on casual inspection and may be somewhat surprising.

We start with the obvious goal probes $\langle\emptyset, c\rangle$ and $\langle\emptyset, a\rangle$. The tool reports that $\langle\emptyset, c\rangle$ is detectable (i.e., c must be true in A_0), but that $\langle\emptyset, a\rangle$ is opaque. For the latter analysis, it also reports that only one minimal witness exists, which includes the clause $a \leftarrow d$. And indeed, the goal probe $\langle\{d\}, a\rangle$ is found to be detectable, hence we can infer that $A_0 \cup \{d\} \vdash a$. (The example is small enough that the interested reader may manually retrace the steps of the algorithm to verify these results.) $\square$

As Theorem 2 indicates, traversing the entire transition system would be infeasible even for small examples. We devised and implemented a number of optimization methods for effectively pruning the search tree. For lack of space, we describe them only very briefly. Full descriptions and proofs of correctness are found in [5].

Order independence. The order in which the probes in Π^+ are processed is irrelevant, since the constructed witness extension A'' is independent of the current witness candidate; it only depends on the currently considered probe. Therefore, we can fix a particular order for Π^+ in an initial state, thereby reducing the search space by a factor of $|\Pi^+|!$ for the search branch starting from that initial state.

Redundant probes. The sets Π^+ and Π^- in an initial state often contain many pairs of probes $\pi_1 = \langle A_1, \varphi_1 \rangle$, $\pi_2 = \langle A_2, \varphi_2 \rangle$ such that $\pi_1 \subseteq \pi_2$ (i.e., $\varphi_1 = \varphi_2 \wedge A_1 \subseteq A_2$). For example, we may have $\pi_1 = \langle \{a.\}, z \rangle$ as well as $\pi_2 = \langle \{a., b.\}, z \rangle$ in Π^+. By monotonicity of $\vdash$ and of the query z, the larger query π_2 is redundant, since any witness candidate that makes π_1 positive also makes π_2 positive. A similar argument can be made for the probes in Π^-. In general, we can first transform initial states $\langle \Pi_0^+, \Pi_0^-, A \rangle \in \mathbf{Init}$ into potentially much smaller states $\langle \Pi_1^+, \Pi_1^-, A \rangle$, where

$$\Pi_1^+ = \{\pi \in \Pi_0^+ \mid \neg \exists \pi' \in \Pi_0^+ : \pi' \subsetneq \pi\}, \text{ and } \Pi_1^- = \{\pi \in \Pi_0^- \mid \neg \exists \pi' \in \Pi_0^- : \pi \subsetneq \pi'\}.$$

These reduced states are then used as initial states.

Conflicting probes. Any initial state $\sigma_0 = \langle \Pi^+, \Pi^-, A \rangle$ in which there exist $\pi_1 \in \Pi^+, \pi_2 \in \Pi^-$ such that $\pi_1 \subseteq \pi_2$ can be discarded straight away, as there are no transitions from σ_0.

Experimental results. To gain a better understanding of the scalability of the opacity checking algorithm, with and without optimization methods, we performed a number of performance tests. We only briefly summarize our findings and refer to the technical report [5] for reproducible details on all test cases and performance plots.

The performance tests are based on the policy from Section 4. The policy is arguably small; however, this fact does not weaken our results, as it is easy to see that the computation time is essentially independent of the size of the policy A_0. The significant parameter with respect to computation time is **Avail**.

We found that the computation time doubles with each irrelevant, trivially positive probe $\langle \{p_i\}, p_i \rangle$ being added to **Avail**, which is predicted by Theorem 2. Successively adding irrelevant, trivially negative probes $\langle \{p_i\}, z \rangle$ only caused a linear increase ($+1.3$ ms per additional probe). This is also to be expected, as negative probes do not cause any branching.

We then explored several variations of the basic scenario from Section 4, running each test case with different combinations of the optimization methods enabled. (The order independence method was enabled in all runs.)

We found that enabling the optimization methods improved performance in all cases, apart from those where **Avail** was manually reduced to only the relevant probes. Even in the latter cases, enabling optimization did not add any significant overhead. The automated optimization methods led to dramatic improvements that were particularly noticeable in the more complex test cases, with speedup factors between 126 and 280. Furthermore, they also significantly improved scalability; for example, increasing the size of **Avail** from 16 to 128 increased the computation time by a factor of 1130 in the unoptimized case, but only by a factor 8 with the optimizations enabled.

Not surprisingly, manually picking only the relevant probes was the most effective strategy for improving performance, with speedup factors between 150 and 19,000. This suggests that significant performance gains can be expected from more sophisticated pruning methods.

The size of **Avail** varied between 16 and 128. All test cases completed in less than one second (with all optimizations enabled, on a standard workstation), apart from the most complex one, which took 7.2 s (vs. 15 min unoptimized) and 150,000 (vs. 7 million) Datalog query evaluations. In the latter test case, the probe queries contained negation, which led to more than 16,000 initial states.

Our results suggest that checking opacity using our tool is feasible in many practical cases, given that the analysis is almost independent of the size of the policy (which may well have millions of clauses), and that it seems reasonable to restrict analysis to adversaries that have no more than about a hundred different probes to their disposal.

7   Discussion

To recapitulate, we first presented a general framework of probing attacks, defining abstract notions of policy, probe, and adversary characterized by available probes, and based on these, notions of observational equivalence, opacity and detectability. We instantiated this framework to Datalog, a language on which many existing policy languages are based.

It has been an open question whether the problem of opacity in Datalog policies is decidable [3]. We answered this question in the positive by presenting a complete decision procedure for opacity. It works by attempting to construct opacity witnesses, i.e., policies that masquerade as the original policy, but falsify the input probe. We also devised a number of optimization strategies for pruning the search space. Our experimental results show that these methods are highly effective.

Opacity is a *possibilistic* information flow property. The mere possibility of the existence of an opacity witness suffices to deem an input probe opaque, no matter how unlikely these witnesses may be. But our algorithm for deciding opacity provides richer information, as it does not merely prove the existence of a witness, but actually enumerates all minimal witnesses. The set of minimal witnesses is a finite representation of the infinite set of all witnesses. Our prototype includes an interface that lets the security analyst browse and inspect the witnesses, thereby enabling her to informally judge the likelihood of opacity or detectability. We believe that this is a more useful approach in practice than ascribing numerical probabilities to the witnesses.

References

1. Anderson, A.: Web Services Profile of XACML (WS-XACML) Version 1.0. OASIS TC Working Draft (2006)
2. Anderson, A., Lockhart, H.: SAML 2.0 Profile of XACML v2. 0. OASIS Standard (2005)
3. Becker, M.Y.: Information flow in credential systems. In: IEEE Computer Security Foundations Symposium, pp. 171–185 (2010)
4. Becker, M.Y., Fournet, C., Gordon, A.D.: Design and semantics of a decentralized authorization language. In: IEEE Computer Security Foundations (2007)
5. Becker, M.Y., Koleini, M.: Information leakage in datalog-based trust management systems. Technical Report MSR-TR-2011-11, Microsoft Research (2011)
6. Becker, M.Y., Sewell, P.: Cassandra: Flexible trust management, applied to electronic health records. In: IEEE Computer Security Foundations, pp. 139–154 (2004)
7. Blaze, M., Feigenbaum, J., Lacy, J.: Decentralized trust management. In: IEEE Symposium on Security and Privacy, pp. 164–173 (1996)
8. Bonatti, P., Kraus, S., Subrahmanian, V.: Foundations of secure deductive databases. IEEE Transactions on Knowledge and Data Engineering 7(3), 406–422 (1995)
9. Ceri, S., Gottlob, G., Tanca, L.: What you always wanted to know about Datalog (and never dared to ask). IEEE Transactions on Knowledge and Data Engineering 1(1), 146–166 (1989)
10. Damianou, N., Dulay, N., Lupu, E., Sloman, M.: The ponder policy specification language. In: Sloman, M., Lobo, J., Lupu, E.C. (eds.) POLICY 2001. LNCS, vol. 1995, pp. 18–38. Springer, Heidelberg (2001)
11. Detreville, J.: Binder, a logic-based security language. In: IEEE Symposium on Security and Privacy, pp. 105–113 (2002)
12. Farkas, C., Jajodia, S.: The inference problem: a survey. ACM SIGKDD Explorations Newsletter 4(2), 6–11 (2002)
13. Gurevich, Y., Neeman, I.: DKAL: Distributed-knowledge authorization language. In: IEEE Computer Security Foundations Symposium (CSF), pp. 149–162 (2008)
14. Gurevich, Y., Neeman, I.: DKAL 2 – a simplified and improved authorization language. Technical Report MSR-TR-2009-11, Microsoft Research (2009)
15. Jajodia, S., Meadows, C.: Inference problems in multilevel secure database management systems. In: Information Security: An Integrated Collection of Essays (1995)
16. Jim, T.: SD3: A trust management system with certified evaluation. In: Proceedings of the 2001 IEEE Symposium on Security and Privacy, pp. 106–115 (2001)
17. Li, N., Mitchell, J.C., Winsborough, W.H.: Design of a role-based trust management framework. In: Symposium on Security and Privacy, pp. 114–130 (2002)
18. OASIS. eXtensible Access Control Markup Language (XACML) Version 2.0 core specification (2005)
19. Sabelfeld, A., Myers, A.: Language-based information-flow security. IEEE Journal on Selected Areas in Communications 21(1), 5–19 (2003)
20. Winsborough, W., Li, N.: Safety in automated trust negotiation. ACM Transactions on Information and System Security (TISSEC) 9(3) (2006)
21. Winsborough, W.H., Li, N.: Towards practical automated trust negotiation. In: IEEE International Workshop on Policies for Distributed Systems and Networks (2002)
22. Winsborough, W.H., Seamons, K.E., Jones, V.E.: Automated trust negotiation. In: DARPA Information Survivability Conference and Exposition, vol. 1 (2000)

On the Inference-Proofness of Database Fragmentation Satisfying Confidentiality Constraints[*]

Joachim Biskup[1], Marcel Preuß[1], and Lena Wiese[2]

[1] Technische Universität Dortmund, Dortmund, Germany
{biskup,preuss}@ls6.cs.tu-dortmund.de
[2] National Institute of Informatics, Tokyo, Japan
wiese@nii.ac.jp

Abstract. Confidentiality of information should be preserved despite the emergence of data outsourcing. An existing approach is supposed to achieve confidentiality by vertical fragmentation and without relying on encryption. Although prohibiting unauthorised (direct) accesses to confidential information, this approach has so far ignored the fact that attackers might infer sensitive information logically by deduction. In this article vertical fragmentation is modelled within the framework of Controlled Query Evaluation (CQE) allowing for inference-proof answering of queries. Within this modelling the inference-proofness of fragmentation is proved formally, even if an attacker has some a priori knowledge in terms of a rather general class of semantic database constraints.

Keywords: Database Security, Information Dissemination Control, Inference-Proofness, First-Order Logic, Outsourcing, Fragmentation.

1 Introduction

In these days information has become one of the most important resources, which has to be protected. In order to protect information from undesired disclosures, confidentiality requirements are declared by setting up a confidentiality policy. According to such a confidentiality policy a system should enforce the declared confidentiality requirements autonomously as for example surveyed in [3].

Moreover, there is an increasing need for storing data cost-efficiently in our economy-driven society. One approach to achieve this goal is called "database as a service" paradigm and leads to third party service providers specialized on hosting database systems and offering the use of these database systems to their customers via Internet in return for payment of rent [12]. These customers may save money because they are freed from purchasing expensive hard- and software and dealing with difficult administrative and maintenance tasks such as upgrading hard- and software or eliminating technical malfunctions.

---

[*] This work has been partially supported by the DFG (SFB 876/A5), and a postdoctoral research grant of the German Academic Exchange Service (DAAD).

X. Lai, J. Zhou, and H. Li (Eds.): ISC 2011, LNCS 7001, pp. 246–261, 2011.

Patient	SSN	Name	DoB	ZIP	Illness	Doctor
	12345	Hellmann	03.01.1981	94142	Hypertension	White
	98765	Dooley	07.10.1953	94141	Obesity	Warren
	24689	McKinley	12.02.1952	94142	Hypertension	White
	13579	Ripley	03.01.1981	94139	Obesity	Warren

Fig. 1. Example of a relational instance containing sensitive associations

Obviously, there is a goal conflict between the discussed "database as a service" paradigm and confidentiality requirements because the service provider cannot be restrained from reading all cleartext information stored in its systems. One natural approach to cope with that conflict lies in encrypting all outsourced data on the user side [12,13]. But, unfortunately, such an approach often makes the efficient evaluation of queries on the server side impossible [2,9].

The benefit of encryption of data lies in making these data – and also the information contained in these data – illegible. But often, in relational database systems single pieces of information are not confidential per se. Due to the storage of data according to some (static) relational schema, semantic associations between different pieces of information are represented and often only these associations are confidential [9]. For example, in a hospital the list of illnesses cured and the list of patients are both not particularly sensitive per se. In contrast, an association between a patient's name and a specific illness is very sensitive and has to be protected. An example adapted from [9] of a relational instance containing this sensitive association among others is given in Fig. 1.

To achieve this protection, some authors suggest to break sensitive associations by splitting relational instances vertically, which is referred to as vertical fragmentation. There are several different approaches to achieving confidentiality based on vertical fragmentation surveyed in [13] and for each of these approaches the corresponding authors describe how fragments of an original relational instance can be outsourced so that unauthorised (direct) accesses to confidential information are prohibited. But it is *not* shown that confidential information cannot be inferred by employing inferences, which may offer the possibility to infer confidential information based on the knowledge of non-confidential information [11]. Moreover, it is *not* considered that an attacker often has some a priori knowledge, which might enable him to infer confidential information [6].

In contrast, there are several approaches to so-called Controlled Query Evaluation (CQE) surveyed in [4] and for each of these approaches it is proven that a declared confidentiality policy is enforced so that any harmful inferences are avoided. "Inference-proofness" is achieved by limiting a user's information gain so that this user cannot infer protected information reliably based on his a priori knowledge and the (possibly distorted) answers to his queries.

The main novel contribution of this article consists of a formal analysis of a specific approach to vertical fragmentation – splitting a relational instance into one externally stored part and one locally-held part – w.r.t. its inference-proofness. More specifically, based on the seminal ideas proposed in [7,8], a formalisation of this approach to vertical fragmentation is developed in Sect. 2. After introducing the framework of CQE briefly in Sect. 3, a logic-oriented

F_o	tid	SSN	Name	DoB		F_s	tid	ZIP	Illness	Doctor
	1	12345	Hellmann	03.01.1981			1	94142	Hypertension	White
	2	98765	Dooley	07.10.1953			2	94141	Obesity	Warren
	3	24689	McKinley	12.02.1952			3	94142	Hypertension	White
	4	13579	Ripley	03.01.1981			4	94139	Obesity	Warren

Fig. 2. Possible fragmentation of the instance given in Fig. 1

modelling of the approach to vertical fragmentation presented in Sect. 2 within the framework of CQE is introduced in Sect. 4 and subsequently analysed w.r.t. its inference-proofness in Sect. 5. Thereby an attacker's a priori knowledge in terms of a rather general class of semantic database constraints is respected.

2   Confidentiality by Fragmentation

Now, the approach to vertical fragmentation (in the following simply referred to as fragmentation) presented in [7,8] is extended. In this approach all data is stored in a single relational instance r over a relational schema $\langle R|A_R|SC_R\rangle$ with relational symbol R and the set $A_R = \{a_1, \ldots, a_n\}$ of attributes. Moreover, the set SC_R contains some semantic (database) constraints, which must be satisfied by each relational instance constructed over this schema. Note that semantic constraints are not considered in [7,8] (hence there $SC_R = \emptyset$).

The approach considered is built on the assumption of a client-server architecture, in which the server is managed by a third party service provider. This third party service provider is not considered to be trustworthy in terms of confidentiality and might actively monitor all queries processed and all data stored on its server. But it is assumed to be guaranteed that this service provider does not manipulate data maliciously so that all data received from the server is always correct in terms of integrity. The client used in this architecture is assumed to be completely trustworthy and also has the ability to store (a limited amount of) data locally. But as this local storage is assumed to be more expensive than the external storage, it is desirable to store as much data as possible on the server.

The idea for achieving confidentiality despite outsourcing (some) data lies in splitting the original instance r over schema $\langle R|A_R|SC_R\rangle$ into two fragment instances f_o and f_s stored instead of r. While f_s may be outsourced to an external server, f_o can only be stored locally on the client.[1] To build f_o and f_s, the attribute set A_R of schema $\langle R|A_R|SC_R\rangle$ is partitioned into two sets $\bar{A}_{F_o}$ and $\bar{A}_{F_s}$ (items (i), (iii) of Def. 1). Then fragment instance f_o (f_s, respectively) is in essence the projection of r on $\bar{A}_{F_o}$ ($\bar{A}_{F_s}$) (item (a) of Def. 1). Obviously, in terms of confidentiality no sensitive information or association is allowed to be contained in fragment instance f_s. Such a fragmentation of the instance given in Fig. 1 (e.g., breaking the name-illness association) is depicted in Fig. 2.

As an authorised user having access to the client as well as to the server should be able to query all information contained in the original instance r, the reconstructability of r based on the fragments f_o and f_s must be guaranteed. For

---

[1] Index s is for server and index o is for owner-side (local) storage.

this purpose each tuple in f_o and f_s is extended by a tuple ID (item (i) of Def. 1) so that in f_o and f_s exactly those two tuples (one tuple per fragment instance) which together correspond to a tuple of r share the same tuple ID (item (c) of Def. 1) being unique in f_o as well as in f_s (item (b) of Def. 1). Strategies for processing SQL queries referring to the original instance on the corresponding fragment instances efficiently are discussed in [7].

Based on the seminal ideas proposed in [7,8] a formalisation of this concept of fragmentation is developed in this article as follows:

Definition 1 (Fragmentation). *Given a relational schema* $\langle R|A_R|SC_R\rangle$, *a vertical fragmentation* $\mathcal{F}$ *of* $\langle R|A_R|SC_R\rangle$ *is a set*

$$\mathcal{F} = \{\langle F_o|A_{F_o}|SC_{F_o}\rangle, \langle F_s|A_{F_s}|SC_{F_s}\rangle\}$$

in which $\langle F_o|A_{F_o}|SC_{F_o}\rangle$ *and* $\langle F_s|A_{F_s}|SC_{F_s}\rangle$ *are relational schemas called* fragments *of* $\mathcal{F}$. *Moreover, for* $i \in \{o, s\}$, *it holds that*

(i) $A_{F_i} := \{a_{tid}\} \cup \bar{A}_{F_i}$ *with* $a_{tid} \notin A_R$ *and* $\bar{A}_{F_i} \subseteq A_R$,
(ii) $SC_{F_i} := \{a_{tid} \to \bar{A}_{F_i}\}$ *with* $a_{tid} \to \bar{A}_{F_i}$ *being a functional dependency*,
(iii) $\bar{A}_{F_o} \cup \bar{A}_{F_s} = A_R$ *and* $\bar{A}_{F_o} \cap \bar{A}_{F_s} = \emptyset$.

Given a relational instance r *over* $\langle R|A_R|SC_R\rangle$, *the fragment instances* f_o *and* f_s *over* $\langle F_o|A_{F_o}|SC_{F_o}\rangle$ *and* $\langle F_s|A_{F_s}|SC_{F_s}\rangle$ *are created by inserting both the tuple* ν_o *into* f_o *and the tuple* ν_s *into* f_s *for each tuple* $\mu \in r$. *Thereby, for* $i \in \{o, s\}$:

(a) $\nu_i[a] = \mu[a]$ *for each attribute* $a \in \bar{A}_{F_i}$,
(b) $\nu'[a_{tid}] \neq \nu''[a_{tid}]$ *for tuples* $\nu', \nu'' \in f_i$ *with* $\nu' \neq \nu''$,
(c) $\nu_o[a_{tid}] = \nu_s[a_{tid}]$ *for attribute* $a_{tid} \in A_{F_o}, A_{F_s}$.

Other tuples do not exist in f_o *and* f_s.

Note that even for two different tuples of r which are equal w.r.t. all attributes of $\bar{A}_{F_s}$ ($\bar{A}_{F_o}$, respectively) there are also two different tuples in f_s (f_o) which are equal w.r.t. all attributes of $\bar{A}_{F_s}$ ($\bar{A}_{F_o}$) because of the existence of unique tuple IDs. Hence, for each tuple μ of r there is *exactly* one tuple ν_s in f_s as well as one tuple ν_o in f_o. If there were no tuple IDs, all duplicates of tuples in f_s (f_o) would be removed. In terms of the example in Fig. 2 the first and the third tuple of f_s would be consolidated without the existence of tuple IDs.

As the goal is to achieve confidentiality by fragmentation, a formal declaration of confidentiality requirements is indispensable. In [7,8] this is obtained by defining a set of so-called confidentiality constraints on the schema level.

Definition 2 (Confidentiality Constraint). *Let* $\langle R|A_R|SC_R\rangle$ *be a relational schema. A* confidentiality constraint c *over* $\langle R|A_R|SC_R\rangle$ *is a subset* $c \subseteq A_R$.

Semantically a confidentiality constraint c claims that each combination of values allocated to the set $c \subseteq A_R$ of attributes in an instance r over schema $\langle R|A_R|SC_R\rangle$ should not be contained completely in f_s. In f_o such a combination of values may be contained completely since f_o is only stored locally.

$$c_1 = \{\texttt{SSN}\} \qquad c_3 = \{\texttt{Name}, \texttt{Illness}\}$$
$$c_2 = \{\texttt{Name}, \texttt{DoB}\} \qquad c_4 = \{\texttt{DoB}, \texttt{ZIP}, \texttt{Illness}\}$$

Fig. 3. Set $\mathcal{C}$ of confidentiality constraints over *Patient*

Definition 3 (Confidentiality of Fragmentation). *Let $\langle R|A_R|SC_R \rangle$ be a relational schema, $\mathcal{F}$ a fragmentation of $\langle R|A_R|SC_R \rangle$ according to Def. 1 and $\mathcal{C}$ a set of confidentiality constraints over $\langle R|A_R|SC_R \rangle$ according to Def. 2. Fragmentation $\mathcal{F}$ is confidential w.r.t. $\mathcal{C}$ iff $c \not\subseteq A_{F_s}$ for each $c \in \mathcal{C}$.*

Note that in case of a singleton constraint $c = \{a_i\}$ a fragmentation can only be confidential if the column of values allocated to a_i in an instance r over $\langle R|A_R|SC_R \rangle$ is not contained in f_s. So, a singleton constraint states that the values allocated to an attribute are sensitive per se. In case of a non-singleton constraint c at least one attribute $a_i \in c$ must not be contained in A_{F_s} and as a consequence of that the corresponding column of an instance r over $\langle R|A_R|SC_R \rangle$ is not in f_s. But in terms of Def. 3 it is irrelevant which of the attributes in c is chosen for not being in A_{F_s} and hence only associations between values allocated to the attributes of c in an instance r over $\langle R|A_R|SC_R \rangle$ are protected.

An example of a set of confidentiality constraints in terms of the running example introduced in Fig. 1 is given in Fig. 3. The fragmentation depicted in Fig. 2 is confidential w.r.t. this set of confidentiality constraints.

In terms of Def. 3 one trivial but feasible solution always is to store all data locally on the client. But since as much data as possible should be stored externally, an optimization problem proven to be NP-hard in [8] has to be solved. To achieve that, an approximation algorithm and several metrics to compare the qualities of computed solutions are presented in [8].

3 Controlled Query Evaluation

In the remainder of this article the inference-proofness of fragmentation is discussed based on a logic-oriented modelling of fragmentation within the framework of Controlled Query Evaluation (CQE). This framework comprises several inference-proof approaches to CQE, which are all based on the same classes of components. These classes and the general procedures of CQE will be introduced briefly on a fairly abstract level based on [4] now.

CQE is a framework with a server hosting a database instance. Although answering queries sent by users, one of the goals of the CQE system is to limit a user's information gain – even by considering information that a user could possibly obtain by employing logical inferences – according to some confidentiality policy. This is achieved by determining each piece of information a (rational) user can possibly infer based on his knowledge before interacting with this user.

To be able to do so, the monitoring of raw data a user receives as answers to his queries is not sufficient. The information contained in these data has to be extracted and represented suitably so that it can be processed. For that purpose the information system considered is assumed to be logic-based in the sense that its database instance is represented by a set of (closed) formulas of a well-defined language of logic (e.g., first-order logic) and the semantics of query evaluation is founded on the (well-defined) notions of validity and implication defined in the context of the semantics of this language.

As inferences can be often drawn by combining several pieces of knowledge, the computation of all inferences a user can employ (based on logical implications) presupposes that the CQE system needs to be aware of the complete knowledge this user has. In case of dynamic CQE – which aims at controlling a user's information gain at runtime – this obviously means that all answers a user receives in response to his queries have to be recorded as a set of formulas in order to be able to decide whether this knowledge combined with the (correct) answers to subsequent queries provides a basis for drawing harmful inferences.

Regardless of using dynamic or static CQE, a user's a priori knowledge expressed as a set of formulas always has to be considered. This a priori knowledge comprises knowledge a user has independently of answers given by the information system considered (e.g., semantic constraints declared for the schema of a relational database or knowledge about the world in general). Although not being harmful per se, such a priori knowledge combined with (uncontrolled) answers to his queries might enable a user to draw some harmful inferences.

To express the knowledge to be kept secret from a specific user, a confidentiality policy in terms of a set of potential secrets is set up for each user. A potential secret Ψ is a formula expressing that the pertinent user must not be able to infer that the information embodied in Ψ is true in the database considered. So, regardless of whether Ψ is actually true in this database or not, from the point of view of this user (established by his a priori knowledge and answers to his queries) it must always be possible that Ψ is *not* true. The conservative approach that a user is aware of the policy set up for him is usually followed.

As already hinted above, there are two general modes of inference control. The dynamic mode controls each answer to a user's query at runtime and therefore the CQE system has to check whether the user's (assumed) knowledge combined with the (correct) answer to his query could enable him to infer some knowledge declared as confidential (i.e., knowledge embodied in a potential secret is implied logically). If this is the case, the system has to distort the answer suitably by lying (i.e., giving a wrong answer) or by refusing an answer at all. The combined usage of both techniques is possible, too.

In static mode the CQE system precomputes an alternative database instance for each user, which is inference-proof according to the confidentiality policy set up for the pertinent user. Although being as close as possible to the original database instance, the alternative instance is distorted by lies or refusals (i.e., missing values) so that the user can query it freely without receiving knowledge enabling him to draw harmful inferences. So, corresponding to the idea of fragmentation that the server knows the externally stored fragment completely, the user's knowledge may comprise the complete alternative instance.

4   A Logic-Oriented View on Fragmentation

The goal of this article is to discuss the inference-proofness of the approach to fragmentation presented in Sect. 2. As CQE is known to be inference-proof, the main idea is to model fragmentation within the framework of CQE. As CQE relies on a logic-oriented view on databases, the approach to fragmentation discussed in Sect. 2 has to be modelled logic-orientedly, too.

For that purpose a language $\mathscr{L}$ of first-order logic with equality, which is suitable for modelling fragmentation logic-orientedly, is presented now. As the externally stored fragment instance f_s over $\langle F_s | A_{F_s} | SC_{F_s} \rangle$, which is assumed to be known to an attacker, must be modelled in $\mathscr{L}$, the set $\mathcal{P}$ of predicate symbols of $\mathscr{L}$ contains the predicate symbol $F_s \in \mathcal{P}$ with arity $k + 1 = |A_{F_s}|$ (including the additional tuple ID attribute plus k original attributes (cf. Fig. 4)). As security should not rely on obscurity, it is assumed that an attacker is aware of the process of fragmentation and knows both the computed fragmentation $\mathcal{F}$ and the schema $\langle R | A_R | SC_R \rangle$ over which the original instance r – being the target of his attacks – is built. To be able to model an attacker's knowledge about r based on these assumptions, language $\mathscr{L}$ also contains a predicate symbol $R \in \mathcal{P}$ with arity $n = |A_R|$. Moreover, there is a distinguished binary predicate symbol $= \notin \mathcal{P}$ for expressing equality. A predicate symbol F_o is *not* needed since an attacker does not have access to the client by assumption.

The language $\mathscr{L}$ also comprises the set Dom of constant symbols, which will be employed for the universe of interpretations for $\mathscr{L}$ as well. In compliance with other approaches to CQE (e.g., [5]) this set is assumed to be fixed and infinite. Further, $\mathscr{L}$ includes an infinite set $Var = \{X_1, X_2, \ldots\}$ of variables.

All formulas contained in $\mathscr{L}$ are constructed inductively in the natural fashion using the quantifiers $\forall$ and $\exists$ and the connectives $\neg$, $\wedge$, $\vee$ and $\Rightarrow$. Thereby each term is either a constant or a variable (functions are not allowed) and each variable is quantified (only closed formulas are in $\mathscr{L}$).

This syntactic specification has to be complemented with an appropriate semantics in which the characteristics of databases are reflected. Such a semantics is established by a so-called DB-Interpretation according to [5]:

Definition 4 (DB-Interpretation). *Given the language $\mathscr{L}$ of first-order logic with a fixed infinite set of constant symbols Dom and a finite set $\mathcal{P}$ of predicate symbols, an interpretation $\mathcal{I}$ over a universe $\mathcal{U}$ is a DB-Interpretation for $\mathscr{L}$ iff*

(i) $\mathcal{U} = Dom$,
(ii) $\mathcal{I}(v) = v \in \mathcal{U}$ holds for every constant symbol $v \in Dom$,
(iii) every $P \in \mathcal{P}$ with arity m is interpreted by a finite relation $\mathcal{I}(P) \subset \mathcal{U}^m$,
(iv) the predicate symbol $= \notin \mathcal{P}$ is interpreted by $\mathcal{I}(=) = \{(v, v) \mid v \in \mathcal{U}\}$.

The semantics of satisfaction of formulas in $\mathscr{L}$ by a DB-Interpretation is the same as in usual first-order logic. A set $\mathcal{S} \subset \mathscr{L}$ of formulas implies a formula $\Phi \in \mathscr{L}$ (written as $\mathcal{S} \models_{DB} \Phi$) iff each DB-Interpretation $\mathcal{I}$ satisfying $\mathcal{S}$ (written as $\mathcal{I} \models_M \mathcal{S}$) also satisfies Φ (written as $\mathcal{I} \models_M \Phi$).

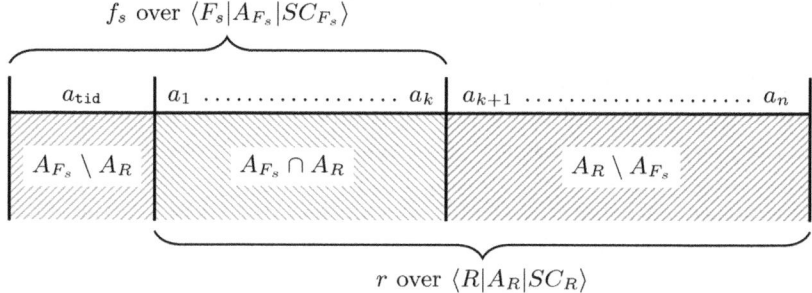

Fig. 4. Rearrangement of columns of r

$$db_{f_s}^+ = \{\ F_s\ (\ 1,\ 94142,\ \text{Hypertension},\ \text{White}\ ),$$
$$F_s\ (\ 2,\ 94141,\ \text{Obesity},\ \ \ \ \ \ \ \text{Warren}\ ),$$
$$F_s\ (\ 3,\ 94142,\ \text{Hypertension},\ \text{White}\ ),$$
$$F_s\ (\ 4,\ 94139,\ \text{Obesity},\ \ \ \ \ \ \ \text{Warren}\ )\ \}$$

Fig. 5. Positive knowledge of f_s in a logic-oriented model

From now on suppose w.l.o.g. that the columns of a relational instance r under investigation are rearranged so that the first k columns of r correspond to the projection of f_s on $A_{F_s} \cap A_R$. This convention is visualised in Fig. 4.

As an attacker is supposed to know the outsourced fragment instance f_s, the knowledge contained in f_s obviously has to be modelled within the logic-oriented view representing an attacker's knowledge. The positive knowledge in terms of the tuples explicitly recorded in f_s can be simply modelled logic-orientedly by adding an atomic formula $F_s(\nu[a_{\text{tid}}], \nu[a_1], \ldots, \nu[a_k])$ for each tuple $\nu \in f_s$. Regarding the fragmentation of Fig. 2 such a set of formulas is given in Fig. 5

But as the original instance r – and so its fragment instance f_s – is assumed to be complete[2], each piece of information *not* contained in r (f_s, respectively) is considered to be *not* valid by Closed World Assumption (CWA). Hence, an attacker knows that each combination of values $(v_{\text{tid}}, v_1, \ldots, v_k) \in Dom^{k+1}$ not contained in any tuple of f_s is not true. This implicitly induces information expressed as $\neg F_s(v_{\text{tid}}, v_1, \ldots, v_k)$. But as Dom is infinite, there is also an infinite number of such combinations not contained in the finite instance f_s.

As this negative knowledge is not explicitly enumerable, it is expressed implicitly by a so-called completeness sentence (cf. [5]) having a universally quantified variable X_j for each attribute $a_j \in A_{F_s}$. This completeness sentence is constructed so that it is satisfied by a DB-Interpretation $\mathcal{I}$ iff $\mathcal{I}$ satisfies each formula $\neg F_s(v_{\text{tid}}, v_1, \ldots, v_k)$ with $(v_{\text{tid}}, v_1, \ldots, v_k) \in Dom^{k+1}$ being a constant combination (substituting the universally quantified variables $X_{\text{tid}}, X_1, \ldots X_k$ of the completeness sentence) which is *not* contained in any tuple of f_s.

---

[2] As there are no statements about the completeness of r or f_s in [7,8] this article relies on the assumption of complete instances.

In terms of the running example the knowledge implicitly known to be *not* valid by CWA can be expressed as the following completeness sentence:

$$(\forall X_t)(\forall X_Z)(\forall X_I)(\forall X_D)\,[$$

$$
\begin{array}{llll}
(X_t = 1 \;\wedge\; X_Z = 94142 \;\wedge\; X_I = \text{Hypert.} \;\wedge\; X_D = \text{White}) & \vee \\
(X_t = 2 \;\wedge\; X_Z = 94141 \;\wedge\; X_I = \text{Obesity} \;\wedge\; X_D = \text{Warren}) & \vee \\
(X_t = 3 \;\wedge\; X_Z = 94142 \;\wedge\; X_I = \text{Hypert.} \;\wedge\; X_D = \text{White}) & \vee \\
(X_t = 4 \;\wedge\; X_Z = 94139 \;\wedge\; X_I = \text{Obesity} \;\wedge\; X_D = \text{Warren}) & \vee \\
\neg F_s(X_t, X_Z, X_I, X_D) & \qquad]
\end{array}
$$

Based on this insight an attacker's knowledge about the fragment instance f_s can be formalised logic-orientedly as follows:

Definition 5 (Logic-Oriented View on f_s). *Given a fragment instance f_s over $\langle F_s | A_{F_s} | SC_{F_s} \rangle$ according to Def. 1 with $A_{F_s} = \{a_{tid}, a_1, \dots, a_k\}$, the positive knowledge contained in f_s is modelled in $\mathscr{L}$ by the set of formulas*

$$db_{f_s}^+ := \{F_s(\nu[a_{tid}], \nu[a_1], \dots, \nu[a_k]) \mid \nu \in f_s\}\;. \tag{1}$$

The implicit negative knowledge contained in f_s is modelled in $\mathscr{L}$ by the singleton set $db_{f_s}^-$ containing the completeness sentence

$$(\forall X_{tid})\dots(\forall X_k)\left[\bigvee_{\nu\in f_s}\left(\bigwedge_{a_j\in A_{F_s}}(X_j = \nu[a_j])\right) \vee \neg F_s(X_{tid}, X_1, \dots, X_k)\right].\tag{2}$$

Moreover the functional dependency $a_{tid} \to \{a_1, \dots, a_k\} \in SC_{F_s}$ is modelled in $\mathscr{L}$ by the singleton set fd_{F_s} containing the formula

$$(\forall X_{tid})\,(\forall X_1)\dots(\forall X_k)\,(\forall X_1')\dots(\forall X_k')\,[\,F_s(X_{tid}, X_1, \dots, X_k)\,\wedge$$
$$F_s(X_{tid}, X_1', \dots, X_k') \Rightarrow (X_1 = X_1') \wedge \dots \wedge (X_k = X_k')]\tag{3}$$

Overall the logic-oriented view on f_s in $\mathscr{L}$ is $db_{f_s} := db_{f_s}^+ \cup db_{f_s}^- \cup fd_{F_s}$.

As already stated above, an attacker is assumed to know the process of fragmentation. This allows him to know that for each tuple $\nu \in f_s$ there is also a tuple $\mu \in r$ which is equal to ν w.r.t. the values allocated to the attributes of $A_{F_s} \cap A_R$. Moreover, being aware of both $\langle F_s | A_{F_s} | SC_{F_s} \rangle$ and $\langle R | A_R | SC_R \rangle$, an attacker knows that the values allocated to the attributes of $A_R \setminus A_{F_s}$ are kept hidden from him in each tuple of r. Regarding the logic-oriented modelling of an attacker's knowledge the ignorance of these values can be stated by using an existentially quantified variable for each term representing such a value.

Moreover – because of the completeness of r and f_s – an attacker knows that for each combination of values $(v_{tid}, v_1, \dots, v_k) \in Dom^{k+1}$ not contained in any tuple of f_s, there is no tuple $\mu \in r$ with $\mu[a_j] = v_j$ for each $j \in \{1, \dots, k\}$. Otherwise there would be a tuple $\nu \in f_s$ containing this combination of values because of the process of fragmentation. So, equivalently, for each tuple $\mu \in r$ there exists a tuple $\nu \in f_s$ with $\nu[a_j] = \mu[a_j]$ for each $j \in \{1, \dots, k\}$.

As already mentioned in Sect. 2, there is *exactly* one tuple in f_s for each tuple of r because of the existence of unique tuple IDs in f_s. So, if there are two different tuples $\nu_1, \nu_2 \in f_s$ being equal w.r.t. the values allocated to the attributes of $A_{F_s} \cap A_R$, an attacker can reason that there are also two tuples $\mu_1, \mu_2 \in r$ which are equal w.r.t. the values allocated to $A_{F_s} \cap A_R$, but differ in at least one of the values allocated to $A_R \setminus A_{F_s}$. Otherwise r would have two equal tuples $\mu_1 = \mu_2$ which is not possible in relational instances.

For now neglecting semantic constraints of the schema of r (i.e., $SC_R = \emptyset$), a logic-oriented view on the (hidden) original instance r based on the knowledge of the outsourced fragment instance f_s can be modelled as follows:

Definition 6 (Logic-Oriented View on r). *Let $\langle F_s | A_{F_s} | SC_{F_s} \rangle$ with $A_{F_s} = \{a_{tid}, a_1, \ldots, a_k\}$ be the outsourced fragment of a fragmentation $\mathcal{F}$ of a relational schema $\langle R | A_R | SC_R \rangle$ with $A_R = \{a_1, \ldots, a_k, \ldots, a_n\}$ and let f_s be a fragment instance over $\langle F_s | A_{F_s} | SC_{F_s} \rangle$ w.r.t. a relational instance r over $\langle R | A_R | SC_R \rangle$. The knowledge about r received from f_s is expressed by*

$$(\forall X_1) \ldots (\forall X_k) \, [\, (\exists X_{tid}) \, F_s(X_{tid}, X_1, \ldots, X_k) \Leftrightarrow \\ (\exists X_{k+1}) \ldots (\exists X_n) \, R(X_1, \ldots, X_k, X_{k+1}, \ldots, X_n) \,] \tag{4}$$

and the knowledge received from preserving duplicates in f_s is expressed by

$$(\forall X_1) \ldots (\forall X_k) \, [\, (\exists X_{tid}) \, (\exists X'_{tid}) \, [\, F_s(X_{tid}, X_1, \ldots, X_k) \wedge \\ F_s(X'_{tid}, X_1, \ldots, X_k) \wedge (X_{tid} \neq X'_{tid}) \,] \Rightarrow \\ (\exists X_{k+1}) \ldots (\exists X_n) \, (\exists X'_{k+1}) \ldots (\exists X'_n) \, [\, R(X_1, \ldots, X_k, X_{k+1}, \ldots, X_n) \wedge \tag{5} \\ R(X_1, \ldots, X_k, X'_{k+1}, \ldots, X'_n) \wedge \bigvee_{j=k+1}^{n} (X_j \neq X'_j) \,] \,] \,.$$

This view on r is referred to as the set of formulas db_r containing (4) and (5).

Before the inference-proofness of fragmentation can be analysed formally, the confidentiality policy according to which a fragmentation is computed has to be modelled logic-orientedly, too. A confidentiality constraint $c \subseteq A_R$ claims that each combination of values allocated to the attributes of c should not be revealed to an attacker completely. To specify this semantics more precisely, it is assumed[3] that c only protects those combinations of values which are explicitly allocated to the attributes of c in a tuple of r. In contrast, an attacker may get to know that a certain combination of values is *not* allocated to the attributes of c in any tuple of r. This semantics complies with the semantics of potential secrets known from the CQE framework (cf. Sect. 3).

The wish to protect a certain combination of values $(v_{i_1}, \ldots, v_{i_\ell}) \in Dom^{|c|}$ can be modelled as a potential secret $(\exists \boldsymbol{X}) \, R(t_1, \ldots, t_n)$ in which $t_j := v_j$ holds for each $j \in \{i_1, \ldots, i_\ell\}$. All other terms are existentially quantified variables.

---

[3] This assumption is needed because the semantics of confidentiality constraints on the instance level is not defined as exactly in [7,8].

But simply modelling each combination of values allocated to the pertinent attributes in r as a potential secret is not sufficient because an attacker is supposed to be aware of the confidentiality policy and could consequently read all sensitive information directly from the policy. This is prevented by protecting *each* combination of values possible according to Dom, regardless of whether it is contained in a tuple of r or not. But enumerating all of these combinations explicitly is not manageable since Dom is infinite.

Equivalently to the enumeration of all combinations of values, free variables $X_{i_1}, \ldots, X_{i_\ell}$ can be used in a potential secret instead of the constants $v_{i_1}, \ldots, v_{i_\ell}$. Then, one single potential secret per confidentiality constraint is sufficient. For that purpose the extended language $\mathscr{L}^f \supset \mathscr{L}$ of first-order logic expanding $\mathscr{L}$ by free variables is introduced.

Definition 7 (Confidentiality Policy). *Let $\mathcal{C}$ be a set of confidentiality constraints over schema $\langle R | A_R | SC_R \rangle$ according to Def. 2. Considering a confidentiality constraint $c_i \in \mathcal{C}$ with $c_i = \{a_{i_1}, \ldots, a_{i_\ell}\} \subseteq \{a_1, \ldots, a_n\} = A_R$ and the set $A_R \setminus c_i = \{a_{i_{\ell+1}}, \ldots, a_{i_n}\}$, constraint c_i can be modelled as a potential secret*

$$\Psi_i(\boldsymbol{X_i}) := (\exists X_{i_{\ell+1}}) \ldots (\exists X_{i_n}) R(X_1, \ldots, X_n)$$

in the extended language $\mathscr{L}^f$. Thereby $\boldsymbol{X_i} = (X_{i_1}, \ldots, X_{i_\ell})$ is the vector of free variables contained in $\Psi_i(\boldsymbol{X_i})$. The set containing exactly one potential secret $\Psi_i(\boldsymbol{X_i})$ constructed as above for every $c_i \in \mathcal{C}$ is called pot_sec$(\mathcal{C})$.

To show the inference-proofness of fragmentation, it has to be proven that none of the potential secrets is implied by the set of formulas representing an attacker's knowledge. This proof cannot be produced (directly) for potential secrets containing free variables because DB-Interpretations are only defined for the language $\mathscr{L}$ not containing free variables. But as free variables of $\mathscr{L}^f$ represent constants of Dom, a so-called expansion of such formulas substituting free variables with constants can be constructed to enable the proof.

Definition 8 (Expansion of Formulas). *Let $\Psi(\boldsymbol{X}) \in \mathscr{L}^f$ be a formula containing the vector $\boldsymbol{X} = (X_1, \ldots, X_\ell)$ of free variables. $\Psi(\boldsymbol{X}) \in \mathscr{L}^f$ is expanded to the set of formulas $\mathrm{ex}(\Psi(\boldsymbol{X})) \subset \mathscr{L}$ by substituting the free variables $\boldsymbol{X}$ with every constant combination $\boldsymbol{v} = (v_1, \ldots, v_\ell) \in Dom^\ell$, thereby creating a formula $\Psi(\boldsymbol{v}) \in \mathscr{L}$. The expansion of a set $\mathcal{S} \subset \mathscr{L}^f$ is $\mathrm{ex}(\mathcal{S}) := \bigcup_{\Psi(\boldsymbol{X}) \in \mathcal{S}} \mathrm{ex}(\Psi(\boldsymbol{X}))$.*

5   Inference-Proofness of Fragmentation

Until now the logic-oriented model only comprises knowledge an attacker has by knowing the outsourced fragment instance. The a priori knowledge an attacker might have has been completely neglected. But as shown in the following example, an attacker might generally employ this knowledge to draw harmful inferences. In terms of the running example, suppose an attacker knows that `Ripley` is the *only* patient who is treated by doctor `Warren` and lives in a small town with zip code 94139. By knowing the set of formulas $db_{f_s}^+$ (see Fig. 5) and moreover knowing the relationship between f_s and its original instance described by formula (4) of Def. 6, an attacker might reason that Ripley suffers from Obesity, thereby violating confidentiality constraint c_3 of Fig. 3.

In this article an attacker is supposed to have a priori knowledge about the original instance r in terms of the set SC_R of semantic constraints declared for schema $\langle R|A_R|SC_R \rangle$. Here, SC_R is a set of arbitrary unirelational and typed semantic constraints as long as they belong to the rather general classes of so-called Equality Generating Dependencies (EGDs) or Tuple Generating Dependencies (TGDs), which together comprise nearly all semantic constraints (cf. [1]).

Intuitively expressed, an EGD claims that the presence of some tuples in r implies that certain components of these tuples are equal and a TGD claims that the presence of some tuples in r implies the existence of certain other tuples in r. Moreover, a constraint is unirelational if it refers to only one relational schema, and it is typed if there is an assignment of variables to column positions preventing the claim for equality of values being in different columns of r [1]. In the case of non-typed constraints, for example, an attacker might infer sensitive information based on a non-typed EGD stating that in some tuple of r a (non-hidden) value allocated to an attribute of $\bar{A}_{F_s}$ is equal to a (hidden) value allocated to an attribute of $\bar{A}_{F_o}$. A well known example for a unirelational and typed EGD is a functional dependency and an example for a unirelational and typed TGD is a join dependency.

According to [1,10] unirelational and typed EGDs and TGDs can be formalised as follows:

Definition 9 (Unirelational Typed TGDs/EGDs). *Let $\langle R|A_R|SC_R \rangle$ be a relational schema. Each unirelational EGD/TGD contained in SC_R can be expressed in $\mathscr{L}$ by a formula $(\forall \boldsymbol{X})\,[\alpha \Rightarrow \beta]$, in which*

(i) α is a conjunction of atomic formulas of the kind $R(X_1,\ldots,X_n)$ with $X_1,\ldots,X_n$ being variables of $\boldsymbol{X}$ and every variable of $\boldsymbol{X}$ appears in α,

(ii) in case of a/an

  – EGD, β is an atomic formula of the kind $(X' = X'')$ with X' and X'' being distinct variables of $\boldsymbol{X}$

  – TGD, β is a formula $(\exists \boldsymbol{Y})\,\gamma$, in which γ is a conjunction of atomic formulas of the kind $R(X_1,\ldots,X_n)$ with $X_1,\ldots,X_n$ being variables of $\boldsymbol{X}$ and $\boldsymbol{Y}$.

A unirelational EGD/TGD is typed iff the set Var of Variables of $\mathscr{L}$ can be partitioned into n disjoint classes so that, for each atomic formula of the kind $R(X_{i_1},\ldots,X_{i_n})$ of α or β, for $1 \le j \le n$, the variable X_{i_j} belongs to class j, and for each atomic formula of the kind $(X' = X'')$ both variables X' and X'' belong to the same class.

For $n = 3$, for example, $(\forall \boldsymbol{X})\,[\,R(X_1,X_2,X_3) \Rightarrow (X_1 = X_3)\,]$ is a non-typed EGD and $(\forall \boldsymbol{X})\,[\,R(X_1,X_2,X_3) \Rightarrow R(X_1,X_3,X_2)\,]$ is a non-typed TGD. In contrast, $(\forall \boldsymbol{X})\,[\,R(X_1,X_2,X_3) \wedge R(X_1,X_2',X_3') \Rightarrow (X_3 = X_3')\,]$ is a typed EGD and replacing $(X_3 = X_3')$ with $R(X_1,X_2',X_3)$ results in a typed TGD.

Now, the inference-proofness of fragmentation can be proved formally based on the logic-oriented modelling of the view an attacker is supposed to have on the original instance r by knowing the externally stored fragment instance f_s and employing his a priori knowledge. Intuitively expressed, it is shown that a rational attacker always has to consider the existence of an alternative instance r' possible which is – from his point of view constituted by his knowledge – indistinguishable from r and does not violate a potential secret.

Theorem 1 (Inference-Proofness). *Let r be a relational instance over a relational schema $\langle R|A_R|SC_R \rangle$ with $A_R = \{a_1, \ldots, a_n\}$ and let $\mathcal{F}$ be a fragmentation of $\langle R|A_R|SC_R \rangle$ according to Def. 1, which is – according to Def. 3 – confidential w.r.t. a set $\mathcal{C}$ of confidentiality constraints constructed in terms of Def. 2. Moreover, let f_s be the fragment instance over fragment $\langle F_s|A_{F_s}|SC_{F_s} \rangle \in \mathcal{F}$ with $A_{F_s} = \{a_{tid}, a_1, \ldots, a_k\}$ created w.r.t. instance r. It holds that*

$$\text{for all } \Psi(\boldsymbol{v}) \in \text{ex}(pot\_sec(\mathcal{C})): \; db_{f_s} \cup db_r \cup prior_{SC_R} \not\models_{DB} \Psi(\boldsymbol{v}) \qquad (6)$$

with $\text{ex}(pot\_sec(\mathcal{C}))$ being the expansion (Def. 8) of $pot\_sec(\mathcal{C})$ constructed according to Def. 7 and db_{f_s} and db_r being constructed according to Def. 5 and Def. 6. Moreover, $prior_{SC_R}$ is a set of unirelational typed TGDs and EGDs contained in SC_R, which are constructed in terms of Def. 9 and satisfied by r.

Proof. To prove formula (6) of Theorem 1, it has to be shown that for an arbitrary $\tilde{\Psi}(\boldsymbol{v}) \in \text{ex}(pot\_sec(\mathcal{C}))$ with $\boldsymbol{v} = (v_{i_1}, \ldots, v_{i_\ell})$ there is a DB-Interpretation $\mathcal{I}^*$ which satisfies db_{f_s}, db_r and $prior_{SC_R}$ and does *not* satisfy $\tilde{\Psi}(\boldsymbol{v})$.

As $\tilde{\Psi}(\boldsymbol{v})$ with $\boldsymbol{v} = (v_{i_1}, \ldots, v_{i_\ell})$ is in $\text{ex}(pot\_sec(\mathcal{C}))$, there has to be the potential secret $\tilde{\Psi}(\boldsymbol{X}) \in pot\_sec(\mathcal{C})$ containing the vector $\boldsymbol{X} = (X_{i_1}, \ldots, X_{i_\ell})$ of free variables and therefore, by construction of $pot\_sec(\mathcal{C})$, there also exists a confidentiality constraint $c = \{a_{i_1}, \ldots, a_{i_\ell}\} \in \mathcal{C}$. Due to $\mathcal{F}$ being confidential by assumption, $c \not\subseteq A_{F_s}$ holds (see Def. 3) and as a consequence of that there is an attribute $a_m \in c$ with $a_m \notin A_{F_s}$. Hence, respecting the rearrangement of the columns of r (see Fig. 4), both $m \notin \{1, \ldots, k\}$ and $m \in \{k+1, \ldots, n\}$ hold.

As a first step towards the construction of $\mathcal{I}^*$, the interpretation of the predicate symbol F_s – that is $\mathcal{I}^*(F_s)$ – is defined as

$$\mathcal{I}^*(F_s) := \{ (\nu[a_{tid}], \nu[a_1], \ldots, \nu[a_k]) \mid \nu \in f_s \} \qquad (7)$$

and obviously this interpretation satisfies all formulas of $db_{f_s}^+$ as well as the closed world assumption contained in $db_{f_s}^-$. Moreover, fd_{F_s} is satisfied because, by assumption, f_s satisfies the functional dependency contained in SC_{F_s} and hence also (3) is satisfied by $\mathcal{I}^*(F_s)$. So, $\mathcal{I}^* \models_M db_{f_s}$ already holds.

Continuing the construction of $\mathcal{I}^*$, the set $\mathcal{I}^*(R)$ is defined as

$$\mathcal{I}^*(R) := \{ (\mu[a_1], \ldots, \mu[a_{m-1}], \varphi_m(\mu[a_m]), \mu[a_{m+1}], \ldots, \mu[a_n]) \mid \mu \in r \} \qquad (8)$$

in which $\varphi_m : \mathcal{U}_m \to \mathcal{U} \setminus \{v_m\}$ is an *injective* function having the finite domain $\mathcal{U}_m := \{ \mu[a_m] \mid \mu \in r \}$ and the infinite range $\mathcal{U} \setminus \{v_m\}$ with $v_m \in \boldsymbol{v}$ and $\mathcal{U}$ being the universe of $\mathcal{I}^*$ (cf. Def. 4). Note that φ_m can always be constructed because of $|\mathcal{U} \setminus \{v_m\}| > |\mathcal{U}_m|$. Moreover, $\mathcal{I}^* \not\models_M \tilde{\Psi}(v)$ holds as $v_m \in (v_{i_1}, \ldots, v_{i_\ell})$ is excluded from the range of φ_m and $\mathcal{I}^*$ can only satisfy $\tilde{\Psi}(v)$ if there is a tuple $(u_1, \ldots, u_m, \ldots, u_n) \in \mathcal{I}^*(R)$ for which $u_j = v_j$ holds for each $j \in \{i_1, \ldots, i_\ell\}$.

Next, it is shown that $\mathcal{I}^* \models_M db_r$ holds by proving that $\mathcal{I}^*$ satisfies the formulas (4) and (5) of Def. 6. To prove the if-part of the equivalence, assume that $(\exists X_{\mathtt{tid}}) \, F_s(X_{\mathtt{tid}}, X_1, \ldots, X_k)$ of (4) is satisfied by $\mathcal{I}^*$ under a constant substitution $(X_1/u_1), \ldots, (X_k/u_k)$ which is feasible according to Dom. Then, according to (7), there is a tuple $(w_{\mathtt{tid}}, u_1, \ldots, u_k) \in \mathcal{I}^*(F_s)$ with $w_{\mathtt{tid}} \in \mathcal{U}$ implying the existence of a tuple $\nu \in f_s$ with $\nu[a_j] = u_j$ for all $j \in \{1, \ldots, k\}$. As f_s is a fragment instance of r (see Def. 1) and the columns of r are rearranged as described above, there is a tuple $\mu \in r$ with $\mu[a_j] = \nu[a_j]$ for all $j \in \{1, \ldots, k\}$. According to (8) and because of $m \notin \{1, \ldots, k\}$ there is a tuple $(u_1, \ldots, u_k, w_{k+1}, \ldots, w_n) \in \mathcal{I}^*(R)$ satisfying the conclusion. To finally establish the equivalence, the only-if-part can be proved by applying the argumentation presented above backwards.

To prove formula (5) of Def. 6, assume that the premise of (5) is satisfied by $\mathcal{I}^*$ under a constant substitution $(X_1/u_1), \ldots, (X_k/u_k)$ which is feasible according to Dom. Then there are two tuples $(w_{\mathtt{tid}}, u_1, \ldots, u_k)$ and $(w'_{\mathtt{tid}}, u_1, \ldots, u_k)$ in $\mathcal{I}^*(F_s)$ and $w_{\mathtt{tid}} \neq w'_{\mathtt{tid}}$ holds for $w_{\mathtt{tid}}, w'_{\mathtt{tid}} \in \mathcal{U}$. Because of the construction of $\mathcal{I}^*(F_s)$ described in (7) there are two different tuples $\nu, \nu' \in f_s$ with $\nu[a_j] = \nu'[a_j] = u_j$ for all $j \in \{1, \ldots, k\}$. Due to the existence of *exactly* one tuple in f_s for each tuple in r (cf. Sect. 2), it can be reasoned that there are also two tuples $\mu, \mu' \in r$ with $\mu[a_j] = \mu'[a_j] = u_j$ for each $j \in \{1, \ldots, k\}$. As relational instances cannot contain two identical tuples, $\mu[a_p] \neq \mu'[a_p]$ must hold for some $p \in \{k+1, \ldots, n\}$ and according to (8) there are two tuples $(u_1, \ldots, u_k, w_{k+1}, \ldots, w_n)$ and $(u_1, \ldots, u_k, w'_{k+1}, \ldots, w'_n)$ in $\mathcal{I}^*(R)$. In the case of $p \neq m$, obviously $w_p \neq w'_p$ holds, and otherwise $w_m \neq w'_m$ holds because of φ_m being injective. Hence, the conclusion of (5) is satisfied by $\mathcal{I}^*$, too.

As a last step $\mathcal{I}^* \models_M prior_{SC_R}$ has to be proved. This is prepared by constructing a temporary DB-Interpretation $\mathcal{I}_t$ for r as a set

$$\mathcal{I}_t(R) := \{ \, (\mu[a_1], \ldots, \mu[a_m], \ldots, \mu[a_n]) \mid \mu \in r \, \} \tag{9}$$

and as (by assumption) r satisfies all constraints of SC_R, all formulas of $prior_{SC_R}$ are satisfied by $\mathcal{I}_t$, too.

As there are no constants in formulas of $prior_{SC_R}$ since all terms of these formulas are quantified variables (cf. Def. 9), an *arbitrary* DB-Interpretation $\mathcal{I}$ satisfying $prior_{SC_R}$ does not need to contain tuples with specific combinations of values corresponding to combinations of constants in formulas of $prior_{SC_R}$. Hence, $\mathcal{I}$ still satisfies $prior_{SC_R}$ if values in tuples of $\mathcal{I}$ are exchanged by other values of $\mathcal{U}$ so that all equalities (to satisfy equalities between variables in formulas of $prior_{SC_R}$) and diversities (to prevent the creation of further equalities resulting in more implications that need to be satisfied) between values of $\mathcal{I}$ are preserved. Moreover, as formulas of $prior_{SC_R}$ are typed, values of $\mathcal{I}$ can be exchanged as long as all equalities and diversities between values of $\mathcal{I}$ are preserved within each column of $\mathcal{I}$.

Obviously, there is a tuple $(u_1, \ldots, u_m, \ldots, u_n) \in \mathcal{I}_t(R)$ iff there is a tuple $(u_1, \ldots, \varphi_m(u_m), \ldots, u_n) \in \mathcal{I}^*(R)$ and as φ_m is injective, two values u'_m and u''_m of $\mathcal{U}$ are equal iff $\varphi_m(u'_m) = \varphi_m(u''_m)$. Hence, $\mathcal{I}^* \models_M prior_{SC_R}$ holds.   □

6 Conclusion and Future Work

Motivated by the wish to achieve confidentiality of information hosted by third party service providers without the usage of encryption, the approach to fragmentation presented in Sect. 2 is developed in [7,8]. In these articles the protection of information is discussed only in terms of direct accesses to data. It is *not* shown that confidential information cannot be inferred based on the knowledge of non-confidential information contained in the externally stored fragment instance f_s and a priori knowledge an attacker might possibly have.

This desirable result is presented in this article under the supposition that an attacker only has a priori knowledge in terms of Equality Generating Dependencies and Tuple Generating Dependencies which are all unirelational and typed. Regarding the possibilities to express knowledge about semantic constraints declared for the schema of an original instance r, this supposition is not very restrictive as most of the semantic constraints commonly used (e.g., functional dependencies, join dependencies) belong to these classes of constraints [1]. Moreover, reconsidering the proof of Theorem 1, it can be seen easily that the inference-proofness still holds if the restriction that semantic constraints have to be typed is replaced by the weaker restriction that the set of semantic constraints considered does not impose that any value of one of the columns $k+1, \ldots, n$ of r is equal to a value of one of the columns $1, \ldots, k$ of r.

Additionally to the knowledge about semantic constraints an attacker might also have some a priori knowledge about the world in general (e.g., a set of facts and inference rules) which cannot be expressed as a set of formulas complying with the restrictions stated above. But as shown in the introductory example of Sect. 5, the inference-proofness of fragmentation *cannot* be guaranteed under arbitrary a priori knowledge – even if no sensitive information can be inferred solely based on this a priori knowledge. So, further research on (weak) syntactic restrictions for modelling a priori knowledge without violating confidentiality requirements might lead to even more expressive languages for that purpose.

As there are other approaches to achieving confidentiality by vertical fragmentation than the one treated in this article (see e.g. [2,9]), another idea for future work might be to analyse the inference-proofness of these approaches. As these approaches free the client from storing data locally by resorting to encryption if necessary, the logic-oriented modelling of an attacker's knowledge has to be adapted suitably to reflect these circumstances. Moreover, approaches based on vertical fragmentation might be combined with approaches based on horizontal fragmentation, which partition the set of tuples of an original instance r with the help of selection criteria into several instances declared over the same set of attributes as r. An approach to achieving inference-proofness based on horizontal fragmentation is presented in [14]. This kind of "hybrid" fragmentation promises a higher amount of outsourced data, but it raises several confidentiality issues (like meta-inferences), which must be analysed with scrutiny.

References

1. Abiteboul, S., Hull, R., Vianu, V.: Foundations of Databases. Addison-Wesley, Reading (1995)
2. Aggarwal, G., Bawa, M., Ganesan, P., Garcia-Molina, H., Kenthapadi, K., Motwani, R., Srivastava, U., Thomas, D., Xu, Y.: Two can keep a secret: A distributed architecture for secure database services. In: 2nd Biennial Conference on Innovative Data Systems Research, CIDR 2005, pp. 186–199 (2005)
3. Biskup, J.: Security in Computing Systems – Challenges, Approaches and Solutions. Springer, Heidelberg (2009)
4. Biskup, J.: Usability confinement of server reactions: Maintaining inference-proof client views by controlled interaction execution. In: Kikuchi, S., Sachdeva, S., Bhalla, S. (eds.) DNIS 2010. LNCS, vol. 5999, pp. 80–106. Springer, Heidelberg (2010)
5. Biskup, J., Bonatti, P.A.: Controlled query evaluation with open queries for a decidable relational submodel. Annals of Mathematics and Artificial Intelligence 50(1-2), 39–77 (2007)
6. Biskup, J., Embley, D.W., Lochner, J.: Reducing inference control to access control for normalized database schemas. Information Processing Letters 106(1), 8–12 (2008)
7. Ciriani, V., De Capitani di Vimercati, S., Foresti, S., Jajodia, S., Paraboschi, S., Samarati, P.: Enforcing confidentiality constraints on sensitive databases with lightweight trusted clients. In: Gudes, E., Vaidya, J. (eds.) Data and Applications Security XXIII. LNCS, vol. 5645, pp. 225–239. Springer, Heidelberg (2009)
8. Ciriani, V., De Capitani di Vimercati, S., Foresti, S., Jajodia, S., Paraboschi, S., Samarati, P.: Keep a few: Outsourcing data while maintaining confidentiality. In: Backes, M., Ning, P. (eds.) ESORICS 2009. LNCS, vol. 5789, pp. 440–455. Springer, Heidelberg (2009)
9. Ciriani, V., De Capitani di Vimercati, S., Foresti, S., Jajodia, S., Paraboschi, S., Samarati, P.: Combining fragmentation and encryption to protect privacy in data storage. ACM Transactions on Information and System Security 13(3) (2010)
10. Fagin, R.: Horn clauses and database dependencies. Journal of the ACM 29(4), 952–985 (1982)
11. Farkas, C., Jajodia, S.: The inference problem: A survey. ACM SIGKDD Explorations Newsletter 4(2), 6–11 (2002)
12. Hacıgümüs, H., Mehrotra, S., Iyer, B.R.: Providing database as a service. In: Proceedings of the 18th International Conference on Data Engineering, ICDE 2002, pp. 29–40. IEEE Computer Society, Los Alamitos (2002)
13. Samarati, P., De Capitani di Vimercati, S.: Data protection in outsourcing scenarios: Issues and directions. In: Feng, D., Basin, D.A., Liu, P. (eds.) ACM Symposium on Information, Computer and Communications Security, ASIACCS 2010, pp. 1–14. ACM, New York (2010)
14. Wiese, L.: Horizontal fragmentation for data outsourcing with formula-based confidentiality constraints. In: Echizen, I., Kunihiro, N., Sasaki, R. (eds.) IWSEC 2010. LNCS, vol. 6434, pp. 101–116. Springer, Heidelberg (2010)

Round-Efficient Oblivious Database Manipulation

Sven Laur[2], Jan Willemson[1,3], and Bingsheng Zhang[1,2]

[1] Cybernetica, Ülikooli 2, Tartu, Estonia
[2] Institute of Computer Science, University of Tartu, Liivi 2, Tartu, Estonia
[3] Software Technology and Applications Competence Center, Ülikooli 2, Tartu, Estonia

Abstract. Most of the multi-party computation frameworks can be viewed as oblivious databases where data is stored and processed in a secret-shared form. However, data manipulation in such databases can be slow and cumbersome without dedicated protocols for certain database operations. In this paper, we provide efficient protocols for oblivious selection, filtering and shuffle—essential tools in privacy-preserving data analysis. As the first contribution, we present a 1-out-of-n oblivious transfer protocol with $O(\log \log n)$ rounds, which achieves optimal communication and time complexity and works over any ring $\mathbb{Z}_N$. Secondly, we show how to construct round-efficient shuffle protocols with optimal asymptotic computation complexity and provide several optimizations.

Keywords: Secure multi-party computation, oblivious transfer, verifiable shuffle, oblivious filtering.

1 Introduction

Privacy issues often arise when sensitive data is gathered from individuals and organizations. Such threats are commonly addressed with organizational and physical methods; however, on some occasions cryptographic methods can provide better alternatives. In this paper, we will concentrate on the methods based on secret sharing and multi-party computations. Such an approach guarantees by design that the computing parties will have no access to the underlying data provided that the number of collaborating malicious parties is small enough. On the other hand, such a guarantee comes with a price of increased computational complexity and generic secure computation techniques [35,7,15] have a prohibitively large overhead.

However, for particular problems it is possible to design protocols that outperform the general-purpose ones. In this paper, we will concentrate on some specific management tasks to be applied on secret shared databases. Since the whole idea behind keeping a database in secret shared form is to prevent data leaks, corresponding database operations should not reveal anything about the underlying dataset (except for what can be concluded from the desired output). Our target is to develop protocols for oblivious database access, filtering, text categorisation and encoding. We consider security in the client-server setting, i.e., the client should obtain correct answer to the query and nothing else, whereas the database holders should not learn anything besides the type of the query. Note that in the context of secret shared databases, client privacy is only guaranteed if the share holders collude below the tolerated threshold.

X. Lai, J. Zhou, and H. Li (Eds.): ISC 2011, LNCS 7001, pp. 262–277, 2011.

In data analysis, one often needs to filter data according to a specific criterion. The corresponding oblivious filtering procedure should produce a new shared database that contains only the records which satisfy the criterion. The share holders should learn nothing beyond the number of records. In particular, they should not learn which database rows we included. As such, the oblivious filtering reduces the database size and thus can remarkably reduce the overall complexity of the remaining steps.

To understand the requirements for data encoding, consider the case of privacy-preserving questionnaire analysis. It is simple to design cryptographically secure data aggregation mechanisms for questions with multiple choice answers, as the responses can be encoded as integers. Questions with free text answers are much more challenging. First, secure text-processing is inherently slower than secure integer arithmetic, as it relies on arithmetic operations. Second, the answers must often be interpreted by human operators in order to extract relevant information. As a trade-off between privacy and efficiency, we present a protocol for oblivious database access. The protocol assures that the human operator who reads text entries and encodes them to standard attributes, cannot link replies to particular responders nor to non-disclosed fields.

Related work. Even though the generic methods of secure multi-party computations have been known for decades [35,7,15], practical implementations of the respective frameworks have emerged only recently, e.g. FairPlayMP [6], VIFF [23], SEPIA [11], SecureSCM [3], VMcrypt [28], TASTY [25] and SHAREMIND [8]. Oblivious transfer has been commonly studied in the two-party setting [10,4,22] where security guarantees are inherently computational. Existing results in the multi-party setting mostly hold for private information retrieval [17,5] where the database is replicated in several sites and only the secrecy of the query index i is protected. Similarly, the secure shuffle problem has been mostly studied in computational setting. Most solutions are given in the context of e-voting with mix-nets [16,34] and onion-routing [12,29]. These solutions propagate encrypted messages through a network of servers to achieve unlinkability at the endpoint. In our context, we need multi-party computation protocols that shuffle secret shared values. As there are no public key operations, such protocols are much more efficient and naturally fit into the framework of share computing environments, such as SHAREMIND [8] and VIFF [20]. Hence, one should view our work in the long line of works [19,8,11,14] that uses addition and multiplication protocols in black-box manner to build secure implementations for standard data processing operations.

Road map. The paper is organized as follows. In Section 2, we describe general frameworks for share computing. Section 3 gives generic constructions for oblivious database manipulation, namely selection, filtering and general read-write access. The described generic protocols use oblivious shuffle as an important building block and three possible instantiations with several round-communication complexity trade-offs for it are presented in Section 4. For severe space constraints, we have also omitted some proofs. These proofs and further performance tweaks are discussed in the full version of the paper available in the Cryptology ePrint Archive [27].

2   Frameworks for Share Computing

General setup. A typical privacy-preserving data analysis application involves three types of entities: data donors, computing parties (referred to as *miners*), and clients. Data donors own sensitive data, miners gather the data and clients want to find answers for various statistical questions. In such a setting, privacy issues can be addressed with the client-server model formalized by Damgård and Ishai [21]. In this model, data donors submit their data in a secret shared form to the miner nodes, which later use share computing protocols to carry out the computations requested by the clients. Since data is secret shared, individual records are protected as long as the miners do not form non-tolerated coalitions, i.e., they as a group follow the restrictions of share-computing protocols. Clients and data donors are not trusted and can arbitrarily violate protocol specifications. Depending on the underlying primitives and protocols, the framework can tolerate either semihonest or malicious corruption of miners.

Share computing. A typical framework for multi-party computations is based on a secret sharing scheme and a set of protocols for manipulating the shares. A secret sharing scheme is specified by randomised sharing and recovery algorithms. The sharing algorithm splits a secret value $x \in \mathbb{Z}_N$ into shares $x_1, \ldots, x_m$ that must be securely transferred to the miners $\mathcal{P}_1, \ldots, \mathcal{P}_m$, respectively. To recover the shared value, miners must together execute the reconstruction algorithm that takes in all shares and outputs the corresponding secret value. For example, the additive secret sharing scheme splits a secret x into shares such that the secret can be reconstructed by adding all the shares, i.e. $x \equiv x_1 + x_2 + \cdots + x_m \mod N$. The vector of shares $(x_1, x_2, \ldots, x_m)$ is commonly denoted by $[\![x]\!]$. Security properties of a secret sharing scheme are defined through a list of tolerable adversarial coalitions. Let $\mathcal{P}_{i_1}, \ldots, \mathcal{P}_{i_k}$ form a tolerable coalition, then the corresponding shares should leak nothing about the secret. More formally, the distributions $(x_{i_1}, \ldots, x_{i_k})$ and $(y_{i_1}, \ldots, y_{i_k})$ must coincide for any inputs $x, y \in \mathbb{Z}_N$.

Share computing protocols enable miners to obliviously compute with shares. For instance, miners can obtain a valid sharing of $x + y$ by locally adding their additive shares of x and y. Similarly, multiplying shares of x locally by a constant $\alpha \in \mathbb{Z}_N$ gives us shares of αx. A secret sharing scheme satisfying these two constraints is referred to as a *linear secret sharing scheme*. As any function can be represented as a Boolean circuit, a share computing framework can be built on top of a linear secret sharing scheme by specifying a protocol for multiplication. For instance, the VIFF framework [1] uses standard solution based on Shamir secret sharing [32], whereas SHAREMIND uses a tailor-suited multiplication protocol for additive secret sharing [8,9].

Although protocols for secure share addition and multiplication are sufficient to achieve Turing completeness, the corresponding generic constructions are not efficient enough for practical applications. In most cases, the effect of the network delay is several orders of magnitudes larger than the time needed to deliver protocol messages. Thus, protocols with minimal round complexity are the most efficient. However, the round count is not an absolute measure, as the delivery time becomes dominant for large messages. Hence, the optimal solution may vary. Table 1 depicts round complexity of various share computing operations implemented in SHAREMIND, where ℓ is the bit length of modulus N. Precise description of the protocols together with empirical

Table 1. Round complexity of common share-computing operations

Operation	Round count	Complexity	Operation	Round count	Complexity
Multiplication	τ_{mul}	$O(1)$	Coin-tossing	τ_{ct}	$O(1)$
Smaller than or equal	τ_{st}	$O(\log \ell)$	Strictly less	τ_{sl}	$O(\log \ell)$
Equality test	τ_{eq}	$O(\log \ell)$	Bit-decomposition	τ_{bd}	$O(\log \ell)$

benchmarking can be found in [8,9]. It is worth noting that logarithmic complexity in the bit size of the modulus N is asymptotically sub-optimal, as theoretical construction by Damgård *et al.* [19] provides a constant round solution. However, the corresponding round count is larger than $\log \ell$ for all practical residue rings $\mathbb{Z}_N$.

For binary operations, we will use a shorthand $[\![x]\!] \circledast [\![y]\!]$ to denote the outputs of a share computing protocol that securely computes $x \circledast y$ from the shares of x and y.

Adversarial model. For clarity, we consider only the static corruption model where adversary specifies parties to be corrupted before the protocol starts, although most protocols can resist more advanced corruption models. Although the list of tolerated adversarial coalitions can be arbitrary, share computing systems can achieve information theoretical security only if the condition Q2 is satisfied in the semihonest model and the condition Q3 is satisfied in the malicious model [26]. Recall that the condition Q2 means that any union of two tolerated adversarial coalitions is not sufficient to corrupt all parties and the condition Q3 means that any union of three tolerated adversarial sets is not sufficient. In the case of threshold corruption, the conditions Q2 and Q3 imply that the number corrupted parties is strictly below $\frac{m}{2}$ and $\frac{m}{3}$, respectively.

Universal composability. As formal security proofs are rather technical, security proofs are often reduced to the security properties of sub-protocols. More specifically, one can deduce security of a compound protocol without delving into details only if all sub-protocols are *universally composable*. Although the formal definition of universal composability is rather complex, the intuition behind it is simple. Let $\varrho\langle \cdot \rangle$ be a global context (often named as environment) that uses the functionality of a protocol π. Let π° be an idealised implementation, where all computations are done by a trusted third party who privately gathers all inputs and distributes the resulting outputs. Then we can compare real and ideal world protocols $\varrho\langle \pi \rangle$ and $\varrho\langle \pi^\circ \rangle$. A protocol π is *universally composable* if for any real world adversary $\mathcal{A}$ there exist an adversary $\mathcal{A}^\circ$ against $\varrho\langle \pi^\circ \rangle$ with comparable complexity and success rate. That is, the joint distribution of all outputs in the real and ideal world must coincide for all input distributions. As a result, a compound protocol consisting of several instances of π° preserves security if we replace π° by π. The latter means that we combine universally composable sub-protocols without any usage restrictions, e.g., execute them in parallel.

We refer to the standard treatments [13,31] for further details. Achieving universal composability in the semihonest model is rather straightforward, most share computing protocols satisfy this including the protocols used in SHAREMIND and VIFF [8,1]. Theoretical constructions for malicious model do exist [20], but these are not widely used in practical systems, yet.

Data-gathering phase
    1. Data donors submit the shares of their inputs x_i to the miner nodes.
Query phase
    A client submits shares of i to the miner nodes.
Processing phase
    1. For $j \in \{1, \ldots, n\}$, miners evaluate in parallel: $[\![y_i]\!] \leftarrow [\![x_j]\!] \cdot ([\![i]\!] = j)$.
    2. Miners compute the shares of the reply: $[\![z]\!] \leftarrow [\![y_1]\!] + \cdots + [\![y_n]\!]$.
Reconstruction phase
    Miners send the shares of z to the client who reconstructs and outputs z.

Protocol 1. Generic oblivious transfer protocol GENOT for the message domain $\mathbb{Z}_N$

3 Oblivious Database Manipulation

In this section, we describe various database operations that are often required in data analysis and show how these can be implemented using protocols for oblivious transfer and shuffle. The objective of the random shuffle protocol is to permute the elements of the underlying database according to a uniformly chosen permutation π, which is oblivious to all miners. Depending on the protocol, the client can either learn the permutation or not. We discuss how to implement the basic protocols in Sections 3.1 and 4.

3.1 Generic Construction for Oblivious Selection

Many data-mining algorithms select and later process a particular sub-sample from the entire data. In the simplest case, the client wants to retrieve a single data record without revealing the index. This problem is commonly referred to as *oblivious transfer*. More formally, let $\boldsymbol{x} = (x_1, \ldots, x_n)$ be a database of ℓ-bit strings and let i be the desired index. Then miner nodes should learn nothing about the client's input and the client should learn nothing beyond x_i. Far more often, the client is not interested in the value itself but needs shares of x_i for further processing. At the end of *oblivious selection* protocol, miner nodes obtain shares of x_i and neither client or miners learn anything new. Note that an oblivious selection protocol becomes an oblivious transfer protocol if miners send the obtained shares securely to the client. Secondly, the client can construct a sub-sample of the original database by executing several oblivious selection operations. For example, assume that the database entries contain private information about individuals, however, it is public which database row belongs to which individual. Then the client can form a group of relevant persons based on private or public information about these individuals. With oblivious selection, the client can select only relevant rows without leaking the selected identities.

    The generic construction for oblivious selection described below works under the assumption that both the database elements x_i and indices i can be considered as elements of a ring $\mathbb{Z}_N$. By the definition of secret sharing, we achieve client privacy by sharing the index i at the client side and just transferring the shares $[\![i]\!]$ to the miners. Then the miners can compute the output shares as follows

$$\llbracket x_i \rrbracket = \sum_{j=1}^{n} (\llbracket i \rrbracket = j) \cdot \llbracket x_j \rrbracket \qquad (1)$$

where ($\llbracket i \rrbracket = j$) denotes the output shares of a secure comparison protocol, see the full paper for further details [27]. The result is correct as comparisons yield a zero-one indicator vector. Protocol 1 depicts the corresponding oblivious transfer protocol, where the output shares are sent back to the client. The GENOT protocol is as secure as the weakest sub-protocol. In particular, as sub-protocols are universally composable, so is the GENOT protocol. If sub-protocols are secure against active corruption, so is the GENOT protocol. Since now this is what we mean by stating that assumptions of share computing are fulfilled.

Theorem 1. *If the assumptions of share computing protocols are fulfilled, the* GENOT *protocol is secure against malicious data donors and clients. The corresponding round complexity is* $\tau_{eq} + \tau_{mul} + 1$ *where* τ_{mul} *and* τ_{eq} *are round complexities of multiplication and equality test protocols.*   □

Remarks. First, note that miners can use any shared value as $\llbracket i \rrbracket$ in the protocol. In particular, same shares can be used to select or fetch elements form different databases. As protocols can be run in parallel, we can obliviously select database records of arbitrary format without increase in the round complexity. Moreover, miners can also assure that the inputs are in a certain range. In particular, miners can assure that $x_i \in \{0,1\}^{\ell}$ by setting $\llbracket x_i \rrbracket \leftarrow \llbracket x_i \rrbracket \cdot (\llbracket x_i \rrbracket < 2^{\ell})$ in the data gathering stage. Second, note that the efficiency of the protocol depends mainly on the efficiency of equality testing. Hence, we must pay special attention to this subprotocol and it is done in the full version of the paper [27]. Third, note that the communication complexities of data donors and clients are optimal up to a multiplicative factor. The miners' computational workload is linear in the database size, which is again optimal, since each database element must be touched.

3.2   Generic Construction for Oblivious Database Filtering

In data analysis tasks, one often needs to separate a certain sub-sample form the entire data. For instance, we might constrain our sample to female patients who are over 65 and have high blood pressure. When the data is secret shared due to privacy reasons, the desired sub-sample must be formed obliviously. The procedure can be split into two phases. First, we must compute shares of an indicator vector f that is set to one when the inclusion criterion is met. Standard share computing frameworks can easily handle predicates consisting of arithmetic and comparison operators, such as the constraint $(\llbracket sex \rrbracket = 1) \cdot (\llbracket age \rrbracket \geq 65) \cdot (\llbracket hbp \rrbracket = 1)$. Second, we must perform the oblivious filtering step. Let x be the vector of database records and let f be a zero-one indicator vector. Then during oblivious filtering protocol miners use shares of x and f to construct a new database of shared records y such that x_i is included into y iff $f_i = 1$. Miners learn nothing except the size of y during the protocol. The latter is unavoidable, if we want to reduce the workload in later processing stages – gains in efficiency unavoidably leak information about the size of the sub-sample y.

---

1. Miners apply oblivious shuffling to permute the shares of the database $(\boldsymbol{x}||\boldsymbol{f})$.
2. Miners reconstruct the permuted vector $\pi(\boldsymbol{f})$ of the resulting database $(\pi(\boldsymbol{x})||\pi(\boldsymbol{f}))$.
3. Miners keep the shares of $\boldsymbol{x}_{\pi(i)}$ for which $f_{\pi(i)} = 1$ as shares of $\boldsymbol{y}$.

Protocol 2. Generic oblivious filtering protocol GENOF

For clarity, let $(\boldsymbol{x}||\boldsymbol{f})$ denote a record-wise concatenation of databases $\boldsymbol{x}$ and $\boldsymbol{f}$ consisting of pairs (x_i, f_i). For any permutation π, let $\pi(\boldsymbol{z})$ denote a reordered database $z_{\pi(1)}, \ldots, z_{\pi(n)}$. Then Protocol 2 depicts a generic construction that reduces oblivious filtering task to oblivious shuffling. As a consequence, efficiency of many data analysis algorithms is determined by the efficiency of oblivious shuffling. See Section 4 for the description of three oblivious shuffle protocols with different trade-offs between round complexity τ_{os} and communication complexity.

Theorem 2. *If assumptions of the oblivious shuffle protocol are fulfilled, the* GENOF *protocol is secure. The round complexity is* $\tau_{os} + 1$ *where* τ_{os} *is the round complexity of oblivious shuffle.*

3.3   Oblivious Read-Write Access

Oblivious database access is a powerful tool in data processing. For instance, we can used it for post-processing secret shared free-text fields. Assume that a human operator obtains all the text entries so that they could not be linked to the responders. Then the operator can extract necessary information form free-text entries for better encoding of answers. Next, the operator must be able to obliviously update the corresponding database fields to complete. The latter can be achieved with the GENOWR protocol, where (1) miners apply oblivious shuffling to permute the shares of the database $\boldsymbol{x}$; (2) the client reads and writes fields of the shuffled database $\pi(\boldsymbol{x})$; (3) miners apply oblivious shuffling to permute the shares of the updated database $\hat{\boldsymbol{x}}$. The protocol does not provide full privacy, as the client learns some database fields and can link field updates with accessed values. Still, such a level of privacy is sufficient for many applications.

Theorem 3. *The overhead of the* GENOWR *protocol is* $2\tau_{os}$ *where* τ_{os} *is the round complexity of the oblivious shuffle protocol. If the assumptions of the oblivious shuffle protocol are fulfilled and the client does not access the same field twice for reading or writing, the miners learn only the number of read and write operations and which operations were performed on the same database record. The client learns the values of accessed database fields and can link updates of database entries with the values of accessed database fields.*

Remarks. The default implementation of GENOWR protocol produces the database where the entries are shuffled. The original order can be restored if the second shuffle implements the inverse of π, or if we add the shared row numbers to the original database and open them after the second shuffle. As another extension, note that the

GENOWR protocol can be used for implementing adaptive oblivious transfer, where the client wants to retrieve several elements from a database, given that the client learns the permutation π. After the database is shuffled, the client can use π to query the required elements, whereas the miners will not know the correspondence. Hence, a single shuffle is sufficient for adaptive oblivious transfer of many elements. If clients do not query the same element more than once, miners will learn only the number of queries and nothing else. The same construction is valid for oblivious selection.

4 Protocols for Oblivious Shuffle

In the following, we describe three oblivious shuffle protocols. The first protocol presented in Section 4.1 based on permutation matrices has constant round complexity. However, the communication complexity is rather high ($\Theta(n^3)$ or $\Theta(n^2)$ depending on the variation). The resharing-based solution presented in Sections 4.3 and 4.4 achieves communication complexity asymptotically optimal in database size ($\Theta(n \log n)$), but is exponential with respect to the number of miners. In many applications this is not a major issue, since the number of miners is usually very limited (e.g. three for the current implementation of SHAREMIND). However, for larger requirements we can propose a good sorting-based trade-off in Section 4.2, as its communication complexity is $\Theta(n \operatorname{polylog} n)$ and it scales well to a larger number of miners.

All three protocols are composed of consecutive shuffling phases which are designed to hide permutations from different sets of miners. In the first and the second protocol, only a single miner knows the permutation of each phase and thus $O(m)$ phases are sufficient for security. In the third protocol, the permutation is known to a large set of miners and thus $\Theta(2^m/\sqrt{m})$ phases are needed in the worst case.

Similarly to oblivious transfer, we only consider the case where all database elements belong to $\mathbb{Z}_N$. As any collection of records can be represented as a matrix where each column is either a data field or part of it, a vector shuffle protocol is sufficient provided that several protocol runs can share the same hidden permutation. It is easy to see that all protocols presented in this section have this property.

4.1 Oblivious Shuffle Based on Permutation Matrices

Recall that for any n-element permutation π there exists a zero-one matrix M_π such that $\pi(\boldsymbol{x}) = M_\pi \boldsymbol{x}$ for all vectors $\boldsymbol{x}$ of size n. Hence, if a miner $\mathcal{P}_i$ generates and shares a permutation matrix M_{π_i}, the database can be shuffled by multiplying it with M_{π_i}. As none of the miners should know the permutation, miners must execute several such shuffle phases. More precisely, let t be the maximal size of a tolerated adversarial coalition. Then it is sufficient if $t+1$ miners permute the database, as at least one permutation remains oblivious to the adversarial coalition. Also, note that the final outcome is indeed a shuffle, as at least one permutation is chosen uniformly. As each permutation phase can be implemented with $\Theta(n^2)$ multiplications and $t = \Theta(m)$ in standard multiparty frameworks, the resulting shuffle protocol contains $O(mn^2)$ multiplications, which can be performed in $O(m\tau_{\mathrm{mul}})$ rounds. The number of rounds can be reduced to $O(\log_2 m\tau_{\mathrm{mul}})$ with the cost of $O(mn^3)$ multiplications if miners first compute the matrix product $M_{\pi_1} \cdots M_{\pi_{t+1}}$ in a balanced manner.

In the malicious model, miners have to additionally verify that each M_{π_i} is a permutation matrix, i.e., all entries are either zeroes or ones and the sums of all the rows and columns of M_{π_i} are ones. Each such test can be expressed as a share computing procedure yielding one if the input M_{π_i} is correct. Moreover, these additional zero-knowledge proofs do not change the complexity estimates. As a result, security follows from the correctness and universal composability of share computing protocols.

4.2 Oblivious Shuffle Based on Sorting

Let $y_1, \ldots, y_n \in \mathbb{Z}_N$ be a random permutation of the set $\{1, \ldots, n\}$. Then we can implement shuffle by obliviously sorting the pairs (x_i, y_i) according to the second element. In such a shuffle phase, a miner $\mathcal{P}_i$ generates and shares values $y_1, \ldots, y_n$ and then all miners use a modified sorting network consisting of compare-exchange gates

$$\text{CompEx}((x_i, y_i), (x_j, y_j)) = \begin{cases} (x_i, y_i), (x_j, y_j) & \text{if } y_i \leq y_j , \\ (x_j, y_j), (x_i, y_i) & \text{if } y_i > y_j , \end{cases}$$

for oblivious relocation of elements. The CompEx-gate can be securely implemented by combining secure multiplication and comparison operations, e.g. we can define the first output element as $([\![y_i]\!] \leq [\![y_j]\!]) \cdot [\![x_i]\!] + ([\![y_i]\!] > [\![y_j]\!]) \cdot [\![x_j]\!]$. Hence, each CompEx block requires $O(\tau_{\text{st}} + \tau_{\text{mul}})$ rounds. Similarly to the first protocol, several shuffle phases are needed to hide the permutation from potentially adversarial miners.

The efficiency of the resulting shuffle protocol depends on the complexity of the sorting network. For instance, randomised shell sort [24] has $O(\log n)$ layers and $O(n \log n)$ CompEx blocks. Thus, the shuffle protocol has $\Theta(m(\tau_{\text{mul}} + \tau_{\text{sl}}) \log n)$ rounds and communication complexity $\Theta(mn \, \text{poly}(\log n))$, where the term $\text{poly}(\log n)$ captures the asymptotic growth of communication in base protocols.

In the full version of the paper [27], we discuss more round-efficient alternatives that can be used when $n \leq \sqrt{N}$. This condition is commonly satisfied in practice.

4.3 Resharing Based Oblivious Shuffle for Semihonest Setting

The suffle phase can be viewed as a hide and seek game, where the aim of the hider set $\mathcal{C}$ is to shuffle the database in a such way that the seeker set $\mathcal{A}$ learns nothing about the permutation. Protocol 3 depicts a setting where the seekers first transfer their shares to the hiders so that the database is secret shared only between the members of $\mathcal{C}$. Next, hiders agree on a permutation π and reorder their shares accordingly. Then the secret sharing is extended to all miners. The protocol is secure and achieves its goal provided that: (a) corrupted parties in the hider set $\mathcal{C}$ cannot recover secrets; (b) shares are extended so that seekers learn nothing about shared values. By repeating this shuffle phase for every maximal tolerable adversarial coalition $\mathcal{A}$, we get a secure oblivious shuffle provided that all sub-protocols remain secure.

Lemma 1. *Let $\mathcal{A}$ be such a coalition that the complement set cannot be corrupted. If assumptions of secret sharing are fulfilled, then the protocol VerShf is secure in the semihonest model.*

Inputs: A database of shares $[\![x_1]\!], \ldots, [\![x_n]\!]$ and a potential adversarial coalition $\mathcal{A}$.
Output: A database of shares $[\![x_{\pi(1)}]\!], \ldots, [\![x_{\pi(n)}]\!]$ for a random permutation π unknown to $\mathcal{A}$.
Clarifying remark: Steps starting with $\otimes$ should be omitted in the semihonest setting.

Mixing phase. Let $\mathcal{C} = \{\mathcal{P}_1, \ldots, \mathcal{P}_m\} \setminus \mathcal{A}$ and let $[\![x_k]\!]_i$ denote the ith share of x_k.

1. Each miner $\mathcal{P}_i \in \mathcal{A}$ additively secret shares its share vector between the miners of $\mathcal{C}$, i.e., each miner $\mathcal{P}_j \in \mathcal{C}$ gets a share u_{kj} such that the shares $(u_{kj})_{j \in \mathcal{C}}$ sum up to $[\![x_k]\!]_i$.
   $\otimes$ $\mathcal{P}_i$ commits $(u_{ij})_{j \in \mathcal{C}}$ to all. CTP, CMP and CSP are executed to verify correctness and consistency of the performed actions.
2. All miners in $\mathcal{C}$ locally compute additive shares for the database elements $x_1, \ldots, x_n$, i.e., each $\mathcal{P}_j \in \mathcal{C}$ holds a share v_{kj} for element x_k, s.t. shares $(v_{kj})_{j \in \mathcal{C}}$ sum up to x_k.
   $\otimes$ All miners use linearity to compute locally commitments to $(v_{kj})_{j \in \mathcal{C}}$.
3. All miners in $\mathcal{C}$ agree on a random permutation π and reorder their shares according to π.
4. Each miner $\mathcal{P}_j \in \mathcal{C}$ uses original secret sharing to share the shuffled shares between all miners. As a result, miners obtain a matrix of shares $([\![v_{\pi(k),j}]\!])_{j \in \mathcal{C}, k \in \{1,2,\ldots,n\}}$.
   $\otimes$ Zero-knowledge proofs for shuffle correctness are run to verify that all parties $\mathcal{P}_i \in \mathcal{C}$ followed the protocol. CSP are executed to commit shares of $(v_{\pi(k),j})_{j \in \mathcal{C}, k \in \{1,2,\ldots,n\}}$.
5. All miners locally add shares $([\![v_{\pi(k),j}]\!])_{j \in \mathcal{C}}$ to obtain $[\![x_{\pi(1)}]\!], \ldots, [\![x_{\pi(n)}]\!]$.

Protocol 3. Verifiable shuffle protocol VERSHF that is oblivious for a coalition $\mathcal{A}$

Proof (Sketch). As the complement group cannot be entirely corrupted, the first step in the mixing phase reveals no information and can be easily simulated. The second, third and fifth step create no communication and thus are trivial to simulate. The fourth step is simulatable due to the properties of the original secret sharing. The claim follows, as the output is guaranteed to be correct in the semihonest model. □

Theorem 4. *If a secret sharing scheme satisfies the Q2 condition, there exists a secure oblivious shuffle protocol with $O(2^m/\sqrt{m})$ rounds and communication complexity $O(2^m m^{3/2} n \log n)$ in the semihonest model.*

Proof. Let us repeat the VERSHF protocol sequentially for each maximal adversarial coalition $\mathcal{A}$. Then each protocol instance is secure as the condition Q2 guarantees that the complement of $\mathcal{A}$ cannot be corrupted. As at least one permutation remains hidden from an adversarial coalition, the protocol indeed implements oblivious shuffle. The round complexity estimate follows from the consideration that maximal adversarial coalitions form an antichain in the partially ordered set of all the subsets of the m-element set of miners. According to Sperner's classical result form 1928, there are up to $\binom{m}{\lfloor m/2 \rfloor} = \Theta(2^m/\sqrt{m})$ elements in such an antichain. The claim concerning the communication complexity follows, since each sub-protocol requires $\Theta(n)$ resharing operations and $O(n \log n)$ bits to generate a random permutation and each resharing operation requires $O(m^2)$ communication for fixed N. □

The protocol described above is asymptotically optimal if we consider only the asymptotic dependency on the database size. Sampling a random permutation requires $\Theta(n \log n)$ bits and thus the communication cannot be decreased further. Moreover, for small number of parties, say $m \leq 10$, the protocol is very efficient as there are no expensive multiplication and comparison operations.

Efficiency tweaks and implementation results. The first possible optimization can be obtained noting that it is not always necessary to consider the full complement of an adversary set as the set of hiding miners. For example, consider the threshold adversary setting, where the maximal adversary coalitions have t members with m strictly larger than $2t + 1$. In this case it is enough to select hiding sets having $t + 1$ elements which would then work against several different adversary sets, allowing us to achieve a reduced number of rounds. For example, if we have $t = 2$ then with $m = 5$ miners we would need 10 phases of computations. At the same time, increasing the number of miners to $m = 6$, six phases are sufficient, since the hider sets can be chosen to be $\{\mathcal{P}_1, \mathcal{P}_2, \mathcal{P}_4\}, \{\mathcal{P}_2, \mathcal{P}_3, \mathcal{P}_5\}, \{\mathcal{P}_3, \mathcal{P}_4, \mathcal{P}_6\}, \{\mathcal{P}_4, \mathcal{P}_5, \mathcal{P}_1\}, \{\mathcal{P}_5, \mathcal{P}_6, \mathcal{P}_2\}, \{\mathcal{P}_6, \mathcal{P}_1, \mathcal{P}_3\}$. It is easy to verify that for no adversary set of up to two elements they have access to all of the permutations and that 6 active sets (i.e. 6 phases) is the minimum achievable for $t = 2$ and $m = 6$. Finding optimal round and communication complexity for any t, m remains an open combinatorial problem.

In practice, communication channels between miner nodes are commonly implemented using authenticated encryption and thus information-theoretical security is unachievable. Hence, the security level does not decrease if the group $\mathcal{C}$ agrees on a short random seed and later uses a secure pseudorandom generator to stretch locally into $O(n \log n)$ bits needed for a random permutation. In particular, we can generate an array $f_{sk}(1), \ldots, f_{sk}(n)$ by applying a pseudorandom permutation f indexed by a seed sk. By sorting the array, we get a permutation that is computationally indistinguishable from a random permutation. We implemented the corresponding shuffle protocol for three miners by using 128-bit AES as f. For each mixing phase, $\mathcal{P}_a$ and $\mathcal{P}_b$ forming the group $\mathcal{C}$ exchanged 128-bit random sub-keys sk_a and sk_b and set $sk \leftarrow sk_a \oplus sk_b$. The protocol was implemented into SHAREMIND framework using C++ programming language. The computing parties and the controller node ran on servers having two Intel Xeon X5670 2.93GHz processors and 48GB of RAM each. The servers were using Debian OS and were connected by gigabit Ethernet. Table 2 shows the times required for oblivious shuffle of databases consisting of $10^3, \ldots, 10^7$ additively shared 32-bit integers.

4.4   Resharing Based Oblivious Shuffle for Malicious Setting

The protocol described above can be used also in the malicious model provided that we can force universal consistency checks: (a) all resharing steps are correct; (b) the permutation π is indeed randomly sampled; (c) hiders permute their shares according to the permutation. To achieve such kind of protection, we follow the standard two-level secret sharing technique [18], where all original shares are secret shared. That is, any share $[\![x]\!]_i$ owned by $\mathcal{P}_i$ is always secret shared between all miners. Such a setup simplifies zero-knowledge correctness proofs, since $\mathcal{P}_i$ can more easily prove that he or she computed output shares $[\![v]\!]_i$ from the input shares $[\![u]\!]_i$. The second level secret sharing is commonly referred to as commitment, as it satisfies both perfect binding and hiding properties. As shown in [18], security against an active adversary can be achieved with

Table 2. Performance of oblivious shuffle in three-party setting

Number of elements	10^3	10^4	10^5	10^6	10^7
Time (ms)	70	102	312	2543	24535

three auxiliary protocols: commitment transfer protocol (CTP), commitment sharing protocol (CSP) and commitment multiplication protocol (CMP). CTP allows to transfer a commitment of a secret from one party to another, CSP allows to share a committed secret in a verifiable way such that the parties will be committed to their shares, and CMP allows to prove that three committed secrets a, b and c satisfy the relation $c = ab$. In order to achieve security against active adversaries, share computing frameworks use two-level secret sharing by default. Moreover, for any robust secret sharing scheme, the second layer can be added on demand. Miners just have to commit their shares. Although a maliciously corrupted miner $\mathcal{P}_i$ can provide incorrect sharings of $[\![x]\!]_i$, the number of correctly shared shares is sufficient to correctly recover the original secret x. If needed, miners can emulate recovery procedure with commitments to detect which shares were incorrectly committed.

Correctness proofs for the resharing steps. As $\mathcal{P}_i$ commits values $(u_{kj})_{j \in C}$ in the first step of Protocol 3, miners can compute $[\![x_k]\!]_i - \sum_{j \in C}[\![u_{kj}]\!]$ and open it. If the result is zero, the shares were correctly formed. The result leaks no information, as the output is always zero for honest miners. By employing CTP, these commitments can be transferred to the intended recipients who are forced to correctly recommit u_{kj}. Verification of the second step is straightforward, since the reconstruction coefficients are public and all parties can locally manipulate shares of shares to get shares of the corresponding linear combinations.

Unbiased sampling of randomness. As the condition Q3 is not satisfied for the set C, hiders cannot agree on the random permutation without outsiders. Hence, all miners must engage in a secure coin-tossing protocol to generate necessary random bits for the permutation π. For instance, a random element of $\mathbb{Z}_N$ can be generated if all parties share random elements of $\mathbb{Z}_N$ and the resulting shares are added together. Next, everybody broadcasts their shares to the set of hiding participants C. To sample a random permutation, the same protocol can be run in parallel to generate enough random bits.

Correctness proof for the local shuffle step. The correctness proof hinges on the fact that miners have commitments to all shares of $\mathcal{P}_j$. After the second step, miners obtain commitments to additive shares $v_{1j}, \ldots, v_{nj}$ and during the fourth step miners receive commitments to permuted shares $v_{\pi(1)j}, \ldots, v_{\pi(n)j}$. Hence, if all miners $\mathcal{P}_j \in C$ prove that the shares were indeed obtained by applying the permutation π, the CSP protocol assures that the resulting double-level sharing of $v_{\pi(1)j}, \ldots, v_{\pi(n)j}$ is also correct.

Protocol 4 depicts a zero-knowledge protocol for a slightly abstracted setting where $\mathcal{P}_j$ has to prove that she has secret shared databases $\boldsymbol{x}$ and $\boldsymbol{y}$ so that $\boldsymbol{y} = \pi(\boldsymbol{x})$ for a permutation π known to the hider set C. At the end of the protocol, honest parties from both sets learn whether the sharing was correct. The soundness error of the protocol can

---

Prover: A prover is a miner $\mathcal{P}_j$ in the hider set knowing $x_1, \ldots, x_n$ and $y_1, \ldots, y_n$.
Helpers: Miners in the hider set $\mathcal{C}$ know a permutation π such that $y_1 = x_{\pi(1)}, \ldots, y_n = x_{\pi(n)}$.
Common inputs: Miners have shares $[\![x_1]\!], \ldots, [\![x_n]\!], [\![y_1]\!], \ldots, [\![y_n]\!]$.

1. All miners engage secure coin-tossing protocol to fix a random permutation σ.
   The resulting bits are revealed to all miners in the hider set so that they can learn σ.
2. The prover $\mathcal{P}_j$ computes $z_1 = y_{\sigma(1)}, \ldots, z_n = y_{\sigma(n)}$ and shares them.
3. All parties engage in a secure coin-tossing protocol to get a public random bit b.
4. (a) If $b = 0$ then all miners in the hider set broadcast σ. Each miner reconstructs σ.
       All miners compute shares of $[\![z_1]\!] - [\![y_{\sigma(1)}]\!], \ldots, [\![z_n]\!] - [\![y_{\sigma(n)}]\!]$ and broadcast the results.
   (b) If $b = 1$ then all miners in the hider set broadcast $\pi \circ \sigma$. Each miner reconstructs $\pi \circ \sigma$.
       All miners compute shares of $[\![z_1]\!] - [\![x_{\pi(\sigma(1))}]\!], \ldots, [\![z_n]\!] - [\![x_{\pi(\sigma(n))}]\!]$ and broadcast
       the results.
5. Proof fails if any of the revealed values is non-zero, otherwise miners accept the proof.

Protocol 4. Zero-knowledge proof ZKSHF for shuffle correctness

be efficiently amplified: any soundness error ε can be achieved by running $\lceil \log_2 1/\varepsilon \rceil$ instances of the protocol in parallel as the protocol is universally composable.

Theorem 5. *Let $\mathcal{A}$ be a maximal adversarial set. If a secret sharing scheme satisfies the Q3 condition then the protocol ZKSHF is a perfect five round zero-knowledge proof with soundness error $\frac{1}{2}$.*

Proof (Sketch). The protocol has five rounds, since each protocol step can be implemented in a single round. For correctness note that if $\mathcal{P}_j$ is honest then $z_i = y_{\sigma(i)} = x_{\pi(\sigma(i))}$ for $i \in \{1, \ldots, n\}$. As the set $\mathcal{A}$ is maximal adversarial set and the condition Q3 holds, the condition Q2 must hold in the hiding set $\mathcal{C}$. Consequently, each miner can correctly reconstruct permutations σ or $\pi \circ \sigma$ in the fourth step. By the condition Q2, honest coalition is strictly larger than any adversarial set in $\mathcal{C}$. As the replies sent by honest parties coincide, each party can detect the correct answer. As the adversarial coalition cannot alter the reconstructed values by changing their shares, the revealed values will always be zeroes. For the soundness claim, assume that the prover cheats and there exists $y_j \neq x_{\pi(j)}$. Let $i = \sigma^{-1}(j)$. Then $z_i \neq y_j$ or $z_i \neq x_{\pi(j)}$ cannot simultaneously succeed. As malicious parties cannot alter the end results, the prover is bound to fail with probability at least $\frac{1}{2}$. For the simulatability, note that the adversarial coalition cannot bias the outputs of coin-tossing protocols. Hence, the permutations σ and $\pi \circ \sigma$ are random permutations in isolation. When no hider is corrupted, we can just send a random permutation to simulate the beginning of the fourth step. Otherwise, we have to extract π form the inputs of the corrupted hider and simulate permutation generation according to the protocol. The public opening of the shares is trivial to simulate, as the result is known to be zero. As the simulator has access to the input shares of the corrupted parties (i.e., the shares of $\boldsymbol{x}$ and $\boldsymbol{y}$), it can augment them to be shares of zero. By emulating the protocol with these dummy shares, the share distribution in the opening phase is guaranteed to coincide with the real protocol. $\square$

Final security claim. As in the semihonest setting, the final shuffle protocol can be obtained by repeating the VERSHF protocol for every maximal tolerable adversarial coalition $\mathcal{A}$. The correctness proofs assure that honest miners detect all deviations form the protocol and catch the corresponding culprit. After each incident, the culprit must be excluded form the computations and the corresponding shuffle phase must be restarted. As the number of malicious parties is limited and the exclusion does not invalidate Q3 condition, honest miners can always complete the protocol.

Theorem 6. *If a secret sharing scheme satisfies the condition Q3, there exists a secure oblivious shuffle protocol with* $O(2^m/\sqrt{m})$ *rounds and communication complexity* $O(2^m m^{3/2} n \log n)$ *in the malicious model.*

5   Conclusions and Future Work

In this paper, we have proposed several round-efficient protocols for oblivious database manipulation over secure MPC in both semihonest and malicious model, including oblivious transfer and oblivious shuffle of secret shared databases. We also presented an oblivious filtering protocol and selection protocols as possible applications of these techniques.

For oblivious transfer, we have presented several versions with round complexity between $O(\log \log n)$ and $O(\log \log \log n)$. For oblivious shuffle, we have proposed three alternative implementations with different trade-offs between the round and communication complexity. Which one of them is the fastest for a given application and deployment scenario, is a non-trivial question requiring further research and benchmarking. We have also provided several possible tweaks for optimization of the presented primitives, but several questions are left open. For example, finding out optimal hider set constructions for possible adversarial structures is an interesting combinatorial problem of its own. As a useful by-product of the tweaking, we obtained an efficient protocol for equality testing based on double recursion with bit decomposition.

In the current paper, we presented initial performance results proving that oblivious database shuffle is efficient for relatively large databases (roughly 25 seconds for 10^7 elements). Further improvements can be achieved by optimizing the implementation, however, these improvements remain the subject for future development as well.

Acknowledgments. This research was supported by Estonian Science Foundation grant #8058, the European Regional Development Fund through the Estonian Center of Excellence in Computer Science (EXCS), and Estonian ICT doctoral school.

References

1. VIFF documentation, http://viff.dk/doc/index.html
2. Proceedings of the Twentieth Annual ACM Symposium on Theory of Computing, Chicago, Illinois, USA, May 2-4. ACM, New York (1988)
3. SecureSCM. Technical report D9.1: Secure Computation Models and Frameworks (July 2008), http://www.securescm.org

4. Aiello, W., Ishai, Y., Reingold, O.: Priced oblivious transfer: How to sell digital goods. In: Pfitzmann, B. (ed.) EUROCRYPT 2001. LNCS, vol. 2045, pp. 119–135. Springer, Heidelberg (2001)

5. Beimel, A., Stahl, Y.: Robust Information-Theoretic Private Information Retrieval. J. Cryptology 20(3), 295–321 (2007)

6. Ben-David, A., Nisan, N., Pinkas, B.: FairplayMP: a system for secure multi-party computation. In: CCS 2008: Proceedings of the 15th ACM Conference on Computer and Communications Security, pp. 257–266. ACM, New York (2008)

7. Ben-Or, M., Goldwasser, S., Wigderson, A.: Completeness Theorems for Non-Cryptographic Fault-Tolerant Distributed Computation (Extended Abstract). In: STOC [2], pp. 1–10

8. Bogdanov, D., Laur, S., Willemson, J.: Sharemind: A Framework for Fast Privacy-Preserving Computations. In: Jajodia, S., Lopez, J. (eds.) ESORICS 2008. LNCS, vol. 5283, pp. 192–206. Springer, Heidelberg (2008)

9. Bogdanov, D., Niitsoo, M., Toft, T., Willemson, J.: Improved protocols for the SHAREMIND virtual machine. Research report T-4-10, Cybernetica (2010), http://research.cyber.ee

10. Brassard, G., Crépeau, C., Robert, J.M.: All-or-Nothing Disclosure of Secrets. In: Odlyzko, A.M. (ed.) CRYPTO 1986. LNCS, vol. 263, pp. 234–238. Springer, Heidelberg (1987)

11. Burkhart, M., Strasser, M., Many, D., Dimitropoulos, X.: SEPIA: Privacy-Preserving Aggregation of Multi-Domain Network Events and Statistics. In: USENIX Security Symposium, Washington, DC, USA, pp. 223–239 (2010)

12. Camenisch, J., Lysyanskaya, A.: A Formal Treatment of Onion Routing. In: Shoup [33], pp. 169–187

13. Canetti, R.: Universally Composable Security: A New Paradigm for Cryptographic Protocols. Cryptology ePrint Archive, Report 2000/067 (2000), http://eprint.iacr.org/

14. Catrina, O., Saxena, A.: Secure Computation with Fixed-Point Numbers. In: Sion, R. (ed.) FC 2010. LNCS, vol. 6052, pp. 35–50. Springer, Heidelberg (2010)

15. Chaum, D., Crépeau, C., Damgård, I.: Multiparty Unconditionally Secure Protocols (Extended Abstract). In: STOC [2], pp. 11–19

16. Chaum, D.L.: Untraceable Electronic Mail, Return Addresses, and Digital Pseudonyms. Communications of the ACM 24(2), 84–90 (1981)

17. Chor, B., Kushilevitz, E., Goldreich, O., Sudan, M.: Private Information Retrieval. J. ACM 45(6), 965–981 (1998)

18. Cramer, R., Damgård, I., Maurer, U.: General Secure Multi-party Computation from any Linear Secret-Sharing Scheme. In: Preneel, B. (ed.) EUROCRYPT 2000. LNCS, vol. 1807, pp. 316–334. Springer, Heidelberg (2000)

19. Damgård, I., Fitzi, M., Kiltz, E., Nielsen, J.B., Toft, T.: Unconditionally Secure Constant-Rounds Multi-party Computation for Equality, Comparison, Bits and Exponentiation. In: Halevi, S., Rabin, T. (eds.) TCC 2006. LNCS, vol. 3876, pp. 285–304. Springer, Heidelberg (2006)

20. Damgård, I., Geisler, M., Krøigaard, M., Nielsen, J.B.: Asynchronous Multiparty Computation: Theory and Implementation. In: Jarecki, S., Tsudik, G. (eds.) PKC 2009. LNCS, vol. 5443, pp. 160–179. Springer, Heidelberg (2009)

21. Damgård, I., Ishai, Y.: Constant-Round Multiparty Computation Using a Black-Box Pseudorandom Generator. In: Shoup [33], pp. 378–394

22. Garay, J.A.: Efficient and Universally Composable Committed Oblivious Transfer and Applications. In: Naor [30], pp. 297–316

23. Geisler, M.: Cryptographic Protocols: Theory and Implementation. PhD thesis, Aarhus University (February 2010)

24. Goodrich, M.T.: Randomized shellsort: A simple oblivious sorting algorithm. In: Charikar, M. (ed.) Proceedings of the Twenty-First Annual ACM-SIAM Symposium on Discrete Algorithms, pp. 1262–1277 (2010)
25. Henecka, W., Kögl, S., Sadeghi, A.-R., Schneider, T., Wehrenberg, I.: TASTY: tool for automating secure two-party computations. In: CCS 2010: Proceedings of the 17th ACM Conference on Computer and Communications Security, pp. 451–462. ACM, New York (2010)
26. Hirt, M., Maurer, U.M.: Player Simulation and General Adversary Structures in Perfect Multiparty Computation. Journal of Cryptology 13(1), 31–60 (2000)
27. Laur, S., Willemson, J., Zhang, B.: Round-efficient Oblivious Database Manipulation. Cryptology ePrint Archive, Report 2011/429 (2011), http://eprint.iacr.org/
28. Malka, L., Katz, J.: VMCrypt – Modular Software Architecture for Scalable Secure Computation. Cryptology ePrint Archive, Report 2010/584 (2010), http://eprint.iacr.org/
29. McLachlan, J., Tran, A., Hopper, N., Kim, Y.: Scalable onion routing with torsk. In: Al-Shaer, E., Jha, S., Keromytis, A.D. (eds.) ACM Conference on Computer and Communications Security, pp. 590–599. ACM, New York (2009)
30. Naor, M. (ed.): TCC 2004. LNCS, vol. 2951. Springer, Heidelberg (2004)
31. Pfitzmann, B., Schunter, M., Waidner, M.: Secure Reactive Systems. Technical Report 3206 (#93252), IBM Research Division, Zürich (May 2000)
32. Shamir, A.: How to share a secret. Commun. ACM 22(11), 612–613 (1979)
33. Shoup, V. (ed.): CRYPTO 2005. LNCS, vol. 3621. Springer, Heidelberg (2005)
34. Wikström, D.: A Universally Composable Mix-Net. In: Naor [30], pp. 317–335
35. Yao, A.C.-C.: Protocols for Secure Computations (Extended Abstract). In: FOCS, pp. 60–164 (1982)

A Privacy-Preserving Join on Outsourced Database

Sha Ma*, Bo Yang, Kangshun Li, and Feng Xia

Department of Information, South China
Agricultural University, Guangzhou, Guangdong, China
{martin_deng}@163.com

Abstract. Outsourced database provides a solution for data owners who want to delegate the task of answering database queries to service provider. Of essential concern in such framework is data privacy. The data owner may want to keep the database hidden from service provider. Simultaneously, potential clients may desire a mean of privacy-preserving computations on outsourced encrypted database. We present a solution of privacy-preserving join query with computational privacy and low overheads. The primary goal of this paper is to provide a set of security notion for such a system as well as a construction which is secure under the newly introduced security notions.

Keywords: searching on encrypted data, join query, outsourced database, data privacy.

1 Introduction

1.1 Motivation

The outsourced database model is a new computing paradigm that has emerged recently [1, 2] which consists of the following entities:

1. Data owner: This is the side that produces data and owns it. It is assumed to have some limited computational resources and storage capabilities but for less than the server.
2. Server: The server stores and manages the data generated by data owners.
3. User: A user can only access data of the data owner.

To save cost, data storage and management are outsourced to database service providers. In other words, highly sensitive data is now stored in locations that is not under the data owner's control, such as leased space and partners' sites. This can put data confidentiality at risk. Therefore, it is desirable to store data in encrypted form to protect sensitive information.

---

* This work is supported by the National Natural Science Foundation of China under Grants 60773175, 60973134 and 70971043, the Foundation of National Laboratory for Modern Communications under Grant 9140C1108020906, and the Natural Science Foundation of Guangdong Province under Grants 10351806001000000 and 9151064201000058.

X. Lai, J. Zhou, and H. Li (Eds.): ISC 2011, LNCS 7001, pp. 278–292, 2011.

Consider the following real-life scenario: A hospital database is required not only to store each patient's general information, for instance, *PatientID, name, age, address, telephone, allergic drugs*, but also to record all medical histories for each patient by doctors, including the description of the disease and the diagnosis information. So the logical schema of the hospital database contains at least two tables, for instance, denoted as *patient*(*PatientID, name, age, address, telephone, allergicdrugs*) and *diagnosis*(*PatientID, disease_desc, diagnosis, DoctorID, date*). Data confidentiality alone can be achieved by encrypting the outsourced content so that even if a database is compromised by an intruder, data remains protected even in the event that a database is successfully attacked or stolen. However, once encrypted, the data cannot be easily processed by the server. Suppose that the outsourced database stores the encrypted table *patient'* and *diagnosis'* according to the plaintext table *patient* and *diagnosis*. Equijoin of *patient'* and *diagnosis'* describes which doctor is responsible for the diagnosis of which patient. It maybe become a powerful witness for the medical malpractice. However, once all attributes are encrypted by a probabilistic encryption algorithm, the server will have much difficulty in processing this query primitive. In addition, the server is not allowed to execute join without the permission from users. This is usually related to the application environment. For example, if the server is able to evaluate join of *patient'* and *diagnosis'* by itself, it might face an expensive lawsuit from the patient because the patient's disease information can be disclosed.

The problem of set-intersection functionality of secure two-party computation [3] sounds very similar to our concerns. However, they are essentially different. Securely computing set-intersection functionality consists of two parties trying to compute their intersection but without disclosing their inputs. In our scenarios, there are three parties participated in the protocol. The encrypted value x and y sent by the data owner are stored in the server, which is desired to do the comparison of x and y under the permission of the user without decryption key and do not obtain any information after the computation.

Thus, it is important to provide mechanisms for server-side data processing that are correctness, privacy of the storage for the data owner, controllability and join privacy(These notions will be explained in section 3). This is particularly relevant in relational settings. This paper presents a vision for how process equijoin on outsourced database in privacy-preserving style. The main contributions of this paper include:(i)a general framework of privacy-preserving join, (ii)the definition of correctness, privacy of the storage for the data owner, controllability and join privacy, (iii)a construction for equijoin, (iv)security proofs of implementation based on security definitions, (v)comparison with the method proposed in reference [4].

1.2   Related Work

Various methods have been proposed recently for secure database in various settings. In particular, encryption is an important technique to protect sensitive data. With data stored in an encrypted form, a crucial question is how to perform queries. There are four classes:

1. Using indices: Relational database use tables to store the information. Rows of the table correspond to records and column to field. Often hidden field or even complete table are added which act as an index. Instead of searching on the encrypted data itself, the actual search is performed in an added index. The index contains for example the hashes of the encrypted records [5, 6] and a Bloom Filter for each encrypted record [4]. This paper focuses on the comparison with reference [4] and analyzes their own advantages and disadvantages.

2. Using trapdoor encryption: Trapdoor encryption makes it possible to give user a way to perform some operations on the encrypted data without decryption key. Many studies focus on the trapdoor encryption to allow user to search for a particular keyword [7,8]. However, few studies solve the problems of comparison query on the encrypted database except order-preserving symmetric encryption [9,10] and hidden-vector encryption [11]. The former one is view as a tool like a blockcipher rather than a full-fledged encryption scheme itself due to the very large size of ciphertext-space. The latter one is required to get the knowledge of data's location in a finite set before the encryption, which is not suitable for database encryption because the computation of the data's location information for each value in the domain, for instance, $char(10)$, is a non-trivial task. However, our construction is not based on any prior information.

3. Using secret sharing: In cryptography, secret sharing means that a secret is split over several parties in such a way that no single party can retrieve the secret. After data can be stored securely by distributing it over several servers, the data is queried by using a secure protocol between the client and the servers. A typical usuage of secret sharing is Private Information Retrieval(PIR) [12,13]. PIR aims at letting a user query the database without leaking to the database which data was required. The idea behind PIR is to replicate the data among several non-communicating servers. A client can hid his query by asking all servers for a part of the data in such a way that no server will learn the whole query by itself.

4. Using homomorphic encryption: Some encryption functions give the ability to perform some simple operations directly on the encrypted data without the need to decrypt the data first. Homomorphic encryption is a type of encryption with a special property. More precisely, an encryption function E is called homomorphic if there exists two(possibly the same) operation ($\oplus$ and $\otimes$), such that $E(a \oplus b) = E(a) \otimes E(b)$. Although this property makes it possible to calculate with encrypted values [6], we need the usage of homomorphic encryption in a delicate manner. The reasons are (1)deterministic homomorphic encryption, for instance, RSA, can not be used in outsourced database scenario because of its weak security. (2)general probabilistic homomorphic encryption makes it difficult to process queries by the server. In essence, our construction is based on a particular probabilistic homomorphic encryption.

Compared with traditional method using homomorphic encryption, the main difference is that we generate trapdoors for the database server to support queries on outsourced database.

2 Preliminaries

2.1 Boneh-Goh-Nissim Encryption System

The public key encryption algorithm of Boneh, Goh and Nissim is proposed in 2005, which resembles the Pailier and the Okamoto-Uchiyama encryption schemes [14]. We describe it as $\mathcal{G}(s)$ below on input a security parameter $s \in \mathbb{Z}^+$.

1. Key generation.
   (a) Choose two random s-bit primes p and q.
   (b) Generate two multiplicative groups $\mathbb{G}$ and $\mathbb{G}_1$ of order $n = pq$ and a bilinear map $e : \mathbb{G} \times \mathbb{G} \to \mathbb{G}_1$ such that for all $u, v \in \mathbb{G}$ and $a, b \in \mathbb{Z}$, we have that $e(u^a, v^b) = e(u, v)^{ab}$. It is also required that if g is a generator of group $\mathbb{G}$ then $e(g, g)$ is a generator of group $\mathbb{G}_1$.
   (c) Choose two random generators $g, u \in_R \mathbb{G}$.
   (d) Calculate the generator $h = u^q$ of a subgroup of $\mathbb{G}$ of order p.
   (e) Publish public key$(n, \mathbb{G}, \mathbb{G}_1, e, g, h)$ and keep private key p secret.
2. Encryption
   (a) Choose a random $r \in_R \{0, \ldots, n-1\}$.
   (b) The encryption of a message m is $c = g^m h^r \in \mathbb{G}$.
3. Decryption
   To decrypt the ciphertext c first compute $c^p = (g^m h^r)^p = (g^p)^m = \hat{g}^m$ and then use Pollard's ρ-method to calculate the discrete log to retrieve m.

Boneh-Goh-Nissim encryption is clearly additively homomorphic because $E(m_1) \cdot E(m_2) = g^{m_1} h^{r_1} \cdot g^{m_2} h^{r_2} = g^{m_1+m_2} h^{r_1+r_2} = E(m_1 + m_2)$.

Theorem 1. *The public key system of Boneh-Goh-Nissim encryption is semantically secure assuming $\mathcal{G}$ satisfies the subgroup decision assumption [14].*

Note: The Boneh-Goh-Nissim encryption is limited in the size of message space due to the need to compute discrete logarithms during decryption [14]. Since in the database application message space is usually defined as a small domain, this restriction does not affect the usefulness of our protocol.

2.2 Bloom Filters

Bloom Filters [15] offer a compact representation of a set of data items, allowing for fast set inclusion tests. A Bloom Filter is an array B of m bits, initialized to zero. It requires a set of n independent hash functions H_i that produce uniformly distributed output in the range [0,m-1] over all possible inputs.

To add an entry W to the filter, calculate

$$b_1 = H_1(W)$$
$$b_2 = H_2(W)$$
$$\cdots$$
$$b_n = H_n(W)$$
$$\forall i, 1 \leq i \leq n, \quad \text{set } B[b_i] = 1 \tag{1}$$

To check if W is in the database, the same b_i are calculated and bits $B[b_i]$ are examined. If any of the bits are 0, the entry does not exists; On the other hand, if all are 1, the record probably does exist. It means that false positives, in which a appears to be in S but actually is not, occur because each location may have also been set by some element other than a.

Eu-Jin Goh firstly quantifies the probability of a false positive occurring in a Bloom Filter [16] . After using n hash functions to insert r distinct elements into an array of size m, the probability of a false positive is $(1 - (1 - (\frac{1}{m}))^{rn})^r \approx (1 - e^{-\frac{rn}{m}})^r$. For a certain number r of inserted elements, there exists a relationship that determines the optimal number of hash functions $h_0 = \frac{m}{r}\ln2 \approx 0.7\frac{m}{r}$ which yields a false positive probability of $p = (\frac{1}{2})^{h_0} = (\frac{1}{2})^{\frac{m}{r}\ln2} \approx 0.62^{\frac{m}{r}}$.

For a Bloom Filter BF, we denote $BF.insert(v)$ the insertion operation and $BF.contains(v)$ the set inclusion test(returning *true* if it contains value v, *false* otherwise).

2.3   Pseudo-Random Generators

Intuitively, a pseudo-random generator outputs strings that are computationally indistinguishable from random string. More precisely, we say that a function $G : \{0,1\}^n \rightarrow \{0,1\}^m$ where $m > n$ is a (t,ε)-*pseudo-random generator* if
1. G is efficiently computable by a deterministic algorithm.
2. For all t time probabilistic algorithms $\mathcal{A}$,

$$|\Pr[A(G(s)) = 0|s \xleftarrow{R} \{0,1\}^n] - \Pr[A(r) = 0|r \xleftarrow{R} \{0,1\}^m]| < \epsilon \tag{2}$$

2.4   Discrete Logarithm Assumption

Let $\mathbb{G}$ be a finite field of size p prime and order q and let g be a generator of $\mathbb{G}$, the Discrete Logarithm assumption(DL):

Definition 1. *Given $g, v \in \mathbb{G}$, it is intractable to find $r \in \mathbb{Z}_q$ such that $v = g^r$ mod p.*

3   Definitions

We begin by defining a general frame work for a general join with preserving privacy on outsourced database. Let X be a data owner, Y be a user, S be a server provider and F be a trusted party. We have described the data owner, the server provider and the user in 1.1. Next, we will explain F. F is usually integrated with client site to process auxiliary operations securely. In reference [5], this role takes responsibility for query translator, meta data storage and result filter. In this paper, F also undertakes the task of storing random numbers.

Definition 2. *A public key storage with join query consists of the following probabilistic polynomial time algorithms and protocols:*

1. *KeyGen(1^s) takes as input a security parameter and outputs a public key PK and a secret key SK.*
2. *Encrypt$_{X,S}(PK, A)$ is a non-interactive two party protocol that allows X to send the messages $A = \{(a_i)_{i=1}^m\}$ (There are m values in the column A)encrypted using the public key PK and generate a auxiliary information Ω_A to be send to the server S. Meanwhile, the random number generated during the encryption is inserted into the list of trusted third party F. We will use Ω_A to support join query where predicate contains A.*
3. *Trapdoor$_{Y,F}(SK, Pre, List)$ is executed by F to take as input a secret key SK, the description of a join predicate Pre and a list of random numbers $List$. It outputs a trapdoor T_{Pre}.*
4. *Join$_{Y,S}(E_A, E_B, T_{Pre}, \Omega_A)$ is a non-interactive two party protocol between the user Y and server S to take a token T_{Pre} for some join predicate, the ciphertext E_A, E_B and Ω_A as input. It outputs true or false. Roughly speaking, if the plaintext A and B satisfy the Pre then algorithm outputs true and outputs false otherwise. The precise requirement is captured in the query correctness property below.*
   Note: Join$_{Y,S}(E_A, E_B, T_{Pre}, \Omega_A)$ is different with Join$_{Y,S}(E_A, E_B, T_{Pre}, \Omega_B)$ despite the predicate is the same because the former uses A's auxiliary information Ω_A to execute join while the latter uses B's auxiliary information Ω_B to execute join.

We now define correctness and security for such a system.

Definition 3. *Let $A_{public}, A_{private} \longleftarrow$ KeyGen(1_s). Fix a finite sequence of messages $\{(a_i)_{i=1}^m\}$ for column A and $\{(b_i)_{j=1}^n\}$ for column B. Suppose that the protocol Encrypt$_{X,S}(PK, A)$ and Encrypt$_{X,S}(PK, B)$ are executed by X and S. After generating T_{Pre} using Trapdoor$_{Y,S}(SK, Pre, List)$ protocol, Y receives the set of message denoted by R_{Pre} after the execution of Join$_{Y,S}(E_A, E_B, T_{Pre}, \Omega_A)$. Then, a privacy-preserving join on outsourced database is said to be correct on the sequence $\{(a_i)_{i=1}^m\}$ and $\{(b_i)_{j=1}^n\}$:*

1. *If $Pre(a_i, b_j) = true$, then $Join(E(A), E(B), T_{Pre}, \Omega_A) = true$.*
2. *If $Pre(a_i, b_j) = false$, then $Pr[Join(E(A), E(B), T_{Pre}, \Omega_A) = \perp] > 1 - \epsilon(s)$ where $\epsilon(s)$ is negligible function, where the probability is taken over all internal randomness used in the protocol Encrypt, Trapdoor and Join.*

A privacy-preserving join on outsourced database is said to be correct if it is correct on all finite sequences.

For privacy, there are several parties involved, and hence there will be several definitional components.

Definition 4. *For sender-storage-privacy, consider the following game between an adversary $\mathcal{A}$ and a challenger $\mathcal{C}$. $\mathcal{A}$ will play the role of outsourced database S and $\mathcal{C}$ will play the role of the message sender X. The game consists of the following steps:*

1. *KeyGen(1^s) is executed by $\mathcal{C}$ who sends the output PK to $\mathcal{A}$.*
2. *$\mathcal{A}$ asks queries of the form (M) where M is a value of a column; $\mathcal{C}$ answers by executing the protocol Encrypt$_{X,S}(PK, M)$ with $\mathcal{A}$.*
3. *$\mathcal{A}$ chooses two pairs(M_0, M_1) and sends this to $\mathcal{C}$, where the messages are of equal size.*
4. *$\mathcal{C}$ picks a random bit $b \in_R \{0, 1\}$ and executes Encrypt$_{X,S}(PK, M_b)$ with $\mathcal{A}$.*
5. *$\mathcal{A}$ outputs a bit $b' \in \{0, 1\}$.*

We define the adversary's advantage as $Adv_{\mathcal{A}}(1^s) = |Pr[b = b'] - \frac{1}{2}|$. We say that a privacy-preserving join on outsourced database is sender-storage-privacy if, for all $\mathcal{A} \in PPT$, we have that $Adv_{\mathcal{A}}(1^s)$ is a negligible function.

Next, we define the controllability, which means the server should not be able to evaluate join predicates on the initial received data without permission from the client. In more detail, without executing the Trapdoor protocol for the join predicate the server can perform join processing by itself, which means out of the control of the client. It may be a leakage of information. We suggest the reader to review the previous example in section 1.1. Since the controllability is a newly concept present in relevant work, we provide a formal description about it:

Definition 5. *For controllability, consider the following game between an adversary $\mathcal{A}$ and a challenger $\mathcal{C}$. Firstly, we define a special case of Join algorithm: Join$_{Y,S}(E_A, E_B, \Box, \Omega_A)$ where $\Box$ denotes the server S does not get the trapdoor from the receiver Y:*

Join$_{Y,S}(E_A, E_B, \Box, \Omega_A)$ is a two party protocol between the user Y and server S for some join predicate, the ciphertext E_A, E_B and Ω_A as input. It outputs true or false.

$\mathcal{A}$ will play the role of outsourced database S and $\mathcal{C}$ will play the role of the receiver Y. The game consists of the following steps:

1. *KeyGen(1^s) is executed by $\mathcal{C}$ who sends the output PK to $\mathcal{A}$.*
2. *$\mathcal{A}$ asks queries of the form Pre, where Pre is a predicate;*
3. *$\mathcal{A}$ chooses two pairs(Pre_0, Pre_1) and sends this to $\mathcal{C}$. We define Pre_0 stands for $Pre_0(a_i, b_j)$=false and Pre_1 stands for $Pre_1(a_i, b_j)$=true where a_i and b_j are values we choose from two columns deliberately.*
4. *$\mathcal{C}$ picks a random bit $b \in_R \{0, 1\}$ for Pre.*
5. *Encrypt$_{X,S}(PK, a_i)$ and Encrypt$_{X,S}(PK, b_j)$ are executed by $\mathcal{C}$ who send the output $E(a_i)$ and $E(b_j)$ to $\mathcal{A}$.*
6. *$\mathcal{C}$ answers by executing the protocol Join$_{Y,S}(E_A, E_B, \Box, \Omega_A)$ with $\mathcal{A}$ for Pre_b.*
7. *$\mathcal{A}$ outputs a decision $b' \in \{0, 1\}$.*

We define the adversary's advantage as $Adv_A(1^s) = |Pr[b = b'] - \frac{1}{2}|$. We say that a privacy-preserving join on outsourced database is controllability if, for all $A \in PPT$, we have that $Adv_A(1^s)$ is a negligible function.

Note: The reader may ask the question: why do not consider the results of $Pre_0(a_i, b_j)$ and $Pre_1(a_i, b_j)$ are the same, which is either true or false. The reason is that if we guarantee the server can not distinguish a pair of predicates where the results of the same data item are different, the distinction of a pair of predicates where the results of the same data item are same is a more difficult task.

Next, we define security for a join query in the sense of semantic-security. We need to ensure that the attacker does not distinguish between a predicate containing column A_0 and B_0 with a predicate containing column A_1 and B_1 from Join protocol even given the trapdoor T_{Pre}.

Definition 6. *For join privacy, consider the following game between an adversary A and a challenger C. A will play the role of outsourced database S and C will play the role of the message receiver Y. The game consists of the following steps:*

1. *KeyGen(1^s) is executed by C who sends the output PK to A.*
2. *Suppose that Pre_0 holds on the column A_0 and B_0; Pre_1 holds on the column A_1 and B_1; A_0, B_0, A_1 and B_1 are of equal size. The attacker A sends the challenger Pre_0, Pre_1 on which it wishes to be challenged. The challenger picks a random $b \in \{0,1\}$ and executes the protocol Join$_{Y,S}(E_{A_b}, E_{B_b}, T_{Pre_b}, \Omega_{A_b})$.*
3. *A output a bit $b' \in \{0,1\}$.*

We define the adversary's advantage as $Adv_A(1^s) = |Pr[b = b'] - \frac{1}{2}|$. We say that a privacy-preserving join on outsourced database is an adaptive chosen predicate attack if for all $A \in PPT$, we have that $Adv_A(1^s)$ is a negligible function.

4   Main Construction

4.1   Outsourced Equijoin

We define the equijoin solution to be the following probabilistic polynomial time algorithms:

1. Setup(s).
    (a) A probabilistic algorithm that takes as input a security parameter $s \in \mathbb{Z}^+$, run $\mathcal{G}(s)$ to obtain the public key PK=$(n, \mathbb{G}, \mathbb{G}_1, e, g, h)$ and the private key SK=p.
    (b) Choose a pseudo-random function $f: \{0,1\}^n \times \{0,1\}^s \to \{0,1\}^s$ to generate a value r_A for each column A and a value r_B for each column B in database D and to generate ℓ values $\{r_1, r_2, \ldots, r_\ell\}$ for a BF viewed as a string of m bits. Keep all values secret.

2. Encrypt(PK, A)
   (a) For each $a_i \in A$, encrypt the plaintext a_i using the public key PK, producing $E_{PK}(a_i) = g^{a_i} h^{r_A} \in \mathbb{G}$.
   (b) Initiate all the bits of a Bloom Filter BF to 0. For each value $r_i (i \in (1, \ldots, \ell))$ in $\{r_1, r_2, \ldots, r_\ell\}$, compute $\rho(r_i) = h^{r_i}$ and insert them into the Bloom Filter (BF.(insert($\rho(r_i)$))))
   (c) Finally, output the values $E_{PK}(a_i)$ for each element of all columns in D and a Bloom Filter BF for column A, which can be viewed as Ω_A.
3. Trapdoor(SK, Pre)
   Given the secret key SK and the description of a predicate Pre for two column name A,B and a list of random numbers hold by F, output a trapdoor for Pre as $T_{Pre_{AB}} = h^{(r_B - r_A + r_\pi)} \bmod n (\pi \in \{1, \ldots, \ell\})$.
4. Join(A, B, T_{Pre}, BF)
   For each element $b_j \in B$, compute

$$\rho_A(a_i, b_j) = \frac{E_{PK}(a_i)}{E_{PK}(b_j)} \times T_{Pre_{AB}} \tag{3}$$

For each element $a_i \in A$, iff $BF.contains(\rho_A(a_i, b_j))$ return the tuple $\langle E_{PK}(a_i), E_{PK}(b_j) \rangle$.

Theorem 2. *The privacy-preserving join on outsourced database from the preceding construction is correct according to definition 3.*

Proof. If $Pre(a_i, b_j) = true$, it means $a_i = b_j$. The correctness of the protocol is easy to verify:

$$\frac{E_{PK}(a_i)}{E_{PK}(b_j)} \times T_{Pre_{AB}} = \frac{g^{a_i} h^{r_A}}{g^{b_j} h^{r_B}} \times h^{r_B - r_A + r_\pi} = g^{a_i - b_j} h^{r_\pi} = h^{r_\pi} \tag{4}$$

h^{r_π} is a member of the set for the BF, so $BF.contains(h^{r_\pi})$ returns true. The client can obtain $\langle E_{PK}(a_i), E_{PK}(b_j) \rangle$ as a part of query results.

If $Pre(a_i, b_j) = false$, it means $a_i \neq b_j$. Suppose that $g = u^\omega$

$$\frac{E_{PK}(a_i)}{E_{PK}(b_j)} \times T_{Pre_{AB}} = g^{a_i - b_j} h^{r_\pi} = (u^\omega)^{a_i - b_j} h^{r_\pi} = h^{\frac{\omega(a_i - b_j)}{q}} h^{r_\pi} = h^{\frac{\omega(a_i - b_j)}{q} + r_\pi} \tag{5}$$

In the case of a join, the false positive rate of query implies that a small percentage of the resulting joined tuples do not match the predicate the join has been executed for. It means that $h^{\frac{\omega(a_i - b_j)}{q} + r_i}$ is considered as a member of the set for the BF and $BF.contains(h^{\frac{\omega(a_i - b_j)}{q} + r_\pi})$ returns true. The occurrence can be divided into two cases:

1. The false positive is caused by Bloom Filters. we have discussed it in 2.2, which is $(\frac{1}{2})^{\frac{\phi}{\ell} \ln 2}$ with respect to a optimal number of hash functions used in the BF, denoted as neg_1. For example, suppose $\phi = 1024$ and $\ell = 100$, the minimal false positive rate is 0.7%. In the case of a join, a small percentage of the resulting joined tuples will then be pruned by the client, so the minimal false positive rate is enough for a practical application.

2. The false positive is caused by executing equation 3 . $h^{\frac{\omega(a_i-b_j)}{q}+r_\pi}\ mod\ n$ is a real member of the set for the BF ($\varpi \in \{1, \ldots, \ell\}$):

$$h^{\frac{\omega(a_i-b_j)}{q}+r_\pi} \equiv h^{r_\varpi}\ mod\ n \qquad (6)$$

$$\frac{\omega(a_i-b_j)}{q} \equiv r_\varpi - r_\pi\ mod\ (p-1)(q-1) \qquad (7)$$

We firstly observe the left side of equation 7. ω and q are constant numbers after Setup phrase. $a_i - b_j$ is the difference of the two compared values. After the database physical structure is defined, the difference of any two values is contained in a finite domain. For example, suppose that the schema of a table *student* is *(ID char(6), name vchar(10), age int)* and the schema of a table *transcript* is *(ID char(6), course vchar(20), grade int)*, we want to perform natural joining between *student* and *transcript*. According to equation 7, the difference of any two values coming from *student.ID* and *transcript.ID* is between [-999999,999999]. Next, we observe the right side of equation 7. $r_\varpi - r_\pi$ is a random number whose domain $\mathbb{Z}$ is infinite. So, the probability of equation 7 is negligible, denoted as neg_2.

Therefore, we can conclude that if $Pre(a_i, b_j) = false$, $Pr[Join(A, B, T_{Pre}, BF) = \bot] = (1 - neg_1)(1 - neg_2) = 1 - (neg_1 + neg_2 - neg_1 neg_2)$. Since neg_1 and neg_2 are both negligible, $Pr[Join(A, B, T_{Pre}, BF) = \bot] > 1 - \epsilon(s)$ where $\epsilon(s)$ is negligible function.

4.2 Proof of Security

Theorem 3. *Assuming semantic-security of the underlying cryptosystem, the privacy-preserving join on outsourced database from the above construction is sender-storage-privacy, according to definition 4.*

Proof. Suppose that there exists an adversary $\mathcal{A} \in$ PPT that can succeed in breaking the security game, from definition 4, with some non-negligible advantage. So, under those conditions, $\mathcal{A}$ can distinguish the distribution of $\mathsf{Encrypt}_{X,S}$ (PK, M_0) from the distribution of $\mathsf{Encrypt}_{X,S}(PK, M_1)$, where the word "distribution" refers to the distribution of the transcript of the interaction between the parties. A transcript of $\mathsf{Encrypt}_{X,S}(PK, M)$ essentially consists of just $E_{PK}(M)$ and a Bloom Filter BF_M for M. We assume that there exists an adversary $\mathcal{A}$ that can distinguish these two distributions. Hence, the encrypted data or the Bloom Filter cannot be computationally indistinguishable. So, there exists an adversary $\mathcal{A}' \in$ PPT that can distinguish between $E_{PK}(M)$, BF_M or the correlation between $E_{PK}(M)$ and BF_M .

1. If $\mathcal{A}' \in$ PPT distinguishes the encrypted column M_0 and M_1, it has distinguished $E_{PK}(M_0)$ from $E_{PK}(M_1)$ which violates our assumption of theorem 1.

2. If $\mathcal{A}' \in$ PPT distinguishes the Bloom Filter, it has distinguished BF_{M_0} from BF_{M_1}. We observe that the construction of BF consists of a serial of random numbers. Although the worst case exists that the hash functions of BF are the same for M_0 and M_1, the BF_{M_0} and BF_{M_1} are indistinguishable due to (t, ε)-*pseudo-random generator* in section 2.3 .

3. If $\mathcal{A}' \in$ PPT distinguishes the correlation between $E_{PK}(M)$ and BF_M. Obviously, there is no correlation between $E_{PK}(M)$ and BF_M because the construction of BF_M for M only consists of random numbers without being related to M

So we conclude that no such $\mathcal{A}$ exists in the first place, and hence the system is sender-storage-privacy according to definition 4.

Theorem 4. *The privacy-preserving join on outsourced database is controllability, according to definition 5.*

Proof. Suppose that there exists an adversary $\mathcal{A} \in$ PPT that can succeed in breaking the security game, from definition 5, with some non-negligible advantage. So, under those conditions, $\mathcal{A}$ can correctly decide whether the equality of a_i and b_j with non-negligible advantage without the trapdoor for the join predicate of the column A and B. We suppose that $\mathcal{A}$ forges a trapdoor, denoted as T'_{Pre}, used by $\mathcal{C}$. Next, there must exists an adversary $\mathcal{A}' \in$ PPT that can make the right decision with non-negligible advantage by executing Join protocol using $T_{Pre'}$. Suppose that $T'_{Pre_{AB}} = h^{r_x}$ and r_x is a random number.

$$\rho'_A(a_i, b_j) = \frac{E_{PK}(a_i)}{E_{PK}(b_j)} \times T_{Pre'_{AB}} = \frac{g^{a_i} h^{r_A}}{g^{b_j} h^{r_B}} \times h^{r_x} = g^{a_i - b_j} h^{r_A - r_B + r_x} \qquad (8)$$

If the $Pre(a_i, b_j) = true$, $\rho'_A(a_i, b_j) = h^{r_A - r_B + r_x}$. Iff $BF.contains(\rho'_A(a_i, b_j))$ return the tuple $\langle E_{PK}(a_i), E_{PK}(b_j) \rangle$. It occurs owing to two cases:

1. $h^{r_A - r_B + r_x}$ is a member of the set for BF_A, denoted by h^{r_π}.

$$h^{r_A - r_B + r_x} \equiv h^{r_\pi} \ mod \ n \qquad (9)$$

$$r_x \equiv r_\pi - r_A + r_B \ mod \ (p-1)(q-1) \qquad (10)$$

Because r_π, r_A and r_B are all random numbers hidden from $\mathcal{A}$, $Adv_{\mathcal{A}'} = Pr[r_x \equiv r_\pi - r_A + r_B]$ is negligible.

2. $BF.contains(\rho'_A(a_i, b_j))$ although $\rho'_A(a_i, b_j)$ is not a member of the set for BF_A. We have discussed the controllable rate of false positive for set inclusion test. So we can choose optimal parameters to obtain a desired small false positive. It means that $BF.contains(\rho'_A(a_i, b_j))$ is negligible.

So we conclude that no such $\mathcal{A}$ exists in the first place, and hence the system is controllability according to definition 5.

Theorem 5. *The privacy-preserving join on outsourced database is join privacy, according to definition 6.*

Proof. Suppose a polynomial time algorithm $\mathcal{A}$ breaks the join security of the system. We construct an algorithm $\mathcal{B}$ that breaks the discrete logarithm assumption. Given $(n, \mathbb{G}, \mathbb{G}_1, e, h)$ as input, algorithm $\mathcal{B}$ works as follows:

1. $\mathcal{B}$ gives algorithm $\mathcal{A}$ the public key $(n, \mathbb{G}, \mathbb{G}_1, e, g, h)$.
2. Algorithm $\mathcal{A}$ outputs two join predicates Pre_0 and Pre_1 to which $\mathcal{B}$ responds with the trapdoor $T_{Pre} = h^{r_{B_b} - r_{A_b} + r_{\pi}}$ for a random $b \xleftarrow{R} \{0, 1\}$ and $v = h^{r_{\pi}}$ is a member of the set of BF. Suppose that Pre_0 holds on the column A_0 and B_0; Pre_1 holds on the column A_1 and B_1; A_0, B_0, A_1 and B_1 are of equal size.
3. Algorithm $\mathcal{A}$ outputs its guess $b' \in \{0, 1\}$ for b. If $b = b'$ algorithm $\mathcal{B}$ outputs 1, which means to find $x \in n$ such that $v = h^x$; otherwise, $\mathcal{B}$ outputs 0.

   Suppose algorithm $\mathcal{B}$'s success probability is ϵ, algorithm $\mathcal{A}$'s success probability is ϵ'. We define algorithm $\mathcal{A}$'s success means $\mathcal{A}$ gets the knowledge of r_A and r_B since r_A and r_B is hidden from the third party F. It is easy to see that $Pr[b = b'] = \frac{\epsilon}{n^2}$ when $Pr[r'_{B_b} = r_{B_b}] = \frac{1}{n}$ and $Pr[r'_{A_b} = r_{A_b}] = \frac{1}{n}$. On the other hand, $Pr[b = b'] = \epsilon' \times \frac{\frac{l}{n}}{1-\alpha}$ where α is the false positive of BF. So we derive the following equation:

$$\frac{\epsilon}{n^2} = \epsilon' \times \frac{\frac{l}{n}}{1 - \alpha} \tag{11}$$

$$\epsilon = \frac{l n \epsilon'}{1 - \alpha} \tag{12}$$

Hence, $\mathcal{B}$ breaks the discrete logarithm assumption with the advantage $\frac{l n \epsilon'}{1-\alpha}$. Since it's well known that the advantage of breaking discrete logarithm assumption is negligible, the advantage of breaking join privacy of our system is negligible.

So we conclude that no such $\mathcal{A}$ exists in the first place, and hence the system is join privacy according to definition 6.

4.3   Discussion

We compare this construction with the method proposed by Bogdan Carbunar et al. [4]. Both solutions have their own advantages and disadvantages.

Bogdan Carbunar et al. Bogdan Carbunar et al. encrypt each column of a relational database. For each element a_i of a column, encrypt a_i with any encryption algorithm, compute an obfuscation $O(a_i)$ of a_i and generate a Bloom Filter containing all possible values associate with a_i satisfying the predicate. Hence, element $a_i \in A$ is stored on the server as $[E_K(a_i), O(a_i), BF(a_i)]$. For each element $b_j \in B$, compute $e_A(b_j) = r_{AB}^{O(b_j)} \bmod p$. For each element $a_i \in A$, iff $BF(a_i)$ contains $e_A(b_j)$ return the tuple $< E_K(a_i), E_K(b_j) >$.

1. Advantages
   (a) Each element of a column can be encrypted by any encryption algorithm, for instance, RSA, ElGamal. Essentially, join is executed through an obfuscation version of a_i and BF, which are similar as a index.
   (b) The encryption method supports both equijoin and non-equijoin, for instance, $|a_i - b_j| < 50$.
2. Disadvantages
   (a) The computational complexity of join operation is modular exponentiation. When the size of database is large, the execution of join is time-consuming.
   (b) The scale of valid payload is small because an obfuscation version and a BF are produced for each elements of a column, for example, we deploy 1024-bit Bloom Filter, 1024-modular using 1024-bit RSA encryption. The scale of valid payload is 33.3%.
   (c) Because executing the same query repeatedly will generate the same trapdoor, Bogdan Carbunar's method may leak the user's query information.

Our Method

1. Advantages
   (a) The computational complexity of equijoin is modular multiplication. Compared to modular exponentiation, the performance of our method is fast.
   (b) The scale of valid payload is large. Besides encrypted data itself, the server only stores a BF for each column but not for each elements of each column.
   (c) Since executing the same query repeatedly will generate different trapdoors, our method provides more secure query service.
2. Disadvantages
   (a) Our method only supports equijoin, which is only a special class of join type.
   (b) Our method is based on the Boneh-Goh-Nissim encryption system. So the security is only semantically secure since Boneh-Goh-Nissim encryption system is semantically secure assuming the subgroup decision assumption.
   (c) The trusted front F needs to reserve some storage space for storing random numbers.

5   Extension

In this paper, we focus on a special case of equivalent join predicate $A = B$ since this type of predicate is frequently used in database. Actually, our construction is also applied to more general equivalent join, for instance, allowing the evaluation of predicate $p(A, B) : (xA = yB + z)$. Next, we will explain it simply to modify the protocol of Trapdoor and Join in the previous construction.

3. Trapdoor(SK, Pre)

   Given the secret key SK and the description of a predicate Pre for two column name A,B, output a trapdoor for Pre as $T_{Pre_{AB}} = h^{(yr_B+r_z-xr_A+r_\pi)} \mod n$ ($\pi \in \{1, \dots, \ell\}$).

4. Join(A, B, T_{Pre}, BF) For each element $b_j \in B$, compute

$$\rho_A(xa_i, yb_j, z) = \frac{[E_{PK}(a_i)]^x}{[E_{PK}(b_j)]^y \cdot E_{PK}(z)} \times T_{Pre_{AB}} \qquad (13)$$

For each element $a_i \in A$,iff $BF.contains(\rho_A(xa_i, yb_j, z))$ return the tuple $\langle E_{PK}(a_i), E_{PK}(b_j) \rangle$.

If $Pre(a_i, b_j) = true$, it means that $xa_i = yb_j + z$. The correctness of the modified protocol is easy to verify:

$$\frac{E_{PK}(xa_i)}{E_{PK}(yb_j) \cdot E_(PK)(z)} \times T_{Pre_{AB}} = \frac{[E_{PK}(a_i)]^x}{[E_{PK}(b_j)]^y \cdot E_{PK}(z)} \times T_{Pre_{AB}}$$

$$= \frac{g^{xa_i} h^{xr_A}}{g^{yb_j} h^{yr_B} \cdot g^z h^{r_z}} \times h^{yr_B+r_z-xr_A+r_\pi}$$

$$= g^{xa_i-yb_j-z} h^{r_\pi} = h^{r_\pi} \qquad (14)$$

h^{r_π} is a member of the set for the BF, so $BF.contains(h^{r_\pi})$ returns true. The client can obtain $\langle E_{PK}(a_i), E_{PK}(b_j) \rangle$ as a part of query results.

Note: Our protocol is "asymmetric". "asymmetric" means that the evaluation of predicate relies on the meta data of one of the two columns, for example, the Bloom Filter for A, not for B. It's not difficult to propose a "symmetry" protocol based on the above construction. However, in database realization, the "asymmetric" style usually appears, for example, a nested loop join method chooses one of two tables as a driving table, so our construction is easier to apply to this application.

6 Conclusion

In this paper, we propose a simple privacy-preserving join on outsourced database. We focus on computing equijoin including the predicate form of $A = B$ and $xA = yB + z$. The main goal is to perform privacy-preserving join on outsourced database server with the permission of users. To the best of our knowledge, there exists few solutions for private joins on encrypted data in this scenarios. Proposed mechanisms for executing join on outsourced database is computational privacy and low overhead. We provide security analysis that our method achieves privacy for the data owner, controllability and join privacy. We also compare it with reference [4], describing their advantages and disadvantages respectively.

References

1. Lehner, W., Sattler, K.U.: Database as a Service (DBaaS). In: 2010 IEEE 26th International Conference on Data Engineering (ICDE 2010), March 1-6, pp. 1216–1217. IEEE, Piscataway (2010)
2. Agrawal, D., El Abbadi, A., Emekci, F., Metwally, A.: Database management as a service: challenges and opportunities. In: IEEE 25th International Conference on Data Engineering, ICDE 2009, USA, March 29-April 2. IEEE, Piscataway (2009)
3. Freedman, M.J., Nissim, K., Pinkas, B.: Efficient private matching and set intersection, pp. 1–19. Springer, Heidelberg (2004)
4. Carbunar, B., Sion, R.: Joining Privately on Outsourced Data. In: Jonker, W., Petković, M. (eds.) SDM 2010. LNCS, vol. 6358, pp. 70–86. Springer, Heidelberg (2010)
5. Hacigumus, H., Iyer, B., Li, C., Mehrotra, S.: Executing SQL over encrypted data in the database-service-provider model. In: ACM SIGMOD 2002 Proceedings of the ACM SIGMOD International Conference on Managment of Data, Jun 3-6, pp. 216–227. Association for Computing Machinery, Madison (2002)
6. Hacıgümüş, H., Iyer, B., Mehrotra, S.: Efficient Execution of Aggregation Queries over Encrypted Relational Databases. In: Lee, Y., Li, J., Whang, K.-Y., Lee, D. (eds.) DASFAA 2004. LNCS, vol. 2973, pp. 125–136. Springer, Heidelberg (2004)
7. Boneh, D., Di Crescenzo, G., Ostrovsky, R., Persiano, G.: Public key encryption with keyword search. In: Cachin, C., Camenisch, J.L. (eds.) EUROCRYPT 2004. LNCS, vol. 3027, pp. 506–522. Springer, Heidelberg (2004)
8. Song, D.X., Wagner, D., Perrig, A.: Practical techniques for searches on encrypted data. In: 2000 IEEE Symposium on Security and Privacy, pp. 44–55. IEEE, Berkeley (2000)
9. Agrawal, R., Kiernan, J., Srikant, R., Xu, Y.: Order preserving encryption for numeric data. In: Proceedings of the ACM SIGMOD International Conference on Management of Data, pp. 563–574. Association for Computing Machinery, Paris (2004)
10. Boldyreva, A., Chenette, N., Lee, Y., O'Neill, A.: Order-preserving symmetric encryption. In: Joux, A. (ed.) EUROCRYPT 2009. LNCS, vol. 5479, pp. 224–241. Springer, Heidelberg (2009)
11. Boneh, D., Waters, B.: Conjunctive, subset, and range queries on encrypted data. In: Vadhan, S.P. (ed.) TCC 2007. LNCS, vol. 4392, pp. 535–554. Springer, Heidelberg (2007)
12. Cachin, C., Micali, S., Stadler, M.: Computationally private information retrieval with polylogarithmic communication. In: Stern, J. (ed.) EUROCRYPT 1999. LNCS, vol. 1592, pp. 402–414. Springer, Heidelberg (1999)
13. Chor, B., Goldreich, O., Kushilevitz, E., Sudan, M.: Private information retrieval. In: Proceedings of the 36th Annual Symposium on Foundations of Computer Science, pp. 41–50. IEEE Computer Society, Milwaukee (1995)
14. Boneh, D., Goh, E.-J., Nissim, K.: Evaluating 2-DNF formulas on ciphertexts. In: Kilian, J. (ed.) TCC 2005. LNCS, vol. 3378, pp. 325–341. Springer, Heidelberg (2005)
15. Bloom, B.H.: Space/time trade-offs in hash coding with allowable errors. Communications of the ACM 13(7), 422–426 (1970)
16. Goh, E.J.: Secure indexes. Cryptology ePrint Archive (2003), http://eprint.iacr.org/2003/216/

APPA: Aggregate Privacy-Preserving Authentication in Vehicular Ad Hoc Networks

Lei Zhang[1], Qianhong Wu[2,3], Bo Qin[2,4], and Josep Domingo-Ferrer[2]

[1] Software Engineering Institute, East China Normal University, Shanghai, China
leizhang@sei.ecnu.edu.cn
[2] UNESCO Chair in Data Privacy, Dept. of Comp. Eng. and Maths
Universitat Rovira i Virgili, Tarragona, Catalonia
{qianhong.wu,bo.qin,josep.domingo}@urv.cat
[3] Key Lab. of Aerospace Information Security and Trusted Computing
Ministry of Education, School of Computer, Wuhan University, China
[4] Dept. of Maths, School of Science, Xi'an University of Technology, China

Abstract. Most security- and privacy-preserving protocols in vehicular *ad hoc* networks (VANETs) heavily rely on time-consuming cryptographic operations which produce a huge volume of cryptographic data. These data are usually employed for many kinds of decisions, which poses the challenge of processing the received cryptographic data fast enough to avoid unaffordable reaction delay. To meet that challenge, we propose a vehicular authentication protocol referred to as APPA. It guarantees trustworthiness of vehicular communications and privacy of vehicles, and enables vehicles to react to vehicular reports containing cryptographic data within a very short delay. Moreover, using our protocol, the seemingly random cryptographic data can be securely and substantially compressed so that the storage space of a vehicle can be greatly saved. Finally, our protocol does not heavily rely on roadside units (RSUs) and it can work to some extent even if the VANET infrastructure is incomplete. These features distinguish our proposal from others and make it attractive in various secure VANET scenarios.

Keywords: Traffic Security, VANETs, Privacy, Protocol Design, Data Compression.

1 Introduction

With the fast development of mobile networks and information processing technologies, vehicular *ad hoc* networks (VANETs) have attracted in recent years particular attention in both industry and academia. A VANET mainly consists of vehicles and properly distributed roadside units (RSUs), both equipped with on-board sensory, processing, and wireless communication modules. The DSRC standard [1] is suggested to support wireless communication in VANETs. In addition to messages routed to/from RSUs, vehicles can also broadcast safety messages concerning accidents, dangerous road conditions, sudden braking, lane changing, etc., in a one-hop or multi-hop fashion. With these mechanisms, VANETs are

X. Lai, J. Zhou, and H. Li (Eds.): ISC 2011, LNCS 7001, pp. 293–308, 2011.
© Springer-Verlag Berlin Heidelberg 2011

expected to enhance the driving experience by improving road safety and traffic efficiency, as well as supporting value-added applications such as automatic toll collection, infotainment, context-oriented personalized services and so on.

Despite the great potential benefits of VANETs, many challenges arise around this technology, especially in what regards security and privacy. VANETs aim at a safer driving environment by allowing vehicle-to-vehicle (V2V) and vehicle-to-RSU (V2R) communications. However, selfish vehicles can also exploit this mechanism to send fraudulent messages for their own profit. Malicious vehicles may impersonate innocent ones to launch attacks without being caught. To address the security requirements in VANETs, digital signatures are usually employed so that the receiving vehicles can verify that these messages have been originated by authentic sources and have not been modified during transmission. Driving privacy is another critical concern in VANETs. Although employing signatures can mitigate the security challenge in VANETs, the presence of signatures in vehicle-generated messages might allow attackers to identify who generated those messages. Since vehicular messages contain speed, location, direction, time and other driving information, a lot of private information about the driver can be inferred, which jeopardizes vehicle and driver privacy.

To address security and privacy concerns, a large body of proposals have striven to secure VANETs (e.g., [5,15,17,18,23,25]). Most proposals heavily rely on time-consuming cryptographic operations which produce a huge volume of cryptographic data. These data are usually employed for many kinds of decisions. Indeed, according to DSRC [1], a vehicle will broadcast a (signed) message to nearby vehicles and/or RSUs every few hundreds of milliseconds. Hence, a vehicle or an RSU may receive hundreds of messages in a short period of time, all of those should be verified in real time. Otherwise, the delay caused by verification of a bulk of signatures may radically impair transmission throughput and system scalability. This poses the challenge of processing the received cryptographic data fast enough to avoid an unaffordable reaction delay, which might result in traffic jams and even accidents.

Furthermore, if some signed messages are later found to be false and have misguided other vehicles into accidents, the message generators and endorsers should be traceable. This implies that the signed vehicle-generated messages have to be stored by the receiving vehicles. However, vehicular messages, especially their appended cryptographic data (being almost random), grow linearly with time. Hence, It's preferable if the vehicular communication protocol can securely compress those cryptographic data.

1.1 Our Results

We propose a security- and privacy-preserving protocol referred to as APPA. The APPA protocol is built on our new notion of one-time identity-based aggregate signature (OTIBAS). This notion incorporates the desirable properties of identity-based cryptography, aggregate signature and one-time signature. In an OTIBAS scheme, a user can compute a signature on a message only if the user has obtained a secret key from a trusted authority, where the secret key is associated with the user's identity. The signature can be verified by anyone who knows the user's identity. Since the user's identity works as her public key, no certificate is needed on the public key, unlike in the conventional PKI setting, which avoids the certificate management overhead. As a one-time signature, the signer cannot use one secret key (corresponding to one identity of the user) to generate no more than one signature, and accordingly, the user's identity should change for every signature and can be viewed as a one-time pseudonym. This feature is well matched to the anonymity requirement of vehicles in VANETs. As an aggregate signature scheme, the signatures on n messages by n signers can be aggregated into one signature which can be verified as if it had been generated by a single signer. This feature caters for the requirements of fast response in VANETs and can be used to save storage space in a vehicle. By exploiting bilinear pairings, we propose an efficient OTIBAS scheme. The security of the proposed OTIBAS scheme is formally proven under the computational co-Diffie-Hellman (co-CDH) assumption. With our provably secure OTIBAS scheme and the *multiplicative secret sharing* technique introduced by Kiltz and Pietrzak [6], we realize the APPA protocol which is efficient and practical to secure vehicular communications.

The APPA protocol exhibits a number of attractive features. The proposal preserves privacy for honest vehicles. However, if a vehicle authenticates a bogus message, then it will be caught for penalty. This mechanism guarantees trustworthiness of vehicular communications and it is a deterrent to malicious vehicles. Furthermore, our protocol enables vehicles to react to vehicular messages within a very short delay and the seemingly random cryptographic data can be securely and substantially compressed: any number of signatures can be compressed into a single group element of about 21 bytes without degrading security. Finally, our protocol does not heavily rely on RSUs, a part of the VANET infrastructure, and it can work to some extent even if the VANET infrastructure is incomplete, *i.e.*, still under construction or partially destroyed by, *e.g.*, an earthquake. These features facilitate deployment in various environments to secure VANETs.

1.2 Related Work

A large number of papers deal with the security and privacy challenges in VANETs [5,15,17,18,23,25]. Among them, pseudonym-based authentication is a popular research line. In [3,17,18], digital signatures for message integrity, authentication and non-repudiation are combined with short-lived anonymous certificates to guarantee the security requirements for the vehicular nodes. However, in this approach, each vehicle needs to pre-load a huge pool of anonymous certificates to achieve privacy of vehicles, and the trusted authority also needs to maintain all the anonymous certificates of all the vehicles, which incurs a heavy burden of certificate management.

To relieve from the certificate management burden, several proposals [12,13] exploit group signatures [4] and present conditional privacy-preserving vehicular authentication protocols. Subsequent efforts have been made to improve the trustworthiness of vehicle-generated messages [23] or to achieve robustness and scalability in VANETs [27]. However, due to the difficulty of dealing with revoked and compromised signers in group signatures (an open question in cryptography), these systems may degrade in performance as the number of revoked/compromised vehicles grows with time.

Another approach to avoid certificate management is to exploit identity-based cryptography [20]. In [25], Zhang *et al.* designed an efficient conditional privacy-preserving protocol for vehicular communications using identity-based cryptography. Their protocol requires the master secret key of the system to be stored in an *idealized* tamper-proof device embedded into vehicles. This device is assumed secure against any attempt of compromise in any circumstance and an attacker cannot extract any data stored in the device. This assumption seems too strong to be met in practice. For instance, without probing into the device, an attacker might collect some side-channel information leaked by cryptographic operations in the device. This kind of attacks are known as side-channel attacks [11,21] and they seem attractive for organized criminals: if the tamper-proof device is compromised, the criminals obtain full control of the system.

Compared to the intensive attention received by security and privacy issues, few efforts have been made to aggregate vehicle-generated messages and cryptographic witnesses to save response time and storage space for vehicles. Picconi *et al.* proposed a PKI-based authentication scheme in VANETs [16]. Their scheme focuses on aggregating messages, rather than aggregating cryptographic witnesses; aggregating the latter is more challenging, because they are almost random. As noted by themselves, their solution suffers from some limitations. In [28], Zhu *et al.* apply short signatures in the PKI setting to aggregate *emergency messages* in VANETs. More recently, Wasef *et al.* [22] employed aggregation technologies to enable each vehicle to simultaneously verify signatures and their certificates in the PKI scenario. Since these schemes are implemented in the PKI setting, certificate management remains a problem. Furthermore, although some general privacy ideas have been discussed in these proposals, it is unclear how they can technically achieve reasonable vehicle privacy.

1.3   Paper Organization

The rest of the paper is organized in the following way. In Section 2, we describe the system architecture and the goals of our design. Section 3 proposes an identity-based aggregate one-time signature scheme as the building block. Section 4 proposes our APPA protocol. We evaluate the new protocol in Section 5. Finally, Section 6 concludes the paper.

2   System Architecture and Design Goals

2.1   System Architecture

As shown in Figure 1, the system architecture in our proposal consists of a trusted authority (TA), a number of RSUs and numerous vehicles.

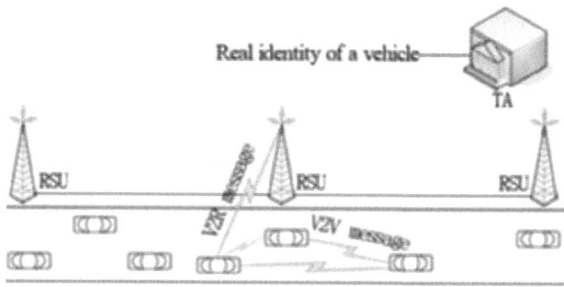

Fig. 1. System architecture

- TA: It generates the system parameters. When an authenticated message is found false, TA might be asked to trace the vehicle having generated that cheating message.
- RSUs: They are distributed along the road side as part of the infrastructure of VANETs. RSUs are equipped with on-board sensory, processing, and wireless communication modules. In our system, they are mainly used to collect/forward data from vehicles or distribute global broadcast from TA to vehicles. We enforce minimum security-related operations on RSUs so that the system can work even if the infrastructure is incomplete at the early stage of VANETs or it is destroyed by some disaster.
- Vehicles: They are equipped with on-board sensory, processing, tamper-proof devices and wireless communication modules. Vehicles move along the roads, and periodically exchange messages with nearby vehicles and RSUs within their communication range.

2.2 Design Goals

Our APPA protocol has the following design goals.

- **Global security.** The vehicle-generated messages should be authenticated to guarantee that they are from real sources and have not been tampered with during transmission. If a vehicle authenticates a bogus message, it must be caught and it must be unable to deny authorship on the cheating message.
- **Individual privacy.** If a vehicle behaves honestly and follows the APPA protocol, its privacy should be guaranteed against attackers who can eavesdrop communication in VANETs.
- **Easy deployment.** The protocol should allow vehicles to quickly react to received messages. It is preferable if there is no heavy overhead incurred by certificate management and the storage requirement of digital signatures is minimized. The protocol should work to some extent in case that the infrastructure is incomplete.

3   The Building Block: OTIBAS

3.1   Modeling OTIBAS

An OTIBAS scheme consists of the following efficient algorithms: Setup, Extract, Sign, Aggregate, and Verify. Setup takes as input a security parameter and outputs the global system parameters including TA's public key. Extract takes as input TA's master secret key and a signer's identity, and outputs a private key for the signer. Sign takes as input a signer's private key and any message, and outputs a signature on the message. The constraint here is that a private key corresponding to a specific identity can be used to generate only one signature. Aggregate takes as input n message-signature pairs generated in the Sign procedure, and outputs an aggregate signature. Verify takes as input the n messages, the aggregate signature, the n identities corresponding to the n message-signature pairs and TA's public key, and it outputs a bit 1 or 0 to represent whether the original n message-signature pairs are valid or not.

   An OTIBAS scheme should be correct in the sense that, if each party honestly follows the scheme, then Verify always outputs 1. An OTIBAS scheme should be also secure. Informally (a more formal definition can be found in Section 3.3), an OTIBAS scheme is said to be secure if any polynomial-time attacker not requesting a private key corresponding to an identity ID^* cannot forge a signature corresponding to ID^* that is aggregated so that Verify outputs 1.

3.2   An OTIBAS Scheme

Our OTIBAS scheme is implemented in bilinear groups [7,14]. Bilinear groups have been widely employed to build versatile cryptosystems [8,24,26].

   Let $\mathbb{G}_1, \mathbb{G}_2$ be two cyclic groups of prime order q and $\mathbb{G}_T$ be a multiplicative cyclic group of the same order. Let g_1 denote a generator of $\mathbb{G}_1$, g_2 a generator of $\mathbb{G}_2$, ψ a computable isomorphism from $\mathbb{G}_2$ to $\mathbb{G}_1$, with $\psi(g_2) = g_1$. A map $\hat{e} : \mathbb{G}_1 \times \mathbb{G}_2 \to \mathbb{G}_T$ is called a bilinear map if $\hat{e}(g_1, g_2) \neq 1$ and $\hat{e}(g_1^\alpha, g_2^\beta) = \hat{e}(g_1, g_2)^{\alpha\beta}$ for all $\alpha, \beta \in Z_q^*$. By exploiting bilinear groups, what follows implements our OTIBAS scheme.

Setup: TA runs this algorithm to generate the system parameters as follows:

1. Choose $q, \mathbb{G}_1, \mathbb{G}_2, \mathbb{G}_T, g_1, g_2, \psi, \hat{e}$.
2. Pick $\kappa \in Z_q^*$ as its master secret key, and compute $y = g_2^\kappa$ as its master public key.
3. Select cryptographic hash functions $H_0(\cdot) : \{0,1\}^* \to \mathbb{G}_1$ and $H_1(\cdot) : \{0,1\}^* \to \mathbb{Z}_q^*$.
4. Publish the system parameter $\Psi = (\hat{e}, q, \mathbb{G}_1, \mathbb{G}_2, \mathbb{G}_T, g_1, g_2, y, H_0(\cdot), H_1(\cdot))$.

Extract: Taking as input κ and a signer's identity ID_i, this algorithm outputs the private key for the signer as follows:

1. Compute $id_{i,0} = H_0(ID_i, 0), id_{i,1} = H_0(ID_i, 1)$.
2. Compute $s_{i,0} = id_{i,0}^\kappa, s_{i,1} = id_{i,1}^\kappa$.
3. Set $s_i = (s_{i,0}, s_{i,1})$ as the private key of the signer.

Sign: To sign a message m_i, a signer with identity ID_i and private key $s_i = (s_{i,0}, s_{i,1})$ computes

$$h_i = H_1(m_i, ID_i), \sigma_i = s_{i,0}s_{i,1}^{h_i}.$$

The signer outputs σ_i as the signature on message m_i. Notice that a signer needs different temporary identities to sign multiple messages, as implied by the name of one-time identity based aggregate signature.

Aggregate: This publicly computable algorithm aggregates n signatures into a single signature. For a set of n users with identities $\{ID_1, \cdots, ID_n\}$, and corresponding message-signature pairs $\{(m_1, \sigma_1), \cdots, (m_n, \sigma_n)\}$, this algorithm outputs

$$\Omega = \prod_{i=1}^{n} \sigma_i$$

as the resulting aggregate signature.

Verify: To verify an aggregate signature Ω on messages $\{m_1, ..., m_n\}$ under identities $\{ID_1, ..., ID_n\}$, the verifier performs the following steps:

1. For $1 \leq i \leq n$, compute $h_i = H_1(m_i, ID_i)$ and $id_{i,0} = H_0(ID_i, 0), id_{i,1} = H_0(ID_i, 1)$.
2. Check

$$\hat{e}(\Omega, g_2) = \hat{e}(\prod_{i=1}^{n} id_{i,0}id_{i,1}^{h_i}, y).$$

Output 1 if the equation holds; else output 0.

3.3 Correctness and Security

The correctness of the OTIBAS scheme in Section 3.2 follows from a direct verification.

In general, the security of an OTIBAS scheme is modeled via the following EUF-OTIBAS-CMA (existential universal forgery under adaptive chosen-message attack) game which is based on the security model of Gentry-Ramzan [8] and takes place between a challenger C and an adversary A. The game has the following three stages:

Initialize: C runs the Setup algorithm to obtain a master secret key and the system parameters. C then sends the system parameters to A while keeping secret the master secret key.

Attack: A can perform a polynomially bounded number of the following types of queries in an adaptive manner.

- Extract *queries*: A can request the private key of an entity with identity ID_i. In response, C outputs the private key of the entity.

- Sign *queries*: $\mathcal{A}$ can request an entity's (whose identity is ID_i) signature on a message M_i. On receiving a query on (M_i, ID_i), $\mathcal{C}$ generates a valid signature σ_i on message M_i under identity ID_i, and replies with σ_i.

Forgery: $\mathcal{A}$ outputs a set of n identities $L_{ID}^* = \{ID_1^*, ..., ID_n^*\}$, a set of n messages $L_M^* = \{M_1^*, ..., M_n^*\}$ and an aggregate signature σ^*.

We say that $\mathcal{A}$ wins the above game, iff

1. σ^* is a valid aggregate signature on messages $\{M_1^*, ..., M_n^*\}$ under identities $\{ID_1^*, ..., ID_n^*\}$.
2. At least one of the identities, without loss of generality say $ID_1^* \in L_{ID}^*$, was not submitted in the Extract queries, and the signature of ID_1^* can be queried at most only once, and (M_1^*, ID_1^*) was never queried during the Sign queries.

We can now define the security of an OTIBAS scheme in terms of the above game.

Definition 1. *An OTIBAS scheme is secure, i.e., secure against existential forgery under adaptive chosen-message attack, iff the success probability of any polynomially bounded adversary in the above EUF-OTIBAS-CMA game is negligible.*

We next recall the co-CDH assumption on which the security of the OTIBAS scheme in Section 3.2 is based.

Definition 2 (co-CDH Assumption). *The co-CDH assumption in two cyclic groups $\mathbb{G}_1$ and $\mathbb{G}_2$ of prime order q equipped with bilinearity states that, given (g_1^a, g_2^b) for randomly chosen $a, b \in Z_q^*$, it is hard for any polynomial-time algorithm to compute g_1^{ab}.*

Regarding the security of our OTIBAS scheme, we have the following claim.

Theorem 1. *Assume there exists an adversary such that: i) it has an advantage ϵ in forging a signature of the OTIBAS scheme of Section 3.2 in an attack modeled by the above EUF-OTIBAS-CMA game, within a time span τ; ii) it can make at most q_{H_i} times $H_i(\cdot)$ $(i = 0, 1)$ queries, q_E times Extract queries, q_S times Sign queries. Then there exists an adversary who can solve the co-CDH problem with probability $\epsilon' \geq \frac{1}{e(q_E + q_S + n + 1)} \epsilon$ within time $\tau' = \tau + \Theta(4q_{H_1} + q_S)\tau_{G_1}$, where τ_{G_1} is the time to compute a point exponentiation in $\mathbb{G}_1$ and n is the size of the aggregating set.*

Due to page limitation, the proof will be presented in the full version of this paper.

4 The APPA Protocol

4.1 Intuition behind Our Protocol

Our APPA protocol is built on the above OTIBAS scheme. The OTIBAS notion incorporates the desirable features of identity-based cryptography [20], aggregate signature [8] and one-time signature. In an OTIBAS scheme, any identity string can be the public key of a signer and the signer can only generate a signature after obtaining the private key corresponding to the identity of the signer. This guarantees the security of the VANETs deployment and eliminates the need for an extra certificate for each identity. Since OTIBAS allows a signer to compute only one signature under one identity, the signer's temporary identity changes for each signature and anonymity is naturally achieved for vehicles/signers. However, since TA knows the secret used by the signer, TA can trace a misbehaving vehicle with the signatures generated by the vehicle. Furthermore, OTIBAS can aggregate n signatures on n (distinct or not) messages from n signers into a single signature. This greatly reduces the time to verify n signatures in order and speeds up the reaction of the verifying vehicles to received messages. Hence, the APPA protocol meets our design goals very well.

Our protocol also requires each vehicle to be equipped with a *practical* tamper-proof device. This is used to allow each vehicle to locally generate its temporary identities (as pseudonyms), without frequently contacting TA, and also to relieve from the reliance on RSUs. However, unlike the protocol in [25] storing the system master secret key in the *idealized* tamper-proof device (assumed secure against any attempt of compromise in any circumstance), we just require the tamper-proof device to store a secret identity of the vehicle and some auxiliary secret information, to enable the vehicle to generate its one-time identity (as a pseudonym) and the private key corresponding to this one-time identity. The secret identity is computed by TA from the real identity of the vehicle and TA's secret. That is, the secret identity in the tamper-proof device of each vehicle is different. If a vehicle does not renew its secret identity, it may leave the opportunity to an attacker to recover this secret. In our protocol, each vehicle can update its secret identity information in its tamper-proof device to counter possible side-channel attacks. This is very different from the protocol in [25] in which the secret in the tamper-proof device cannot be updated.

4.2   The Concrete Protocol

The APPA protocol consists of the following five stages.

[System Setup]

At this stage, TA initializes the system-wide parameters as follows:

1. Generate $q, \mathbb{G}_1, \mathbb{G}_2, \mathbb{G}_T, g_1, g_2, \psi, \hat{e}$, where $q - 1$ has a large prime factor (say $q - 1 = 2q'$ for a prime q').
2. Pick $\kappa \in Z_q^*$ as its master secret key, and compute $y = g^\kappa$ as its master public key.
3. Select cryptographic hash functions $H_0(\cdot) : \{0,1\}^* \to \mathbb{G}_1$, $H_1(\cdot) : \{0,1\}^* \to \mathbb{Z}_q^*$ and $H_2^{key}(\cdot) : \{0,1\}^* \to \{0,1\}^l$, where $H_2^{key}(\cdot)$ is a keyed hash.
4. Choose Λ as a hash key of $H_2^{key}(\cdot)$.
5. Pre-load the system parameters $\Psi = (\hat{e}, q, \mathbb{G}_1, \mathbb{G}_2, \mathbb{G}_T, g_1, g_2, y, H_0(\cdot), H_1(\cdot), H_2^{key}(\cdot))$ in each vehicle and RSU.

The TA also maintains a member list ML which is kept secret. We will define this list later.

[Vehicle Join]

Each vehicle is equipped with a tamper-proof device. Before a vehicle $\mathcal{V}_i$ joins a VANET, its tamper-proof device should be initialized. The tamper-proof device of $\mathcal{V}_i$ is preloaded with the system parameters Ψ and two secret values (α_i, β_i), where α_i and β_i satisfy $\kappa = \alpha_i \beta_i$. Furthermore, an internal pseudo-identity (IPID) and a hash key λ_i are also preloaded to the tamper-proof device. In the following, we show how to set IPID and λ_i.

Assume that each vehicle is associated with a real identity, *e.g.*, the driving license number. The real identity is used by TA to generate IPIDs for the vehicle. Suppose the real identity of a vehicle $\mathcal{V}_i$ is $ID_{\mathcal{V}_i}$. To generate an IPID for $\mathcal{V}_i$, TA concatenates the real identity and a validity period VP_i, *e.g.*, "01.01.2011-01.02.2011", which is associated with and used to compute the IPID $IPID_{\mathcal{V}_i} = H_2^{\Lambda}(ID_{\mathcal{V}_i} \| VP_i)$. TA chooses a hash key λ_i, and stores $IPID_{\mathcal{V}_i}, \lambda_i$ in the tamper-proof device. $(ID_{\mathcal{V}_i}, VP_i, \lambda_i)$ is added to the member list ML.

We notice that, if a vehicle does not update its IPID, its privacy is exposed to potential side-channel attacks. Hence, we suggest each IPID to have a limited life time, and we require the vehicle to periodically renew $(IPID_{\mathcal{V}_i}, \lambda_i)$ before the current IPID expires.

[Vehicle Sign]

At this stage, with the help of the embedded tamper-proof device, a vehicle $\mathcal{V}_i$ computes a signature on a message. This stage has four steps: *Generate public pseudo-identity*, *Extract one-time signing key*, *Sign message* and *Randomize secret values*. The description of each step is as follows:

Generate public pseudo-identity: At this step, $\mathcal{V}_i$ uses its internal pseudo-identity $\overline{IPID_{\mathcal{V}_i}}$ to generate its public pseudo-identity (PPID). It generates a PPID $PPID_{i,j}$ by computing $PPID_{i,j} = H_2^{\lambda_i}(IPID_{\mathcal{V}_i}, \tau)$, where τ is a timestamp.

Extract one-time signing key: At this step, $\mathcal{V}_i$ generates the private signing key corresponding to the PPID $PPID_{i,j}$. Assuming that the current secret values of $\mathcal{V}_i$ are $\alpha_{i,j}$ and $\beta_{i,j}$, $\mathcal{V}_i$ generates the private key associated with the PPID $PPID_{i,j}$ as follows:

1. Compute $pid_{i,j,0} = H_0(PPID_{i,j}, 0), pid_{i,j,1} = H_0(PPID_{i,j}, 1)$.
2. Compute $s_{i,j,0} = pid_{i,j,0}^{\alpha_{i,j}} pid_{i,j,0}^{\beta_{i,j}}, s_{i,j,1} = pid_{i,j,1}^{\alpha_{i,j}} pid_{i,j,1}^{\beta_{i,j}}$.
3. Set $s_{i,j} = (s_{i,j,0}, s_{i,j,1})$ as the one-time signing key of the vehicle.

Sign message: At this step, $\mathcal{V}_i$ computes and outputs the signature $\sigma_{i,j} = s_{i,j,0} s_{i,j,1}^{h_i}$, where $h_i = H_1(m_i, PPID_{i,j})$.

Randomize secret values: To counteract potential side channel attacks, the vehicle's secret values in the tamper-proof device should be updated. The *multiplicative secret sharing* technique introduced by Kiltz and Pietrzak [6] is employed to achieve the goal. Assuming that the current secret values of the tamper-proof device are $\alpha_{i,j}, \beta_{i,j}$, the device generates the new secret values as follows:

1. Choose a random $r \in \mathbb{Z}_q^*$.
2. Compute $\alpha_{i,j+1} = r\alpha_{i,j}$ and $\beta_{i,j+1} = r^{-1}\beta_{i,j}$.
3. Set $(\alpha_{i,j+1}, \beta_{i,j+1})$ as the new secret values.

[Message Verification and Signature Storage]

Suppose that a vehicle or an RSU receive n message-signature pairs $\{(m_1, \sigma_1), \cdots, (m_n, \sigma_n)\}$ from n vehicles with public pseudo identities $\{PPID_{1,j_1}, ..., PPID_{n,j_n}\}$, respectively. The verifier computes the aggregate signature $\Omega = \prod_{i=1}^{n} \sigma_i$. To verify the aggregate signature Ω, the verifier performs the following steps:

1. For $1 \le i \le n$, compute $h_i = H_1(m_i, PPID_{i,j_i})$, $pid_{i,j_i,0} = H_0(PPID_{i,j_i}, 0)$, $pid_{i,j_i,1} = H_0(PPID_{i,j_i}, 1)$.
2. Output 1 if $\hat{e}(\Omega, g_2) = \hat{e}(\prod_{i=1}^{n} pid_{i,j_i,0}pid_{i,j_i,1}^{h_i}, y)$. Else output 0.

After verifying the aggregate signature, a vehicle or an RSU may store

$$(m_1||...||m_n; PPID_{1,j_1}||...||PPID_{n,j_n}; \Omega)$$

in its local database. One may notice that the last field is of constant size, (the length of one group element, which is argued to be 21 bytes in Section 5.2).

[Trace]

If a message has passed the verification procedure but is found false, TA should be able to trace the real identity of the message originator. Assume the false message is m and the corresponding public pseudo-identity is $PPID_{\mathcal{V}_{?,?}}$. To recover the real identity corresponding to $PPID_{\mathcal{V}_{?,?}}$, TA extracts the timestamp τ from m. According to τ, TA may know the valid period of the internal pseudo-identity of the message sender. To find who is the real sender, TA tests $H_2^{\lambda_i}(H_2^\Lambda(ID_{\mathcal{V}_i}||VP_i), \tau) \stackrel{?}{=} PPID_{\mathcal{V}_{?,?}}$, where $(ID_{\mathcal{V}_i}, VP_i, \lambda_i)$ is on the member list ML. If the equation holds, TA outputs $ID_{\mathcal{V}_i}$.

5   Evaluation

5.1   Security and Privacy

It is easy to see that our protocol satisfies the *global security* requirement of Section 2.2, by noting that: i) the employed IDAOTS scheme is shown to be secure; ii) the real identity of a vehicle can be traced as in the Trace stage. Below we analyze whether our protocol satisfies the *individual privacy* requirement. In our protocol, the only information that can be used by an eavesdropper to trace a vehicle is the public pseudo-identities. However, it is hard for anyone (except TA and the vehicle itself) to know the vehicle's internal pseudo-identity IPID: computing IPID implies finding the inverse of PPID (*i.e.*, the keyed hash output), which is impossible since the keyed hash is one-way. Furthermore, since the keyed hash outputs are computationally indistinguishable, that is, it is hard for an attacker to determine whether two different public pseudo-identities (*i.e.*, two outputs of the keyed hash) are computed from the inputs of the same internal pseudo-identity, different time stamps and a secret key or the inputs of different internal pseudo-identities, different time stamps and a secret key. This implies that our protocol is unlinkable in the sense that an attacker cannot know whether two pseudonyms are linked with the same vehicle or not. Therefore, the *privacy* goal is achieved.

It is also worth noticing that our protocol is resistant to side-channel attacks. In our protocol, there are two kinds of secrets stored in the tamper-proof device. The first one is $(IPID_{\mathcal{V}_i}, \lambda_i)$ which is related to a vehicle $\mathcal{V}_i$'s privacy. If $\mathcal{V}_i$ does not renew $(IPID_{\mathcal{V}_i}, \lambda_i)$, it may leave the opportunity to an attacker to recover this secret, so that the attacker can trace $\mathcal{V}_i$. However, one may notice that, in practice, the attacker can only launch a side-channel attack occasionally. Therefore, in most cases, before the attacker collects enough side-channel information to recover the current secret $(IPID_{\mathcal{V}_i}, \lambda_i)$, the vehicle has already updated it, noting that APPA suggests a vehicle to renew $(IPID_{\mathcal{V}_i}, \lambda_i)$ periodically. Furthermore, in the worst case, even if the attacker recovers $(IPID_{\mathcal{V}_i}, \lambda_i)$, the attacker can only track the vehicle for a short period of time, *i.e.*, before $(IPID_{\mathcal{V}_i}, \lambda_i)$ is renewed. The second kind of secrets are the secret values $(\alpha_{i,j}, \beta_{i,j})$. In our protocol, these secret values are used once and then a random value is chosen to blind them (the *multiplicative secret sharing* technique). This technique is introduced by Kiltz and Pietrzak [6], and, can be used to convert a scheme to be leakage resilient one. With this technique, the attacker cannot collect enough information about these secret values. Because the precondition of side-channel attacks is that the same secret value be involved in a large number of cryptographic operations so that enough side-channel information about the secret value can be collected to perform a statistical analysis.

5.2   Transmission and Storage Overhead

Table 1 compares the transmission and storage overhead incurred by the security and privacy mechanisms with up-to-date protocols in the literature. For fairness, we consider protocols which, as our APPA protocol, do not require pre-storing a large number of anonymous certificates/pseudo identities.

According to [2], the length of a point in group $\mathbb{G}_1$ is 171 bits (about 21 bytes). In addition, the length of an identity is 20 bytes. For our protocol, it is easy to see that the length of a signature and an IPID is about $21 + 20$ bytes. However, the signatures in our protocol can be aggregated into a single point in $\mathbb{G}_1$. Hence, the length of the aggregate signature will not increase with the number of messages received and the total overhead is $21 + 20n$ bytes. This length is about $1/2$ of that in [25]. Furthermore, the transmission and storage overhead of our protocol is much lower than that in [12] and [27].

5.3   Impact of Signature Verification on Response Time

Table 2 compares the computational overhead of signature verification with the same protocols considered in Table 1. It is sufficient to only consider the most

Table 1. Comparison of protocols in terms of Transmission and Storage Overhead

	Send/Store a single message	Store n messages
Protocol 1 [12]‡	192 bytes	$192n$ bytes
Protocol 2 [25]†	$21 + 41$ bytes	$21 + 41n$ bytes
Protocol 3 [27]‡	368 bytes	$368n$ bytes
Our Protocol†	$21 + 20$ bytes	$21 + 20n$ bytes

†: Identity-based signature based
‡: Group signature based

Table 2. Comparison of protocols in terms of Computational Overhead

	Verify a single signature	Verify n signatures
Protocol 1 [12]	$5\tau_{bp} + 12\tau_{pe}$	$5n\tau_{bp} + 12n\tau_{pe}$
Protocol 2 [25]	$3\tau_{bp} + \tau_{pe} + \tau_{mh}$	$3\tau_{bp} + n\tau_{pe} + n\tau_{mh}$
Protocol 3 [27]	$2\tau_{bp} + 14\tau_{pe}$	$2\tau_{bp} + \frac{14n}{4.8}\tau_{pe}$
Our Protocol	$2\tau_{bp} + \tau_{pe} + 2\tau_{mh}$	$2\tau_{bp} + n\tau_{pe} + 2n\tau_{mh}$

costly operations, *i.e.*, pairing, point exponentiation and map-to-point hash (*e.g.*, $H_0(\cdot)$) operations.

The processing time for one bilinear pairing operation is about $\tau_{bp} = 1.87$ ms, the time for one point exponentiation operation is about $\tau_{pe} = 0.49$ ms and the computational cost of one map-to-point hash is about 0.22 ms [9,10]. In Figure 2, one may notice that the signature verification cost in our protocol is much more efficient than that in [12] and slightly more efficient than that in [27]. Further, when the number of signatures to be verified is small, our protocol is more efficient than that in [25]. When the number of signatures to be verified grows, our protocol has comparable efficiency with that in [25]. However, the signature length in our protocol is about 1/2 of that in [25]. Furthermore, unlike the protocol in [25], our protocol is resistant to side-channel attacks.

5.4   Tracing Efficiency

As shown in Section 4, to find the real identity of a message sender, TA tests

$$H_2^{\lambda_i}(H_2^{\Lambda}(ID_{\mathcal{V}_i} \| VP_i), \tau) \stackrel{?}{=} PPID_{\mathcal{V}_{?,?}},$$

where $(ID_{\mathcal{V}_i}, VP_i, \lambda_i)$ is on the list ML. If the equation holds, TA outputs $ID_{\mathcal{V}_i}$. The cost of recovering the real identity of a sender may seem very high, because there may be millions of tuples in ML. However, we notice that TA does not need to test all the tuples on ML. Instead, TA only needs to test the tuples that contain VPs that match τ. Therefore, for a VANET with 1 million vehicles, TA only needs to test 0.5 million times in average. We use the popular keyed SHA-1 to estimate the tracing efficiency of our protocol. It only takes 0.001 ms on

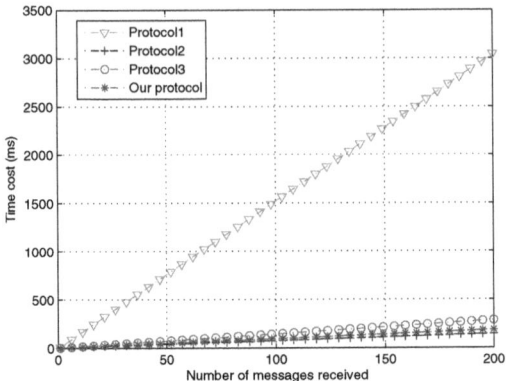

Fig. 2. Verification cost

average for TA to perform a SHA-1 hash operation [19]. Therefore, for a VANET with 1 million vehicles, our protocol only needs about one second to find the real identity of the message originator.

6   Conclusion

In this paper, we proposed the APPA protocol to secure VANETs. The protocol allows aggregate privacy-preserving authenticated vehicular communications. The protocol guarantees trustworthiness of vehicle-generated messages and privacy of vehicles. With APPA, vehicles can react to received messages within a very short delay. The digital signatures for the authentication purpose are securely and substantially compressed. The APPA protocol does not heavily rely on roadside units, which implies that the protocol can work even if the VANET infrastructure is incomplete. These features seem desirable and allow our protocol to be deployed in various secure VANET scenarios.

Acknowledgments and disclaimer. This work is partly supported by the Nature Science Foundation of China through projects 61021004, 11061130539, 60970114, 60970115, 60970116 and 61003214, by the European commission under FP7 project "DwB", by the Spanish Government under projects TSI2007-65406-C03-01 "E-AEGIS", TIN2009-11689 "RIPUP", "eVerification" TSI-020100-2009-720, SeCloud TSI-020302-2010-153 and CONSOLIDER INGENIO 2010 CSD2007-00004 "ARES", by the Fundamental Research Funds for the Central Universities of China through projects 3103004, 78210044, by the Scientific Research Program Funded by Shaanxi Provincial Education Department (Program No. 2010JK727), by the Shanghai Natural Science Foundation under Grant No. 11ZR1411200 and by the Beijing Municipal Natural Science Foundation to Project 4112052. The forth author is partially supported as an ICREA-Acadèmia researcher by the Catalan Government. The views of the author with the UN-ESCO Chair in Data Privacy do not necessarily reflect the position of UNESCO nor commit that organization.

References

1. Dedicated Short Range Communications (DRSC) home,
   http://www.leearmstrong.com/Dsrc/DSRCHomeset.htm
2. Boneh, D., Boyen, X., Shacham, H.: Short group signatures. In: Franklin, M. (ed.) CRYPTO 2004. LNCS, vol. 3152, pp. 41–55. Springer, Heidelberg (2004)
3. Calandriello, G., Papadimitratos, P., Hubaux, J.-P., Lioy, A.: Efficient and robust pseudonymous authentication in vanet. In: ACM VANET 2007, pp. 19–28. ACM Press, New York (2007)
4. Chaum, D., van Heyst, E.: Group signatures. In: Davies, D.W. (ed.) EUROCRYPT 1991. LNCS, vol. 547, pp. 257–265. Springer, Heidelberg (1991)
5. Daza, V., Domingo-Ferrer, J., Sebé, F., Viejo, A.: Trustworthy privacy-preserving car-generated announcements in vehicular ad hoc networks. IEEE Transactions on Vehicular Technology 58(4), 1876–1886 (2009)
6. Kiltz, E., Pietrzak, K.: Leakage resilient elGamal encryption. In: Abe, M. (ed.) ASIACRYPT 2010. LNCS, vol. 6477, pp. 595–612. Springer, Heidelberg (2010)
7. Frey, G., Rück, H.-G.: A remark concerning m-divisibility and the discrete logarithm in the divisor class group of curves. Mathematics of Computation 62(206), 865–874 (1994)
8. Gentry, C., Ramzan, Z.: Identity-Based Aggregate Signatures. In: Yung, M., Dodis, Y., Kiayias, A., Malkin, T. (eds.) PKC 2006. LNCS, vol. 3958, pp. 257–273. Springer, Heidelberg (2006)
9. Icart, T.: How to hash into elliptic curves. In: Halevi, S. (ed.) CRYPTO 2009. LNCS, vol. 5677, pp. 303–316. Springer, Heidelberg (2009)
10. Jiang, Y., Shi, M., Shen, X., Lin, C.: BAT: A robust signature scheme for vehicular networks using binary authentication trees. IEEE Transactions on Wireless Communications 8(4), 1974–1983 (2009)
11. Kocher, P.: Timing attacks on implementations of Diffie-Hellman, RSA, DSS, and other systems. In: Koblitz, N. (ed.) CRYPTO 1996. LNCS, vol. 1109, pp. 104–113. Springer, Heidelberg (1996)
12. Lin, X., Sun, X., Ho, P., Shen, X.: GSIS: A secure and privacy preserving protocol for vehicular communications. IEEE Transactions on Vehicular Technology 56(6), 3442–3456 (2007)
13. Lu, R., Lin, X., Zhu, H., Ho, P., Shen, X.: ECPP: Efficient conditional privacy preservation protocol for secure vehicular communications. In: IEEE INFOCOM 2008, pp. 1229–1237. IEEE Computer Society Press, Los Alamitos (2008)
14. Menezes, A., Okamoto, T., Vanstone, S.A.: Reducing elliptic curves logarithms to logarithms in a finite field. IEEE Transactions on Information Theory 39(5), 1639–1646 (1993)
15. Papadimitratos, P., Gligor, V., Hubaux, J.: Securing vehicular communications - Assumptions, requirements, and principles. In: ESCAR 2006 (2006)
16. Picconi, F., Ravi, N., Gruteser, M., Iftode, L.: Probabilistic validation of aggregated data in vehicular ad hoc networks. In: ACM VANET 2006, pp. 76–85. ACM Press, New York (2006)

17. Raya, M., Hubaux, J.: The security of vehicular ad hoc networks. In: ACM SASN 2005, pp. 11–21. ACM Press, New York (2005)
18. Raya, M., Hubaux, J.: Securing vehicular ad hoc networks. Journal of Computer Security 15(1), 39–68 (2007)
19. Satizábal, C., Martínez-Peláez, R., Forné, J., Rico-Novella, F.: Reducing the computational cost of certification path validation in mobile payment. In: López, J., Samarati, P., Ferrer, J.L. (eds.) EuroPKI 2007. LNCS, vol. 4582, pp. 280–296. Springer, Heidelberg (2007)
20. Shamir, A.: Identity-Based Cryptosystems and Signature Schemes. In: Blakely, G.R., Chaum, D. (eds.) CRYPTO 1984. LNCS, vol. 196, pp. 47–53. Springer, Heidelberg (1985)
21. Standaert, F., Malkin, T., Yung, M.: A unified framework for the analysis of side-channel key recovery attacks. In: Joux, A. (ed.) EUROCRYPT 2009. LNCS, vol. 5479, pp. 443–461. Springer, Heidelberg (2009)
22. Wasef, A., Shen, X.: ASIC: Aggregate signatures and certificates verification scheme for vehicular networks, http://www.engine.lib.uwaterloo.ca
23. Wu, Q., Domingo-Ferrer, J., Gonzalez-Nicolas, U.: Balanced trustworthiness, safety, and privacy in vehicle-to-vehicle communications. IEEE Transactions on Vehicular Technology 59(2), 559–573 (2010)
24. Wu, Q., Mu, Y., Susilo, W., Qin, B., Domingo-Ferrer, J.: Asymmetric group key agreement. In: Joux, A. (ed.) EUROCRYPT 2009. LNCS, vol. 5479, pp. 153–170. Springer, Heidelberg (2009)
25. Zhang, C., Lu, R., Lin, X., Ho, P., Shen, X.: An efficient identity-based batch verification scheme for vehicular sensor networks. In: IEEE INFOCOM 2008, pp. 246–250. IEEE Computer Society Press, Los Alamitos (2008)
26. Zhang, L., Wu, Q., Qin, B., Domingo-Ferrer, J.: Identity-based authenticated asymmetric group key agreement protocol. In: Thai, M.T., Sahni, S. (eds.) COCOON 2010. LNCS, vol. 6196, pp. 510–519. Springer, Heidelberg (2010)
27. Zhang, L., Wu, Q., Solanas, A., Domingo-Ferrer, J.: A scalable robust authentication protocol for secure vehicular communications. IEEE Transactions on Vehicular Technology 59(4), 1606–1617 (2010)
28. Zhu, H., Lin, X., Lu, R., Ho, P., Shen, X.: AEMA: An aggregated emergency message authentication scheme for enhancing the security of vehicular ad hoc networks. In: IEEE ICC 2008, pp. 1436–1440. IEEE Computer Society Press, Los Alamitos (2008)

Assessing Location Privacy in Mobile Communication Networks

Klaus Rechert[1], Konrad Meier[1], Benjamin Greschbach[2],
Dennis Wehrle[1], and Dirk von Suchodoletz[1]

[1] Faculty of Engineering, Albert-Ludwigs University Freiburg i. B., Germany
[2] School of Computer Science and Communication, KTH - Royal Institute of
Technology, Stockholm, Sweden

Abstract. In this paper we analyze a class of location disclosure in
which location information from individuals is generated in an automated way, i.e. is observed by a ubiquitous infrastructure. Since such
information is valuable for both scientific research and commercial use,
location information might be passed on to third parties. Users are usually aware neither of the extent of the information disclosure (e.g. by
carrying a mobile phone), nor how the collected data is used and by
whom.

In order to assess the expected privacy risk in terms of the possible
extent of exposure, we propose an adversary model and a privacy metric that allow an evaluation of the possible privacy loss by using mobile
communication infrastructure. Furthermore, a case study on the privacy
effects of using GSM infrastructure was conducted with the goal of analyzing the side effects of using a mobile handset. Based on these results
requirements for a privacy-aware mobile handheld device were derived.

1 Introduction

Mobile communication systems as well as location-based services are now both
well established and well accepted by users. The combination of location determination, powerful mobile devices (so-called Smartphones) and ubiquitous
network communication options provide a lot of useful new applications but also
bring new challenges to the users' privacy especially when it comes to location
disclosure.

There are various occasions and motives for location disclosure. In general, one
can classify location disclosure types into two different categories based on trust
relationship between involved communication peers (cf. Fig. 1). A very common
situation today is when users exchange their whereabouts with location-based
service providers for tailored and context-sensitive information. Exchanging position information within groups through social network services (SNS) is also
gaining in popularity. These services usually involve informed users who are
aware that they are disclosing their location data. Service providers as well as
social peers are considered as partially trusted, at least for the specific communication context, as communication is voluntary and communication peers are
known.

X. Lai, J. Zhou, and H. Li (Eds.): ISC 2011, LNCS 7001, pp. 309–324, 2011.

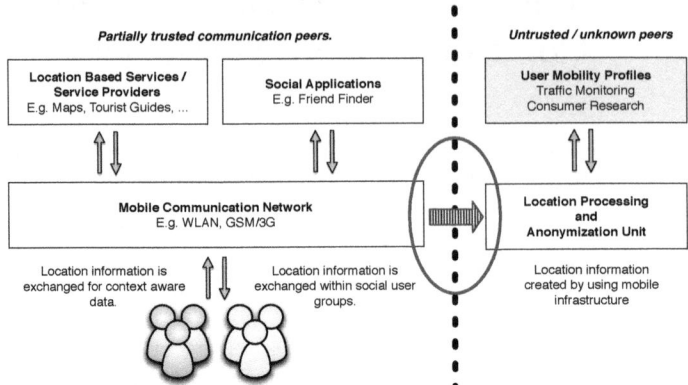

Fig. 1. Transfer of location information from (partially-)trusted peers to untrusted peers

Nowadays there is typically mobile communication involved. The communication infrastructure is usually also considered as partly trusted, i.e., there is a service agreement between the user and provider. Due to the specific nature of mobile communication networks, the location of mobile subscribers is known to the underlying infrastructure.

Hence, there is a second class of location disclosure, where location information from individuals is generated automatically or is observed by the infrastructure. Such information is valuable for scientific research [1] but also for commercial use (e.g. traffic monitoring[1] or location-aware advertising [2]). Therefore, location information might be passed on to third parties in an anonymized and / or aggregated way. In this case users are usually aware neither of the extent of their information disclosure (e.g. by carrying a mobile phone), nor how the collected data is used and by whom.

A lot of research has been done (e.g. [3]) on protecting a user's anonymity. However, when using mobile infrastructure users face two difficulties. First, due to regulations, the quality of service, but also technical conditions, location privacy protection measures like anonymity and obfuscation techniques seem inadequate or difficult to employ. Second, users suffer from limited and asymmetric knowledge. They have no knowledge on the nature, accuracy and amount of location data they have generated by using mobile communication infrastructure so far. As well, they cannot make a judgment about the level and quality of anonymization if the location data is exploited for various services. Since the data might be de-anonymized (e.g. [4]) or, even worse, if raw data leaks for whatever reason, users bear a latent privacy risk.

Thus, disclosing location data always conflicts with the user's privacy, since position information or movement history might lead to the user's identity. Col-

---

[1] For instance A1 Traffic (http://www.a1.net/business/a1traffictechnologie [12/1/2010]) or Vodafone HD Traffic (http://www.vodafone.de/business/firmenkunden/verkehrsinfo-hd-traffic.html [5/1/2011]).

lected location data can become a quasi-identifier, similar to a fingerprint [5]. Hence, by using external knowledge the identity of a specific user can be determined (e.g. [6]). Furthermore, through observing and evaluating a user's movements, his preferences and other possible sensitive information might be revealed. Such sensitive location-related data contains places a user visits frequently or at certain times and thus has special interest in. With the location data of other individuals his social relations become visible.

In order to assess the expected privacy risk in terms of the possible extent of exposure, first an adversary model has to be defined. Based on this model a simple privacy metric is proposed in order to assess the privacy loss in mobile communication networks and provide the groundwork for developing requirements for a privacy-aware communication device. As an example, we analyzed the impact on the user's privacy of different network configurations found at four GSM telephony infrastructure providers.

2   Related Work

Recent analysis of mobile phone call data records (CDR) showed that even sporadic anonymous location data with coarse spatial resolution contain sensitive information and could lead to possible identification. Humans tend to move in very regular patterns. A study using six months of call data records showed that humans stay more than 40% of the time at the same two places [1,7]. In another study on anonymized aggregated call data records, the movement patterns of commuters in two cities were compared [8]. Similar studies were conducted on tourist movement patterns in New York and Rome [9,10].

Sohn et al. analyzed GSM data to determine a user's movement mode based on radio signal fingerprints. The authors were able to distinguish between walking, driving and stationary profiles with a success rate of about 85% [11]. De Mulder et al. conducted a study on the possibilities of re-identification of individual mobile phone subscribers based on available cell data. In their study they evaluated a Markovian model and a model based on the sequence of cell-IDs. They report a success rate of about 80 % for the latter method [12].

From the aforementioned studies one could conclude that using a mobile communication network (e.g. GSM) is a threat to a user's privacy. However, from a user's perspective the question remains how much knowledge a network provider has on his movement patterns and which of the available networks pose the least threat for his privacy. Lee et al. dealt with location privacy in GSM networks only on the protocol layer in the relation between mobile station, VLR and HLR. However, location data as a possible quasi-identifier was not discussed [13]. Ardagna et al. introduce a scenario with a semi-trusted (mobile) network provider and propose a multi-path communication approach to achieve k-anonymity for the initiating sender of a message [14]. Their approach relies on a hybrid network infrastructure where subscribers are able to form ad-hoc networks. However, such an approach protects only the relation between sender and final recipient (e.g. LBS-provider).

In order to evaluate a privacy metric an adversary model is required. A popular model is an adversary that observes in some way generalized location data and tries to reconstruct this data to connected traces of a single individual. In a second step the adversary may re-identify the traced individual through his workplace or home by incorporating external knowledge (e.g. [15]) For instance, Shorki et al. defined a location privacy metric that measures the (in)ability of an adversary to accurately track a mobile user over space and time [16].

A different method for measuring location privacy is to make use of the uncertainty of an adversary in order to assign a new observation to a trace of a specific individual, e.g. by assigning probabilities to movement patterns and thus compensating changed pseudonyms [17]. A similar measurement was proposed as *time-to-confusion* metric, the tracking time of an individual until the adversary cannot determine the next position with sufficient certainty [15]. The (un)certainty measure is based on the entropy of the observed position and the probability of the expected or calculated next location. Sending a variety of locations for each query also increases the ambiguity and thus the level of privacy [18].

However, in most mobile communication relations the anonymity assumption seems inadequate. Furthermore, the aforementioned privacy metrics usually require full insight into the set of all users to determine the level of privacy for a single user within this set.

3   Location Privacy in Mobile Telephony Networks

When using mobile communication, location data is generated, accumulated and stored, as a technical and possibly legal[2] necessity. As an example we discuss the GSM infrastructure, because it is widely deployed and recently software and analysis tools have become available [3,4]. Its successors UMTS (3G) and LTE (4G) still share most of its principal characteristics. The goal is to uncover hidden privacy risks posed by the network's background communication and the effects of different network configurations on the user's privacy. In contrast to active usage (e.g. phone calls, texting), location data is not always provided voluntarily.

A typical GSM network is structured into cells, served by a single base station (BTS) and larger cell-compounds called location area (LA). In idle state no dedicated channel is assigned to the mobile station (MS). It only listens to the common control channel (CCCH) and to the broadcast control channel (BCCH) [19,20] and is otherwise in standby mode to save energy. Through System Information Messages on the BCCH the MS receives periodically a list of neighboring cells from the serving BTS and performs signal strength measurements on these base stations. This way the MS can always select the BTS with good received

---

[2] For instance EU Directive 2006/24/EC (Data retention),
`http://eur-lex.europa.eu/LexUriServ/`
`LexUriServ.do?uri=CELEX:32006L0024:en:NOT`, [5/21/2011]
[3] Open Source GSM Baseband implementation, `http://bb.osmocom.org` [05/19/2011]
[4] Open Base Station Controller OpenBSC, `http://openbsc.osmocom.org` [05/19/2011]

signal strength in order to maintain network attachment. To establish a connection to the MS in case of an incoming connection, the network has to know if the MS is still connected to the network and in which LA the MS is currently located. To accomplish that, the location update procedure was introduced. Either periodically or when changing the LA, a location update (LU) procedure is performed. Within this procedure the phone starts an active communication with the network infrastructure, sending a so-called measurement report to the base station. This report consists of the received signal strength of up to six of the strongest neighboring cells and the received signal strength of the serving base station. The range between the periodic location updates may vary among the infrastructure providers.

3.1   Locating Mobile Phones

Due to regulatory requirements [5] but also driven by commercial opportunities locating mobile phones gained the attention of research and industry. There is a variety of possibilities for determining a mobile station's location from the view point of the infrastructure, e.g., by Cell Origin with timing advance (TA) and Uplink Time Difference of Arrival (U-TDOA) for GSM [21] [6]. While the latter method requires sophisticated network infrastructure, Cell Origin and TA are available in any network setup. However, all these methods work without special requirements for the mobile station and achieve a positioning accuracy of up to 50 m in urban areas (TDOA) [23].

Another (nonstandard) method to determine a MS's location makes use of measurement results. Usually based on databases built from signal propagation models used during the planning phase of the infrastructure, this data can be used to create a look-up table for signal measurements to determine the MS's location. Based on the cell, TA and received signal strength of the serving cell as well as the six neighboring cells, Zimmermann et al. achieved positioning accuracy of below 80 m in 67% and 200 m in 95% in an urban scenario [24]. With a similar method but more generic setup, Peschke et al. report a positioning accuracy of 124 m in 67% [25]. While the mobile phone is in idle mode, network-assisted positioning is not possible. The network either has to wait for the next active period of the MS (e.g. phone call, location update) or has to initiate MS activity. This can be done by transmitting a *silent text message* to the MS in order to force an active communication, but without raising the user's awareness. The procedure is used for instance by law enforcement authorities or by location-based services based on GSM positioning.

3.2   Privacy Threats

Providing a proper (especially technical) definition of location privacy has proven to be a difficult task. Many different definitions were published, all covering

---

[5] e.g. FCC Enhanced 911 Wireless Service,
   http://www.fcc.gov/pshs/services/911-services/enhanced911 [5/15/2011]
[6] Location determination options for UTRAN [22]

specific aspects. One abstract definition, first formulated by Westin [26] and modified by Duckham & Kulik [27], describes location privacy as:

> "[...] a special type of information privacy which concerns the claim of individuals to determine for themselves *when*, *how*, and to *what extent* location information about them is communicated to others."

According to this definition the user should be in control of the dissemination of his location information. Thus, the user's privacy is threatened (according to the aforementioned definition) because of technical necessities frequent location information is generated and possibly stored for different reasons (e.g., for technical network monitoring and improvement, regulation and law enforcement requirements). The user's mobile station collects and transmits location data without notification or consent. Furthermore, besides using the mobile device for active communication the user is unaware when location data is generated and transmitted (e.g. location updates, silent text messages, etc.). Furthermore, the user's privacy is threatened because of monetization of available location information. Even though this data is usually aggregated or anonymized, users are not aware of the technique used and thus bear a risk of re-identification (which they can't assess) (cf. [4]). Neither the final data consumer nor the intended use of the location data is known. Finally, users are not aware of the extent of the information they share. Usually one can assume that a single location datum does not reveal much information to a ubiquitous observer. In contrast to trusted communication peers such generic observers do not have appropriate background knowledge on a single individual and thus have difficulties deriving the user's real life context or current activity, especially for service providers with a subscriber base. However, by the accumulation of location observations knowledge about a user can be easily created. In order to improve the privacy situation in mobile communication networks, any location disclosure has to be made transparent and a privacy metric is required in order to evaluate the extent of location disclosures.

3.3 Adversary Model

In order to measure the extent of location disclosure in mobile communication networks and to assess the effects on the user's privacy the adversary has to be modeled. From a user's perspective, there is no assured knowledge on the capabilities of the observing / listening adversary, especially how disclosed or observed location data is used and what kind of conclusion the adversary is able to make based on the information gained. In general, the user's knowledge is limited to common knowledge about the technical and architectural characteristics of the mobile communication technology he or she uses (e.g. communication infrastructure service, etc.) as well as to a general estimation of the location determination abilities, limited by technical or physical factors. Furthermore, the user is able to monitor her own usage patterns by logging her exposure to the network, and has knowledge about the surrounding landscape, i.e., map knowledge.

Hence, the adversary model is limited to information an adversary may have gained during a defined observation period. We assume that an adversary A has a memory $O = \{o_1, \ldots, o_m\}$ of observations on the user's movement history based on time-stamped location observations $o_t = (c, \varepsilon)_t \in \mathbb{O}$, which are tuples of a geographic coordinate $c \in \mathbb{C}$ and a spatiotemporal error estimate $\varepsilon \in \mathbb{E}$ of this coordinate. The index t is a timestamp describing when the location observation was made, with o_m being the latest observation. The function $loc : \mathbb{O} \rightarrow \mathbb{C}$ extracts the location information from the tuple and $err : \mathbb{O} \rightarrow \mathbb{E}$ returns the error estimate.

In our scenario we assume that the adversary's utility (denoted as U_A) is negatively correlated with the user's privacy level in a communication relation with adversary A denoted as $P^A \in [0, -\infty)$, with $P^A = 0$ as the maximal achievable privacy level:

$$U_A(O) \simeq -P^A(O). \tag{1}$$

For instance if the user does not disclose any location information, her privacy is maximal but also the adversary's utility is zero. Henceforward, there is a utility gain if the adversary extends his knowledge either on the user's preferences or on his (periodic) behavior. This utility gain might be due to technical reasons (e.g. efficient network planning) or to the reuse of the data for commercial purposes. We can also assume that the adversary's utility is not decreased through any location data as long as the data is accurate, i.e., the user is not lying. Accordingly, $U_A(O') \geq U_A(O)$, with $O' := O \cup o'$, iff. o' reveals previously unknown information to the adversary A. Hence the user's privacy w.r.t. adversary A can only decrease by disclosing additional information: $P^A(O') \leq P^A(O)$.

Furthermore, the adversary's utility as well as the user's privacy depends on the nature and magnitude of the error estimate ε. First, with more accurate information possibly more information might be disclosed and thus, $err(o') < err(o) \Rightarrow U_A(o') \geq U_A(o)$ whereas the actual information gain is dependent, e.g., on landscape characteristics or additional knowledge on the user's context. Second, depending on the adversary and the kind of observation, the error value ε for a given location sample is evaluated differently. If the adversary determines the location by direct observation (o^{adv}), e.g., through WiFi/GSM/3G infrastructure, the adversary knows the size and distribution of the expected spatial error for the observed location sample. Furthermore, the temporal error component can be ignored. In contrast, if location information is given by the user (o^{usr}), the adversary has no information about the quality and thus the magnitude of the error ε of the observed sample. The user might have altered the spatial and/or temporal accuracy of the location information before submission. In general we can assume that $err(o^{adv}) \leq err(o^{usr})$ and therefore $U_A(o^{adv}) \geq U_A(o^{usr})$, since a robust error estimation reduces the adversary's uncertainty and thus increases the potential information gain. But more importantly the adversary chooses time and frequency of location observations.

3.4   Determining an Adversary's Knowledge Level

In order to reflect the duration, density, and quality of observation, a model of all past disclosures (i.e. history or knowledge (K)) w.r.t. a given adversary is required. The user's privacy is threatened by the discovery of his regular behavior and preferences (i.e. movement pattern). Since a user cannot change the knowledge an adversary already has, the user may evaluate the level of completeness of an adversary's information and the information gain or privacy loss involved in disclosing a further location sample.

Based on the adversary's utility function, we require that the knowledge gain $\Delta K_A(O, o') = K_A(O, o') - K_A(O \setminus \{o_m\}, o_m) \geq 0$ for any o'. If no new information is released, $\Delta K_A = 0$, and thus no privacy loss is experienced by the user. For a user it is important to know what extra information the disclosure of a single location sample o' gives to each listening or observing adversary A w.r.t. his history.

In a study on movement patterns of mobile phone users, Gonzalez et al. found a characteristic strong tendency of humans to return to places they visited before. Furthermore, the probability of returning to a location depends on the number of location samples for that location. A rough estimation can be denoted as $Pr(l_k) \sim k^{-1}$ where k is the rank of the location l based on the number of observations [1]. In a similar study it was shown that the range in the number of significant places is limited ($\approx$ 8-15). At these places a user spends about 85% of the time. However, there is a long tail area with several hundred places which were visited less than 1% but covered about 15% of the user's total observation time [7]. For the proposed privacy model we concentrate on the top L popular places (with L being in the range of about 8-15), as these places are likely to be revisited and therefore are considered as significant places in a user's routine.

If we assume that the attacker's a-priori knowledge on the observed location sample o' is limited to the generic probability distribution describing human mobility patterns and the accumulated knowledge so far, then we can model the adversary's knowledge as the uncertainty assigning the observed location information to a top L place. Entropy can be used to express the uncertainty of the adversary and therefore the user's privacy. Using entropy to quantify privacy was already used in different settings (e.g [28]). In the following we consider a location $l \in \mathbb{C}^*$ to be an arbitrarily shaped area in $\mathbb{C}$ and denote the spatial inclusion of a precise coordinate $c \in \mathbb{C}$ in the area l by writing $c \cong l$. To comply with the characteristics of human mobility patterns as described above, we define the probability of an observed location sample o' belonging to one of the top L locations $(l_i, i \in \{1, \dots, L\})$ as $p_{l_i} := Pr(loc(o') \cong l_i) = \frac{\tau}{i}$ where $\tau \in (0, 1]$ is chosen in a way such that $\left(\sum_{i=1}^{L} p_{l_i}\right) + \gamma = 1$ with $\gamma \in [0, 1)$ representing the summed probability of o' belonging to one of the many seldom visited places in the long tail distribution observed by Bayir et al. [7]. Assuming that the adversary A has already discovered the top k locations of the user (by making use of the previously observed user locations in O), we make a distinction between two cases: (A) o' belongs to a frequently visited location already known to the adversary ($\exists i \in \{1, \dots, k\} : loc(o') \cong l_i$), or (B) the adversary is not able to unambiguously connect the location observation to an already detected top L location.

In case (A) no information about new frequently visited places is revealed (which we denote by $K_A^{L(A)}(O, o') = 0$). For case (B) we measure privacy as the uncertainty (i.e. entropy) of assigning o' to one of the remaining unknown top L locations. We denote with $p_{sk} := \sum_{i=1}^{k} p_{l_i}$ the summed probability for the k top locations *known* to the adversary and accordingly $p_{su} := \sum_{i=k+1}^{L} p_{l_i}$ the summed probability for the *unknown* top locations. Given that o' does not belong to one of the k known places, the probability for the remaining places $l_{k+1} \ldots l_L$ changes to $p_{l_i}^k = p_{l_i} \cdot (1 + \frac{p_{sk}}{p_{su}})$, which yields the following entropy calculation:

$$K_A^{L(B)}(O, o') = -(\sum_{i=k+1}^{L} p_{l_i}^k \log p_{l_i}^k) - \gamma \log \gamma, \qquad (2)$$

where γ denotes the summed probability of location samples which do not belong to the top L locations. The overall uncertainty level of the adversary is the weighted sum of the two cases (A) and (B) described above:

$$K_A^L(O, o') = p_{(A)} \cdot K_A^{L(A)}(O, o') + p_{(B)} \cdot K_A^{L(B)}(O, o'), \qquad (3)$$

where $p_{(A)} = p_{sk}$ is the probability of case (A) and $p_{(B)} = 1 - p_{(A)}$ the probability of case (B). By merging the equations of the two cases, the overall uncertainty of an adversary in assigning o' to a yet unknown top location can be expressed as:

$$K_A^L(O, o') = (1 - p_{sk}) \cdot \left(-(\sum_{i=k+1}^{L} p_{l_i}^k \log p_{l_i}^k) - \gamma \log \gamma \right). \qquad (4)$$

Until now, we assumed a simple binary decision as to whether a location sample belongs to a regular visited place (i.e. cluster) or not, hence $\varepsilon \simeq 0$ and a function $C_O(l) = |\{o \in O \mid loc(o) \cong l\}|$ counting the number of times a user was observed at a given location $l \in \mathbb{C}^*$ (cf. section 3.4 above), making it possible to rank the places by their popularity ($l_1, l_2, \ldots$ where $C_O(l_i) \geq C_O(l_{i+1})$ – which means that l_1 is the most frequently visited location). Location information observed by mobile communication infrastructure is error prone. Depending on the communication infrastructure used, users can make assumptions on the physical limitations of the involved technology and thus can estimate a best case value for ε. In order to model the adversary's uncertainty we introduce p_c as the probability of function C^E assigning o' correctly to a location $l \in \mathbb{C}^*$, taking $\varepsilon = err(o')$ into account (and $\overline{p_c} := 1 - p_c$). As the precise definition of p_c depends on the implementation of C^E we only assume a correlation between the error and this probability: $p_c \sim \varepsilon^{-1}$.

However, modeling the adversary's uncertainty based on ε is in practice both difficult and possibly harmful to the user, since the adversary's capabilities might be underestimated, which may result in a higher and misleading privacy level. Hence, in order to get a robust reflection on a user's frequently visited places, using a clustering approach leads to an efficient but also abstract representation of the user's regular behavior. Several studies (e.g. [15,29]) demonstrate that clustering is an effective tool for identification of a user's significant places.

Thus, instead of modeling the uncertainty of an adversary in assigning an observation to a certain location l_i, the spatial size of a possible location cluster is increased by the estimated spatial error. Thus, the adversary's uncertainty can be translated into the problem of choosing a single location out of all possible (and plausible) locations within the clustered spatial area. This uncertainty can be calculated using map data. Fig. 2 shows the resulting clusters for GPS data (left) and GSM data (right) from a 17-day trace with hourly location observations (GSM) and an estimated error of 250 m (GSM).

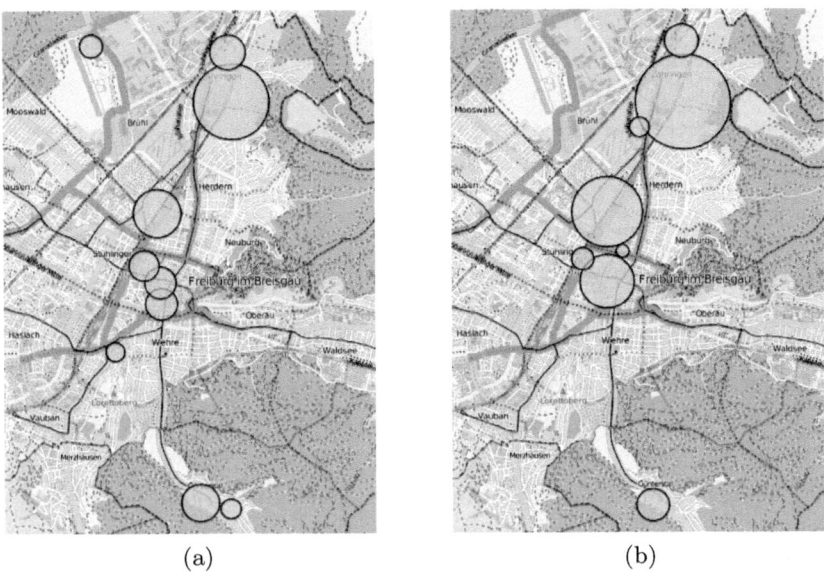

(a)   (b)

Fig. 2. Clusters generated by a 17-day GPS trace (a) and (b) 17 days of hourly location updates (GSM) with an estimated spatial error of 250 m. The radius of each cluster denotes its significance for the user.

3.5 Determining an Adversary's Knowledge Gain

With the uncertainty value before and after disclosure of o', an adversary only gains new information if a new frequently visited location is uncovered and can be calculated as $\Delta K_A^L(O, o') = K_A^L(O, o') - K_A^L(O \setminus \{o_m\}, o_m)$ where o_m is the latest location observation in O (and therefore the direct predecessor of o').

If o' can be assigned to a known location $l_i \in L$, then $\Delta K_A^L = 0$, as by definition no information about new frequently visited places is revealed. However, the weight of already determined frequently visited places may change due to such an observation. Furthermore, people's preferences are not static and hence neither are their preferences as to frequently visited places. For instance, people change employer and/or move from time to time. Such changes in regular behavior disclose private information and thus compromise the user's privacy. To model these changes, the observation horizon can be limited and any information older than a certain amount of time could be discarded.

To model changes in the frequency of the user's top locations and a user's regular behavior, we measure the change in the distribution made by a new observation. The adversary's a-priori knowledge is the distribution of the time spent in all known locations and hence their relative importance to the user. Thus, an adversary gains extra knowledge if the distribution of time spent changes, i.e., the user's preferences change. For every detected location we assume that the true probability $q(O, o', l_i) := Pr_O(loc(o') \cong l_i)$ is the relative observed importance of location l_i derived from the previous observations in O (e.g. $Pr_O(loc(o') \cong l_i) \sim C_O(l)$). We define the information gain as the difference between the observed distribution before and after a disclosure of additional data. One simple method for measuring the information gain is the relative entropy using KL-divergence [30]

$$K_A^C(O, o') = -\sum_{i=1}^{k} q(O, o', l_i) \log \frac{q(O, o', l_i)}{q((O \cup o'), o', l_i)} \quad , \tag{5}$$

where $q(O, o', l_i)$ denotes the probability of returning to l_i before and $q((O \cup o'), o', l_i)$ the new probability after the new observation o'.

Finally, we express the privacy loss as

$$\Delta K_A(O, o') = \Delta K_A^L(O, o') + K_A^C(O, o'). \tag{6}$$

4   Case Study GSM Network

In contrast to previous work with focus on analyzing call data records, our focus was on uncovering the side effects of using a mobile handset, since these are currently the most personal devices we know. The mobile handset is a highly sensitive device not only due to the huge subscriber count, but especially because people hardly do any activity without keeping their mobile phones nearby. The aforementioned studies showed the expressiveness of call data records. However, we are focusing especially on location updates, since these are scheduled periodically and configuration among network providers differs significantly. While one mobile telephony provider requires a client to initiate a LU every 60 minutes, another one only requires updates every 12 hours. The remaining two telephony providers configured their networks requiring four and six hour intervals respectively. From a user's privacy perspective, location updates are especially threatening because of their regularity but also because these events happen without the user noticing.

The information on the network infrastructure and its configuration a user gets through the handset's UI is usually limited to the mobile operator name, signal strength of the serving cell and type of network connection (e.g. GSM or 3G). In order to analyze the user's exposure, we developed a logger device to record any communication between the GSM infrastructure and a mobile phone. This device was carried by test persons; however, the phone was kept in a passive / idle mode, i.e., no phone calls were made or received. The logger device is based on a Nokia 3310 phone. These phones are able to provide raw network data through a specific debug interface [7]. This data can be recorded, decoded and analyzed in a second step. To make this setup mobile, a micro controller writing the data to an SD-Card was attached. Furthermore, a GPS device was added to tag the network data with a time stamp and to record the user's movements. With this method we could not directly determine the knowledge the network infrastructure has. However, we could record each time the user was exposed to the infrastructure, and a network-based determination of his location was possible or location data was generated by the network infrastructure (e.g. measurement of the timing advance). In order to simulate the larger error of GSM positioning, a random radial error of ε was added to the GPS position. For our experiments we implemented a cluster algorithm based on a radius filter. For periodic and gap-based location data (e.g. GSM) such a filter simply reflects the frequency a user was observed at a specific place. Additionally, for GPS data a gap filter was used. With this method we were able to detect the places a user revisited frequently. Throughout the experiments a value of $\tau = 0.3$ was used, which roughly represents the results from the aforementioned studies on human mobility patterns. Furthermore, 12 clusters were expected. We assumed $\varepsilon = 250\,\text{m}$ as the average positioning accuracy of the GSM network.

4.1   Data Analysis

The first analyzed data set was created by a test person carrying the logger device for about 17 days equipped with a SIM-card from a German provider which requires location updates hourly. 17744 GPS points and 312 location updates were recorded. The reason that the number of location updates is lower than anticipated is twofold: the first and the last day were not complete, but there were also signal losses and user operation errors like empty batteries. However, these results should correspond with real life mobile phone usage. During that time a total of 10 clusters could be identified through the GPS data, 8 based on GSM data. The remaining clusters found by the GPS method were not detected. This is due not only to the short evaluation period and lower spatial resolution of the GSM positioning but also to the short length of stay at the remaining two places (i.e. less than one hour). In a test trial with 6 hourly location updates and an equal test period, only three clusters could be detected. Fig. 3a shows the temporal development of the discovery of frequently visited places using different methods and location sample frequency. Fig. 3b shows the adversary's knowledge on the remaining, yet uncovered, places. In a further trial with 12-hour location update intervals, within 8 days only a single cluster could

---

[7] GSM decoding with a Nokia 3310 phone,
  https://svn.berlin.ccc.de/projects/airprobe/wiki/tracelog, [5/15/2011]

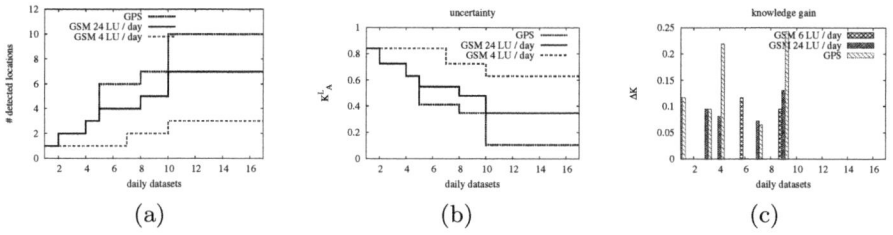

Fig. 3. Modeling an adversary's knowledge gain using GSM and GPS data sources. (a) shows the temporal development of detecting frequently visited places, (b) shows the uncertainty assigning a new observation to a yet unknown frequently visited place and (c) shows the adversary's knowledge gain (ΔK) by disclosing a daily data set.

be determined. One reason was the disadvantageous time points at 7:30 AM and 7:30 PM. However, the time points were chosen by the network. For this configuration a long term trial is pending. Due to long distance traveling, offline phases, and time periods without reception, we expect random shifts for the time point of location updates. Therefore, for a long term observation it seems likely that some (2-3) additional clusters should be detected. The probability of detecting a cluster where a user spends large amounts of time is more likely and thus is likely to be uncovered first. The privacy measurement also implicitly captures the distribution of the observed location samples. If the distribution of location samples is concentrated within certain time spans, fewer clusters will be discovered. The same applies for evenly distributed but sparse samples (e.g. every 12 hours).

4.2 Privacy Improvements

Based on the aforementioned analysis, several enhancements could improve the user's privacy in mobile communication networks. First, one can observe that a simple quantitative privacy policy as offered by network providers, stating only the length of possible data storage is neither meaningful nor helpful for a sub-scriber w.r.t. location privacy. Especially the density of periodic location samples makes a significant difference as to the provider's possible knowledge base and thus the user's present and future privacy risks. Therefore, subscribers also need to know when, how and to what extent location information is generated. With such knowledge the user's awareness as to his privacy loss is raised. In a second step the user should be able to control location dissemination by making informed decisions.

A *privacy aware* mobile phone requires software interfaces to the mobile phone stack controlling and exposing signaling attempts (e.g. detecting silent text messages), measurement reports and the occurrence of location updates. With the help of *osmoconbb* GSM baseband implementations, first steps toward a privacy aware phone were made. The mobile station is able to log location data and expose it to the user, which is sent to the service provider The measurement results sent by the MS during location updates include signal strength measurements from surrounding BTS. The measurement information is used for the handover decision during the connection. Since a LU requires only a very brief communication with the network, a handover between different cells is very unlikely or even impossible. Thus, sending measurements of neighboring stations is technically not always required. If the number of transmitted measurements is reduced or completely omitted, the accuracy of the network's position estimation is significantly decreased. In the best case (no measurements transmitted), the accuracy is decreased to cell origin with timing advance. A further step to decrease the accuracy of the position determination is transmitting modified or false measurement information. To decrease the accuracy of the position estimation further, a MS could send with a slight timing offset. Such offsets have direct impact on the timing advance calculation of the BTS. Consequently, this leads to an incorrect distance estimation between MS and BTS. The combination of manipulating measurement results and timing advance gives the possibility to conceal the actual position of the MS. The rough position of the MS is still given by the BTS used and its covered area.

5   Conclusion

We fully acknowledge that the evaluation is based on too little data to be statistically significant. However, the data clearly indicates that network configuration has an impact on the user's location privacy. Furthermore, the proposed metric gives the user a tool to understand the impact of mobile communication on his privacy without knowing the adversary's capabilities or behavior.

In contrast to other personal digital devices, mobile phones are hardly ever switched off, thus offering unique options for (unobserved) user tracking. The disclosure or detection of any significant place decreases the user's privacy by roughly the same level, independent of the relative importance or rank of the place. The proposed user model and accordingly the privacy metric showed that the user's privacy loss is roughly the same for all detected clusters. Especially for a setting with semi-trusted adversaries, this result reflects the (commercial) importance of lower ranked clusters w.r.t. the completeness of a user's profile. Since lower ranked clusters are harder to detect, the ability to uncover such a place reflects the density and/or the length of observation by an adversary and thus on the user's exposure. Therefore, network configuration is crucial for the individual's privacy. In our tests we saw time ranges for location updates range from 60 minutes to 12 hours with different results for detecting frequently visited places. These different network configurations have a significant impact on the user's location privacy especially if privacy policies only specify the duration of data storage. Finally, countermeasures to improve the user's privacy were proposed. The evaluation of the effectiveness of the proposed actions remains for future work.

References

1. Gonzalez, M.C., Hidalgo, C.A., Barabasi, A.L.: Understanding individual human mobility patterns. Nature 453, 779–782 (2008)
2. Krumm, J.: Ubiquitous advertising: The killer application for the 21st century. IEEE Pervasive Computing 99 (2010)
3. Gruteser, M., Grunwald, D.: Anonymous usage of location-based services through spatial and temporal cloaking. In: MobiSys 2003: Proceedings of the 1st International Conference on Mobile Systems, Applications and Services, pp. 31–42. ACM, New York (2003)
4. Ma, C.Y., Yau, D.K., Yip, N.K., Rao, N.S.: Privacy vulnerability of published anonymous mobility traces. In: Proceedings of the Sixteenth Annual International Conference on Mobile Computing and Networking, MobiCom 2010, pp. 185–196. ACM, New York (2010)
5. Bettini, C., Wang, X.S., Jajodia, S.: Protecting privacy against location-based personal identification. In: Jonker, W., Petković, M. (eds.) SDM 2005. LNCS, vol. 3674, pp. 185–199. Springer, Heidelberg (2005)
6. Golle, P., Partridge, K.: On the anonymity of home/Work location pairs. In: Tokuda, H., Beigl, M., Friday, A., Brush, A., Tobe, Y. (eds.) Pervasive 2009. LNCS, vol. 5538, pp. 390–397. Springer, Heidelberg (2009)
7. Bayir, M., Demirbas, M., Eagle, N.: Discovering spatiotemporal mobility profiles of cellphone users. In: IEEE International Symposium on a World of Wireless, Mobile and Multimedia Networks & Workshops, WoWMoM 2004, pp. 1–9 (2009)
8. Isaacman, S., Becker, R., Cáceres, R., Kobourov, S., Rowland, J., Varshavsky, A.: A tale of two cities. In: HotMobile 2010: Proceedings of the Eleventh Workshop on Mobile Computing Systems; Applications, pp. 19–24. ACM, New York (2010)
9. Girardin, F., Calabrese, F., Dal Fiorre, F., Biderman, A., Ratti, C., Blat, J.: Uncovering the presence and movements of tourists from user-generated content. In: Proceedings of International Forum on Tourism Statistics (2008)
10. Girardin, F., Vaccari, A., Gerber, A., Biderman, A., Ratti, C.: Towards estimating the presence of visitors from the aggragate mobile phone network activity they generate. In: Proceedings of International Conference on Computers in Urban Planning and Urban Management (2009)
11. Sohn, T., Varshavsky, A., LaMarca, A., Chen, M., Choudhury, T., Smith, I., Consolvo, S., Hightower, J., Griswold, W., de Lara, E.: Mobility detection using everyday GSM traces. In: Dourish, P., Friday, A. (eds.) UbiComp 2006. LNCS, vol. 4206, pp. 212–224. Springer, Heidelberg (2006)
12. De Mulder, Y., Danezis, G., Batina, L., Preneel, B.: Identification via location-profiling in gsm networks. In: Proceedings of the 7th ACM Workshop on Privacy in the Electronic Society, WPES 2008, pp. 23–32. ACM, New York (2008)
13. Lee, C.H., Hwang, M.S., Yang, W.P.: Enhanced privacy and authentication for the global system for mobile communications. Wirel. Netw. 5, 231–243 (1999)
14. Ardagna, C., Jajodia, S., Samarati, P., Stavrou, A.: Privacy preservation over untrusted mobile networks. In: Bettini, C., Jajodia, S., Samarati, P., Wang, X. (eds.) Privacy in Location-Based Applications. LNCS, vol. 5599, pp. 84–105. Springer, Heidelberg (2009)
15. Hoh, B., Gruteser, M., Xiong, H., Alrabady, A.: Achieving guaranteed anonymity in gps traces via uncertainty-aware path cloaking. IEEE Transactions on Mobile Computing 9, 1089–1107 (2010)

16. Shokri, R., Freudiger, J., Jadliwala, M., Hubaux, J.P.: A distortion-based metric for location privacy. In: WPES 2009: Proceedings of the 8th ACM Workshop on Privacy in the Electronic Society, pp. 21–30. ACM, New York (2009)
17. Beresford, A., Stajano, F.: Location privacy in pervasive computing. IEEE Pervasive Computing 2, 46–55 (2003)
18. Duckham, M., Kulik, L.: A formal model of obfuscation and negotiation for location privacy. In: Gellersen, H.-W., Want, R., Schmidt, A. (eds.) PERVASIVE 2005. LNCS, vol. 3468, pp. 152–170. Springer, Heidelberg (2005)
19. 3rd Generation Partnership Project (3GPP): TS 24.008 Technical Specification Group Core Network and Terminals; Mobile radio interface Layer 3 specification; Core network protocols; Stage 3 (Release 10) (2010)
20. 3rd Generation Partnership Project (3GPP): TS 45.008 Technical Specification Group GSM/EDGE Radio Access Network; Radio subsystem link control (Release 9) (2010)
21. 3rd Generation Partnership Project (3GPP): TS 43.059 Technical Specification Group GSM/EDGE Radio Access Network; Functional stage 2 description of Location Services (LCS) in GERAN (Release 9) (2009)
22. 3rd Generation Partnership Project (3GPP): TS 25.305 Technical Specification Group Radio Access Network; Stage 2 functional specification of User Equipment (UE) positioning in UTRAN (Release 10) (2010)
23. Sun, G., Chen, J., Guo, W., Liu, K.: Signal processing techniques in network-aided positioning: a survey of state-of-the-art positioning designs. IEEE Signal Processing Magazine 22, 12–23 (2005)
24. Zimmermann, D., Baumann, J., Layh, A., Landstorfer, F., Hoppe, R., Wolfle, G.: Database correlation for positioning of mobile terminals in cellular networks using wave propagation models. In: IEEE 60th Vehicular Technology Conference, VTC 2004-Fall, vol. 7, pp. 4682–4686 (2004)
25. Haeb-Umbach, R., Peschke, S.: A novel similarity measure for positioning cellular phones by a comparison with a database of signal power levels, vol. 56, pp. 368–372 (2007)
26. Westin, A.F.: Privacy and Freedom, 1st edn. Atheneum, New York (1967)
27. Duckham, M., Kulik, L.: Location privacy and location-aware computing, pp. 35–51. CRC Press, Boca Raton (2006)
28. Díaz, C., Seys, S., Claessens, J., Preneel, B.: Towards measuring anonymity. In: Dingledine, R., Syverson, P.F. (eds.) PET 2002. LNCS, vol. 2482, pp. 54–68. Springer, Heidelberg (2003)
29. Ashbrook, D., Starner, T.: Using gps to learn significant locations and predict movement across multiple users. Personal Ubiquitous Comput. 7, 275–286 (2003)
30. Kullback, S., Leibler, R.A.: On information and sufficiency. The Annals of Mathematical Statistics 22, 79–86 (1951)

How Much Is Enough?
Choosing ϵ for Differential Privacy

Jaewoo Lee and Chris Clifton

Department of Computer Science, Purdue University,
West Lafayette, IN 47907
{jaewoo,clifton}@cs.purdue.edu
http://www.cs.purdue.edu

Abstract. Differential privacy is a recent notion, and while it is nice conceptually it has been difficult to apply in practice. The parameters of differential privacy have an intuitive theoretical interpretation, but the implications and impacts on the risk of disclosure in practice have not yet been studied, and choosing appropriate values for them is non-trivial. Although the privacy parameter ϵ in differential privacy is used to quantify the privacy risk posed by releasing statistics computed on sensitive data, ϵ is not an absolute measure of privacy but rather a relative measure. In effect, even for the same value of ϵ, the privacy guarantees enforced by differential privacy are different based on the domain of attribute in question and the query supported. We consider the probability of identifying any particular individual as being in the database, and demonstrate the challenge of setting the proper value of ϵ given the goal of protecting individuals in the database with some fixed probability.

Keywords: Differential Privacy, Privacy Parameter, ϵ.

1 Introduction

As volumes of personal data collected by many organizations increase, the problem of preserving privacy is increasingly important. The potential social benefits of analyzing such datasets drive many organizations to be interested in releasing statistical information about the data. In the field of privacy preserving data analysis, the main goal is to release statistical information about sample databases safely without compromising the privacy of any individuals who's records contribute to the database. These two conflicting objectives pose challenging trade-off between providing useful information about the population and protecting the privacy of any individuals.

Privacy laws typically protect *individually identifiable data*; data that cannot be linked to an individual is not considered a privacy risk. Unfortunately, what it means for data to be *individually identifiable* is not simple to define. Statistical summaries can reveal information about a single individual, particularly if an adversary knows information about other individuals. Differential Privacy [6] provides a strong guarantee of privacy even when the adversary has arbitrary

X. Lai, J. Zhou, and H. Li (Eds.): ISC 2011, LNCS 7001, pp. 325–340, 2011.

external knowledge. Basically, differential privacy hides the presence of an individual in the database from data users by making two output distributions, one with and the other without an individual, be computationally indistinguishable (for all individuals). To achieve this, differential privacy uses an output perturbation technique which adds random noise to the outputs. The magnitude of noise to add, which determines the degree of privacy, depends on the type of computation and it must be large enough to conceal the largest contribution that can be made to the output by one single individual. To be specific, let X be a database to release statistics about and f be a query function. ϵ-differentially private mechanism gives perturbed response $f(X)+Y$ instead of the true answer $f(X)$, where Y is the random noise.

While this seems a perfect solution, the amount of noise needed to achieve indistinguishability between two datasets generally eliminates any useful information. The actual definition is for ϵ-differential privacy (see Definition 1), where the ϵ factor is a difference between the probabilities of receiving the same outcome on two different databases. ϵ becomes a parameter on the degree of privacy provided. ϵ is a relative measure since it bounds the data user's information gain, instead of the absolute amount. Even for the same value of ϵ, the probability of identifying an individual enforced by differential privacy is different depending on the universe.

Unfortunately, ϵ does not easily relate to practically relevant measures of privacy. For example, assume a very simple problem where an adversary wants to determine the value of a binary attribute about an individual - simply "is the individual in the dataset" (such as a research dataset for diabetes, where simply revealing presence in the dataset places an individual at risk of discrimination.) What we would really like is a measure of the risk to an individual – what is the probability that an individual is in the dataset given release of statistical information about the data? If disclosure allows an adversary to calculate too high a probability that the individual is in the dataset, then that individual's privacy (in legal terms, which typically protect "individually identifiable data") is at risk. This is addressed for anonymization in [14], and the problem would seem a perfect match for differentially privacy. The challenge is that choosing an appropriate value of ϵ turns out to be quite challenging.

The problem is that protecting privacy requires knowing not only the data to be protected, but also the entire universe of individuals from which that data might be drawn. This is a known challenge with differential privacy, as calculating the sensitivity of a query is based on all *possible* databases differing by a single value. This may be an inherent problem with protecting privacy; δ-presence [14] faces the same issue (although an approximation based on univariate statistics is given in [13].) What makes this a particular problem for differential privacy is that not only do we need to know the entire universe to use a differentially private mechanism, it is also needed to determine an appropriate value of ϵ. In this paper, we will show that given a goal of controlling probabilistic disclosure of the presence of an individual, the proper of ϵ varies depending on individual values, even for individuals not in the dataset.

To see this, imagine the following (hypothetical) scenario. Purdue University has put together a "short list" of alumni as possible commencement speakers. A local newspaper is writing a story on the value Indiana taxpayers get from Purdue, and would like to know if these distinguished alumni are locals or world travelers. Purdue does not want to reveal the list (to avoid embarrassing those that are not selected, for example), but is willing to reveal the average distance people on the list have traveled from Purdue in their lifetimes. Using a differentially private mechanism to add noise to the resulting average will protect individuals - but how much noise is needed? Outliers in the data, such as Purdue's Apollo astronaut alumni (who have been nearly 400,000km from campus), result in a requirement for a significant amount of noise. More critically, we show that such outliers also change the appropriate value for ϵ. As a result, simply setting parameters to be used for differential privacy is an unsolved problem.

Although differential privacy has been extensively studied in many papers, to our best knowledge, no studies have been conducted toward the issues on the application of differential privacy in practice. In many papers, the value of privacy parameter ϵ is chosen arbitrarily or assumed to be given. This leaves an impression that ϵ can be freely chosen as needed but, in reality, decision on the value of ϵ should be made carefully with considerations of the domain and the acceptable ranges of risk of disclosure. In this paper, we illustrate why the choice of ϵ is important using the perspective of the risk of revealing presence and how an inappropriate value of ϵ can cause a privacy breach. We also show that a value of ϵ that is appropriate for a particular universe of values may lead to a breach with a different set of values.

2   Related Work

The concept of differential privacy was motivated by the impossibility of absolute protection [4] against adversaries with arbitrary external information [5]. In a differentially private mechanism, what a potential adversary can learn from interactions with the mechanism is limited (within a multiplicative factor) no matter what external information the adversary has. Essentially, what can be learned from a dataset with a particular individual also can be learned from a dataset without that individual [9,11]. This definition enables a privacy model that does not need to make assumptions on an adversary's external information, a key limitation of prior work on protecting privacy. A line of research on indistinguishability between two neighboring databases leads to emergence of differential privacy. [2,9,10]

The notion of differential privacy has received much theoretical attention in the privacy community and has been extensively studied in the literature [2,3, 10,8,1]; a recent survey on differential privacy is provided in [7]. However, most research on differential privacy has focused on exploring theoretical properties of the model. The main focus of study has been how to safely release database while preserving privacy for a particular function f. For example, [5] studies how to release count queries and [9] touches on more general query functions such as histograms and linear algebraic functions. The concept of global sensitivity was introduced in [6] and it has been shown that releasing a database with noise proportional to the global sensitivity of the query functions achieves differential privacy. Nissim et al. [15] expanded the framework of differential privacy by introducing *smooth sensitivity*, which reduces the amount of noise added. It is motivated from the observation that, for many types of query functions, the local sensitivity is small while global (worst-case) sensitivity is extremely large. To decide the magnitude of noise, they use a smooth upper bound function S, which is an upper bound on local sensitivity.

There are a few implementations supporting differential privacy. PINQ [12] is an implementation of differential privacy that provides answers to SQL queries in a differentially private way. AIRAVAT [16] is another system that applies differential privacy mechanism for MapReduce computation in a cloud computing environment. Although their system has been built upon differential privacy framework, this doesn't mean that privacy is actually enforced by the system. It is still the responsibility of users who use the system to select the value of ϵ that prohibits any inferences on the dataset beyond what is allowed.

3   Differential Privacy

A database D is a collection of data elements drawn from the universe U. A row in a database corresponds to an individual whose privacy needs to be protected. Each data row consists of a set of attributes $A = A_1, A_2, ..., A_m$. The set of values each attribute can take, attribute domain, is denoted by $dom(A_i)$ where $1 \leq i \leq m$. A mechanism $\mathcal{M} : D \rightarrow \mathbb{R}^d$ is a randomized function that maps database D to a probability distribution over some range and returns a vector of randomly chosen real numbers within the range. A mechanism $\mathcal{M}$ is said to be ϵ-differentially private if adding or removing a single data item in a database only affects the probability of any outcome within a small multiplicative factor. The formal definition of an ϵ-differentially private mechanism is:

Definition 1 (ϵ-differentially private mechanism). *A randomized mechanism $\mathcal{M}$ is ϵ-differentially private if for all data sets D_1 and D_2 differing on at most one element, and all $S \subseteq Range(\mathcal{M})$*

$$Pr[\mathcal{M}(D_1) \in S] \leq \exp(\epsilon) \times Pr[\mathcal{M}(D_2) \in S]$$

This paper considers an interactive privacy mechanism and the same assumptions as in [6]. In an interactive model, users issue queries to the database and receive a noisy response where the magnitude of noise added to the response is determined based on the query function f. Sensitivity of a query function f represents the largest change in the output to the query function which can be made by a single data item.

Table 1. Example database X

Name	School year	Absence days
Chris	1	1
Kelly	2	2
Pat	3	3
Terry	4	10

Definition 2 (Global Sensitivity). *For the given query function* $f : D \to \mathbb{R}^d$, *the global sensitivity of* f *is*

$$\Delta f = \max_{x,y} |f(x) - f(y)|$$

for $\forall x, y$ *differing in at most one element.*

Let $Lap(\lambda)$ be the Laplace distribution whose density function is $h(x) = \frac{1}{2\lambda} \exp(-\frac{|X-\mu|}{\lambda})$ where μ is a mean and $\lambda(> 0)$ is a scale factor. Dwork et al. proved that, for the given query function f and a database X, a randomized mechanism $\mathcal{M}_f$ that returns $f(X) + Y$ as an answer where Y is drawn i.i.d from $Lap(\frac{\Delta f}{\epsilon})$, gives ϵ-differential privacy [9].

4 Example: Mean

In this section, we illustrate how the value of ϵ should be adjusted according to the change of domain (or universe) of the attribute in question to enforce the same level of protection. While essentially the same as the problem described in Section 1, we switch to a different motivation both to show the generality of this problem and to give realistic numbers that are easy to demonstrate.

Consider a database consisting of 4 students registered for a course, which includes each student's name, school year, and number of absence days in the previous semester. Let X denotes the database. In X, there is one student, Terry, who was placed on academic probation in the previous semester. Table 1 shows the example database. Note that Terry's number of absence days is relatively large compared to those of other students. Assume that the school wants to release data on students who have not been on academic probation to support academic success research. Let X' denote the database to be published, i.e., $X' = X - \{Terry\}$. Since knowing if a student has been on academic probation is clearly a privacy breach, the school allows faculty and staff (who may know the year in school and absence days of individual students, but not who has been placed on probation) to query X' only via an ϵ-differentially private mechanism. For the purpose of illustration, let us assume that the goal of the data provider is to hide the presence of data contributors (or, in this case, non-contributors) by keeping the adversary's probability of identifying their presence in the database less than 1/3. Throughout the example, we show that what values of ϵ needed to achieve that goal for queries on mean year and mean absence days, and demonstrate that in spite of the similarity between the data columns (in fact identical for the data in X'), a different value of ϵ is needed depending on which column is queried.

4.1   Achieving Differential Privacy

We now describe an ϵ-differentially private mechanism κ for releasing the mean school year and absence days of students in X'. Informally, the goal of κ is to make the query responses from any databases that differ in only one element be indistinguishable within a factor of e^ϵ, so that the absence of Terry in X' (and thus probation status) isn't revealed. At the same time, the privacy of individuals who participated in the database X' needs to be protected as well. The sensitivity of the mean query function Δf is computed by measuring the maximum change in the query output caused by a single individual. Notice that calculating the sensitivity requires global knowledge on that domain since every possible attribute value that not only presently exists in X' but also could exist needs to be considered. For any possible data instance Y of size 3 and a tuple t of Y, Δf is determined as the maximum value among the results of the following computation:

$$\Delta f(X') = \max_{Y \subset X} |f(Y) - f(Y - t)|\ where\ |Y| = 3$$

From our example, it is calculated as follows:

$$\Delta f = \left| \frac{1 + 2 + 10}{3} - \frac{1 + 2}{2} \right| = \frac{17}{6}$$

For now, let's assume $\epsilon=2$; we later show how this choice of ϵ discloses the information. The random noise drawn from the Laplace distribution with mean 0 and scale factor $\lambda = \frac{\Delta f}{\epsilon}$ is 1.1677. The query response γ is produced as $\gamma = \kappa_f(X') = f(X') + Lap(\frac{\Delta f}{\epsilon}) = 2 + 1.1677 = 3.1677$. Consider the following probabilities:

$$\frac{Pr[\kappa_f(X) > 3.1677]}{Pr[\kappa_f(X') > 3.1677]} = 3.2933 \leq e^2$$

Note that the above value computed from the cumulative density function of the Laplace distribution. The adversary cannot distinguish the response from queries against X and X' within the factor of e^2, so differential privacy is achieved. However, does this also mean that Terry's privacy is protected?

4.2   Adversary Model

In this example (as is the goal of differential privacy), we assume a very strong adversary who has complete knowledge on the universe, i.e., full access to all records in the universe U; thus each attribute value of all records in X is known to the adversary. In other words, the adversary can potentially access the records of every student in this school. The adversary knows everything about the universe except that which individual is missing in the database X' (i.e., who is on academic probation). In addition to the complete knowledge about X, the adversary knows the fact that X' consists of students who have never been on academic probation. Assuming an adversary having complete knowledge about each individual in the database is not unrealistic because differential privacy is supposed to provide privacy given adversaries with arbitrary background knowledge.

In our model, the adversary has a database X consisting of n records, i.e., knowledge of the exact attribute values of each individual in X, and has an infinite computational power. Given a database X' with $n-1$ records sampled from X (i.e., $X' \subset X$ and $|X'| = |X|-1$), the adversary's goal is to figure out absence of a victim individual in X' by using knowledge of X. This is identical to find out other individuals' presences in X'. With respect to our example, a privacy breach is to allow the adversary to guess absence/presence of an individual in X' correctly with high probability.

4.3   Attack Model

To determine membership in X', the adversary maintains a set of tuples $\langle \omega, \alpha, \beta \rangle$ for each possible combination ω of X', where α and β are the adversary's prior belief and posterior belief on $X' = \omega$ given a query response. Let Ψ denote the set of all possible combinations of X'. For simplicity, we assume α is a uniform prior, i.e., $\forall \omega \in \Psi, \alpha(\omega) = \binom{n}{n-1} = \frac{1}{n}$. We refer to each possible combination ω in Ψ as a possible world. The posterior belief β is defined in Definition 3.

Definition 3 (Posterior belief on $X' = \omega$). *Given the query function f and the query response $\gamma = \kappa_f(X')$, for each possible world ω, the adversary's posterior belief on ω is defined as:*

$$\beta(\omega) = P(X' = \omega | \gamma) = \frac{P(\kappa_f(\omega) = \gamma)}{P(\gamma)} = \frac{P(\kappa_f(\omega) = \gamma)}{\sum_{\psi \in \Psi} P(\kappa_f(\psi) = \gamma)}$$

where κ_f is an ϵ-differentially private mechanism for the query function f.

The posterior belief $\beta(\omega)$ represents the adversary's changed belief on each possible world that the underlying database being queried against is ω. To figure out which individuals are in the database, the adversary issues a query against X' and gets a noisy answer. After seeing the query response, the adversary computes the posterior belief for each possible world. Finally, the adversary selects one with the highest posterior belief as a "best guess". The confidence of the adversary's guess is calculated using Definition 4.

Definition 4 (Confidence level). *Given the best guess ω', the adversary's confidence in guessing the missing element is defined as*

$$conf(\omega') = \beta(\omega') - \alpha(\omega')$$

As the adversary's posterior belief on each possible world becomes large, the chances of disclosing any individual's presence in the database also become high, which makes disclosure of the statistics. This has an implication that the adversary's posterior belief on each possible world can be thought of as the risk of disclosure.

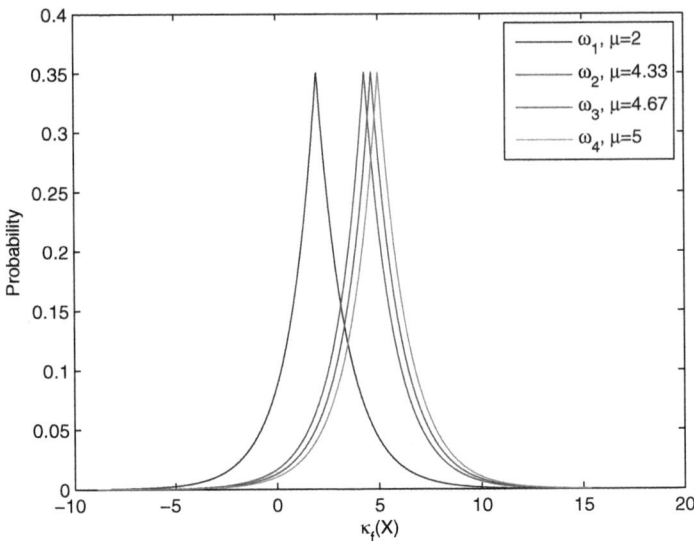

Fig. 1. Query response distributions

Definition 5 (Risk of disclosure Γ). *Given the set of possible worlds Ψ, the risk of disclosing presence/absence of any individual in the database Γ is defined as the adversary's maximum posterior belief.*

$$\Gamma = \max_{\omega \in \Psi} \beta(\omega)$$

4.4   Limitation of Differential Privacy

Basically, the underlying assumption that differential privacy is relying on is that, if two extreme query answers that can be produced from any dataset possible in the universe are indistinguishable, the presence or absence of any individual can be hidden. The difference between those two extreme answers is masked by random noise. However, there is a problem with this approach. Although differential privacy ensures that every possible database of the same size is indistinguishable within some factor ϵ, there always exists a distribution that is more likely than others given the query response. For example, in Figure 1, ω_1 is the most likely to be the true distribution among 4 possible worlds given the response $\gamma=1$. This allows the adversary to improve the belief of each possible world after seeing the response.

In our example, for illustration we assume $U = X$ and, without loss of generality, the response is 2.2013. Recall that the sensitivity Δf of mean query function for the domain of absence days is $\frac{17}{6}$. The adversary's posterior belief of ω_1 when $\epsilon = 2$ is:

Table 2. Posterior belief $\beta(\omega)$

Possible world(ω)	$\epsilon = 5$	$\epsilon = 2$	$\epsilon = 1$	$\epsilon = 0.5$	$\epsilon = 0.1$	$\epsilon = 0.01$
{1, 2, 3}	0.9705	0.5519	0.4596	0.3477	0.2328	0.2482
{1, 2, 10}	0.0159	0.1859	0.2019	0.2305	0.2527	0.2503
{1, 3, 10}	0.0087	0.1463	0.1791	0.2171	0.2558	0.2506
{2, 3, 10}	0.0049	0.1159	0.1594	0.2048	0.2588	0.2509

Table 3. Posterior belief $\beta(\omega)$

Possible world(ω)	$\epsilon = 5$	$\epsilon = 2$	$\epsilon = 1$	$\epsilon = 0.5$	$\epsilon = 0.1$	$\epsilon = 0.01$
{1, 2, 3}	0.9705	0.6825	0.4596	0.3477	0.2680	0.2518
{1, 2, 10}	0.0158	0.1315	0.2017	0.2303	0.2469	0.2497
{1, 3, 10}	0.0088	0.1039	0.1793	0.2172	0.2440	0.2494
{2, 3, 10}	0.0049	0.0821	0.1594	0.2048	0.2411	0.2491

$$\beta(\omega_1) = \frac{P(\kappa_f(\omega) = 2.2013)}{\sum_{i=1}^{4} P(\kappa_f(\omega_i) = 2.2013)}$$

$$= \frac{0.3602}{0.3062 + 0.0784 + 0.0619 + 0.0489} = 0.6180$$

The adversary will come to the conclusion that $X' = \omega_1$. Even though the output, mean of absence days, is released via a differentially private mechanism, the adversary can still make a correct guess on who is absent from the list with high probability and confidence. Consider the probability when the attribute queried is school year. The sensitivity of mean query function for the school year domain is $\frac{5}{6}$. The adversary's posterior belief on ω_1 for this case is 0.3390. Although the same parameter and response values are used for both cases, the resulting adversary's probabilities are significantly different. Notice that the adversary's posterior belief, $\beta(\omega)$, is a random variable and Table 2 and Table 3 show two different instances of it. As shown in Table 2, an adversary's best guess would be $X' = \{\text{Chris}, \text{Kelly}, \text{Pat}\}$, which is correct, with the confidence of 0.7705=0.9705-0.25 when ϵ=5. When ϵ=0.5, the adversary still get it right but the confidence is only 0.0977(0.3477-0.25). When ϵ=0.01, the adversary fails to make the correct guess given the set of query responses.

5   Choice of ϵ

The previous section demonstrated that given sufficiently low ϵ, ϵ-differential privacy does limit an adversary's ability to identify an individual. However, as lowering ϵ reduces the utility of the answer, the question of the proper value of ϵ is still open. We now demonstrate how to choose ϵ to control the adversary's success at identification of an individual in this particular scenario, and demonstrate the difficulties that arise.

5.1   Upper Bound on Adversary's Posterior Belief

Let $X = \{x_1, x_2, \ldots, x_{n-1}, v\}$ and $X' = \{x_1, x_2, \ldots, x_{n-1}\}$. Without loss of generality, assume that elements are sorted in ascending order (i.e., $x_1 < x_2 < \ldots < x_{n-1} < v$). Two databases X and X' are identical except that only one element, v, is missing in X'. For illustration, we impose an ordering to the enumeration of possible worlds. Let ω_i denote the ith possible world maintained by an adversary and x_i^k be the kth smallest element of ω_i. For any $\omega_i, \omega_j \in \Psi$, ω_i is lower than ω_i if $\forall k s.t. 1 \le k \le n, x_i^k \le x_i^k$. For example, $\omega_i = \{x_1, x_2, \ldots, x_{n-1}\}$, $\omega_2 = \{x_1, x_2, \ldots, x_{n-2}, v\}$, $\omega_3 = \{x_1, x_2, \ldots, x_{n-3}, x_{n-1}, v\}$, etc. To get an upper bound on the adversary's probability of a correct guess, we have to assume the worst case in which the correct answer seems to be most likely (i.e., $\gamma = f(\omega_i)$).[1] Given the query response γ, the adversary's posterior probability on $X' = \omega_i$ is $\beta(\omega_i) = \frac{P(\kappa_f(\omega_i)=\gamma)}{\sum_{k=1}^{n} P(\kappa_f(\omega_k)=\gamma)}$. If we divide numerator and denominator of $\beta(\omega_i)$ by $P(\kappa_f(\omega_i) = \gamma)$,

$$\beta(\omega_i) = \frac{1}{1 + \sum_{k=1, k \neq i}^{n} \frac{P(\kappa_f(\omega_k)=\gamma)}{P(\kappa_f(\omega_i)=\gamma)}} \tag{1}$$

$$= \frac{1}{1 + \sum_{k=1, k \neq i}^{n} \frac{e^{-\frac{|\gamma - f(\omega_k)|}{\lambda}}}{e^{-\frac{|\gamma - f(\omega_i)|}{\lambda}}}} \tag{2}$$

$$\le \frac{1}{1 + \sum_{k=1, k \neq i}^{n} e^{-\frac{|f(\omega_i) - f(\omega_k)|}{\lambda}}} \tag{3}$$

$$\le \frac{1}{1 + \sum_{k=1, k \neq i}^{n} e^{-\frac{\Delta v}{\lambda}}} \tag{4}$$

$$= \frac{1}{1 + (n-1)e^{-\frac{\epsilon \Delta v}{\Delta f}}} \tag{5}$$

where $\Delta v = \max_{1 \le i, j \le n} |f(\omega_i) - f(\omega_j)|$ and $i \neq j$.

In (4), the distance between $f(\omega_i)$ and $f(\omega_k)$ is approximated with Δv, the longest distance between $f(\omega_i)$ and $f(\omega_j)$ where $1 \le i, j \le n$. Recall that in our example, the missing value is the largest in the database, which means ω_1 is a true distribution. Under this condition, the distance between $f(\omega_i)(1 \le i \le n)$ and $f(\omega_1)$ has little difference, which makes (3) and (4) approximately the same. Therefore, the upper bound becomes tight. In order to make every possible world look equally likely, the following equation needs to be satisfied:

$$\frac{1}{1 + (n-1)e^{-\frac{\epsilon \Delta v}{\Delta f}}} = \frac{1}{n}$$

---

[1] This "all possible worlds" knowledge is the same information needed to calculate global sensitivity for differential privacy; extending differential privacy to work with more limited information is beyond the scope of this paper, and would face similar challenges to those addressed for generalization-based anonymization in moving from [14] to [13].

The value of ϵ which satisfies the equation is 0. Therefore, to make every possible dataset in the universe to look equally likely, the query results would be pure random noise, providing no utility.

5.2   Determining the Right Value of ϵ

We now show how the proper value of ϵ can be chosen given the goal of hiding any individual's presence (or absence) in the database. Assume that the privacy requirement for our example dataset is to limit the any individual's probability of being identified as present in the database to be no greater than $\frac{1}{3}$. We show two ways of selecting a good choice of ϵ that guarantees the probability of identifying any individual's presence is no greater than the maximum tolerable value δ. One is to use the upper bound presented in Section 5.1; the other is to search for the right value. We first consider how the upper bound on the adversary's probability can be utilized to enforce the requirement. Let ρ be the probability of being identified as present in the database. We need to find ϵ that satisfies the following inequality.

$$\frac{1}{1 + (n-1)e^{-\frac{\epsilon \Delta v}{\Delta f}}} \leq \rho \tag{6}$$

Rearranging yields

$$\epsilon \leq \frac{\Delta f}{\Delta v} \ln \frac{(n-1)\rho}{1 - \rho} \tag{7}$$

Note that the greater n and ρ are, the greater the minimum ϵ needed. Therefore, as the size of database to publish and the probability to bound get larger, less noise need be added.

In our example, the maximum distance between function values of every possible world, Δv, is $f(\omega_4) - f(\omega_1) = 5 - 2 = 3$. Thus, in order to enforce the adversary's probability to be no greater than $\frac{1}{3}$,

$$\frac{1}{1 + (n-1)e^{-\frac{\epsilon \Delta v}{\Delta f}}} \leq \frac{1}{3} \tag{8}$$

$$\epsilon \leq \frac{17}{18} \ln(\frac{3}{2}) \approx 0.3829 \tag{9}$$

Let's consider how this value changes when the attribute to release is the school year rather than absence days. In this case, the sensitivity Δf and the maximum distance between possible answers Δv need to be recalculated since those are the parameter values that are completely dependent on the universe. The recomputed values of Δf and Δv are $\frac{5}{6}$ and 1, respectively. Note that the mean value of school year information is less sensitive than that of absence days, which results in smaller values of both Δf and Δv. The right value of ϵ for the school year domain using the upper bound is $\epsilon \leq \frac{5}{6} \ln \frac{3}{2} \approx 0.3379$. However, even when $\epsilon = 0.5(> 0.3379)$, the risk of disclosure is

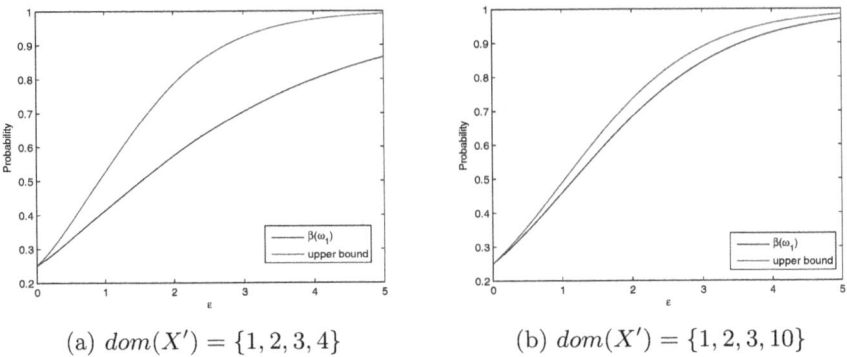

(a) $dom(X') = \{1,2,3,4\}$ (b) $dom(X') = \{1,2,3,10\}$

Fig. 2. Upper bound on the risk of disclosure by varying domains

$$\max_i \beta(\omega_i) = \beta(\omega_1) \tag{10}$$

$$= \frac{1}{1 + e^{-\frac{1}{5}} + e^{-\frac{2}{5}} + e^{-\frac{3}{5}}} \tag{11}$$

$$\approx 0.3292 < \frac{1}{3}, \tag{12}$$

which is still lower than the maximum acceptable level of risk. This means that a more precise value of ϵ can be found. As shown in Figure 2, our upper bound on the risk of disclosure is not tight when the domain does not include outliers. In other words, when the domain of the attribute to be released has low sensitivity, the upper bound of Section 5.1 gives a value of ϵ that may be significantly greater than the actual value of ϵ needed to satisfy the privacy constraint (the actual impact on the amount of noise added follows from the differential privacy literature.) Although the value obtained using the upper bound ensures that any individual's risk of being identified as present in the database is no greater than δ, this might be overkill, especially when there is no value that significantly deviates from the mean of that distribution. In such case, we can perform binary search to determine the maximum ϵ that meets the requirement. Before performing the search, the range within which the value of ϵ will be searched needs to be determined. The minimum would be 0 which means no information can be learned while the maximum can be calculated using the upper bound above. Let ϵ_s and ϵ_f denote the beginning and end of the range to search, respectively. Firstly, compute the risk of disclosure when $\epsilon = \frac{\epsilon_s + \epsilon_f}{2}$. Next, if it is greater than δ, set ϵ_f to the current value of ϵ. Otherwise, set ϵ_s to ϵ. Repeat this search process until the maximum value of ϵ that satisfies the constraint is found.

6 Example: Median

We now show that the appropriate value of ϵ is dependent not only on the data and universe of values, but also on the query to be computed. This section

Table 4. Senstivity of median for the database X

Possible world(ω)	$LS_f(\omega)$
$\{1, 2, 3\}$	0.5
$\{1, 2, 10\}$	4
$\{1, 3, 10\}$	3.5
$\{2, 3, 10\}$	3.5

repeats the previous analysis on the example student database shown in Table 1, but where the query is instead *median*.

Let $f(X) = median(x_1, x_2, ..., x_n)$ where x_i are real numbers. The median of a finite list of numbers is defined to be the middle one when all the observations are arranged from lowest value to highest value. If a dataset has an even number of observations, there may be no single middle value; in this case we define the median to be the mean of the two middle numbers. Without loss of generality, assume $x_1 \leq \cdots \leq x_n$; this gives the following definition for median:

$$f(X) = \begin{cases} x_k & \text{for } n = 2k - 1 \quad (13) \\ \dfrac{x_k + x_{k+1}}{2} & \text{for } n = 2k \quad (14) \end{cases}$$

where k is a positive integer. To calculate sensitivity, let X' be a database obtained by removing one element from X. If X has an odd number of elements (i.e., $n = 2k - 1$), $f(X) = x_k$ and $f(X')$ could be $\frac{x_k + x_{k+1}}{2}$, $\frac{x_{k-1} + x_k}{2}$ or $\frac{x_{k-1} + x_{k+1}}{2}$. On the other hand, if X has an even number of elements (i.e., $n = 2k$), $f(X) = \frac{x_k + x_{k+1}}{2}$ and $f(X')$ is either x_k or x_{k+1}. Thus, the local sensitivity of median for X is

$$LS_f(X) = \begin{cases} \max\left(\frac{x_{k+1} - x_k}{2}, \frac{x_k - x_{k-1}}{2}, \left|\frac{x_{k+1} + x_{k-1}}{2} - x_k\right|\right) & \text{for } n = 2k - 1 \\ \frac{x_{k+1} - x_k}{2} & \text{for } n = 2k \end{cases} \quad (15)$$

and the global sensitivity of median is

$$\Delta f(X) = \max_{\omega \in \Psi} LS_f(\omega)$$

Table 4 shows the sensitivity of median for the absence days attribute of our example database. As with Section 5.2, our target is that the adversary's probability estimate for the value of this attribute should be no greater than $\frac{1}{3}$. In our example, the maximum difference between medians of each possible world is 1 (i.e., $\Delta v = \max_{1 \leq i,j \leq 4} |f(\omega_i) - f(\omega_j)| = 1$). Applying the upper bound gives

$$\frac{1}{1 + (n-1)e^{-\frac{\epsilon \Delta v}{\Delta f}}} \leq \frac{1}{3} \quad (16)$$

$$\frac{1}{1 + 3e^{-\frac{\epsilon}{4}}} \leq \frac{1}{3} \quad (17)$$

$$\epsilon \leq 4\ln(\frac{3}{2}) \approx 1.6219 \quad (18)$$

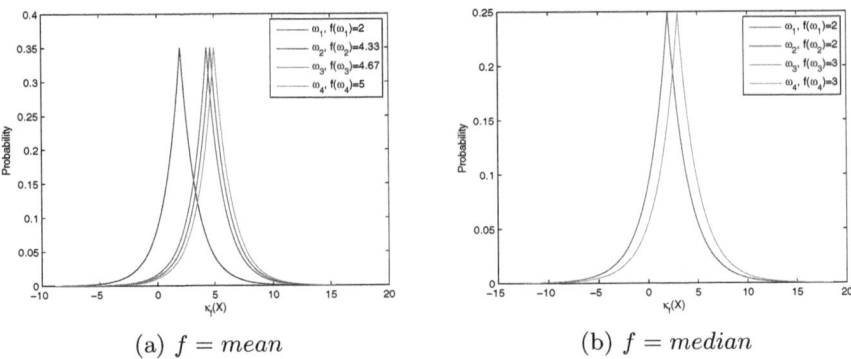

(a) $f = mean$ (b) $f = median$

Fig. 3. Distributions of each possible world for different type of queries

A more precise upper bound can be found by replacing Δv with exact values as follows:

$$\beta(\omega_i) = \frac{1}{1 + \sum_{k=1, k\neq i}^{n} e^{-\frac{|f(\omega_i)-f(\omega_k)|}{\lambda}}} \tag{19}$$

$$= \frac{1}{1 + e^0 + e^{-\frac{1}{\lambda}} + e^{-\frac{1}{\lambda}}} \tag{20}$$

$$= \frac{1}{2 + 2e^{-\frac{1}{\lambda}}} \tag{21}$$

The inequality $\frac{1}{2+2e^{-\frac{\epsilon}{\Delta f}}} \leq \frac{1}{3}$ leads to $\epsilon \leq 4\ln 2 \approx 2.776$.

Recall that the sensitivity of the mean function for the database X' is $\frac{17}{6} \approx 2.83$ and $\epsilon \leq 0.3829$ to limit the adversary's probability no greater than $\frac{1}{3}$. With the same universe, we can allow larger epsilon ($\epsilon \leq 2.776$) to enforce the same degree of privacy for the median whose sensitivity is larger than that of the mean. This is because the type of query affects the distributions of possible world. In Figure 3(a), given the response $\gamma < 3.16$, ω_1 is significantly more likely than others. On the other hand, in Figure 3(b), given any response value of γ, both ω_1 and ω_2 are always equally likely and difference of likelihood between each possible world is relatively small.

7   Conclusion

While the concept of differential privacy has received considerable attention in the literature, there has been little discussion of how to apply it in practice. Although ϵ is the privacy parameter for differential privacy, it does not directly correlate to a practical privacy standard. We have shown that given a practical standard, namely the risk of identifying an individual, it is possible to determine an appropriate value of ϵ. However, this requires knowing the queries to be computed, the universe of data, and the subset of that universe to be queried. While this is not a disabling issue, as such knowledge (except the subset to be queried, presumably known to the data holder) is typically required to construct a differentially private mechanism anyway, it does raise additional research challenges. Succinctly, any discussion of a differentially private mechanism requires a discussion of how to set an appropriate ϵ for that mechanism, a challenge that may be as or more difficult than developing the mechanism itself.

Acknowledgments. Partial support for this work was provided by MURI award FA9550-08-1-0265 from the Air Force Office of Scientific Research.

References

1. Barak, B., Chaudhuri, K., Dwork, C., Kale, S., McSherry, F., Talwar, K.: Privacy, accuracy, and consistency too: a holistic solution to contingency table release. In: Proceedings of the Twenty-Sixth ACM SIGMOD-SIGACT-SIGART Symposium on Principles of Database Systems, pp. 273–282. ACM, New York (2007)
2. Blum, A., Dwork, C., McSherry, F., Nissim, K.: Practical privacy: the SuLQ framework. In: Proceedings of the Twenty-Fourth ACM SIGMOD-SIGACT-SIGART Symposium on Principles of Database Systems, pp. 128–138. ACM, New York (2005)
3. Blum, A., Ligett, K., Roth, A.: A learning theory approach to non-interactive database privacy. In: Proceedings of the 40th Annual ACM Symposium on Theory of Computing, pp. 609–618. ACM, New York (2008)
4. Dalenius, T.: Towards a methodology for statistical disclosure control. Statistik Tidskrift 15(429-444), 2–1 (1977)
5. Dinur, I., Nissim, K.: Revealing information while preserving privacy. In: Proceedings of the Twenty-Second ACM SIGMOD-SIGACT-SIGART Symposium on Principles of Database Systems, pp. 202–210. ACM, New York (2003)
6. Dwork, C.: Differential privacy. In: Bugliesi, M., Preneel, B., Sassone, V., Wegener, I. (eds.) ICALP 2006, Part II. LNCS, vol. 4052, pp. 1–12. Springer, Heidelberg (2006)
7. Dwork, C.: Differential privacy: A survey of results. In: Agrawal, M., Du, D.-Z., Duan, Z., Li, A. (eds.) TAMC 2008. LNCS, vol. 4978, pp. 1–19. Springer, Heidelberg (2008)
8. Dwork, C., Kenthapadi, K., McSherry, F., Mironov, I., Naor, M.: Our data, ourselves: Privacy via distributed noise generation. In: Vaudenay, S. (ed.) EUROCRYPT 2006. LNCS, vol. 4004, pp. 486–503. Springer, Heidelberg (2006)
9. Dwork, C., McSherry, F., Nissim, K., Smith, A.: Calibrating noise to sensitivity in private data analysis. In: Halevi, S., Rabin, T. (eds.) TCC 2006. LNCS, vol. 3876, pp. 265–284. Springer, Heidelberg (2006)
10. Dwork, C., Nissim, K.: Privacy-preserving datamining on vertically partitioned databases. In: Franklin, M. (ed.) CRYPTO 2004. LNCS, vol. 3152, pp. 528–544. Springer, Heidelberg (2004)
11. Kasiviswanathan, S., Smith, A.: A note on differential privacy: Defining resistance to arbitrary side information. Arxiv preprint arXiv:0803.3946 (2008)
12. McSherry, F.: Privacy integrated queries: an extensible platform for privacy-preserving data analysis. In: Proceedings of the 35th SIGMOD International Conference on Management of Data, pp. 19–30. ACM, New York (2009)

13. Nergiz, M.E., Clifton, C.: δ-presence without complete world knowledge. IEEE Transactions on Knowledge and Data Engineering 22(6), 868–883 (2010), http://doi.ieeecomputersociety.org/10.1109/TKDE.2009.125
14. Nergiz, M., Atzori, M., Clifton, C.: Hiding the presence of individuals from shared databases. In: Proceedings of the 2007 ACM SIGMOD International Conference on Management of Data, Beijing, China, June 11-14, pp. 665–676 (2007), http://doi.acm.org/10.1145/1247480.1247554
15. Nissim, K., Raskhodnikova, S., Smith, A.: Smooth sensitivity and sampling in private data analysis. In: Proceedings of the Thirty-Ninth Annual ACM Symposium on Theory of Computing, pp. 75–84. ACM, New York (2007)
16. Roy, I., Setty, S., Kilzer, A., Shmatikov, V., Witchel, E.: Airavat: Security and privacy for MapReduce. In: Proceedings of the 7th USENIX Conference on Networked Systems Design and Implementation, p. 20. USENIX Association (2010)

Non-interactive CDH-Based Multisignature Scheme in the Plain Public Key Model with Tighter Security

Yuan Zhou[1,2], Haifeng Qian[1,*], and Xiangxue Li[1]

[1] Department of Computer Science and Technology,
East China Normal University, China
hfqian@cs.ecnu.edu.cn
[2] Network Emergency Response Technical Team/Coordination Center, China

Abstract. A multisignature scheme allows an ad hoc set of users to sign a message so that the resulting single signature certifies that the users endorsed the message. However, all known multisignatures are either at the price of complexity and additional trust of Certificate Authority (CA), or sacrificing efficiency of computation and communication (including both bandwidth and round). This paper proposes a new multisignature scheme with efficient verification in the plain public key model. Our multisignatures enjoys the most desired features: (1) Our plain public key model-based multisignatures do not impose any impractical key setup or PKI requirements; (2) Our multisignature scheme is non-interactive, which saves computation and communication in signature generation; (3) Through pre-computation, our scheme achieves $\mathcal{O}(1)$ verification in the plain public key model; (4) Provable tighter security under the standard CDH assumption ensures high level of security in both practice and theory. Hence, our non-interactive multisignatures are of great use in authentication of routes in networks.

1   Introduction

Multisignatures extend standard digital signatures to allow an ad hoc set of ℓ users to jointly sign a message which are very useful in applications such as contract signing, authentication of routes in mobile networks, distribution of a certificate authority (CA), aggregation of acknowledgements in multicast and secure vehicular communications (refer to [1,2,3,4,5,6,7,8,9,10] for detailed application scenarios).

Due to many practical reasons for applications, we desire that a multisignature scheme might have the following features: (1) the resulting signature is shorter than ℓ separate signatures, even constant for ℓ; (2) Multisignature generation and verification (even key generation) are very fast; (3) The communication overhead including rounds and messages transmitted in multisignature generation should be as fewer as possible (even reduced to a minimum); (4) Trust on

---

* Corresponding author.

X. Lai, J. Zhou, and H. Li (Eds.): ISC 2011, LNCS 7001, pp. 341–354, 2011.

the trusted third party (e.g., Certificate Authority) should be reduced as less as possible. All these specific aspects are important in the real life applications of multisignatures.

1.1 Rogue Key Attacks

In order to generate multisignatures of constant size, the homomorphic properties of arithmetic operations involved in standard signatures are often used. However, these homomorphic properties that enable aggregation of signatures into multisignatures often incurs rogue key attacks for multisignature schemes. In this attack, the adversary chooses its public key as a function of those of honest signers in such a way that it can forge multi-signatures easily[2,5]. For example, rogue key attack succeeds if the verification key (combination of public keys of signers) for multisignature has a fixed formula as $PK = \prod_i^{\ell} pk_i$ where the adversary can choose $pk_a = g^s \cdot (\prod_i^{\ell-1} pk_i)^{-1}$ and know the corresponding private key s for $PK = pk_a \prod_i^{\ell-1} pk_i = g^s$ (e.g., [11,12,3,13,14] are vulnerable to such attacks in the plain setting). Actually rogue-key attack is considered as a main menace for discrete logarithm based multi-signature schemes.

Early approaches prevent rogue-key attacks for multisignatures at the cost of complexity and expense in the scheme, or using unrealistic and complicate key setup assumptions on the public-key infrastructure (PKI)[1]. These key setup operation assumptions include *dedicated key generation* (DKG), *knowledge of Secret Key* (KOSK), *proof of possesion of private key* (POP) and *the plain public key model* (PPK).

The first effort to prevent rogue key attacks (called DKG) is due to Micali et al. [1]. However, the DKG is impractical because of expensive interactions of key generation, complex and large public keys, and static group of signers. The second approach, KOSK assumption needs a party to prove knowledge of its secret key, during public key registration with a certificate authority (CA). The requirement of handing over the secret keys leads to obtaining simple constructions and proofs of security [3,13]. However, the existing public key infrastructures (PKIs) do not require proofs of knowledge of secret keys [15].

The third one is contributed by [14] and [5,6] where users are required to provide proofs of procession of secret keys (during key registrations) in order to prevent rogue key attacks. Ristenpart and Yilek in [14] showed that the schemes in [3,13] by using the Key Registration (KR) model can be improved more secure without reducing efficiency. A similar idea named the Key Verification (KV) model was later proposed in [5,6], but having the multisignature receiver verify the POP message (together with verification of PKI certificates)[16], instead of the CAs during the key registration.

We note that either the KR model or the KOSK assumption needs non-standard trust on the CAs because it requires the CAs must perform specific verifications. If a CA is corrupted then the adversary can easily forge multisignature through rogue key attacks. On the other hand, the interaction and verification during key registration also causes additional computational burden for the

CAs. While the KV model causes additional computational cost of the verifier (linear to the number of signers) during the verification of a multisignature.

Obviously, it is highly interesting and desirable to provide multisignature schemes which are secure in the plain setting where no special registration process is assumed for public keys registration. Bellare and Neven provided such a scheme in the plain (public-key) model [2], the others are shown in [5,17,18]. In the plain (public-key) model, there is no dedicated key generation (DKG) procedures, or well-formedness proofs accompanied to the public keys and a party can obtain a certificate on any arbitrary key without any proof.

1.2   Multisignatures in the PPK Model

So far as we know, there are only very few multisignature schemes with provable security in the plain public key model and all these multisignatures are proved secure in the random oracle model [19].

The first multisignature in such a model was proposed by Boneh et al [18] (for brevity BGLS), which is implied by the construction so-called aggregate signature [20]. The main drawback of the BGLS scheme is the high cost of verification. In fact verification of a single multisignature of the BGLS scheme requires $\mathcal{O}(\ell)$ pairings where ℓ is the number of signers participating in signing, making it extremely impractical.

Under the DL assumption Bellare and Neven [2], provided a concrete multisignature scheme (denoted BN) with rather an efficient verification in the PPK model. Followed the idea of [2], a lot of *interactive* multisignature schemes are proposed [5,17] in the PPK model. However, the interaction is quite expensive in many important application because communicating even one bit of data may use significantly more power than executing one 32-bit instruction [21] and also in many settings, communication is not reliable, and so the fewer interactions, the better.

The only known non-interactive multisignature scheme with efficient verification in the PPK model (the QX scheme) is proposed by Qian and Xu quite recently in [22]. Comparing with the BGLS scheme, the QX scheme improves verification efficiency by reducing pairing computation complexity $\mathcal{O}(\ell)$ to $\mathcal{O}(1)$. However, security reduction of the QX scheme is even looser than that of the BN scheme[2]. Intuitively, a tight security (proof) means that the scheme is almost as hard to break as the underlying cryptographic problem to solve. Therefore, it is always welcome to find a non-interactive multisignature scheme in the PPK model with more tighter security reduction.

1.3   Our Contributions

We present a non-interactive multisignature scheme, which operates in the plain public-key model and is proven secure based on the standard CDH assumption in the random oracle model. Compared with the BGLS scheme, our scheme minimize the verification cost, by reducing $(\ell + 1)$ pairings to two pairings through

Table 1. Multisignature Comparison

Scheme	Assump.	Rounds	Sign	Comm. Cost	Verify	σ Size
BN [2]	< DL	3	1 exp	$|\mathbb{G}'| + |q| + l_0$	1 $\mathsf{mexp}^{(\ell+1)}$	$|\mathbb{G}'| + |q|$
BGLS [18,20]	$\approx$ CDH	1	1 exp	$|\mathbb{G}|$	$(\ell+1)$ pr	$|\mathbb{G}|$
QX[22]	<< CDH	1	1 exp	$|\mathbb{G}|$	$2\mathsf{pr} + \mathsf{mexp}^{(\ell+1)}$	$|\mathbb{G}|$
Ours	$\approx$ CDH	1	2 exp	$2\,|\mathbb{G}|$	2 pr	$2\,|\mathbb{G}|$

Remark 1. For each scheme, we summarize the underlying cryptographic assumption; number of rounds of the signing protocol ("1" means non-interactive); the computational complexity for each signer (Sign); the communication cost required for the signing (Comm. Cost) ; the computational complexity for verifying a multisignature (Verify); the size of multisignature (σ Size). For CDH-based schemes we assume the symmetric pairing $e : \mathbb{G} \times \mathbb{G} \to \mathbb{G}_T$ for convenient comparison. For DL-based schemes, we assume we work over a 160-bit elliptic-curve (EC) group $\mathbb{G}'$. We assume the order of $\mathbb{G}$, $\mathbb{G}_T$ and $\mathbb{G}'$ are equal (i.e., $p = q$). We denote by exp an exponentiation in group $\mathbb{G}$ (or $\mathbb{G}'$ and $\mathbb{G}_T$) whose order is q (or p), and by $\mathsf{mexp}^{(t)}$ a multi-exponentiation with t exponentiation coefficients (e.g., $\mathsf{mexp}^{(2)}$ corresponds to $g^{k_1} h^{k_2}$ for some g, h, k_1, k_2), by ℓ the number of signers, by l_0 the length of hash value, by pr a bilinear pairing, by $|\mathbb{G}|$ the bit length of the representation for elements in group $\mathbb{G}$, and by $|p|$ the bits length of p. DL stands for Discrete Logarithm, CDH stands for Computational Diffie-Hellman. " << " and " < " means very loose security and loose security, respectively. " $\approx$ " means close security.

pre-computation (cf. Table 1)[1]. The improvement in verification time is substantial because one pairing costs about 6-20 exponentiations [2]. Given ℓ signers, the verification key is fixed (consisting of ℓ partial verification keys that can be derived from the public keys independently), which means that we can compute once and for all, the verification key before signing or verification.

On the other hand, our scheme also enjoys a tighter security proof, compared with both the BN scheme and the QX scheme. Then security parameters of our scheme could be smaller than those of both the BN scheme and the QX scheme, while preserving the same security level. Actually, our security proof shows that an adversary can at most with probability roughly $q_s \cdot \varepsilon$ break our multisignature scheme where ε is the upper bound of probability for breaking the underlying cryptographic problem. While in the BN scheme the corresponding upper bound is roughly $\sqrt{q_h \cdot \varepsilon}$. Let $q_h = 2^{60}$, $q_s = 2^{30}$, $\varepsilon = 2^{-80}$, our scheme ensures 50 bits of security level, while the BN scheme (or the QX scheme) ensures at most 10 bits of security level in practice. Compared with the QX scheme, our scheme also saves the mult-exponentiation in verification of a multisignature. Detailed comparisons amongst our scheme, the BN scheme, the BGLS scheme and the QX scheme are depicted in Table 1.

---

[1] Since the signers' public keys are used as prefixes to the corresponding message in the BGLS scheme, verification of the BGLS multisignature needs $(\ell + 1)$ pairings necessarily.

Our technique for dealing with the rogue key attack in the plain public-key model is quite different from those of [2]. Instead of using a dynamic key with respective to messages [2] as the verification key for the multisignatures, we use a combined key derived from the public keys of signers, irrelevant to messages. Therefore, we can pre-compute the verification key for any group of signers before knowing the signed messages. Such pre-computation could be done when the certificates of public keys are verified.

The structure of our multisignatures are similar to that of the WMS scheme [13] and its variant [14] which enables us to reduces the communication rounds of multisignatures to optimal (non-interactive). The computational cost and communication cost (the amount of data transmit in generating a single multisignature) are the same as the WMS scheme [13] whose security holds under the KOSK assumption, but not secure in the PPK model. Moreover, our multisignature reaches high level of security, in fact our scheme is almost as secure as the Computational Diffie-Hellman (CDH) problem.

2   Preliminaries

We recall the basic definitions for multisignatures, then review the cryptographic complexity assumption in this section. We the notations of [22] in the following.

2.1   Definitions of Multisignatures

The definition of interactive multisignatures in the plain public-key model was first formalized in [2] with ℓ signers, each having as input its own public and private keys as well as the public keys of the other signers. The signers interact via a protocol to generate a multisignature.

In this paper, we consider the following non-interactive variant: Given the same inputs as in an interactive scheme, each signer contributes a partial signature *without* interacting with each other, and the partial signatures can be "assembled" into a multisignature by any one.

A non-interactive multisignature scheme MS = (Setup, Gen, MSign, MVf) consists of three algorithms and one protocol (adapted from [2]):

- Setup(1^λ): This is a randomized algorithm that takes as input a security parameter λ and produces a set of global public parameters pp. (This algorithm should be run by a trusted party and pp can also be viewed as a common reference string.)
- Gen(pp): This is a randomized algorithm that, on input of public parameters pp, outputs (an honest) signer i's private/public key pair (sk_i, pk_i).
- MSign($pp, \{sk_i\}, M, L$): Given public parameters pp, a message M and a (multi)set $L = (pk_1, \ldots, pk_\ell)$ of purported signers (note that $pk_i = pk_j$ for some $i \neq j$ is allowed in the plain public-key model because one can simply claim another's public key as its own), signer i uses its private key sk_i to compute a partial signature σ_i and then broadcasts σ_i. Given σ_j for $j = 1, \ldots, \ell$, any one can obtain a multisignature σ with respect to the public keys on L.

– MVf(pp, M, σ, L): Given $L = (pk_1, \ldots, pk_\ell)$, pp, a message M, a multisignature σ, this deterministic algorithm outputs 0 (reject) or 1 (accept).

We require a multisignature scheme to be *correct*, meaning that every multisignature σ obtained from the partial signatures of legitimate signers (according to
MSign) is always accepted as valid.

Experiment $\mathrm{Exp}^{MS}_{uu.cma}(\mathcal{A})$:

1. $pp \leftarrow \mathsf{Setup}(1^\lambda)$; $(pk^\star, sk^\star) \leftarrow \mathsf{Gen}(pp)$; $\mathcal{M} \leftarrow \phi$ where $\mathcal{M}$ is the set of
   pairs (M, L) previously queried for signatures.
2. Run $\mathcal{A}(pp, pk^\star)$ as follows:
   – $\mathcal{A}$ can choose arbitrary public key for any user, possibly as a function
     of the honest user's public key $pk^\star$.
   – To obtain a multisignature, $\mathcal{A}$ can invoke the execution of
     $\mathsf{MSign}(\cdot, \cdot, \cdot, \cdot)$ (concurrently) by presenting a message M and a
     (multi)set $L = (pk_1, \ldots, pk_n)$ for any n, as long as $pk^\star$ appears at
     least once in L.
   – If the multisignature scheme uses hash functions that are treated as
     random oracles, $\mathcal{A}$ can submit strings and obtain their corresponding
     hash values.
3. Eventually, $\mathcal{A}$ outputs an alleged multisignature $\sigma^\star$ on a message $M^\star$ with
   respect to $L^\star = (pk_1, \ldots, pk_\ell)$. If

   $$\mathsf{MVf}(pp, M^\star, \sigma^\star, L^\star) = 1$$

   and

   $$(pk^\star \in L^\star) \wedge (M^\star, L^\star) \notin \mathcal{M} = 1$$

   then return 1, otherwise 0.

Fig. 1. Experiment for security definition

Intuitively, security of multisignature scheme requires that it is infeasible for
adversary $\mathcal{A}$ to forge a multisignature with respect to a new message which
extends the classical security notion of digital signature scheme known as existential unforgeability under adaptively chosen-message attacks [23]. Following
[2,5], we can assume there is a single honest signer. Then, unforgeability must
hold despite that in the plain public-key model, the adversary can corrupt all
other signers and choose their public keys arbitrarily (even registering the public key of the honest user as their own public keys), and can interact with the
honest signer in any number of concurrent signing instances before outputting
its forgery.

Formally, we define the advantage of $\mathcal{A}$ against multisignature scheme MS as
the probability that the experiment $\mathrm{Exp}^{MS}_{uu.cma}(\mathcal{A})$ in Figure 1 outputs 1.

We say the adversary $(t, q_s, q_h, \ell, \varepsilon)$-breaks multisignature scheme MS, if it in time t, after q_s signature queries or q_s invocations of $\mathsf{MSign}(\cdot, \cdot, \cdot, \cdot)$, and optionally q_h queries to the hash functions (if any) that are treated as random oracles, has an advantage at least ε in forging a multisignature co-signed by ℓ signers, namely

$$\Pr[\mathsf{Exp}^{\mathsf{MS}}_{\mathsf{uu.cma}}(\mathcal{A}) = 1] \geq \varepsilon.$$

If there is no such adversary, we say the multisignature scheme is $(t, q_s, q_h, \ell, \varepsilon)$-secure.

2.2 Cryptographic Complexity Assumption

Let $\mathbb{G}$ and $\mathbb{G}_T$ be two (multiplicative) cyclic groups of prime order p where the group action on $\mathbb{G}$ and $\mathbb{G}_T$ can be computed efficiently, g be a generator of $\mathbb{G}$, $e : \mathbb{G} \times \mathbb{G} \to \mathbb{G}_T$ be an efficiently computable map (i.e., pairing) with the following properties:

- Bilinear: for all $(u, v) \in \mathbb{G} \times \mathbb{G}$ and $a, b \in \mathbb{Z}_p$, $e(u^a, v^b) = e(u, v)^{ab}$;
- Non-degenerate: $e(g, g) \neq 1$.

For specific applications, we recommend the asymmetric setting, namely $\mathbb{G}_1 \neq \mathbb{G}_2$ for bilinear maps (i.e., $e : \mathbb{G}_1 \times \mathbb{G}_2 \to \mathbb{G}_T$), that allows for short signatures without side-effect. Our scheme is also adaptable for such a setting. For more details, refer to [24,25].

We define the computational Diffie-Hellman problem (with pairings) as follow.

Definition 1 (CDH). *Given* $(g, g^a, h) \in \mathbb{G} \times \mathbb{G} \times \mathbb{G}$ *for some* $a \xleftarrow{R} \mathbb{Z}_p$ *and* $h \xleftarrow{R} \mathbb{G}$, *find* $h^a \in \mathbb{G}$.

Define the success probability of an algorithm $\mathcal{A}$ solving the CDH problem as

$$\mathsf{Adv}^{\mathrm{cdh}}_{\mathcal{A}} \stackrel{\mathrm{def}}{=} \Pr\left[h^a \leftarrow \mathcal{A}(g, g^a, h) : g \xleftarrow{R} \mathbb{G}, a \xleftarrow{R} \mathbb{Z}_p, h \xleftarrow{R} \mathbb{G}\right].$$

The probability is taken over the uniform random choice of g from $\mathbb{G}$, of a from $\mathbb{Z}_p$, of h from $\mathbb{G}$, and the coin tosses of $\mathcal{A}$. We say the algorithm $\mathcal{A}$ (t, ε)-solves the CDH problem if $\mathcal{A}$ runs in time at most t and $\mathsf{Adv}^{\mathrm{cdh}}_{\mathcal{A}} \geq \varepsilon$. We say the CDH problem is (t, ε)-intractable if there is no algorithm $\mathcal{A}$ that can (t, ε)-solve it.

3 Our Construction

Our scheme uses the Waters-like signature (e.g., $\sigma = (sk \cdot H(m)^r, g^r)$) to construct multisignatures, however this scheme is different from the WMS multisignature scheme in [13] since security of our scheme is proved in the plain public key model (with random oracles) while the WMS multisignature (whose security is proved in the KOSK model) must impose additional requirement on the traditional PKIs to ensure security (e.g., it requires the CAs and users to perform additional protocols to get public key certificated).

Our scheme does not require the signers to share a common random or commit [2,17] during multisignature generation as well, otherwise it seems impossible to reduce the round of signing to minimum. Although our construction is similar to the WMS scheme [13], security of multisignatures in the plain public key model does not rely on the trust of the CAs because of the plain public key model. While security of the WMS scheme are proved in the KOSK model that means security also relies on the trust of the CAs. Therefore, our scheme reduces the trust of the third party (e.g., CA) because even a malicious CA can't do any harm to the honest users in our system.

Our scheme consists of the following algorithms (or protocols):

Setup(1^λ): On input a security parameter λ, select global public parameters $pp = (\mathbb{G}, \mathbb{G}_T, p, g, e, H, H_m)$, where $H_m : \{0,1\}^* \rightarrow \mathbb{G}$ and $H : \mathbb{G} \rightarrow \mathbb{G}$ are secure hash functions (viewed as random oracles here). This algorithm may be run by a trust party.

Gen(pp): On input pp, an honest user i selects $x_i \stackrel{R}{\leftarrow} \mathbb{Z}_p$ and its private/public key pair is (sk_i, pk_i) where

$$sk_i = H(pk_i)^{x_i}, \quad pk_i = g^{x_i}.$$

MSign($pp, \{sk_i\}, M, L$): On input pp, message M and $L = (pk_1, \ldots, pk_\ell)$, user i ($1 \leq i \leq \ell$) executes the following:

1. Pick a random $r_i \stackrel{R}{\leftarrow} \mathbb{Z}_p$ and compute

$$s_i \leftarrow sk_i \cdot H_m(M||L)^{r_i} \quad \text{and} \quad t_i \leftarrow g^{r_i}.$$

2. Broadcast $\sigma_i = (s_i, t_i)$ as the partial signature for message M (which is a standard signature already).

Given partial signatures $\sigma_1, \ldots, \sigma_\ell$, any one can compute the multisignature for group L as follows:

$$\sigma = \bigotimes_{1 \leq j \leq \ell} \sigma_j = \left(\prod_{1 \leq j \leq \ell} s_i, \prod_{1 \leq j \leq \ell} t_i \right).$$

MVf(pp, M, σ, L): Given pp, $L = (pk_1, \ldots, pk_\ell)$, message M, and an alleged multisignature $\sigma = (s, t)$, a verifier accepts the multisignature if

$$e(s, g) = e(H_m(M||L), t) \cdot \prod_{pk_i \in L} A_i,$$

where $A_i = e(H(pk_i), pk_i)$ and rejects otherwise.

Theorem 1. *Our multisignature scheme is correct. Namely any multisignature generated by legal signers can be verified.*

Proof. The scheme is *correct* because

$$e(s, g) = e\left(\left(\prod_{i=1}^{\ell} H(pk_i)^{x_i}\right) \cdot H_m(M||L)^r, g\right)$$

$$= e(H_m(M||L)^r, g) \cdot \prod_{i=1}^{\ell} e(H(pk_i), g^{x_i})$$

$$= e(H_m(M||L), t) \cdot \prod_{i=1}^{\ell} e(H(pk_i), pk_i)$$

$$= e(H_m(M||L), t) \cdot \prod_{pk_i \in L} A_i,$$

where

$$r = \sum_{i=1}^{\ell} r_i \quad \text{and} \quad t = g^r.$$

Remark 2. In our scheme, each signer generates its own randomness. Without any shared common random generator, we reduce the round of communication to minimum. On the other hand, our multisignature generation is also quite simple and straightforward: each signer produces its Waters-like signature $\sigma = (sk \cdot H(m)^r, g^r)$ on the message, then the multisignature is just the component-wise product of these signatures.

Remark 3. To compute $A_i = e(H(pk_i), pk_i)$ needs one pairing computation, while this computation is once for all and can be finished when checking the validity of the public key certificates of signers. Therefore this computation of pairing can be saved through pre-computation before multisignature verification since A_i is independent from messages. Comparing with those in the plain public key model, our multisignature scheme is the most efficient one in verification since our scheme achieves $\mathcal{O}(1)$-verification (respective to pairing computations).

Remark 4. As in many applications, signers might not know who (included L) are going to sign the message M, it is also interesting to find a proper multisignature scheme that can be applied to this situation. Indeed, we only need replace "$M||L$" by "M" in our signing protocol to yield such a scheme. However, it security only prevents forgery on a new message [13,14], not a pair of message/signers.

4   Security Analysis

We state the result of security for our multisignature scheme in the following theorem.

Theorem 2. *If there is an algorithm $\mathcal{A}$ in the random oracle model that $(t, q_s, q_h, \ell, \varepsilon)$-breaks our scheme, then there is an algorithm $\mathcal{B}$ that (t', ε')-solves the CDH problem, where*

$$t' = t + \mathcal{O}(q_h + 3q_s + \ell + 2)T_e \quad and \quad \varepsilon' = \frac{\varepsilon}{e(q_s + 1)},$$

T_e *is the running-time of exponentiation in $\mathbb{G}$, q_h and q_s are the bound of two hash queries to H_m, H and signature queries, respectively, and the mathematical constant e is the base of the natural logarithm.*

Proof. The strategy is to construct algorithm $\mathcal{B}$ that solves the computational Diffie-Hellman problem, by utilizing algorithm $\mathcal{A}$ which $(t, q_s, q_h, \ell, \varepsilon)$-breaks our multisignature scheme where $q_h = q_H + q_{H_m}$ is the total number of hash queries.

Suppose $\mathcal{B}$ is given $(g, g^a, h) \in \mathbb{G} \times \mathbb{G} \times \mathbb{G}$ for $a \xleftarrow{R} \mathbb{Z}_p$, $h \xleftarrow{R} \mathbb{G}$, and asked to find h^a. Without loss of generality, assume user 1 is the honest signer. $\mathcal{B}$ simulates the random oracles $H_m(\cdot)$ and $H(\cdot)$, the signature oracle $\mathcal{O}_{\mathsf{MSign}}(pp, pk^\star, M, L)$ for providing user 1's partial signatures valid under the public key $pk^\star \stackrel{def}{=} pk_1 = g^{x_1} \stackrel{def}{=} g^a$ with $x_1 \stackrel{def}{=} a$ unknown to $\mathcal{B}$.

Setup. $\mathcal{B}$ gives $\mathcal{A}$ the public key $pk^\star = g^a$ and other public parameters $(\mathbb{G}, \mathbb{G}_T, e, H_m(\cdot), g, H(\cdot))$.

$H_m(M\|L)$-**Queries.** $\mathcal{B}$ responds to queries to random oracle $H_m(\cdot)$ as follows:

1. If there is a tuple $(M\|L, b, r, H_m(M\|L))$ in the H_m-list which is initially empty, return $H_m(M\|L)$; otherwise, execute the following.
2. Choose $r \xleftarrow{R} \mathbb{Z}_p$, a bit $b \xleftarrow{\delta} \{0, 1\}$ such that b takes 1 with probability δ that will be specified later, return $H_m(M\|L) = h^b \cdot g^r$ and add $(M\|L, b, r, H_m(M\|L))$ to the H_m-list.

$H(X)$-**Queries.** $\mathcal{B}$ initializes an H-list which only has $(pk^\star, h \cdot g^k, k)$ for $k \xleftarrow{R} \mathbb{Z}_p$, by setting $H(pk^\star) = h \cdot g^k$ and then executes as follows:

1. If $(X, H(X), k)$ has been defined, return $H(X)$; otherwise do the following.
2. Choose $k \xleftarrow{R} \mathbb{Z}_p$, return $H(X) = g^k$ and add $(X, H(X), k)$ to the H-list.

Note that if the argument of the query cannot be parsed as $X \in \mathbb{G}$, $\mathcal{B}$ simply returns a random element of $\mathbb{G}$, while preserving consistency if the same query has been asked before.

$\mathcal{O}_{\mathsf{MSign}}(pp, pk^\star, M, L)$-**Queries.** $\mathcal{B}$ proceeds as follows:

1. If $pk^\star \notin L$, return $\perp$ and abort; otherwise find $(M\|L, b, r, H_m(M\|L))$ (where $H_m(M\|L) = h^b \cdot g^r$) in the H_m-list. (We assume that $\mathcal{A}$ has asked the corresponding hash values; otherwise $\mathcal{B}$ just acts as if it is responding to the hash queries to $H_m(\cdot)$.)

2. If $b = 0$, abort; otherwise randomly choose α from $\mathbb{Z}_p$ and compute

$$s_1 = (g^a)^{k-r}(hg^r)^\alpha$$
$$= (hg^k)^{x_1}(hg^r)^{\alpha-x_1}$$
$$= H(pk_1)^{x_1} H_m(M||L)^{\alpha-x_1},$$
$$t_1 = g^{\alpha-x_1} = g^\alpha(g^{x_1})^{-1} = g^\alpha(g^a)^{-1}.$$

3. Output $\sigma_1 = (s_1, t_1)$.

Note that $\sigma_1 = (s_1, t_1)$ is valid when we set randomness $r' = \alpha - a$.

Output h^a. Eventually $\mathcal{A}$ outputs a multisignature forgery $\sigma = (s, t)$ on message M^* with respect to $L = (pk^* = pk_1, pk_2, \ldots, pk_\ell)$, where $(M^*, L) \notin \mathcal{M}$ (the set of previously queried messages and the signing group for partial signatures from user 1). Since σ is valid, we know that for some $d \in \mathbb{Z}_p$,

$$s = \left(\prod_{i=1}^\ell H(pk_i)^{x_i} \right) \cdot H_m(M^*||L)^d, \quad t = g^d.$$

Then, $\mathcal{B}$ executes the following to compute h^a:

1. Perform additional queries $H(pk_i)$ for $1 \le i \le \ell$, making sure that $H(pk_i)$ for $2 \le i \le \ell$ is defined.
2. Let J be the index such that $pk_i = pk_1$ and Δ be the number of such keys for $1 \le i \le \ell$.
3. Find $(M^*||L, b^*, r^*, H_m(M^*||L))$ in the H_m-list.
4. If $b^* = 1$, abort; otherwise $b^* = 0$ compute and output

$$h^a = \left(s \cdot t^{-r^*} \cdot \prod_{pk_i \ne pk_1} pk_i^{-k_i} \cdot (g^a)^{-k_1\Delta} \right)^{\Delta^{-1}} \tag{4.1}$$

because

$$s \cdot t^{-r^*} \cdot \left(\prod_{pk_i \ne pk_1} pk_i^{k_i} \right)^{-1} \cdot (g^a)^{-k_1\Delta}$$

$$= s \cdot \left(g^{-r^*} \right)^d \cdot \left(\prod_{pk_i \ne pk_1} (g^{k_i})^{x_i} \right)^{-1} \cdot (g^a)^{-k_1\Delta}$$

$$= s \cdot H_m(M^*||L)^{-d} \left(\prod_{pk_i \ne pk_1} H(pk_i)^{x_i} \right)^{-1} \cdot (g^a)^{-k_1\Delta}$$

$$= \prod_{pk_i = pk_1} H(pk_i)^{x_i} \cdot (g^a)^{-k_1\Delta}$$

$$= \left((hg^{k_1})^{x_1} \right)^\Delta \cdot (g^{x_1})^{-k_1\Delta}$$

$$= h^{a\Delta}.$$

where $H(pk_i) = g^{k_i}$ for $pk_i \ne pk_1$ and $x_1 = a$.

Although $\mathcal{B}$ perfectly simulates the random oracle $H_m(\cdot)$, in the simulation of $\mathcal{O}_{\mathsf{MSign}}(\cdot, \cdot, \cdot, \cdot)$ $\mathcal{B}$ succeeds with probability δ, namely the probability that $b_i = 1$ for each $i = 1, \ldots, q_s$. Thus $\mathcal{B}$ doesn't abort in the simulation of the signature oracle with probability

$$\Pr\left[\bigwedge_{i=1}^{q_s}(b_i = 1)\right] = \prod_{i=1}^{q_s}\Pr\left[b_i = 1\right] = \delta^{q_s}.$$

When $\mathcal{A}$ outputs a valid forgery, $\mathcal{B}$ will abort if $b^\star = 1$. Therefore, $\mathcal{B}$ succeeds with probability that $b^\star = 0$ at that time (i.e., $1 - \delta$). Let $\mathcal{E}$ denote the event that $\mathcal{A}$ outputs a valid forgery, and $\mathcal{E}_{\mathsf{suc}}|\mathcal{E}$ denote the event $\mathcal{B}$ succeeds in solving the CDH problem (i.e., outputting h^a) under the condition that $\mathcal{E}$ happens. We have $\Pr[\mathcal{E}_{\mathsf{suc}}|\mathcal{E}] = (1 - \delta)\delta^{q_s}$. Then,

$$\varepsilon_{\mathcal{B}} = \Pr[\mathcal{E} \bigwedge \mathcal{E}_{\mathsf{suc}}]$$
$$= \Pr[\mathcal{E}] \cdot \Pr[\mathcal{E}_{\mathsf{suc}}|\mathcal{E}]$$
$$= \varepsilon \cdot (1 - \delta)\delta^{q_s}$$

Let $f(\delta) = (1 - \delta)\delta^{q_s}$. Since f is maximal at $\delta_0 = \frac{q_s}{q_s+1}$, we know $f(\delta_0) \geq \frac{1}{e \cdot (q_s+1)}$, where the mathematical constant e is the base of the natural logarithm. Therefore,

$$\varepsilon_{\mathcal{B}} \geq \frac{\varepsilon'}{e \cdot (q_s + 1)}.$$

Finally, for the running-time of $\mathcal{B}$, we take into account the running-time t of $\mathcal{A}$, the exponentiations on hash queries $\mathcal{A}$ made, and the linear number of exponentiations in each signing query and $\ell + 2$ exponentiation on extracting h^a. This takes time at most $t + \mathcal{O}(q_h + q_s + \ell + 2) \cdot T_e$, where T_e is running-time of exponentiation and q_h, q_s are the number of hash queries to H_m and signature queries $\mathcal{O}_{\mathsf{MSign}}(\cdot, \cdot, \cdot, \cdot)$, respectively.

5   Additional Related Work

The BGLS scheme was originally proposed for the purpose of aggregate signatures [18], and was later shown (through a new analysis [20]) to be a secure multisignature scheme as well. However, the BGLS scheme is extremely inefficient in multisignature verification. While in principle sequential aggregate signatures (e.g., [26,27]) can be used to construct multisignatures, this approach has drawbacks such as *interactive* signing (signers cannot contribute their partial signatures independently) and expensive verification time. Other aggregate signatures impose special strong assumption on time *synchronization* [28,29] which also will impose additional operational assumption for multisignatures.

6 Conclusion

We presented an efficient non-interactive multisignature scheme in the plain public key model. Our scheme not only reduces the trust assumption on the third party and but also achieves optimal rounds of communication. This scheme, to our best knowledge enjoys a tighter security proof, comparing with non-interactive construction in the plain public key model. Furthermore, our scheme only needs $\mathcal{O}(1)$ (pairings) in verification through pre-computation. We believe it is amongst the most practical schemes currently available in many realistic application scenarios.

Acknowledgements. This work has been supported by the National Natural Science Foundation of China, Grant No. 60873217, 61172085, 11061130539 and 61021004.

References

1. Micali, S., Ohta, K., Reyzin, L.: Accountable-Subgroup Multisignatures: Extended Abstract. In: Eighth ACM Conference on Computer and Communications Security, pp. 245–254. ACM Press, New York (2001)
2. Bellare, M., Neven, G.: Multisignatures in the Plain Public-Key Model and a General Forking Lemma. In: 13th ACM Conference on Computer and Communications Security, pp. 390–399. ACM Press, New York (2006)
3. Boldyreva, A.: Threshold Signatures, Multisignatures and Blind Signatures Based on the Gap-Diffie-Hellman-Group Signature Scheme. In: Desmedt, Y.G. (ed.) PKC 2003. LNCS, vol. 2567, pp. 31–46. Springer, Heidelberg (2002)
4. Kim, J., Tsudik, G.: Srdp: Securing Route Discovery in DSR. In: Second Annual International Conference on Mobile and Ubiquitous Systems: Networking and Services, pp. 247–260. IEEE Press, Los Alamitos (2005)
5. Bagherzandi, A., Cheon, J., Jarecki, S.: Multisignatures Secure under the Discrete Logarithm Assumption and a Generalized Forking Lemma. In: The 15th ACM Conference on Computer and Communications Security, pp. 449–458. ACM Press, New York (2008)
6. Bagherzandi, A., Jarecki, S.: Multisignatures Using Proofs of Secret Key Possession, as Secure as the Diffie-Hellman Problem. In: Ostrovsky, R., De Prisco, R., Visconti, I. (eds.) SCN 2008. LNCS, vol. 5229, pp. 218–235. Springer, Heidelberg (2008)
7. Castelluccia, C., Jarecki, S., Kim, J., Tsudik, G.: Secure Acknowledgment Aggregation and Multisignatures with Limited Robustness. Comput. Netw. 50, 1639–1652 (2006)
8. Lin, X., Sun, X., Ho, P.H., Shen, X.: Gsis: A Secure and Privacy Preserving Protocol for Vehicular Communications. IEEE Trans. on Vehicular Tech. 56, 3442–3456 (2007)
9. Lu, R., Lin, X., Zhu, H., Ho, P.H., Shen, X.: Ecpp: Efficient Conditional Privacy Preservation Protocol for Secure Vehicular Communications. In: The 27th Conference on Computer Communications IEEE INFOCOM 2008, pp. 14–18. IEEE Press, Los Alamitos (2008)
10. Lu, R., Lin, X., Shen, X.: Spring: A social-based Privacy-Preserving Packet Forwarding Protocol for Vehicular Delay Tolerant Networks. In: The 29th Conference on Computer Communications, IEEE INFOCOM 2010, pp. 14–19. IEEE Press, Los Alamitos (2010)

11. Itakura, K., Nakamura, K.: A Public Key Cryptosystem Suitable for Digital Multisignatures. NEC Research & Development 71, 1–8 (1983)
12. Ohta, K., Okamoto, R.: Multisignature Schemes Secure Against Active Insider Attacks. IEICE Transactions on Fundamentals E82-A, 21–31 (1999)
13. Lu, S., Ostrovsky, R., Sahai, A., Shacham, H., Waters, B.: Sequential Aggregate Signatures and Multisignatures Without Random Oracles. In: Vaudenay, S. (ed.) EUROCRYPT 2006. LNCS, vol. 4004, pp. 465–485. Springer, Heidelberg (2006)
14. Ristenpart, T., Yilek, S.: The Power of Proofs-of-Possession: Securing Multiparty Signatures Against Rogue-Key Attacks. In: Naor, M. (ed.) EUROCRYPT 2007. LNCS, vol. 4515, pp. 228–245. Springer, Heidelberg (2007)
15. Adams, C., Farrell, S., Kause, T., Monen, T.: Internet X.509 Public Key Infrastructure Certificate Management Protocol, cmp (2005)
16. Schaad, J.: Internet X.509 Public Key Infrastructure Certificate Request Message Format (2005)
17. Ma, C., Weng, J., Li, Y., Deng, R.: Efficient Discrete Logarithm Based Multi-Signature Scheme in the Plain Public Key Model. Des. Codes Cryptography 54, 121–133 (2010)
18. Boneh, D., Gentry, C., Lynn, B., Shacham, H.: Aggregate and Verifiably Encrypted Signatures. In: Biham, E. (ed.) EUROCRYPT 2003. LNCS, vol. 2656, pp. 416–432. Springer, Heidelberg (2003)
19. Bellare, M., Rogaway, P.: Random Oracles Are Practical: a Paradigm for Designing Efficient Protocols. In: 10th ACM Conference on Computer and Communications Security, pp. 62–73. ACM Press, New York (1993)
20. Bellare, M., Namprempre, C., Neven, G.: Unrestricted Aggregate Signatures. In: Arge, L., Cachin, C., Jurdziński, T., Tarlecki, A. (eds.) ICALP 2007. LNCS, vol. 4596, pp. 411–422. Springer, Heidelberg (2007)
21. Barr, K., Asanović, K.: Energy-Aware Lossless Data Compression. ACM Trans. Comput. Syst. 24, 250–291 (2006)
22. Qian, H., Xu, S.: Non-Interactive Multisignatures in the Plain Public-Key Model with Efficient Verification. Inf. Process. Lett. 111, 82–89 (2010)
23. Goldwasser, S., Micali, S., Rivest, R.: A Digital Signature Scheme Secure Against Adaptive Chosen-Message Attacks. SIAM Journal of Computing 17, 281–308 (1988)
24. Boneh, D., Lynn, B., Shacham, H.: Short Signatures from the Weil Pairing. In: Boyd, C. (ed.) ASIACRYPT 2001. LNCS, vol. 2248, pp. 514–532. Springer, Heidelberg (2001)
25. Galbraith, S., Paterson, K., Smart, N.: Pairings for Cryptographers. Discrete Applied Mathematics 156, 3113–3121 (2008)
26. Lysyanskaya, A., Micali, S., Reyzin, L., Shacham, H.: Sequential Aggregate Signatures from Trapdoor Permutations. In: Cachin, C., Camenisch, J.L. (eds.) EUROCRYPT 2004. LNCS, vol. 3027, pp. 74–90. Springer, Heidelberg (2004)
27. Neven, G.: Efficient Sequential Aggregate Signed Data. In: Smart, N.P. (ed.) EUROCRYPT 2008. LNCS, vol. 4965, pp. 52–69. Springer, Heidelberg (2008)
28. Gentry, C., Ramzan, Z.: Identity-Based Aggregate Signatures. In: Yung, M., Dodis, Y., Kiayias, A., Malkin, T. (eds.) PKC 2006. LNCS, vol. 3958, pp. 257–273. Springer, Heidelberg (2006)
29. Ahn, J.H., Green, M., Hohenberger, S.: Synchronized Aggregate Signatures: New Definitions, Constructions and Applications. In: 17th ACM Conference on Computer and Communications Security, pp. 473–484. ACM Press, New York (2010)

An Efficient Construction of Time-Selective Convertible Undeniable Signatures

Qiong Huang[1,2], Duncan S. Wong[2], Willy Susilo[3,*], and Bo Yang[1]

[1] South China Agricultural University, Guangzhou, China
csqhuang@alumni.cityu.edu.hk, byang@scau.edu.cn
[2] City University of Hong Kong, Hong Kong S.A.R., China
duncan@cityu.edu.hk
[3] Centre for Computer and Information Security Research
University of Wollongong, NSW 2522, Australia
wsusilo@uow.edu.au

Abstract. A time-selective convertible undeniable signature scheme allows a signer to release a *time-selective converter* which converts undeniable signatures pertaining to or up to a specific time period to publicly verifiable ones but not those in any other time periods. The security of existing schemes relies on a strong and interactive assumption called xyz-DCAA in random oracle model or several relatively new hash function assumptions in the generic group model. For some of them, the converter size for each time period also grows linearly or logarithmically with the number of previous time periods. In this paper, we propose a new construction in which all the converters (i.e. time-selective, selective and universal) are of constant size. In particular, the time-selective converter for each time period is only one group element, no matter how many previous time periods there are already. The security of this new construction is proved in the random oracle model based on non-interactive and falsifiable assumptions.

Keywords: convertible undeniable signature, anonymity, time-selective conversion, random oracle.

1 Introduction

Undeniable signature [6], introduced by Chaum and van Antwerpen, allows a signer to generate a signature such that its validity or invalidity can only be verified by running a confirmation or disavowal protocol, respectively, with the signer. Convertible Undeniable Signature (CUS) [4], further allows the signer to convert an undeniable signature to a publicly verifiable one by releasing an additional information called *converter*. There are two types of converters: (1) a *selective converter* makes a specific undeniable signature publicly verifiable; and

---

* W. Susilo was supported by the Australian Research Council Future Fellowship FT0991397.

X. Lai, J. Zhou, and H. Li (Eds.): ISC 2011, LNCS 7001, pp. 355–371, 2011.

(2) a *universal converter* makes all the undeniable signatures of a signer publicly verifiable. Once the universal converter is released, not only the undeniable signatures that have been generated become publicly verifiable, but also will the signatures to be generated become publicly verifiable. The universal conversion makes all signatures, both past and future, be publicly verifiable. This property may not be desirable in some context and furthermore, in order to continue generating undeniable signatures, the signer needs to re-register a new public/secret key pair (in the PKI setting).

In CT-RSA 2005, Laguillaumie and Vergnaud [13] introduced the notion of *Time-Selective Convertible Undeniable Signature* (TSCUS). In a TSCUS scheme, there is a new type of converters called *time-selective converter*. First, the time is divided into time periods, then each signature is generated with respect to a particular time period. The time-selective converter of the particular time period enables a verifier to check the validity of all signatures corresponding to that particular time period only. Undeniable signatures corresponding to other time periods remain unverifiable and the confirmation or disavowal protocol has to be carried out with the signer for checking their validity. In [13], a concrete TSCUS scheme was also proposed. Time-selective converters and selective converters are of size linear with the number of time periods, which were recently reduced to logarithmic size in [14]. The security of it, specifically, the anonymity relies on the intractability of a strong decisional problem called *xyz-Decisional Collusion Attack Algorithm* (xyz-DCAA)[1] in random oracle model [2]. In both schemes [13,14], the converter for time period t validates not only undeniable signatures with respect to t, but also those with respect to previous periods. El Aimani and Vergnaud [1] proposed a variant of the TSCUS mentioned above, where the converter for a time period converts undeniable signatures with respect to that particular period only. The converter is also constant in size. However, the security of the scheme is only proven in the generic group model [23] and in particular, relies on *new* assumptions on hash functions.

In this paper we propose a new TSCUS scheme. All the converters in the scheme are constant in size. Specifically, the time-selective converter for each time period contains only one element of a bilinear group, no matter how many previous time periods there are already. The security of this new scheme does not rely on any new hash function assumption but on the conventional random oracle abstraction [2], and on some non-interactive and falsifiable number-theoretic assumptions, namely Strong Diffie-Hellman (CDH) assumption [3] and Decisional Double Strong Diffie-Hellman (DDSDH) assumption (see Sec. 3). The scheme is also efficient. The signature generation involves two exponentiations in a bilinear group, which are implemented as two elliptic curve scalar multiplications (ECSM) in an elliptic curve group. Different from [13,14] but similar to [1], a time-selective converter in this new TSCUS scheme is specific to a particular time period only, and time periods are treated as ordinary bit strings, rather

---

[1] The assumption is strong in the sense that after seeing part of the problem instance, the adversary submits a target number to an oracle, and gets back the rest of the instance constructed based on the target number. (See [13] for details.)

Table 1. A brief comparison with [13,14,1]

	SCvt	TSCvt	Assumptions
[13]	$O(t)$	$O(t)$	CDH + xyz-DCAA
[14]	$O(\log t)$	$O(\log t)$	CDH + xyz-DCAA
[1]	$O(1)$	$O(1)$	raPre + rlColl + Pre + Prf
Ours	$O(1)$	$O(1)$	SDH + DDSDH

than integers as in [13,14]. Table 1 shows a comparison between this new scheme and those existing ones [13,14,1]. In the table, we use t to denote the number of time periods. More details will be given in Sec. 4.5.

1.1   Related Work

Since the introduction of (convertible) undeniable signature, a couple of schemes [16,9,8,12,19] with provable security in the random oracle model and others [7,12,19,20,11,22] with security proven in the standard model have been proposed. There are also several related notions to undeniable signature available in the literature. For example, Chaum [5] proposed the notion of *designated confirmer signature* (DCS), in which there is a third party called *confirmer*, who can confirm/disavow signatures on behalf of the signer. Related work of DCS include [18,10,21]. In [13], Laguillaumie and Vergnaud introduced the notion of Time-Selective Convertible Undeniable Signature (TSCUS) and proposed the first concrete one. The signature is short and consists of one group element and one short binary string. The selective converter of a signature and the time-selective converter of a specific time period, however, grows linearly in size with the number of time periods. The time-selective converter for a time period t allows a verifier to check the validity of all the undeniable signatures with respect to time periods t and earlier. Very recently, Laguillaumie and Vergnaud [14] proposed another scheme which reduces the converter size so that it grows logarithmically with the number of time periods. On the security, the unforgeability of both schemes are based on the CDH assumption while the anonymity relies on the intractability of a strong and interactive decisional problem called xyz-DCAA. Both of them were proven in the random oracle model.

In 2007, El Aimani and Vergnaud [1] refined the notion of time-selective conversion in TSCUS. They proposed to time-selectively convert signatures *gradually*. That is, a time-selective converter for time period t only converts signatures pertaining to time period t but not signatures pertaining to any other time periods. They proposed a TSCUS scheme which is based on Michels-Petersen-Horster CUS [15]. The security of the scheme is proven in the generic group model [23] assuming some new properties on the underlying hash functions.

Outline. In the following we describe the definition of TSCUS and security models. This is followed by the definitions of number-theoretic assumptions in Sec. 3. In Sec. 4, we propose a new TSCUS scheme and evaluate its performance as well as a batch verification technique. Security analysis is also given in this section. Finally, we conclude the paper in Sec. 5.

2 Time-Selective Convertible Undeniable Signature

2.1 Definition

Definition 1. *A time-selective convertible undeniable signature (TSCUS) scheme consists of the following probabilistic polynomial-time (PPT) algorithms, and interactive protocols:*

Setup *takes as input 1^k where k is a security parameter, and outputs the public parameters* Params.

Kg *takes as input* Params, *and outputs a key pair* (pk, sk) *for a signer.*

Sign *takes as input* Params, sk, *a message m and a time period $t \in \mathcal{T}$ where $\mathcal{T}$ is the space of time periods, and outputs a signature σ.*

Control *takes as input* Params, sk, m, t *and σ, and outputs 1 if σ is a valid signature on m under* pk *with respect to t, and 0 otherwise.*

SConv *takes as input* Params, sk, m, t *and σ, and outputs a converter* scvt *if* Control(Params, sk, m, t, σ) $= 1$, *and $\perp$ otherwise.*

TSConv *takes as input* Params, sk *and t, and outputs a time-selective converter* tscvt_t.

UConv *takes as input* Params *and* sk, *and outputs a universal converter* ucvt.

SVer *takes as input* Params, m, t, σ, scvt *and* pk, *and outputs 1 for acceptance, or 0 for rejection.*

TSVer *takes as input* Params, m, t, σ, tscvt_t *for time period t and* pk, *and outputs 1 for acceptance, or 0 for rejection.*

UVer *takes as input* Params, m, t, σ, ucvt *and* pk, *and outputs 1 for acceptance, or 0 for rejection.*

Confirm, Disavow *are interactive protocols, in which the signer is the prover. The common input to both the prover and the verifier consists of* Params, m, t, σ *and* pk. *The secret input to the prover is* sk. *At the end of a confirmation (resp. disavowal) protocol run, the verifier outputs 1 for accepting σ as a valid (resp. invalid) signature on m under* pk *with respect to t, and 0 otherwise.*

The definition above differs from that in [13,14] slightly. We consider three types of converters explicitly, i.e. selective converter, time-selective converter and universal converter, while in [13,14], only the first two types are considered. For simplicity, we omit Params when describing the algorithms and schemes from now on.

Correctness. For all (pk, sk) $\leftarrow$ Kg(1^k), $m \in \{0,1\}^*$, $t \in \mathcal{T}$, $\sigma \leftarrow$ Sign(sk, m, t), $\mathsf{tscvt}_t \leftarrow$ TSConv(sk, t) and scvt $\leftarrow$ SConv(sk, m, t, σ), the following equations should hold:

Control(sk, m, t, σ) $= 1$, SVer(m, t, σ, scvt, pk) $= 1$ and TSVer(m, t, σ, tscvt_t, pk) $= 1$.

Let σ' be a signature such that Control(sk, m, t, σ') $= 0$. The confirmation and disavowal protocols should satisfy the following conditions:

Completeness. If signer S and verifier V honestly follow the protocol, then

$$\Pr[\mathsf{Confirm}_{S(\mathsf{sk}),V}(m,t,\sigma,\mathsf{pk}) = 1] = 1, \text{ and}$$
$$\Pr[\mathsf{Disavow}_{S(\mathsf{sk}),V}(m,t,\sigma',\mathsf{pk}) = 1] = 1.$$

Soundness. For any PPT algorithm S', it holds that

$$\Pr[\mathsf{Confirm}_{S'(\mathsf{sk}),V}(m,t,\sigma',\mathsf{pk}) = 1] \leq \epsilon(k), \text{ and}$$
$$\Pr[\mathsf{Disavow}_{S'(\mathsf{sk}),V}(m,t,\sigma,\mathsf{pk}) = 1] \leq \epsilon(k),$$

where $\epsilon(\cdot)$ is a negligible function in k.

Non-transferability. We require that the confirmation and disavowal protocols are zero-knowledge so that no PPT verifier can transfer its conviction of the validity (resp. invalidity) of a signature to others.

2.2   Security Models

Unforgeability. This requires that no efficient adversary can forge a new signature with non-negligible probability. We consider the following game:

1. Challenger C generates Params and $(\mathsf{pk}, \mathsf{sk})$, computes $\mathsf{ucvt} \leftarrow \mathsf{UConv}(\mathsf{sk})$, and invokes adversary $\mathcal{A}$ on input (Params, pk, ucvt).
2. $\mathcal{A}$ issues queries to the following oracle adaptively:
   OSign takes as input a message m and a time period t, and outputs a signature σ.
3. $\mathcal{A}$ outputs a triple (m^*, t^*, σ^*), and wins the game if
   (a) $\mathsf{UVer}(m^*, t^*, \sigma^*, \mathsf{ucvt}, \mathsf{pk}) = 1$, and
   (b) σ^* was not returned by **OSign** on input (m^*, t^*).

Definition 2 (Unforgeability). *A TSCUS scheme is* $(\mathsf{t}, q_s, \epsilon)$*-unforgeable, if there is no* $\mathcal{A}$ *which runs in time* t*, issues at most* q_s *queries to* **OSign***, and wins the game with probability at least* ϵ*.*

In the definition above, the adversary is given the universal converter, while in [13,14] the adversary is given all the time-selective converters. As the universal converter enables verification of all the signatures regardless of which time periods that the signatures are corresponding to, the unforgeability definition above is at least as strong as that in [13,14]. The definition above also captures strong unforgeability, i.e. $\mathcal{A}$ is not able to come up with a new signature on m^* with respect to time period t^* even if **OSign** has been queried with (m^*, t^*).

Invisibility. Given an undeniable signature, the invisibility requires that no verifier can tell if it is valid.

1. Challenger C generates Params and a key pair $(\mathsf{pk}, \mathsf{sk})$, and gives (Params, pk) to adversary $\mathcal{A}$.
2. $\mathcal{A}$ issues queries to the following oracles:

OSign takes as input (m, t) where m is a message and t a time period, and outputs a signature σ generated using sk.

OSConv takes as input (m, t, σ) and outputs selective converter scvt if $\mathsf{SVer}(m, t, \sigma, \mathsf{scvt}, \mathsf{pk}) = 1$, and $\bot$ otherwise.

OTSConv takes as input t and outputs time-selective converter tscvt_t generated using sk.

OConfirm takes as input (m, t, σ) and uses sk to carry out the confirmation protocol with $\mathcal{A}$ if $\mathsf{Control}(\mathsf{sk}, m, t, \sigma) = 1$, and does nothing otherwise.

ODisavow takes as input (m, t, σ), and uses sk to carry out the disavowal protocol with $\mathcal{A}$ if $\mathsf{Control}(\mathsf{sk}, m, t, \sigma) = 0$, and does nothing otherwise.

3. $\mathcal{A}$ submits a message m^* and a time period t^*. C then tosses a coin b. If $b = 0$, C computes $\sigma^* \leftarrow \mathsf{Sign}(\mathsf{sk}, m^*, t^*)$; otherwise, it randomly selects σ^* from the space of the signer's signatures. C returns σ^* to $\mathcal{A}$.

4. $\mathcal{A}$ continues to query as in Step 2. Finally, it outputs a bit b', and wins the game if $b' = b$ and
   (a) $\mathcal{A}$ did not query **OTSConv** on input t^*;
   (b) $\mathcal{A}$ did not query **OSConv** on input m^*, t^* and σ^*; and
   (c) $\mathcal{A}$ did not query **OConfirm** or **ODisavow** on input (m^*, t^*, σ^*).

The advantage of $\mathcal{A}$ is defined as the absolute difference between the probability that $\mathcal{A}$ wins the game and $1/2$.

Definition 3 (Invisibility). *A TSCUS scheme is* $(\mathsf{t}, q_s, q_{sc}, \ q_t, q_c, q_d, \epsilon)$-*invisible, if there is no $\mathcal{A}$ which runs in time* t*, issues at most q_s queries to* **OSign***, q_{sc} queries to* **OSConv***, q_t queries to* **OTSConv***, q_c queries to* **OConfirm** *and q_d queries to* **ODisavow***, wins the game above with advantage at least ϵ.*

Anonymity. Another security property related to the undecidability of undeniable signatures is *anonymity*, introduced by Galbraith and Mao [8]. Given an undeniable signature, anonymity requires that no verifier is able to find out the signer of the signature. It captures the non-self-authenticating property and is formalized in the indistinguishability-based game below.

1. Challenger C generates Params, and two key pairs $(\mathsf{pk}_0, \mathsf{sk}_0)$, $(\mathsf{pk}_1, \mathsf{sk}_1)$ using Kg, and gives $(\mathsf{Params}, \mathsf{pk}_0, \mathsf{pk}_1)$ to adversary $\mathcal{A}$.

2. $\mathcal{A}$ issues queries as in Def. 3, except that now all the oracles take as input an additional bit d, and use sk_d to compute the answer.

3. $\mathcal{A}$ submits a message m^* and a time period t^*. C then tosses a coin b, and computes $\sigma^* \leftarrow \mathsf{Sign}(\mathsf{sk}_b, m^*, t^*)$. It returns σ^* to $\mathcal{A}$.

4. $\mathcal{A}$ continues to query as in Step 2. Finally, it outputs a bit b', and wins the game if $b' = b$ and
   (a) $\mathcal{A}$ did not query **OTSConv** on input t^* and any $d \in \{0, 1\}$;
   (b) $\mathcal{A}$ did not query **OSConv** on input m^*, t^*, σ^* and any $d \in \{0, 1\}$; and
   (c) $\mathcal{A}$ did not query **OConfirm** nor **ODisavow** on input (m^*, t^*, σ^*, d) for any $d \in \{0, 1\}$.

The advantage of $\mathcal{A}$ is defined as the absolute difference between the probability that $\mathcal{A}$ wins the game and $1/2$.

Definition 4 (Anonymity). *A TSCUS scheme is* $(t, q_s, q_{sc}, q_t, q_c, q_d, \epsilon)$-*anonymous, if there is no* $\mathcal{A}$ *which runs in time* t, *issues at most* q_s *queries to* **OSign**, q_{sc} *queries to* **OSConv**, q_t *queries to* **OTSConv**, q_c *queries to* **OConfirm** *and* q_d *queries to* **ODisavow**, *wins the game above with advantage at least* ϵ.

Access to random oracles will also be given to $\mathcal{A}$ if a scheme's security is proven in the random oracle model.

3   Number-Theoretic Assumptions

Admissible Pairings. Let $\mathbb{G}$ and $\mathbb{G}_T$ be two cyclic groups of large prime order p. The mapping $\hat{e} : \mathbb{G} \times \mathbb{G} \to \mathbb{G}_T$ is said to be an *admissible pairing*, if (1. *bilinearlity*) $\forall u, v \in \mathbb{G}$ and $\forall a, b \in \mathbb{Z}$, $\hat{e}(u^a, v^b) = \hat{e}(u, v)^{ab}$; (2. *non-degeneracy*) $\exists u, v \in \mathbb{G}$ such that $\hat{e}(u, v) \neq 1_T$, where 1_T is the identity element of $\mathbb{G}_T$; and (3. *computability*) there exists an efficient algorithm for computing $\hat{e}(u, v)$ for any $u, v \in \mathbb{G}$. Let $\mathcal{IG}(1^k)$ be a PPT which on input 1^k outputs $(\mathbb{G}, \mathbb{G}_T, p, g, \hat{e})$ which are defined as above.

Strong Diffie-Hellman (SDH) Assumption [3]. The SDH assumption (t, q, ϵ)-holds in the context of $(\mathbb{G}, \mathbb{G}_T, p, g, \hat{e})$, if there is no adversary $\mathcal{A}$ which runs in time at most t, and

$$\mathrm{Adv}_{\mathcal{A}}^{\mathrm{SDH}}(k) \overset{\mathrm{def}}{=} \Pr[g \leftarrow \mathbb{G}, x \leftarrow \mathbb{Z}_p, (c, g^{\frac{1}{x+c}}) \leftarrow \mathcal{A}(g, g^x, g^{x^2}, \cdots, g^{x^q})] > \epsilon,$$

where the probability is taken over the random choices of $x \in \mathbb{Z}_p$ and the random coins used by $\mathcal{A}$.

Decisional Double Strong Diffie-Hellman (DDSDH) Assumption. The DDSDH assumption (t, q_1, q_2, ϵ)-holds in the context of $(\mathbb{G}, \mathbb{G}_T, p, g, \hat{e})$, if there is no adversary $\mathcal{A}$ which runs in time at most t, and

$$\mathrm{Adv}_{\mathcal{A}}^{\mathrm{DDSDH}}(k) \overset{\mathrm{def}}{=} \left| \Pr \left[\begin{array}{l} Z_1 \leftarrow \mathbb{G}, x, y \leftarrow \mathbb{Z}_p, Z_0 \leftarrow g^{\frac{1}{x}+\frac{1}{y}}, b \leftarrow \{0,1\}, \\ b' \leftarrow \mathcal{A}(g, g^x, \cdots, g^{x^{q_1}}, g^y, \cdots, g^{y^{q_2}}, Z_b) \end{array} : b'=b \right] - \frac{1}{2} \right| > \epsilon,$$

where the probability is taken over the random choices of $Z_1 \in \mathbb{G}$, $x, y \in \mathbb{Z}_p$, $b \in \{0,1\}$ and the random coins used by $\mathcal{A}$.

This assumption, although being q-type, is still non-interactive and falsifiable [17]. If an algorithm $\mathcal{A}$ solves the (q_1, q_2)-DDSDH problem with advantage ϵ where $q_1, q_2 < o(\sqrt[3]{p})$, it has to perform at least $\Omega(\sqrt{\epsilon p/(q_1 + q_2)})$ generic group operations. Due to the page limit, we defer the intractability analysis of DDSDH problem in the generic group model [23,3] to the full version of this paper.

4   Our TSCUS Scheme

Before proposing the new TSCUS scheme, we briefly review the Zhang-Safavi-Naini-Susilo (ZSS) signature scheme [24] which is existentially unforgeable under chosen message attacks in the random oracle model.

4.1   Zhang-Safavi-Naini-Susilo (ZSS) Signature

Let $(\mathbb{G}, \mathbb{G}_T, p, g, \hat{e})$ be a random output of algorithm $\mathcal{IG}(1^k)$. Zhang-Safavi-Naini-Susilo signature [24] works as below:

Kg. The signer selects at random $x \leftarrow \mathbb{Z}_p$ and computes $X := g^x$. Its public/private key pair is $(\mathtt{pk}, \mathtt{sk}) := (X, x)$.

Sign. Given a message $m \in \{0,1\}^*$, the signer computes $\sigma := g^{\frac{1}{x+\mathtt{H}(m)}}$, where $\mathtt{H} : \{0,1\}^* \to \mathbb{Z}_p$ is a collision-resistant hash function. The signer then returns σ as its signature on the message.

Ver. Given a message m, a purported signature σ and a public key $\mathtt{pk}$, the verifier checks if $\hat{e}(\sigma, X g^{\mathtt{H}(m)}) = \hat{e}(g, g)$. It accepts if the equation holds, and rejects otherwise.

4.2   A New TSCUS Scheme

To support unbounded number of time periods and furthermore, to allow the adversary to obtain time-selective converters for time periods of its choices, we treat the time-selective converters as "signatures" on the time periods and generate them using a fully secure signature scheme. Since the verifier is unable to compute the time-selective converters by itself, if we hide the signer's (standard) signature by the corresponding time-selective converter, the verifier could not verify the resulting signature. On the other side, in order to allow a verifier to verify all signatures with respect to a time period using the corresponding time-selective converter, we employ a deterministic signature scheme (e.g. ZSS signature) to generate time-selective converters. With a time-selective converter, anyone can recover the signer's standard signature on the message from the given undeniable signature. Below are the details of our scheme.

Let $\mathbb{G}, \mathbb{G}$ be cyclic multiplicative groups of prime order p, $\hat{e} : \mathbb{G} \times \mathbb{G} \to \mathbb{G}_T$ be an admissible pairing, and g, h be random generators of $\mathbb{G}$. Let $\lambda(k)$ be a polynomial in the security parameter k. We write it as λ for simplicity. Let $\mathcal{T} \subseteq \{0,1\}^\lambda$ be a polynomial-size set of time periods[2]. Let $\mathtt{H}_1 : \{0,1\}^\lambda \to \mathbb{G}$, $\mathtt{H}_2 : \{0,1\}^* \times \{0,1\}^\lambda \to \mathbb{Z}_p$ and $\mathtt{H}_3 : \{0,1\}^* \times \{0,1\}^\lambda \times \mathbb{G}^4 \times \mathbb{G}_T \to \mathbb{Z}_p$ be hash functions which will be modeled as random oracles in security proofs. The TSCUS scheme is described as follows:

Kg. The signer selects at random $x, y \leftarrow \mathbb{Z}_p$, computes $X := g^x$ and $Y := h^y$, and sets the public/private key pair to $(\mathtt{pk}, \mathtt{sk}) := ((X, Y), (x, y))$.

Sign. Given a message m and a time period t, the signer computes

$$\sigma := g^{\frac{1}{x+\mathtt{H}_1(t)}} h^{\frac{1}{y+\mathtt{H}_2(m,t)}}. \tag{1}$$

The signature on m with respect to time period t is σ.

---

[2] This requirement is actually not necessary for our scheme, as our scheme supports exponentially large sets of time periods. Here we just follow Laguillaumie and Vergnaud's definition [13,14]. Different from [13,14] where they treat the time period as an integer, in this paper, we treat it as a binary string.

Control. Given a message m, a time period t and a signature σ, the signer checks if Eq. (1) holds. It rejects if the check fails, and accepts otherwise.

Confirm. To confirm the validity of an alleged signature σ on a message m with respect to a time period t, if $\mathsf{Control}(\mathsf{sk}, m, t, \sigma) = 0$, the signer does nothing. The validity of σ tells that

$$\hat{e}(\sigma, Xg^{u_1}) = \hat{e}(g, g)\hat{e}(h, Xg^{u_1})^{1/(y+u_2)},$$

where $u_1 = \mathsf{H}_1(t)$ and $u_2 = \mathsf{H}_2(m, t)$. Denote

$$W_1 = \hat{e}(\sigma, Xg^{u_1})/\hat{e}(g, g) \quad \text{and} \quad W_2 = \hat{e}(h, Xg^{u_1}).$$

The signer carries out the following proof of knowledge of y with the verifier:

$$\mathrm{PK}\,\{y : h^y = Y \wedge W_1^y = W_2/W_1^{u_2}\}, \tag{2}$$

which is the equality proof of two discrete logarithms. There are standard protocols for fulfilling this task.

Disavow. To disavow σ with respect to (m, t), if $\mathsf{Control}(\mathsf{sk}, m, t, \sigma) = 1$, the signer does nothing; otherwise, it uses the knowledge of y to carry out the following proof of knowledge with the verifier:

$$\mathrm{PK}\,\{y : h^y = Y \wedge W_1^y \neq W_2/W_1^{u_2}.\} \tag{3}$$

This is the inequality proof of two discrete logarithms, and again there are standard protocols for fulfilling the task.

SConv. To generate a selective converter for σ with respect to (m, t), the signer checks if $\mathsf{Control}(\mathsf{sk}, m, t, \sigma) = 1$. If not, it returns $\bot$; otherwise, it generates the selective converter $\mathsf{scvt} := (c, z) \in \mathbb{Z}_p^2$ by applying Fiat-Shamir heuristic to the confirmation protocol. Specifically, it computes scvt as below:
1. select at random $r \in Z_p$, and set $R_1 = h^r$ and $R_2 = W_1^r$;
2. set $c := \mathsf{H}_3(m, t, \sigma, \mathsf{pk}, R_1, R_2)$, and $z = r + cy$.

TSConv. To time-selectively convert undeniable signatures with respect to a time period t, the signer computes

$$\mathsf{tscvt}_t := g^{\frac{1}{x+\mathsf{H}_1(t)}},$$

and releases the time-selective converter tscvt_t.

UConv. To universally convert undeniable signatures, the signer releases $\mathsf{ucvt} := x$, with which anyone can compute the time-selective converter for any time period by following the **TSConv** algorithm.

SVer. Given m, t, σ, a selective converter $\mathsf{scvt} = (c, z)$ and pk, the verifier V checks if the following equations hold:

$$h^z = R_1 Y^c \quad \text{and} \quad W_1^z = R_2(W_2 W_1^{-u_2})^c.$$

Output 1 if they hold, and 0 otherwise.

TSVer. Given $(m, t, \sigma, \mathsf{tscvt}_t, \mathsf{pk})$, the verifier computes $u_1 = \mathsf{H}_1(t)$ and $u_2 = \mathsf{H}_2(m, t)$, and checks if

$$\hat{e}(\mathsf{tscvt}_t, Xg^{u_1}) = \hat{e}(g, g) \quad \text{and} \quad \hat{e}(\sigma \cdot \mathsf{tscvt}_t^{-1}, Yh^{u_2}) = \hat{e}(h, h)$$

Output 1 if both equations hold, and 0 otherwise.

UVer. Given $(m, t, \sigma, \mathsf{ucvt} = x, \mathsf{pk})$, the verifier computes $u_1 = \mathsf{H}_1(t)$, $u_2 = \mathsf{H}_2(m, t)$ and checks if

$$g^x = X \quad \text{and} \quad \hat{e}(\sigma \cdot g^{-1/(x+u_1)}, Yh^{u_2}) = \hat{e}(h, h).$$

Output 1 if both equations hold, and 0 otherwise.

4.3 Security Analysis

Theorem 1. *The time-selective convertible undeniable signature is strongly unforgeable in the random oracle model, provided that SDH assumption holds.*

Proof. Let $\mathcal{A}$ be a PPT forger against the time selective convertible undeniable signature scheme. Suppose it issues at most q_1 queries to H_1, q_2 queries to H_2, q_3 queries to H_3 and q_s signing queries, and wins the unforgeability game with advantage ϵ. We make use of $\mathcal{A}$ to build an algorithm $\mathcal{B}$ which breaks the $(Q+1)$-SDH assumption, where $Q := q_2 + q_s$.

The input of $\mathcal{B}$ is $(\tilde{g}, \tilde{g}^{\beta}, \tilde{g}^{\beta^2}, \cdots, \tilde{g}^{\beta^{Q+1}})$, and $\mathcal{B}$ is to find a pair $(c, \tilde{g}^{1/(\beta+c)})$. It chooses at random a Q-vector $\boldsymbol{w} = (w_1, \cdots, w_Q) \in \mathbb{Z}_p^Q$. Let $G(\cdot)$ be the function

$$G(\beta) := \prod_{j=1}^{Q} (\beta + w_j).$$

It selects at random $g \in \mathbb{G}$ and $x \in \mathbb{Z}_p$, and computes

$$h := \tilde{g}^{G(\beta)}, \quad X := g^x \quad \text{and} \quad Y := h^{\beta}.$$

Note that the second component y of the secret key is implicitly set to $y := \beta$, which is unknown to $\mathcal{B}$. Let $\mathsf{pk} = (X, Y)$ and $\mathsf{ucvt} = x$. $\mathcal{B}$ then chooses at random $j^* \in [q_2]$, invokes the adversary $\mathcal{A}$ on input $(\mathbb{G}, \mathbb{G}_T, \hat{e}, p, g, h, \mathsf{pk}, \mathsf{ucvt})$, and starts to simulate oracles for $\mathcal{A}$ as below.

H_1. Given a distinct query t to oracle H_1, $\mathcal{B}$ chooses at random string $s \in \mathbb{Z}_p$, returns s to the adversary, and stores (t, s) into a hash table HT_1.

H_2. Given the j-th distinct query (m_j, t_j) to oracle H_2, if $j \neq j^*$, $\mathcal{B}$ retrieves w_j from its memory; otherwise, it selects at random a string $w_j \in \mathbb{Z}_p$. In either case, $\mathcal{B}$ returns w_j to the adversary, and stores $((m_j, t_j), w_j)$ into a hash table HT_2.

H_3. Given a distinct query R, $\mathcal{B}$ selects at random a string $u \in \mathbb{Z}_p$, returns u to the adversary, and stores (R, u) in a hash table HT_3.

Sign. Given a distinct signing query (m,t), $\mathcal{B}$ first retrieves $s = \mathrm{H}_1(t)$ and $w = \mathrm{H}_2(m,t)$ from tables HT_1 and HT_2, respectively. If there is no such entry in the table, $\mathcal{B}$ creates one as above. It then computes

$$\sigma := g^{\frac{1}{x+s}} h^{\frac{1}{y+w}}.$$

Since x was chosen by $\mathcal{B}$ and $w \in \{w_1, \cdots, w_Q\}$, σ could be efficiently computed by $\mathcal{B}$.

Finally, $\mathcal{A}$ outputs its forgery (m^*, t^*, σ^*). If $(m^*, t^*) \neq (m_{j^*}, t_{j^*})$, $\mathcal{B}$ aborts; otherwise, $\mathcal{B}$ retrieves the answer w^* to query (m^*, t^*) from table HT_2. Suppose that $\mathcal{A}$ wins its unforgeability game. We have that $\mathcal{A}$ did not issue a signing query on input (m^*, t^*) and therefore, $\mathcal{B}$ did not compute $h^{1/(y+w^*)}$ itself. The validity of σ^* shows that

$$\hat{e}(\sigma^*/g^{1/(x+\mathrm{H}_1(t^*))}, Y h^{w^*}) = \hat{e}(g, g),$$

which guarantees that $T := \sigma^*/g^{1/(x+\mathrm{H}_1(t^*))}$ is of the form $T = h^{1/(y+w^*)}$. Then we obtain that

$$T = h^{\frac{1}{y+w^*}} = \tilde{g}^{\frac{G(\beta)}{\beta+w^*}} = \tilde{g}^{G'(\beta)} \tilde{g}^{\frac{d}{\beta+w^*}},$$

where $G'(\beta)$ is a polynomial of degree $Q-1$, and d is the constant number such that $G'(\beta)(\beta+w^*)+d = G(\beta)$. $\mathcal{B}$ then computes

$$Z = \left(\frac{T}{\tilde{g}^{G'(\beta)}}\right)^{1/d},$$

and outputs (w^*, Z), which is a valid solution to the given SDH problem instance.

Probability Analysis. The simulation of the random oracles and the signing oracle is perfect, and brings no difference to the adversary's view. According to the simulation above, $\mathcal{B}$ aborts if $\mathcal{B}$'s guess of which query (m,t) to oracle H_2 would be the pair in $\mathcal{A}$'s forgery, is not correct. Denote by **good** the event that $\mathcal{B}$'s guess is correct. Since $\mathcal{B}$ chooses j^* randomly and uniformly from the set $[q_2]$, the probability that the event **good** happens is $1/q_2$. Conditioned on that it does not abort, $\mathcal{B}$ succeeds in breaking the SDH assumption if $\mathcal{A}$ wins the game. Therefore, the advantage of $\mathcal{B}$ in solving the SDH problem is lower bounded by

$$\mathsf{Adv}_{\mathcal{B}}^{\mathrm{SDH}}(k) \geq \Pr[\mathbf{good}]\epsilon = \frac{1}{q_2}\epsilon.$$

If ϵ is non-negligible, so is $\mathsf{Adv}_{\mathcal{B}}^{\mathrm{SDH}}(k)$. $\qquad\qquad\square$

Theorem 2. *The time-selective convertible undeniable signature is invisible in the random oracle model, provided that DDSDH assumption holds.*

Proof. Suppose the adversary $\mathcal{A}$ issues at most q_1 queries to H_1, q_2 queries to H_2, q_3 queries to H_3, q_s signing queries, q_{sc} queries for time-selective converters, and q_c (q_d, respectively) confirmation (disavowal, respectively) queries, and wins the invisibility game with advantage ϵ. We use $\mathcal{A}$ to construct another algorithm $\mathcal{B}$ for solving the Decisional Double Strong Diffie-Hellman problem, namely the (Q_1+1, Q_2+2)-DDSDH problem where $Q_1 := q_1 + q_s$ and $Q_2 := q_2 + q_s$.

Let $(\tilde{g}, \tilde{g}^\alpha, \cdots, \tilde{g}^{\alpha^{Q_1+1}}, \tilde{g}^\beta, \cdots, \tilde{g}^{\beta^{Q_2+2}}, Z)$ be a random instance of (Q_1+1, Q_2+2)-DDSDH problem. $\mathcal{B}$ selects at random a (Q_1+1)-vector $\boldsymbol{s} := (s^*, s_1, \cdots, s_{Q_1}) \in \mathbb{Z}_p^{Q_1+1}$ and a $(Q_2 + 2)$-vector $\boldsymbol{w} := (w^*, w_1, \cdots, w_{Q_2+1}) \in \mathbb{Z}_p^{Q_2+2}$ subject to

$$\prod_{i=1}^{Q_1}(s_i - s^*) = \prod_{j=1}^{Q_2+1}(w_j - w^*), \tag{4}$$

and computes the following two functions

$$F(\alpha) = \prod_{i=1}^{Q_1}((\alpha - s^*) + s_i) \quad \text{and} \quad G(\beta) = \prod_{j=1}^{Q_2+1}((\beta - w^*) + w_j).$$

It sets $g := \tilde{g}^{F(\alpha)}$, $X := g^{\alpha - s^*}$, $h := \tilde{g}^{G(\beta)}$ and $Y := h^{\beta - w^*}$. The public key of the signer is $\mathtt{pk} := (X, Y)$, and the corresponding secret key is $\mathtt{sk} := (x, y) = (\alpha - s^*, \beta - w^*)$ which is unknown to $\mathcal{B}$. Observe that conditioned on the choice of $\boldsymbol{s}$ and the relation (4), polynomial $G(\beta)$ is still of dimension Q_2, which guarantees the independence of h from g and the signatures (to be generated later). Hence, $\mathtt{pk}$ is well distributed. $\mathcal{B}$ then invokes the adversary $\mathcal{A}$ on input $(\mathbb{G}, \mathbb{G}_T, \hat{e}, p, g, h, \mathtt{pk})$, and begins to simulate oracles for $\mathcal{A}$ as below.

$\mathtt{H}_1$. Given the i-th distinct query t_i to oracle $\mathtt{H}_1$, $\mathcal{B}$ returns s_i to the adversary, and stores (t_i, s_i) into a hash table HT_1.

$\mathtt{H}_2$. Given the j-th distinct query (m_j, t_j) to oracle $\mathtt{H}_2$, $\mathcal{B}$ returns w_j to the adversary, and stores $((m_j, t_j), w_j)$ into a hash table HT_2.

$\mathtt{H}_3$. Given a distinct query $(m, t, \sigma, \mathtt{pk}, R_1, R_2)$, $\mathcal{B}$ selects at random a string $c \in \mathbb{Z}_p$, returns c to the adversary, and stores $((m, t, \sigma, \mathtt{pk}, R_1, R_2), c)$ in a hash table HT_3.

Sign. Given a distinct signing query (m, t), $\mathcal{B}$ first retrieves $s = \mathtt{H}_1(t)$ and $w = \mathtt{H}_2(m, t)$ from tables HT_1 and HT_2, respectively. If there is no such entry in the table, $\mathcal{B}$ creates one as above. It then computes

$$\sigma := g^{\frac{1}{x+s}} h^{\frac{1}{y+w}}.$$

Since $s \in \{s_1, \cdots, s_{Q_1}\}$ and $w \in \{w_1, \cdots, w_{Q_2+1}\}$, σ could be efficiently computed by $\mathcal{B}$.

Confirm/Disavow. Given a confirmation/disavowal query (m, t, σ), $\mathcal{B}$ runs the simulator of the confirmation/disavowal protocol to produce the corresponding proof.

SConv. Given (m, t, σ), $\mathcal{B}$ calls the simulator of the non-interactive version of the confirmation protocol to produce a proof $\pi = (c, z)$, and returns it to the adversary. In particular, $\mathcal{B}$ picks at random $(c, z) \in \mathbb{Z}_p^2$, and patches the oracle $\mathtt{H}_3$ with $\mathtt{H}_3(m, t, \sigma, \mathtt{pk}, h^z/Y^c, W_1^z/(W_2 W_1^{-\mathtt{H}_2(m,t)})^c) = c$. If a collision occurs, $\mathcal{B}$ aborts with outputting a random bit.

TSConv. Given a time period t, $\mathcal{B}$ retrieves $s = \mathsf{H}_1(t)$ from table HT_1, computes and returns $\mathsf{tscvt} := g^{1/(x+s)}$.

When the adversary is ready, it submits a message m^* and a time period t^* such that (m^*, t^*) is distinct from all the pairs it submitted to the signing oracle and t^* was not submitted as a **TSConv** query. An honestly generated signature on (m^*, t^*) should be of the form:

$$
\begin{aligned}
\sigma^* &= g^{\frac{1}{x+s^*}} h^{\frac{1}{y+w^*}} = g^{\frac{1}{\alpha-s^*+s^*}} h^{\frac{1}{\beta-w^*+w^*}} \\
&= g^{\frac{1}{\alpha}} h^{\frac{1}{\beta}} = \tilde{g}^{\frac{F(\alpha)}{\alpha}} \tilde{g}^{\frac{G(\beta)}{\beta}} \\
&= \tilde{g}^{\frac{\prod_{i=1}^{Q_1}(\alpha-s^*+s_i)}{\alpha}} \tilde{g}^{\frac{\prod_{j=1}^{Q_2+1}(\beta-w^*+w_j)}{\beta}} \\
&= \tilde{g}^{F'(\alpha)} \tilde{g}^{G'(\beta)} \tilde{g}^{\frac{\prod_{i=1}^{Q_1}(s_i-s^*)}{\alpha} + \frac{\prod_{j=1}^{Q_2+1}(w_j-w^*)}{\beta}},
\end{aligned}
$$

where $F'(\alpha)$ is a polynomial of degree $Q_1 - 1$ and $G'(\beta)$ is a polynomial of degree Q_2. Note that the two polynomials could be efficiently computed by $\mathcal{B}$ from the given input. By Eq. (4) we know that $\prod_{i=1}^{Q_1}(s_i - s^*) = \prod_{j=1}^{Q_2+1}(w_j - w^*)$. Denote it by d. Then we have

$$
\sigma^* = \tilde{g}^{F'(\alpha)} \tilde{g}^{G'(\beta)} (\tilde{g}^{\frac{1}{\alpha}+\frac{1}{\beta}})^d.
$$

Hence, $\mathcal{B}$ prepares the challenge signature σ^* by computing

$$
\sigma^* := \tilde{g}^{F'(\alpha)} \tilde{g}^{G'(\beta)} Z^d.
$$

If $Z = \tilde{g}^{\frac{1}{\alpha}+\frac{1}{\beta}}$, σ^* is a valid signature; if Z is randomly chosen from $\mathbb{G}$, σ^* is also random. Let b be the bit such that $b = 1$ if Z is a random element of $\mathbb{G}$, and $b = 0$ otherwise. $\mathcal{B}$ returns σ^* and continues to simulate the oracles for $\mathcal{A}$. Finally, $\mathcal{A}$ outputs a bit b'. $\mathcal{B}$ outputs b' as well. If $\mathcal{A}$ wins its game, so does $\mathcal{B}$.

Probability Analysis. The simulation of oracles H_1, H_2, **TSConv** and **Sign** are perfect, as long as the numbers of queries $\mathcal{A}$ issued do not exceed the presumed bounds. In the simulation of oracle H_3, $\mathcal{B}$ patches the output of the oracle, and collisions might occur. Denote by col the event that a collision occurs. The probability that col happens is at most $q_3 q_{sc}/p$, which is negligible.

In addition, $\mathcal{B}$ answers the adversary's confirmation and disavowal queries by running the corresponding simulator, which would also bring difference to the adversary's view. However, since the protocols are zero-knowledge, the simulated proofs look almost the same as real ones, up to a negligible difference δ. According to the simulation above, $\mathcal{B}$ succeeds in telling how Z was chosen if $\mathcal{A}$ succeeds in its invisibility game. Therefore, we have that

$$
\begin{aligned}
\mathsf{Adv}_{\mathcal{B}}^{\mathrm{DDSDH}}(k) &\geq \left| \Pr[\mathsf{col}]\frac{1}{2} + \Pr[\neg\mathsf{col}](1-\delta)(\frac{1}{2}+\epsilon) - \frac{1}{2} \right| \\
&\geq \left| \frac{1}{2}\frac{q_3 q_{sc}}{p} + (1 - \frac{q_3 q_{sc}}{p})(1-\delta)(\frac{1}{2}+\epsilon) - \frac{1}{2} \right|
\end{aligned}
$$

$$= \left| (1 - \frac{q_3 q_{sc}}{p})(\epsilon - \delta(\frac{1}{2} + \epsilon)) \right| \geq \left| (1 - \frac{q_3 q_{sc}}{p})(\epsilon - \delta) \right|.$$

The last inequality holds because $\epsilon \leq 1/2$. Since the terms $q_3 q_{sc}/p$ and δ are negligible in the security parameter, it turns out that $\mathrm{Adv}_{\mathcal{B}}^{\mathrm{DDSDH}}(k)$ is negligibly close to ϵ.   □

It is known that invisibility and anonymity of undeniable signature are equivalent to each other if the signer's signature space is indistinguishable from uniform [8]. This is the case in our scheme, i.e. the range of each signer's signature space is $\mathbb{G}$. We have the following corollary:

Corollary 1. *Our time-selective convertible undeniable signature scheme is anonymous in the random oracle model if DDSDH assumption holds.*

4.4   Randomized Signature Generation

In our TSCUS scheme each message has a unique signature per time period. To allow the generation of multiple signatures on a message m w.r.t. a time period t, we may modify the signature generation algorithm as follows. The signer chooses at random $r \leftarrow \mathbb{Z}_p$, and computes $U = g^{1/(x+\mathrm{H}_1(t))} h^{1/(y+\mathrm{H}_2(m,t,r))}$. The signature on m w.r.t. t is $\sigma := (U, r)$. The other part of the scheme can then be modified accordingly.

4.5   Efficiency and Comparison

In Table 2, the key size (both public and private keys), the signature size and converter size (both selective and time-selective) of our scheme are compared with that of existing ones [13,14,1]. The table also shows the differences in security assumptions. In the table, t is the time period. Note that we do not consider the size of Params, while we should note that besides group description, Params of the scheme in [1] also contains the description of two pseudorandom functions. Hence the size of their Params is larger than that of [13,14] and ours. Regarding security, the anonymity of schemes in [13,14] relies on a strong and interactive assumption, xyz-DCAA, and the security of the scheme in [1] relies on some special properties of hash functions, and the pseudorandomness of pseudorandom functions, while the scheme proposed in this paper relies on SDH and DDSDH assumptions.

Table 3 shows the comparison of these schemes in terms of the computational cost. In the table, 'E' and 'E_T' represent the exponentiation operation in $\mathbb{G}$ and $\mathbb{G}_T$, respectively, 'P' represents the bilinear pairing, and 'PRF' refers to the evaluation of a pseudorandom function.

Note that in Laguillaumie and Vergnaud's schemes [13,14], the size of a time-selective converter grows linearly/logarithmcally with the number of time periods. If the converter for the t-th period has already been published, the signer only needs to release one extra group element in the $(t + 1)$-th time period.

Table 2. Comparison with [13,14,1] in size and assumption

	PK	SK	Sig	SCvt	TSCvt	Assumptions
[13]	2G	$2\mathbb{Z}_p$	$\ell + 1$G	tG	tG	CDH + xyz-DCAA
[14]	2G	$2\mathbb{Z}_p$	$\ell + 1$G	$\lceil \log t \rceil$G	$\lceil \log t \rceil$G	CDH + xyz-DCAA
[1]	2G	$2\mathbb{Z}_p$	2G $+ 1\mathbb{Z}_p$	$1\mathbb{Z}_p$	k	raPre + rlColl + Pre + Prf
Ours	2G	$2\mathbb{Z}_p$	1G	$2\mathbb{Z}_p$	1G	SDH + DDSDH

Table 3. Comparison with [13,14,1] in computational cost

	Sign	SConv	SVer	TSConv	TSVer
[13]	1E	t E	1E+$(2t + 2)$ P	t E	1E+$(2t + 2)$ P
[14]	1E	$1.5\lceil \log t \rceil$ E	1E+$(3\lceil \log t \rceil + 2)$ P	$1.5\lceil \log t \rceil$E	1E+$(3\lceil \log t \rceil + 2)$ P
[1]	2E+2PRF	1PRF	1PRF+4E	2PRF	4E
Ours	2E	2E+$1E_T$+3P	3E+$3E_T$+3P	1E	2E+4P

The computation of our scheme can be optimized. If we pre-compute $\hat{e}(g, g)$ and $\hat{e}(h, h)$ and put them into the system parameter, the number of pairing evaluations in **SConv**, **SVer** and **UVer** can be reduced by 1 each, and that in **TSVer** can be reduced by 2. In practice, the verifier only needs to check the validity of the time-selective converter (i.e. the first equation in **TSVer**) once for each time period. Then for each subsequent signature corresponding to the same time period, the number of pairing evaluations in **TSVer** can be further reduced to just one.

5   Conclusion

In this paper we proposed a new TSCUS scheme, which has short and constant-size converters (i.e. universal, selective and time-selective). The scheme has low computational complexity and low communication complexity. The security of the scheme is based on SDH assumption and DDSDH assumption.

The security of the concrete TSCUS scheme relies on the random oracle model. One of the future work is to construct a TSCUS scheme which can be proven secure in the standard model, while having comparable computational complexity and size on keys, signatures and converters.

References

1. El Aimani, L., Vergnaud, D.: Gradually convertible undeniable signatures. In: Katz, J., Yung, M. (eds.) ACNS 2007. LNCS, vol. 4521, pp. 478–496. Springer, Heidelberg (2007)
2. Bellare, M., Rogaway, P.: Random oracles are practical: A paradigm for designing efficient protocols. In: CCS, pp. 62–73. ACM, New York (1993)

3. Boneh, D., Boyen, X.: Short signatures without random oracles. In: Cachin, C., Camenisch, J.L. (eds.) EUROCRYPT 2004. LNCS, vol. 3027, pp. 56–73. Springer, Heidelberg (2004)

4. Boyar, J., Chaum, D., Damgård, I., Pederson, T.P.: Convertible undeniable signatures. In: Menezes, A., Vanstone, S.A. (eds.) CRYPTO 1990. LNCS, vol. 537, pp. 189–205. Springer, Heidelberg (1991)

5. Chaum, D.: Designated confirmer signatures. In: De Santis, A. (ed.) EUROCRYPT 1994. LNCS, vol. 950, pp. 86–91. Springer, Heidelberg (1995)

6. Chaum, D., van Antwerpen, H.: Undeniable signatures. In: Brassard, G. (ed.) CRYPTO 1989. LNCS, vol. 435, pp. 212–216. Springer, Heidelberg (1990)

7. Damgård, I., Pedersen, T.: New convertible undeniable signature schemes. In: Maurer, U.M. (ed.) EUROCRYPT 1996. LNCS, vol. 1070, pp. 372–386. Springer, Heidelberg (1996)

8. Galbraith, S.D., Mao, W.: Invisibility and anonymity of undeniable and confirmer signatures. In: Joye, M. (ed.) CT-RSA 2003. LNCS, vol. 2612, pp. 80–97. Springer, Heidelberg (2003)

9. Gennaro, R., Krawczyk, H., Rabin, T.: RSA-based undeniable signatures. In: Kaliski Jr., B.S. (ed.) CRYPTO 1997. LNCS, vol. 1294, pp. 132–149. Springer, Heidelberg (1997)

10. Gentry, C., Molnar, D., Ramzan, Z.: Efficient designated confirmer signatures without random oracles or general zero-knowledge proofs (extended abstract). In: Roy, B. (ed.) ASIACRYPT 2005. LNCS, vol. 3788, pp. 662–681. Springer, Heidelberg (2005)

11. Huang, Q., Wong, D.S.: Short convertible undeniable signature in the standard model. In: Bao, F., Weng, J. (eds.) ISPEC 2011. LNCS, vol. 6672, pp. 257–272. Springer, Heidelberg (2011)

12. Kurosawa, K., Takagi, T.: New approach for selectively convertible undeniable signature schemes. In: Lai, X., Chen, K. (eds.) ASIACRYPT 2006. LNCS, vol. 4284, pp. 428–443. Springer, Heidelberg (2006)

13. Laguillaumie, F., Vergnaud, D.: Time-selective convertible undeniable signatures. In: Menezes, A. (ed.) CT-RSA 2005. LNCS, vol. 3376, pp. 154–171. Springer, Heidelberg (2005)

14. Laguillaumie, F., Vergnaud, D.: Time-selective convertible undeniable signatures with short conversion receipts. Information Sciences 180(12), 2458–2475 (2010)

15. Michels, M., Petersen, H., Horster, P.: Breaking and repairing a convertible undeniable signature scheme. In: CCS, pp. 148–152. ACM, New York (1996)

16. Michels, M., Stadler, M.: Efficient convertible undeniable signature schemes. In: SAC 1997, pp. 231–244 (1997)

17. Naor, M.: On cryptographic assumptions and challenges. In: Boneh, D. (ed.) CRYPTO 2003. LNCS, vol. 2729, pp. 96–109. Springer, Heidelberg (2003)

18. Okamoto, T.: Designated confirmer signatures and public-key encryption are equivalent. In: Desmedt, Y.G. (ed.) CRYPTO 1994. LNCS, vol. 839, pp. 61–74. Springer, Heidelberg (1994)

19. Phong, L.T., Kurosawa, K., Ogata, W.: New RSA-based (selectively) convertible undeniable signature schemes. In: Preneel, B. (ed.) AFRICACRYPT 2009. LNCS, vol. 5580, pp. 116–134. Springer, Heidelberg (2009)

20. Phong, L.T., Kurosawa, K., Ogata, W.: Provably secure convertible undeniable signatures with unambiguity. In: Garay, J.A., De Prisco, R. (eds.) SCN 2010. LNCS, vol. 6280, pp. 291–308. Springer, Heidelberg (2010)

21. Huang, Q., Wong, D.S., Susilo, W.: A new construction of designated confirmer signature and its application to optimistic fair exchange. In: Joye, M., Miyaji, A., Otsuka, A. (eds.) Pairing 2010. LNCS, vol. 6487, pp. 41–61. Springer, Heidelberg (2010)
22. Schuldt, J.C.N., Matsuura, K.: An efficient convertible undeniable signature scheme with delegatable verification. In: Kwak, J., Deng, R.H., Won, Y., Wang, G. (eds.) ISPEC 2010. LNCS, vol. 6047, pp. 276–293. Springer, Heidelberg (2010)
23. Shoup, V.: Lower bounds for discrete logarithms and related problems. In: Fumy, W. (ed.) EUROCRYPT 1997. LNCS, vol. 1233, pp. 256–266. Springer, Heidelberg (1997)
24. Zhang, F., Safavi-Naini, R., Susilo, W.: An efficient signature scheme from bilinear pairings and its applications. In: Bao, F., Deng, R., Zhou, J. (eds.) PKC 2004. LNCS, vol. 2947, pp. 277–290. Springer, Heidelberg (2004)

Efficient Fail-Stop Signatures from the Factoring Assumption

Atefeh Mashatan and Khaled Ouafi

Ecole Polytechnique Fédérale de Lausanne (EPFL),
CH-1015 Lausanne, Switzerland
{Atefeh.Mashatan,Khaled.Ouafi}@epfl.ch

Abstract. In this paper, we revisit the construction of fail-stop signatures from the factoring assumption. These signatures were originally proposed to provide information-theoretic-based security against forgeries. In contrast to classical signature schemes, in which signers are protected through a computational conjecture, fail-stop signature schemes protect the signers in an information theoretic sense, i.e., they guarantee that no one, regardless of its computational power, is able to forge a signature that cannot be detected and proven to be a forgery. Such a feature inherently introduced another threat: malicious signers who want to deny a legitimate signature.

Many construction of fail-stop signatures were proposed in the literature, based on the discrete logarithm, the RSA, or the factoring assumptions. Several variants of this latter assumption were used to construct fail-sop signature schemes. Bleumer et al. (EuroCrypt '90) proposed a fail-stop signature scheme based on the difficulty of factoring large integers and Susilo et al. (The Computer Journal, 2000) showed how to construct a fail-stop signature scheme from the so-called "strong factorization" assumption. A later attempt by Schmidt-Samoa (ICICS '04) was to propose a fail-stop signature scheme from the $p^2 q$ factoring assumption.

Compared to those proposals, we take a more traditional approach by considering the Rabin function as our starting point. We generalize this function to a new bundling homomorphism while retaining Rabin's efficient reduction to factoring the modulus of the multiplicative group. Moreover, we preserve the efficiency of the Rabin function as our scheme only requires two, very optimized, modular exponentiations for key generation and verification. This improves on older constructions from factoring assumptions which required either two unoptimized or four exponentiations for key generation and either two unoptimized or three modular exponentiations for verifying.

Keywords: Fail-stop signature schemes, Digital signatures, Rabin function, Factoring.

1 Introduction

Digital signatures are one of the main achievements of public key cryptography: they are the main primitive that ensures authenticity of data transmitted

X. Lai, J. Zhou, and H. Li (Eds.): ISC 2011, LNCS 7001, pp. 372–385, 2011.

through an insecure channel. In being so, digital signatures were originally designed to be the "digital" equivalent to handwritten signatures. Their underlying mechanism is simple and elegant: to each person is associated a "public" key that can be used to verify any signature this person has produced for a document or certificate. To each such key, that is publicly known, is associated a so-called "private" key that is known to the signer only and is used to sign electronic messages or documents. Naturally, this private key should not be computable from the public key. A computational assumption believed to be "impossible" to solve is usually used to prevent this scenario.

However, when coming to the practical world, the question of forgeries was quickly raised by lawyers and judges. This case was even more critical in countries like Germany in which digital signatures have a legal value. However, cryptographers initially ignored the problem of forgery and repudiation, that is because the whole security of a digital signature scheme stands on the shoulders of a problem that we trust to be "hard" to solve. So, as long as the computational assumption holds, forgeries do not happen, except with very small probability that we also hope to never occur. Nevertheless, from a legal point of view, forgeries do exist and signers have to be given the ability to defend themselves against them.

Fail-stop signatures were designed to address this problem: they provide the signer with a mean to prove that a signature is a forgery and that the cryptographic assumption holding the security of the scheme has been broken. So, in the case of a dispute, the signer can exhibit the evidence and deny any responsibility on it. Furthermore, such a forgery shows that the scheme has become insecure, and thus has to be stopped, hence the denomination "fail-stop". Since they consider the extreme eventuality of the break of a hard cryptographic problem, fail-stop signature schemes should protect the signers against computationally unlimited adversaries. However, granting to signers the ability to prove forgeries is not without risk. Indeed, a malicious signer may try to prove that a signature he has produced is a forgery. This way, he can free himself from any commitment the signed document may induce. Unfortunately, it is theoretically impossible to ensure security against computationally unlimited malicious signers if we assume that the forgers are so. Instead, the security against a malicious signer is only computational, i.e., it assumes that solving the hard problem for this party is infeasible.

1.1   Previous Work

While the most popular and efficient scheme is based on the discrete logarithm problem [19], few convincing schemes from the factoring assumption were proposed in the literature. Surprisingly, the first fail-stop signature scheme was based on the factorization assumption [3]. More precisely, it was originally related to the notion of claw-free pairs of permutations in [3] but then reformulated in [11] to explicitly relate the security of the scheme to the factoring assumption. This scheme, to which we refer as the QR scheme, is based on the hardness of factoring an RSA modulus $n = pq$ with factors being such that $p \equiv 3 \mod 8$ and $q \equiv 7 \mod 8$. Up to date, it is unclear how to compare the difficulty of factoring such moduli compared to the classical RSA ones. Moreover, the scheme seems to be made so that it works so its structure is not very "natural".

Later, Susilo et al. proposed another scheme [17], that, following [15], we refer to as the order scheme, based on the difficulty of factoring an RSA modulus $n = pq$ when the algorithm is also given an element from a superset $\mathbb{Z}_P^\star$ of $\mathbb{Z}_n^\star$ (such that P is prime and a that $P-1$ is a multiple of n) of order q. This scheme has been shown to be insecure by Schmidt-Samoa [15] but was also repaired in the same paper by reducing the message space form $\mathbb{Z}_n^\star$ to $\mathbb{Z}_p$. This fix affected the efficiency of the scheme as it made signatures around three times larger than their corresponding messages.

In parallel, Schmidt-Samoa proposed a scheme based on the hardness of factoring RSA moduli of the form $n = p^2q$ [15] (we refer to this one as the p^2q scheme). Although it is probably the most elegant scheme of those mentioned before, this scheme inherently requires larger sizes for the modulus compared to the other two. Recently, Susilo proposed a fail-signature scheme from the factoring and discrete logarithm assumption [16]. However, we note that its security proof does not make any reduction to the factoring problem but rather to the discrete logarithm problem in an RSA group. Therefore, we do not consider it based on a factoring assumption.

1.2   Our Contribution

We revisit the design of fail-stop signatures from the factoring assumption. Our starting point is the Rabin function, $f(x) = x^2 \bmod n$, whose invertibility is known to be equivalent to the problem of factoring the integer n. As a consequence of the Chinese Remainder theorem, when $n = pq$, a classical RSA modulus, this function maps four integers to a single one. In this work, we use a generalization of the Rabin function, $f(x) = x^a \bmod n$, for a a that divides the order of the group $\mathbb{Z}_n^\star$. In practice, we will construct such moduli by choosing a p that satisfies $p-1 \equiv 0 \bmod a$. We remark that this type of RSA moduli, with a publicly known, was already used by Benaloh to define a public key encryption scheme [4].

As we show later, the function $f(x) = x^a \bmod n$ enjoys a number of nice properties. At first, using a result from number theory due to Frobenius [6], we prove that this function actually maps a elements of $\mathbb{Z}_n^\star$ to a single a-th residue of the same group. The second property, inherited from the Rabin function is that finding collisions is provably as hard as factoring the modulus with the knowledge of a. The last observation is that both proofs for the results above do not take into consideration the structure of a, but only on its size. Consequently, numbers of the form $2^s + 1$ will be used for a as it allows fast computation of the modular exponentiation in s multiplications using the square-and-multiply method, resulting in a quadratic complexity for computing the function f. All these properties allow us to use this function as a bundling homomorphism and then instantiate Pfitzmann's general construction [12] of fail-stop signatures to yield the most efficient scheme based on a factoring assumption.

The rest of the paper is structured as follows. Section 2 provides the necessary background, definitions of fail-stop signature schemes and the factoring intractability assumption. We then dedicate Section 3 to our instantiation of the fail-stop signature scheme and prove its security. Efficiency analysis is discussed in Section 4.

2 Preliminaries

2.1 Notations and Negligible Probabilities

Throughout this paper, we use the expression $y \leftarrow \mathcal{A}(x)$ to mean that y is assigned the output of algorithm $\mathcal{A}$ running on input x. An algorithm is said to be polynomial if its running time can be expressed as a polynomial in the size of its inputs.

If X denotes a set, $|X|$ denotes its cardinality and $x \in_R X$ expresses that x is chosen from X according to the uniform distribution. If X and Y are sets, then the relative complement of X in Y, denoted $X \setminus Y$, is the set of elements in X, but not in Y.

We also recall the classical notions of negligible function and one-way function.

Definition 1 (Negligible Function). *A function $f : \mathbb{N} \to \mathbb{R}$ is negligible in k if for any positive polynomial function $p(\cdot)$ there exists k_0 such that:*

$$k \geq k_0 \Rightarrow f(k) < \frac{1}{p(k)}.$$

Negligible functions in k are denoted $\mathsf{negl}(k)$.

2.2 Number Theoretic and Factoring Background

We call a prime number p to be a-strong, if $p = 2ap' + 1$ for a prime $p' > 2a$. Note that such prime numbers are called strong by Rivest and Silverman [14] and that for $a = 1$ we get the usual definition of strong primes.

In the following, we consider a probabilistic polynomial-time algorithm Gen that, on input a security parameter 1^k, that picks an integer a and generates a random a-strong prime $p = 2ap' + 1$ and a regular, not a-strong, prime number q, such that $\lfloor \log_2 p \rfloor = \lfloor \log_2 q \rfloor = \ell_F(k)$, where $\ell_F(k)$ denotes a function that represents, for any given security parameter k, the recommended (bit-)size of the RSA modulus $n = pq$. In the end, Gen outputs $n = pq$ and a along with p and q. The following definition formalizes the main assumption that will be used throughout this work.

Definition 2 (The Factorization Assumption with a-strong primes).
Let us consider a probabilistic polynomial-time algorithm $\mathcal{A}_{\mathsf{FACT}}$ who takes as input an RSA modulus $n = pq$ and an odd integer a, such that p, q, and $p - 1/2a$

are all prime numbers, and outputs p and q. The factoring assumption states that for any such $\mathcal{A}_{\mathsf{FACT}}$, we have

$$\Pr\left[(p,q) \leftarrow \mathcal{A}_{\mathsf{FACT}}(1^k, n, a) \,\middle|\, (n, a, p, q) \leftarrow \mathsf{Gen}(1^k, a)\right] = \mathsf{negl}(k).$$

The probability is taken over the random tapes of Gen and $\mathcal{A}_{\mathsf{FACT}}$.

While this type of prime numbers has already been used before by Benaloh to construct a public encryption scheme [4], we refer the reader to the work of Groth [8] which makes a more detailed treatment of the factorization of such numbers by Pollard's rho method [13] and other factoring algorithms such as the general number field sieve.

In the rest of this section, we concentrate on exponentiation to a and a-th residuosity. It is well known that every $x \in \mathbb{Z}_n^\star$ has a unique a-th root if and only if $\gcd(a, \varphi(n)) = 1$, where $\varphi(\cdot)$ denotes the Euler totient function. In fact, this ensures the correctness of the RSA cryptosystem. However, when a and $\varphi(n)$ are not coprime, an element of $\mathbb{Z}_n^\star$ may admit multiple a-th roots as stated by the following theorem.

Theorem 1 (Frobenius [6]). *If a divides the order of a group, then the number of elements in the group whose order divides a is a multiple of a. If the group is cyclic, then this number is exactly a.*

Note that $\mathbb{Z}_n^\star$ is not cyclic when n is an RSA modulus. However, the next theorem proves that the number of a-th roots of any element in $\mathbb{Z}_n^\star$ is *exactly* equal to a in the class of RSA moduli we consider in this paper.

Theorem 2. *Let a be an odd number and $n = pq$ be an RSA modulus such that p is an a-strong prime, and q is a regular prime number that satisfies $\gcd(a, q-1) = 1$. If y is an a-th residue in $\mathbb{Z}_n^\star$, then the equation $g^a = y \mod n$ admits exactly a solutions.*

Proof. Let $\mathsf{CRT} : \mathbb{Z}_n^\star \to \mathbb{Z}_p^\star \times \mathbb{Z}_q^\star$ denote the isomorphism induced by the chinese remainder theorem. By applying CRT, if g is a solution to the equation $g^a = y \mod n$, then $(g_p, g_q) = (g \mod p, g \mod q)$ is a solution to the two equations

$$(g_p)^a = y \mod p, \tag{1}$$
$$(g_q)^a = y \mod q. \tag{2}$$

On one hand, Recalling that $\mathbb{Z}_p^\star$ is cyclic when p is prime, we can apply Thm. 1 to deduce that Eq. (1) has exactly a solutions (remember that a divides $\varphi(p) = ap'$). On another hand, since $\gcd(a, \varphi(q)) = 1$, there must exist a_q' such that $a \cdot a_q' = 1$ (mod $q - 1$) and Eq. (2) can be rewritten as $g_q = y^{a_q'} \mod q$. Hence, Eq. (2) admits one unique solution.

As there exist a tuples of the form (g_p, g_q) that satisfy equations (1) and (2), by applying CRT^{-1}, we deduce that the number of a-th roots of y in $\mathbb{Z}_n^\star$ is exactly a. $\qquad\square$

2.3   Fail-Stop Signatures

We briefly review the basic definition of fail-stop signature schemes and their security properties. A complete and more formal definition can be found in [5].

The idea of fail-stop signatures is to associate, to each possible message m, a number of signatures s that passes the verification test with the public key. These signatures are called acceptable signatures. However, the signer should not be able to construct more than one signature from the secret key. This one signature is called the valid signature. Of course, the set of acceptable signatures must be small in comparison with the signature space, so that it should be difficult for a computationally bounded signer to find an acceptable signature different from his own.

In all cases, an adversary with unlimited computational power can always compute the set of acceptable signatures (one way is to try the verification algorithm with every element of the signature space). So, in order to achieve security against such an adversary, we have to consider an information-theoretic security determined by a security parameter $\sigma \in \mathbb{N}$: an adversary with unlimited computational power should not have enough information to distinguish the valid signature in the set of acceptable ones except with a probability upper-bounded by $2^{-\sigma}$ (the probability to guess it successfully). This property is called the signer's security.

In parallel, verifiers must be secure against signers who, by computing a valid proof of forgery, manage to disavow one of their own legitimate signatures. The security against these signers can only be computational which means that the probability of a signer disavowing his own signature is negligible in k. This property is known as the verifier's security.

Definition 3 (Fail-Stop Signature Schemes). *A fail-stop signature scheme is defined by the following five polynomial-time algorithms:*

- KeyGen$(1^k, 1^\sigma, 1^N) \longrightarrow (sk, pk)$. *This is a probabilistic polynomial-time, possibly interactive, protocol run by the signer and the verifier (or a trusted center) runs on security parameters k, σ and an integer N representing the number of signatures the signer can release. At the end of the protocol, he signer obtains a private key sk which can be used to sign at most N messages (This type of signature schemes are said to be N-times). The other party obtains the corresponding public key pk. For the sake of simplicity, the input 1^N is omitted when the scheme is meant to be used to sign one message only.*
- Sign$(sk, i, m) \longrightarrow s$. *Given a message m, a counter $i \in \{1 \ldots N\}$, incremented at each invocation of this algorithm, and the private key sk, this (probabilistic) polynomial-time algorithm creates a valid signature s for the message m. When the scheme is a one-time signature scheme, i.e., $N = 1$, the input i is omitted.*
- Verify$(pk, m; s) \longrightarrow \{0, 1\}$. *Given a signature s for m and a public key pk, Verify is a deterministic polynomial-time algorithm that outputs 1 if the signature is acceptable, otherwise it outputs 0.*

- ProveForgery$(sk, m, s) \longrightarrow \{\mathsf{pr}, \perp\}$. *On input an acceptable signature s on m, and private key sk, this algorithm outputs a bit-string* pr *or* $\perp$ *in case of failure.*
- VerifyProof$(pk, m, s, \mathsf{pr}) \longrightarrow \{0, 1\}$. *A polynomial-time algorithm that takes on input pk, an acceptable signature s of a message m and a proof of forgery* pr *and outputs either 1, meaning that the proof is valid, or 0, meaning that the proof is invalid.*

Additionally, any fail-stop signature scheme should also satisfy two correctness properties: every signature honestly produced using Sign is acceptable and every proof computed using ProveForgery passes the verification. This is more formally defined hereafter.

Definition 4 (Correctness of a Fail-Stop Signature Scheme). *We say that a fail-stop signature scheme is correct if the two conditions hold:*

1. *Every honestly generated signature is valid, i.e.,*

$$\forall \lambda, N, m \in M : \Pr\left[1 \leftarrow \mathsf{Verify}(pk, m, s) \,\middle|\, \begin{array}{l} (pk, sk) \leftarrow \mathsf{Keygen}(1^\lambda, 1^N), \\ s \leftarrow \mathsf{Sign}(sk, m) \end{array} \right] = 1.$$

2. *Every honestly generated proof is valid, i.e.,*

$$\forall \lambda, N, m \in M, s \in S :$$
$$\Pr\left[1 \leftarrow \mathsf{VerifyProof}(pk, m, s, pr) \,\middle|\, \begin{array}{l} (pk, sk) \leftarrow \mathsf{Keygen}(1^\lambda, 1^N), \\ pr \leftarrow \mathsf{ProveForgery}(sk, m, s), \\ pr \neq \perp \end{array} \right] = 1.$$

Security against malicious signers and powerful forgers is formalized as follow.

Definition 5 (Security of a Fail-Stop Signature Scheme). *A fail-stop signature scheme with security parameters k and* σ *is said to be secure if the two properties hold*

1. **Signer's security.** *The probability that a computationally unlimited adversary knowing pk and the signature of N adaptively chosen messages* $(m_1, s_1), \ldots, (m_N, s_N)$ *outputs a pair* (m, s) *such that* $(m, s) \neq (m_i, s_i), \forall i = 1 \ldots N$, *and* ProveForgery$(sk, m, s) = \perp$ *must be smaller than* $2^{-\sigma}$.
2. **Verifier's security.** *The probability that a polynomially bounded malicious signer generates a public key pk and then outputs a pair* (m, s) *and a proof of forgery* pr *such that* VerifyProof$(pk, m, s, \mathsf{pr}) = 1$ *must be negligible in k.*

A constraining result for fail-stop signatures is that, in order to be able to sign N messages, the signer has to generate at least $(N + 1)(k - 1)$ secret random bits. We stress that these bits do not necessarily need to generated during the key generation as the signature algorithm may be probabilistic and update the internal state according to some random bits. Although this result seems a sever limitation against the practical implementation of fail-stop signature schemes, it has been proved in [18] that it is unavoidable if one hopes to achieve security against unbounded adversaries.

2.4   The General Construction Using Bundling Homomorphisms

A general framework for constructing a fail-stop signature scheme secure for signing one message, on which are built all known secure fail-stop signature schemes, has been proposed by Pfitzmann [12]. It is based on the notion of bundling homomorphism. We note that classical techniques such as Merkle trees [9,10], top-down authentication trees [7], and one-way accumulators [1,2], can be used to extend a one-time fail-stop signature scheme to many messages.

Definition 6 (Family of Bundling Homomorphisms). *Let $\imath \in I$ be an indexing for a family of triples $(h_\imath, G_\imath, H_\imath)$ such that for all possible $\imath \in I$, $(G_\imath, +, 0)$ $(H_\imath, \times, 1)$ are Abelian groups and $h_\imath : G_\imath \to H_\imath$. $(h_\imath, G_\imath, H_\imath)$ is called a family of bundling homomorphisms with degree level 2^τ and collision-resistance security of level k if it satisfies:*

1. *For every $\imath \in I$, $h_\imath$ is an homomorphism.*
2. *There exist polynomial-time algorithms for sampling from I, computing $h_\imath$ and the group operations in $G_\imath$ and $H_\imath$, for every $\imath \in I$*
3. *For every $\imath \in I$, every image $y \in \mathsf{Im}(h_\imath)$ has at least 2^τ preimages in $G_\imath$. We call 2^τ the bundling degree of the homomorphism.*
4. *It is computationally infeasible to find collisions, i.e. for any probabilistic polynomial-time algorithm $\tilde{A}$, we have*

$$\Pr\left[h_\imath(x_1 - x_2) = 1, \quad x_1 \neq x_2 \,\middle|\, \begin{array}{c} \imath \in_R I, \\ (x_1, x_2) \leftarrow \tilde{A}(\imath) \end{array}\right] = \mathsf{negl}(k)$$

where the probability is taken among the random choice of $\imath$ and the random coins of $\tilde{A}$.

The basic idea behind Pfitzmann's construction is to use the high bundling degree of a bundling homomorphism to "hide" the signer's secret key, which lies in the space of the G's, even against powerful adversaries who can invert the homomorphism (note that the collision resistance of bundling homomorphisms stated in point 4 of the definition implies resistance against preimage attacks).

The complete description of the construction is described below. For the sake of simplicity, we assume the case in which the key generation is run by the signer in conjonction with a center trusted by the verifiers (and not necessarily by the signer). This easily generalizes to the case of many verifiers [11].

– KeyGen$(1^k, 1^\sigma)$. As a prekey, the center picks a random index K for the family of bundling homomorphisms $H_\imath$. It sets $h = h_K$, $G = G_K$, $H = H_K$ and also defines the finite message space $M \subset \mathbb{Z}$.
  For prekey verification, the signer has to be convinced that h is a group homomorphism with bundling degree 2^τ for an appropriate τ to be later discussed. Once he is convinced, the signer chooses randomly his private key

$$sk = (sk_1, sk_2) \in_R G^2,$$

and computes the public key

$$pk = (pk_1, pk_2) = (h_K(sk_1), h_K(sk_2)) \in H^2.$$

- Sign(sk, m). For signing a message $m \in M$, compute

$$s = sk_1 + m \cdot sk_2,$$

where $m \cdot sk_2$ denotes the operation of performing m additions of sk_2 in G.
- Verify(pk, m, s). Check whether:

$$h(s) = pk_1 \times pk_2^m.$$

- ProveForgery(sk, m, s'). Given s' a forgery for m that passes the verification, the signer computes $s = \text{Sign}(sk, m)$ and exhibits

$$(m, s, s').$$

- VerifyProof(pk, m, s, pr). Given m and $(s, s') \in G^2$, verify that

$$s \neq s' \quad \wedge \quad h(s) = h(s')$$

For the security analysis of this construction, we will denote SK_{pk} the set of secret keys that correspond to the public key pk, i.e,

$$SK_{pk} = \{sk \in SK : h(sk) = pk\}.$$

Since h has a bundling degree of 2^τ, we must have $|SK_{pk}| \geq 2^\tau$. In consequence, given a public key (pk_1, pk_2) there are at least $2^{2\tau}$ private keys that match the signer's public key. Each of these private keys produces a signature that passes the verification. These keys can be computed by an adversary with unlimited computational power, but he cannot know which of these 2^{2^τ} keys is the signer's key that he used to sign the message. However, because of the equation $sk_1 + m \times sk_2 = s$, it is sufficient to find one key to determine the other, thus the number of possible private keys is reduced to 2^τ.

A signer is able to prove a forgery on a message m' only if his signature differs from the forged signature. To measure the probability that an adversary outputs the signer's signature we must analyze the number of pairwise different signatures that can be produced using the 2^τ private keys. This number is related to the number of possible secret keys that produce a valid signature. Some easy implications show that this number is upper-bounded by the size of the set

$$T = \{d \in G : h(d) = 1 \wedge (m' - m)d = 0\}.$$

In order to prove security, we have to consider the worst case with respect to the choice of the message, i.e., we consider the signature that can be produced by the maximum number of candidate secret keys,

$$T_{max} = \max_{m' \in M \setminus \{0\}} |\{d \in G : h(d) = 1 \wedge m'd = 0\}|.$$

A more detailed proof and analysis of this construction can be found in [11,12].

Theorem 3 (Security of the General Construction [11,12]). *Let a fail-stop signature scheme following the general construction above with security parameters k, τ. We do have*

1. *The scheme provides a level of security k for the verifiers.*
2. *If 2^τ is chosen such that $T_{max} \leq 2^{\tau - \sigma}$, then this scheme provides a level of security of σ for the signer.*

3   New Fail-Stop Signatures from Factoring

3.1   The Construction

Let us consider the set of RSA moduli $n = pq$ for which p is an a-strong prime number and $p - 1$ is co-prime with q. We define the following group homomorphism

$$H_a : \mathbb{Z}_n^\star \longrightarrow \mathbb{Z}_n^\star$$
$$x \longmapsto x^a \mod n.$$

And using Pfitzmann's general construction, the following fail-stop signature scheme comes naturally:

KeyGen. On input σ, k the center chooses a σ-bit odd integer a and two equally sized primes p, q such that $p - 1/2a$ is also a prime number and that $\gcd(q-1, a) = 1$. The prekey of the scheme then consists of $n = pq$ and a.

To verify that the prekey is correctly generated, the signer has to be provided with a zero-knowledge proof that a indeed divides $\varphi(n)$ (such a proof can easily be constructed from general zero-knowledge proofs). From that point, he generates the key material as follow

$$(sk_1, sk_2) \in_R \mathbb{Z}_n^{\star 2}, \qquad (pk_1, pk_2) = (sk_1^a \mod n, sk_2^a \mod n).$$

Sign. The message space is defined to be $M = \{1, \ldots, \varphi(n) - 1\} \setminus \{x > 1 \mid \gcd(a, x) \neq 1\}$. To sign a message $m \in M$, the signer computes

$$s = sk_1 \times sk_2^m \mod n$$

Verify. An element $s \in \mathbb{Z}_n^\star$ is an acceptable signature on $m \in M$ if and only if the following equality holds

$$s^a = pk_1 \times pk_2^m \mod n.$$

ProveForgery. On the happening of a forgery $(m, s^\star)$, the proof consists of, as states the general construction, the signer's signature s on the message m using his secret key (sk_1, sk_2).

VerifyProof. Given two signatures s and $s^\star$ on the same message m, the proof of forgery is valid if both signatures are different and $s^a = s^{\star a} \mod n$ (It will be shown in the next section that this equation leads to the factorization of n).

3.2 Security Analysis

To prove that h_n, described above, is a family of bundling homomorphisms, we first need to prove this lemma.

Lemma 1. *Let $n = pq$ and a be an odd number such that p, q are prime number and a divides $p-1/2$, $p-1/2a$ is prime and $\gcd(q-1, a) = 1$. If there exists a successful polynomial-time algorithm $\tilde{A}$ who succeeds in finding x_1 and x_2 such that $h_n(x_1) = h_n(x_2)$ and $x_1 \neq x_2$, then there exists a successful polynomial-time algorithm against the factorization problem of Def. 2.*

Proof. We construct the algorithm against factoring as follows. First, it calls upon $\tilde{A}$ and gets x_1 and x_2 such that

$$x_1^a = x_2^a \pmod{n}.$$

Since p and q are of the same size, it must be that $q > a$. Hence, $a \mod q - 1 = a$. So,

$$x_1^a = x_2^a \pmod{q}.$$

Since $\gcd(a, q-1) = 1$, a is invertible modulo $q-1$ and it results that

$$x_1 = x_2 + k \cdot q, \quad \text{for } 1 \leq k < p$$

Hence, we conclude that $\gcd(x_1 - x_2, q) = q$.

So it is sufficient for A_{FACT} to compute $q = \gcd(x_1 - x_2, n)$ and then deduce $p = n/q$. Note that A_{FACT} runs in polynomial-time if $\tilde{A}$ does so. Moreover, it has the same success probability of $\tilde{A}$. $\qquad\square$

The following theorem proves that h_n, described above, is a family of bundling homomorphisms of degree 2^σ:

Theorem 4. *Under the Factorization Assumption of Def. 2, the construction above is a family of bundling homomorphisms with bundling degree 2^σ.*

Proof. It is trivial to see that h is an homomorphism.

Regarding its bundling degree, we recall Thm. 2 which states that the number of solutions of the equation $x^a = y \mod n$ is a when y is an a-th residue of $\mathbb{Z}_n^\star$. Setting $y = 1$ trivially leads the kernel of the homomorphism. The kernel is thus of size a which is lower-bounded by 2^σ. $\qquad\square$

At this point, we state the following theorem, which asserts the security of our construction.

Theorem 5. *The fail-stop signature scheme defined in Sec. 3.1 is secure under the assumption that factoring integers $n = pq$ with the knowledge of a divisor of $\varphi(n)$ is hard.*

Proof. As it has already been show in Thm. 4 that h_n is a family of bundling homomorphisms, we only need to prove security for the signer. For that sake, and according to Thm. 3, we analyse the size of the following set (Here the neutral element is 1):

$$
\begin{aligned}
T_{max} &= \max_{m' \in M \setminus \{1\}} \left| \{d \in \mathbb{Z}_n^\star : h_n(d) = 1 \wedge d^{m'} = 1\} \right| \\
&= \max_{m' \in M \setminus \{1\}} \left| \{d \in \mathbb{Z}_n^\star : d^a = 1 \wedge d^{m'} = 1\} \right| \\
&= \max_{m' \in M \setminus \{1\}} \left| \{d \in \mathbb{Z}_n^\star : \mathrm{ord}_{\mathbb{Z}_n^\star}(d) | a \wedge \mathrm{ord}_{\mathbb{Z}_n^\star}(d) | m'\} \right| \\
&= \max_{m' \in M \setminus \{1\}} \left| \{d \in \mathbb{Z}_n^\star : \mathrm{ord}_{\mathbb{Z}_n^\star}(d) | \gcd(a, m')\} \right| \\
&= 1
\end{aligned}
$$

since $\gcd(m', a) = 1$ by definition of M.

Hence we conclude that taking a bundling homomorphism with degree 2^σ is sufficient to achieve security against powerful adversaries.   □

4   Efficiency Analysis

Among the five algorithms of the scheme, the key generation is certainly the most expensive operation to perform. Generating strong primes of our specified form is done in time $O\left(\ell_F^4(\lambda) + (1 + \sigma + \ell_F(\lambda))^4\right)$. However, this is not a critical issue as it is run only once. Note that this complexity is in fact very close to the key generation of RSA with strong primes [14].

Using some simple optimization techniques, we can drastically reduce the complexity of modular exponentiations to the power of a by choosing values for a with very low (or very high) Hamming weight. Indeed, the security proof of our scheme is based the *size* of a and not its Hamming weight. According to the state of the art of factoring algorithms such that ECM and NFS, assuming low Hamming weight for a does not help the adversary factoring. This way, one can use the classical square-and-multiply technique and set $a = 2^\alpha \pm 1$, for some $\alpha \geq \sigma$. This trick allows us to reduce the complexity of all exponentiations to a from $O(\ell_F^3(\lambda))$ to $O(\sigma \ell_F^2(\lambda))$. For practical applications in which the one-time key has to be regenerated at each signature, such a feature is clearly a benefit. Verification also benefit from this optimisations as only one modular exponentiation (to m) remains to be done.

For typical values $k = \sigma = 80$, we need to generate a 80-bit odd a, e.g., $a = 2^{80} + 1$, and a 1024-bit RSA modulus composed of one a-strong prime number. Key generation is then performed using $40 + 1 = 41$ modular multiplications. Verification is also performed using 41 modular multiplications and one exponentiation.

As our scheme has its operations performed in a smaller group with optimized exponents, it clearly outperforms the p^2q scheme in terms of efficiency and signature length. Since this latter is more efficient than the order scheme [15], we omit

Table 1. Comparaison of efficiency parameters of the most efficient factorization-based fail-stop signature schemes. For better readability and easier notations, we let k denote the length of the RSA moduli and define $\rho = \tau - \sigma$. Note that we set $a = 2^\sigma + 1$ in our scheme so that it does not need to be explicitly included in the public-key.

	Size of sk	Size of pk	Size of m	Size of s	Sign (# mult.)	Verify (# mult.)
QR Scheme	$2(\rho + \sigma + k)$	$2k$	ρ	$\rho + \sigma + k$	ρ	$< 2\rho + \sigma$
Our scheme	$2k$	$2k$	σ	$2k$	σ	$< 2\sigma$

the order scheme from the comparison. Regarding the QR scheme, we provide Table 1 to put a clear comparison with our scheme. It turns out that in term of size of the keys and the signatures, our scheme behaves better than the QR scheme. Regarding the efficiency of the other parameters and algorithms, having a direct comparison is more difficult as the parameter ρ of the QR scheme can be arbitrarily chosen whereas setting σ to a too large value in our scheme would require to increase the modulus size to guard against $p-1$ factoring methods. However, for practical values of comparison $\rho = \sigma = 160$, we can still use 1024-bit moduli. In such a scenario, our scheme still outperforms the QR one.

References

1. Barić, N., Pfitzmann, B.: Collision-Free Accumulators and Fail-Stop Signature Schemes without Trees. In: Fumy, W. (ed.) EUROCRYPT 1997. LNCS, vol. 1233, pp. 480–494. Springer, Heidelberg (1997)
2. Benaloh, J.C., de Mare, M.: One-Way Accumulators: A Decentralized Alternative to Digital Signatures. In: Helleseth, T. (ed.) EUROCRYPT 1993. LNCS, vol. 765, pp. 274–285. Springer, Heidelberg (1994)
3. Bleumer, G., Pfitzmann, B., Waidner, M.: A Remark on Signature Scheme Where Forgery Can Be Proved. In: Damgård, I.B. (ed.) EUROCRYPT 1990. LNCS, vol. 473, pp. 441–445. Springer, Heidelberg (1991)
4. Clarkson, J.B.: Dense probabilistic encryption. In: Workshop on Selected Areas of Cryptography, pp. 120–128 (1994)
5. Damgård, I., Pedersen, T.P., Pfitzmann, B.: On the existence of statistically hiding bit commitment schemes and fail-stop signatures. J. Cryptology 10(3), 163–194 (1997)
6. Frobenius, G.: Über einen Fundamentalsatz der Gruppentheorie, II. Sitzungsberichte der Preussischen Akademie Weissenstein (1907)
7. Goldwasser, S., Micali, S., Rivest, R.L.: A digital signature scheme secure against adaptive chosen-message attacks. SIAM J. Comput. 17(2), 281–308 (1988)
8. Groth, J.: Cryptography in subgroups of Z_n^*. In: Kilian, J. (ed.) TCC 2005. LNCS, vol. 3378, pp. 50–65. Springer, Heidelberg (2005)
9. Merkle, R.C.: Protocols for public key cryptosystems. In: IEEE Symposium on Security and Privacy, pp. 122–134 (1980)
10. Merkle, R.C.: A certified digital signature. In: Brassard, G. (ed.) CRYPTO 1989. LNCS, vol. 435, pp. 218–238. Springer, Heidelberg (1990)

11. Pedersen, T.P., Pfitzmann, B.: Fail-stop signatures. SIAM J. Comput. 26(2), 291–330 (1997)
12. Pfitzmann, B.: Digital Signature Schemes, General Framework and Fail-Stop Signatures. LNCS, vol. 1100. Springer, Heidelberg (1996)
13. Pollard, J.M.: A monte carlo method for factorization. BIT Numerical Mathematics 15(3), 331–334 (1975)
14. Rivest, R., Silverman, R.: Are 'strong' primes needed for RSA? Cryptology ePrint Archive, Report 2001/007 (2001), http://eprint.iacr.org/
15. Schmidt-Samoa, K.: Factorization-based fail-stop signatures revisited. In: López, J., Qing, S., Okamoto, E. (eds.) ICICS 2004. LNCS, vol. 3269, pp. 118–131. Springer, Heidelberg (2004)
16. Susilo, W.: Short fail-stop signature scheme based on factorization and discrete logarithm assumptions. Theor. Comput. Sci. 410(8-10), 736–744 (2009)
17. Susilo, W., Safavi-Naini, R., Gysin, M., Seberry, J.: A new and efficient fail-stop signature scheme. Comput. J. 43(5), 430–437 (2000)
18. van Heijst, E., Pedersen, T.P., Pfitzmann, B.: New Constructions of Fail-Stop Signatures and Lower Bounds. In: Brickell, E.F. (ed.) CRYPTO 1992. LNCS, vol. 740, pp. 15–30. Springer, Heidelberg (1993)
19. van Heyst, E., Pedersen, T.P.: How to Make Efficient Fail-Stop Signatures. In: Rueppel, R.A. (ed.) EUROCRYPT 1992. LNCS, vol. 658, pp. 366–377. Springer, Heidelberg (1993)

Author Index

Becker, Moritz Y.   229
Biskup, Joachim   246

Chen, Huiyu   1
Chen, Jiageng   32
Chen, Liqun   119
Chen, Songqing   135
Chen, Yu   119
Chen, Zhong   79
Clifton, Chris   325

Domingo-Ferrer, Josep   293
Dürholz, Ulrich   47

Eidenbenz, Stephan   135
Emura, Keita   102

Fang, Hui   168
Fischer-Hübner, Simone   152
Fischlin, Marc   47
Fukushima, Kazuhide   63

Gadaleta, Francesco   213
Greschbach, Benjamin   309
Gu, Dawu   182

Henricksen, Matt   63
Huang, Qiong   355
Huang, Yin   168

Joosen, Wouter   213

Kasper, Michael   47
Kiayias, Aggelos   17
Kiyomoto, Shinsaku   63
Koleini, Masoud   229
Kong, Fanyu   95
Kontaxis, Georgios   197

Laur, Sven   262
Lee, Jaewoo   325
Li, Juanru   182
Li, Kangshun   278

Li, Xiangxue   341
Liu, Peng   1

Ma, Sha   278
Mao, Bing   1
Markatos, Evangelos P.   197
Mashatan, Atefeh   372
Meier, Konrad   309
Miyaji, Atsuko   32, 102

Nakano, Yuto   63
Nikiforakis, Nick   213

Omote, Kazumasa   102
Onete, Cristina   47
Ouafi, Khaled   372

Pehlivanoglu, Serdar   17
Polychronakis, Michalis   197
Preuß, Marcel   246

Qian, Haifeng   341
Qin, Bo   293

Rechert, Klaus   309

Susilo, Willy   355

von Suchodoletz, Dirk   309

Wang, Shuhong   168
Wang, Xinche   1
Wehrle, Dennis   309
Wiese, Lena   246
Willemson, Jan   262
Wong, Duncan S.   355
Wu, Lei   95
Wu, Qianhong   79, 293
Wu, Yongdong   168

Xia, Feng   278
Xie, Li   1
Xin, Zhi   1
Xiong, Hu   79

Yan, Guanhua   135
Yang, Bo   278, 355
Yap, Wun-She   63
Younan, Yves   213
Yu, Jia   95
Yu, Ran   182

Zhang, Bingsheng   262
Zhang, Ge   152
Zhang, Lei   293
Zhao, Ruoxu   182
Zhou, Yuan   341
Zhu, Sencun   1